CW00470588

# Ion Channels of Nociception

# Ion Channels of Nociception

Special Issue Editor

**Rashid Giniatullin**

MDPI • Basel • Beijing • Wuhan • Barcelona • Belgrade • Manchester • Tokyo • Cluj • Tianjin

*Special Issue Editor*
Rashid Giniatullin
A.I. Virtanen Institute,
University of Eastern Finland
Finland

*Editorial Office*
MDPI
St. Alban-Anlage 66
4052 Basel, Switzerland

This is a reprint of articles from the Special Issue published online in the open access journal *International Journal of Molecular Sciences* (ISSN 1422-0067) (available at: https://www.mdpi.com/journal/ijms/special_issues/nociception).

For citation purposes, cite each article independently as indicated on the article page online and as indicated below:

LastName, A.A.; LastName, B.B.; LastName, C.C. Article Title. *Journal Name* **Year**, *Article Number*, Page Range.

**ISBN 978-3-03936-549-4 (Hbk)**
**ISBN 978-3-03936-550-0 (PDF)**

# Contents

# About the Special Issue Editor

**Rashid Giniatullin** is working at the University of Eastern Finland. He graduated from Kazan Medical University in 1977, where he received a PhD in Physiology in 1981 and the degree of Doctor of Medical Sciences in 1992. He worked as Professor of Physiology at Kazan Medical University from 1993 to 2000. From 2000 to 2008, he was a Visiting Professor at SISSA, Trieste, Italy. In 2008, he was elected as Professor of Cell Biology at the University of Eastern Finland. His main research interests are molecular and cellular mechanisms of pain, migraine pathophysiology, synaptic transmission, ligand-gated receptors, desensitization, and signalling via reactive oxygen species

International Journal of
*Molecular Sciences*

*Editorial*

# Ion Channels of Nociception

**Rashid Giniatullin**

A.I. Virtanen Institute, University of Eastern Finland, 70211 Kuopio, Finland; Rashid.Giniatullin@uef.fi;
Tel.: +358-403553665

Received: 13 May 2020; Accepted: 15 May 2020; Published: 18 May 2020

**Abstract:** The special issue "Ion Channels of Nociception" contains 13 articles published by 73 authors from different countries united by the main focusing on the peripheral mechanisms of pain. The content covers the mechanisms of neuropathic, inflammatory, and dental pain as well as pain in migraine and diabetes, nociceptive roles of P2X3, ASIC, Piezo and TRP channels, pain control through GPCRs and pharmacological agents and non-pharmacological treatment with electroacupuncture.

**Keywords:** pain; nociception; sensory neurons; ion channels; P2X3; TRPV1; TRPA1; ASIC; Piezo channels; migraine; tooth pain

---

Sensation of pain is one of the fundamental attributes of most species, including humans. Physiological (acute) pain protects our physical and mental health from harmful stimuli, whereas chronic and pathological pain are debilitating and contribute to the disease state.

Despite active studies for decades, molecular mechanisms of pain—especially of pathological pain—remain largely unaddressed, as evidenced by the growing number of patients with chronic forms of pain. There are, however, some very promising advances emerging. A new field of pain treatment via neuromodulation is quickly growing, as well as novel mechanistic explanations unleashing the efficiency of traditional techniques of Chinese medicine. New molecular actors with important roles in pain mechanisms are being characterized, such as the mechanosensitive Piezo ion channels [1].

Pain signals are detected by specialized sensory neurons, emitting nerve impulses encoding pain in response to noxious stimuli. Many of these nociceptive neurons are equipped with a rich repertoire of specific ion channels which serve as pain transducers. These ion channels are located at the peripheral terminals of dorsal root or trigeminal ganglia neurons as well as in sensory neurons of the viscera (Figure 1). Pain transducers react to a variety of chemical or physical stimuli (algogens) by opening the ion channels and inducing neuronal depolarization known as the generator potential. These pain transducers include ATP-gated P2X3, classical heat/capsaicin-sensitive TRPV1 and cold/redox-sensitive TRPA1 channels, acid sensitive ion channels (ASICs), and mechanosensitive Piezo, to name just a few (Figure 1, for further details see the classical review [2]). Sufficiently high generator potential, assisted by voltage-gated sodium channels, initiates the propagating action potentials (spikes). Some voltage-gated channels are expressed exclusively in nociceptive neurons, for instance, sodium Nav1.8 and Nav1.9 channels [3]. The peripheral pain signals travel via the multi-synaptic network of the nociceptive system to the higher pain centers to be perceived as a feeling of pain.

A remarkable property of nociceptive neurons is sensitisation (enhanced responsiveness to triggers of pain). This phenomenon, a type of neuronal plasticity, can be mediated by a plethora of metabotropic receptors for different classes of pain modulators: classical mediators, neurotrophins, cytokines, and other active molecules (Figure 1). Normally, many pain transducers are in "sleeping" or low activity mode, but they can become very active during the action of these endogenous modulators triggering neuronal sensitization.

The fundamental approach: "treat pain at the source" provides a strategic rationale to diminish pain by counteracting its peripheral mechanisms. The knowledge of the function and structure of pain transducers and associated ion channels is essential to develop new medicines. The fact that the molecular structures of most ion channels implicated in nociception are known facilitates the development of new anti-nociceptive (analgesic) medicines. Direct targeting of pain transducers by specific blockers provides an opportunity to block pain mediated by these ion channels. However, this is a challenging task since the lifetime of channels on the membrane is limited because of the continuous traffic and renewal as it was shown for the nociceptive P2X3 receptors [4].

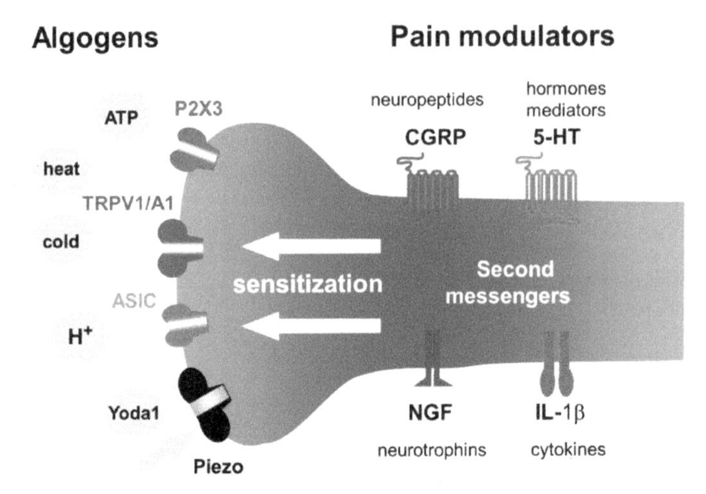


**Figure 1.** Chemical and physical stimuli (algogens) activating ion channel of nociception and pain modulators acting at the nerve terminal of the nociceptive neuron. The nerve terminal expresses specific ion channels of nociception such as ATP-gated P2X3, heat/capsaicin-sensitive TRPV1 and cold/redox-sensitive TRPA1 channels, acid-sensitive ion channels (ASICs) and mechanosensitive Piezo channels. The activity of these pain transducers can be enhanced leading to the neuronal sensitization, via metabotropic receptors, by different classes of pain modulators including neuropeptides, hormones, classical mediators, neurotrophins, cytokines, and other endogenous pro-nociceptive molecules.

This special issue "Ion Channels of Nociception" contains 13 articles published by 73 authors from different countries united by the main focusing on the peripheral mechanisms of pain. The content covers the mechanisms of neuropathic, inflammatory, and dental pain as well as pain in migraine and diabetes, nociceptive roles of P2X3, ASIC, Piezo and TRP channels, pain control through GPCRs and pharmacological agents and non-pharmacological treatment with electroacupuncture.

Not surprisingly, the majority of papers in this collection are devoted to the functioning of TRP channels, which are the best-studied transducers of noxious stimuli but still a puzzling issue for the researchers. In the large family of TRP channels, TRPV1 and TRPA1 are best known for their leading role in key pain mechanisms. TRPV1 receptors are often co-expressed with TRPA1 in nociceptive neurons and probably can interact by means of the protein TMEM100. The details of such interactions are discussed in the well-illustrated review by Takayama and co-workers [5] summarizing the role TRP channels play in the pain states including acute pain, inflammatory pain, migraine pain, and other disorders. Much attention in this review is paid to the ANO1 protein, a calcium-activated chloride channel expressed in nociceptive neurons. The authors propose that TRP channels and ANO1 can collaborate to generate a strong pain signal in primary sensory neurons. This is an interesting "marriage" between cation and anion permeable ion channels to control the neuronal excitability.

By focusing on TRPA1 channels, Feng and colleagues [6], using electrophysiological recordings from brainstem slices, detected activation of presynaptic polymodal TRPA1 ion channels in glutamatergic terminals synapsing on the caudal nucleus of the solitary tract neurons. The enhanced glutamatergic synaptic neurotransmission onto the second-order sensory neurons was activated by TRPA1 agonists. These findings were supported by data from the TRPA1 knockout mice where TRPA1 agonists failed to alter synaptic efficacy. The study suggests that in the caudal brainstem the input from visceral noxious stimuli can be targeted by multiple endogenous TRPA1 ligands including reactive oxygen species and probably by the classic analgesic paracetamol via its derivatives [6].

The original view on the role of ion channels, including TRP and other unrelated channels in nociception, was presented by Ciotu with co-workers [7]. As stated by the authors, they describe "more novel and less known features" of ion channels. In particular, they summarize not commonly considered ion channel properties such as a memory of the previous voltage and chemical stimulation, alternative ion conduction pathways, cluster formation, and role of the silent accessory subunits. This fresh view on the channels function in a realistic membrane environment should stimulate the interest to these little appreciated phenomena and can suggest new ideas for many researchers studying the pain mechanisms.

One study in this special issue, reports the expression of TRPA1 channels not only in neurons but also in non-neuronal cells. Thus, Kameda with colleagues [8], studied the expression and activity of the pro-nociceptive TRPA1 and TRPV1 in the intervertebral disc (IVD) as its degeneration is associated with inflammatory pain. As the object of the study, they analyzed human fetal, healthy, and degenerated IVD tissues. They found that the TRPA1 agonist AITC activated inflamed IVD cells, induced expression of IL-8, but reduced disintegrin and metalloproteinase with thrombospondin motifs 5 (ADAMTS5). By comparing knockout TRPA1 versus TRPV1 mice, they further confirmed the leading role of TRPA1 in control of inflammation in IVD cells.

The experimental study by Demartini with collaborators [9], has a significant translational impact as they describe a new treatment approach for neuropathic pain. Notably, neuropathic pain is a severe disabling and often intractable condition and new therapeutic targets are of high interest. They used a drug candidate compound ADM-12 blocking the nociceptive TRPA1 receptors in the trigeminal nerve fibers. The authors used the well-established model of the neuropathic pain based on the chronic constriction injury of the infraorbital nerve. Apart from the antagonism of TRPA1 receptors, they found also other benefits of this treatment including reduced expression of TRPA1/V1 receptors and pro-nociceptive neuropeptides and cytokines, collectively contributing to diminished mechanical allodynia, a leading symptom of neuropathic pain.

Whereas the trigeminal neuralgia is a relatively rare type of head pain, migraine headaches are very common. Migraine headache is characterized by a severe pulsatile (throbbing) type of pain also transmitted by trigeminal nerve fibers. Review by Della Pietra et al. [10] provides a new explanation for the throbbing pain in migraine by proposing a rhythmic activation, by pulsatile blood flow, of mechanosensitive Piezo channels in trigeminal fibers in meninges. Piezo ion channels (presented by Piezo1 and Piezo2 isoforms) with the highest sensitivity to mechanical stimuli were recently discovered [1]. The Piezo 1 subtype sensitive to mechanical and chemical agonists such as Yoda1, is found in trigeminal neurons along with Piezo2 subtype [11]. Della Pietra and co-authors describe the function and modulation of Piezo mechanotransducers in the trigeminovascular system as sensors generating the rhythmic migraine pain signals. The emerging field of Piezo currently attracts much attention as a new way for efficient control of Piezo-related diseases, including migraine and chronic pain.

The review by Tardiolo et al. [12] expands our knowledge of current and novel therapeutic approaches to migraine. The particular focus in this review is on the pharmacological targets for novel drugs based on 5-HT receptor agonists (ditans), CGRP receptor antagonists (new generation of gepants), and CGRP receptor or ligand antagonists such as monoclinal antibodies (mAbs). Of note, this review contains a detailed description of these and other new emerging treatments in migraine. Further in this review, the authors present the animal models of migraine including the dural application of the inflammatory soup or high potassium to induce CSD, nitroglycerin, and transgenic migraine models.

Apart from the clear role in migraine of the neuropeptide CGRP, another neuropeptide, substance P, often co-expressed with CGRP in nociceptive neurons, contributes to other types of pain. For decades, substance P was considered a putative promoter of pain. A review included in this collection by Chang with colleagues [13] highlights a new paradoxical role of substance P in anti-nociception, in particular, in muscle pain. The authors suggest that the anti-nociception by the substance P in muscle pain is mediated via the enhancement of the M-channel outward currents in local sensory neurons. The review contains a detailed description of ion channels modulated by substance P. Elucidating the dual role of substance P in pain control would further improve our understanding of the biological functions of this neuropeptide for better development of anti-nociceptive treatments.

The nociceptive neurons express also a high level of P2X3 receptors activated by extracellular ATP, which is a powerful algogenic substance (Figure 1). Xiang and co-workers [14] uncovered the role of ATP-gated P2X3 receptors in analgesia by electroacupuncture in CFA-induced inflammatory pain. P2X3 receptors, opening a cationic ion channel after ATP binding, are primarily implicated in the inflammatory type of pain. The authors found that the short-term or long-term application of 100 Hz electroacupuncture increased the paw withdrawal threshold and reversed the elevation of P2X3 receptors in sensory DRG neurons.

Mustăciosu with collaborators [15] analyzed the expression of the neuron-specific Elav-like Hu RNA-binding proteins in sensory DRG neurons in mice with the streptozotocin (STZ)-induced diabetes. These proteins presented with three isoforms HuB, HuC, and HuD, typically play a role in neurogenesis and neuronal plasticity. As the original approach, the authors compared STZ-sensitive to STZ-resistant mice with high or low glycemia, respectively. With thermal pain testing, they found that the hypoalgesia was observed only in the diabetic mice. This effect was associated with HuD downregulation and HuB upregulation, which might be related to the altered post-transcriptional control of RNAs involved in the regulation of thermal hypoalgesia.

In their comprehensive review, Salzer and co-workers [16] discuss how the members of the big family of G-protein coupled receptors (GPCRs) control the function of different types of ion channels implicated in nociception (see also Figure 1). Notably, these CGRPs not only detect the endogenous active molecules such as opioids or cannabinoids operating as modulators of pain transducing ion channels but also represent the most important targets for various analgesic therapeutics. The authors provide in-depth analysis of GPCRs-mediated modulation of the main subtypes of ligand- (TRPs, ASIC, CaCCs), mechano- (K2P and Piezo) as well as voltage-gated (sodium, calcium, and potassium) channels implicated in nociception.

Lee and collaborators [17] present an interesting review on the mechanisms of the dental pain, which is often extremely severe and based on the unique anatomical structure of the tooth. Along with other chemical stimuli, they paid much attention to the heat-/cold-induced nociceptive signaling via TRP type ion channels. Although thermal stimuli are the primary signals to trigger tooth pain, this cannot explain the sudden and intense tooth pain elicited by innocuous mechanical stimuli. Moreover, similar to migraine headaches, dental pain often has a pulsating character. The latter can be activated by mechanical stimulation from the rhythmic movement of dentinal fluid or deformation of tooth microstructure. Logically, as in the migraine paper by Della Pietra et al, Lee and colleagues have considered the role of the professional Piezo ion channels as detectors of these mechanical forces. As Piezo2 channels are expressed mainly in low threshold mechanoreceptive neurons, the authors proposed that, in the tooth, these neurons serve as nociceptors. Interestingly, Piezo2 were also detected in odontoblastic processes in dentinal tubules, suggesting the complex role of these mechanosensitive channels in dental pain.

Shteinikov and co-workers [18] report the results of the experimental study on the action of the hydrophobic amines and their guanidine analogs on activation and desensitization of acid-sensing ion channels (ASIC). The ASIC3 subtype, studied here, with the highest acid sensitivity is primarily expressed in nociceptive neurons and likely implicated in various types of chronic and inflammatory pain associated with tissue acidosis. By testing a series of hydrophobic monoamines on CHO cells expressing rat ASIC3 channels, they found an interesting combination of two opposite effects of these potentially analgesic agents. This finding can explain previous contradictory results obtained with these ASIC modulators. They showed the inhibition of ASIC3 activation due to the acidic shift of proton sensitivity but they also detected the reduced desensitization of ion channels, which is normally very fast and limits the activity of these membrane proteins.

In summary, this collection of articles provides an overview of different aspects of peripheral pain mechanisms. These papers are extending our understanding of the role of ion channels in situ, ion channel interactions, functional role of sensitization-desensitization and ion channel inactivation, endogenous modulators, and other important aspects of the functioning of excitable nociceptive neurons partnering with non-excitable cells. We believe that this issue will provide new insights into the remarkable field of pain research.

**Author Contributions:** Writing, review, and editing, R.G.

**Funding:** The author is supported by the Finnish Academy (grant number 325392).

**Conflicts of Interest:** The author declares no conflict of interest.

## References

1. Coste, B.; Mathur, J.; Schmidt, M.; Earley, T.J.; Ranade, S.; Petrus, M.J.; Dubin, A.E.; Patapoutian, A. Piezo1 and Piezo2 Are Essential Components of Distinct Mechanically Activated Cation Channels. *Science* **2010**, *330*, 55–60. [CrossRef] [PubMed]
2. Julius, D.; Basbaum, A.I. Molecular mechanisms of nociception. *Nature* **2001**, *413*, 203–210. [CrossRef] [PubMed]
3. Bennett, D.L.; Clark, A.; Huang, J.; Waxman, S.G.; Dib-Hajj, S.D. The Role of Voltage-Gated Sodium Channels in Pain Signaling. *Physiol. Rev.* **2019**, *99*, 1079–1151. [CrossRef] [PubMed]
4. Pryazhnikov, E.; Fayuk, D.; Niittykoski, M.; Giniatullin, R.; Khiroug, L. Unusually Strong Temperature Dependence of P2X3 Receptor Traffic to the Plasma Membrane. *Front. Cell. Neurosci.* **2011**, *5*. [CrossRef] [PubMed]
5. Takayama, Y.; Derouiche, S.; Maruyama, K.; Tominaga, M. Emerging Perspectives on Pain Management by Modulation of TRP Channels and ANO1. *Int. J. Mol. Sci.* **2019**, *20*, 3411. [CrossRef] [PubMed]
6. Feng, L.; Uteshev, V.V.; Premkumar, L.S. Expression and Function of Transient Receptor Potential Ankyrin 1 Ion Channels in the Caudal Nucleus of the Solitary Tract. *Int. J. Mol. Sci.* **2019**, *20*, 2065. [CrossRef] [PubMed]
7. Ciotu, C.I.; Tsantoulas, C.; Meents, J.; Lampert, A.; McMahon, S.B.; Ludwig, A.; Fischer, M.J. Noncanonical Ion Channel Behaviour in Pain. *Int. J. Mol. Sci.* **2019**, *20*, 4572. [CrossRef] [PubMed]
8. Kameda, T.; Zvick, J.; Vuk, M.; Sadowska, A.; Tam, W.K.; Leung, V.Y.-L.; Bölcskei, K.; Helyes, Z.; Applegate, L.A.; Hausmann, O.; et al. Expression and Activity of TRPA1 and TRPV1 in the Intervertebral Disc: Association with Inflammation and Matrix Remodeling. *Int. J. Mol. Sci.* **2019**, *20*, 1767. [CrossRef] [PubMed]
9. DeMartini, C.; Greco, R.; Zanaboni, A.M.; Francesconi, O.; Nativi, C.; Tassorelli, C.; Deseure, K. Antagonism of Transient Receptor Potential Ankyrin Type-1 Channels as a Potential Target for the Treatment of Trigeminal Neuropathic Pain: Study in an Animal Model. *Int. J. Mol. Sci.* **2018**, *19*, 3320. [CrossRef] [PubMed]
10. Della Pietra, A.; Mikhailov, N.; Giniatullin, R. The Emerging Role of Mechanosensitive Piezo Channels in Migraine Pain. *Int. J. Mol. Sci.* **2020**, *21*, 696. [CrossRef]
11. Mikhailov, N.; Leskinen, J.; Fagerlund, I.; Poguzhelskaya, E.; Giniatullina, R.; Gafurov, O.; Malm, T.; Karjalainen, T.; Gröhn, O.; Giniatullin, R. Mechanosensitive meningeal nociception via Piezo channels: Implications for pulsatile pain in migraine? *Neuropharmacology* **2019**, *149*, 113–123. [CrossRef] [PubMed]
12. Tardiolo, G.; Bramanti, P.; Mazzon, E. Migraine: Experimental Models and Novel Therapeutic Approaches. *Int. J. Mol. Sci.* **2019**, *20*, 2932. [CrossRef] [PubMed]
13. Chang, C.-T.; Jiang, B.-Y.; Chen, C.-C. Ion Channels Involved in Substance P-Mediated Nociception and Antinociception. *Int. J. Mol. Sci.* **2019**, *20*, 1596. [CrossRef] [PubMed]
14. Xiang, X.; Wang, S.; Shao, F.; Fang, J.; Xu, Y.; Wang, W.; Sun, H.; Liu, X.; Du, J.; Fang, J. Electroacupuncture Stimulation Alleviates CFA-Induced Inflammatory Pain Via Suppressing P2X3 Expression. *Int. J. Mol. Sci.* **2019**, *20*, 3248. [CrossRef] [PubMed]

15. Mustaciosu, C.C.; Banciu, A.; Rusu, C.M.; Banciu, D.D.; Savu, D.; Radu, M.; Radu, B.M. RNA-Binding Proteins HuB, HuC, and HuD are Distinctly Regulated in Dorsal Root Ganglia Neurons from STZ-Sensitive Compared to STZ-Resistant Diabetic Mice. *Int. J. Mol. Sci.* **2019**, *20*, 1965. [CrossRef] [PubMed]
16. Salzer, I.; Ray, S.; Schicker, K.; Boehm, S. Nociceptor Signalling through ion Channel Regulation via GPCRs. *Int. J. Mol. Sci.* **2019**, *20*, 2488. [CrossRef] [PubMed]
17. Lee, K.; Lee, B.-M.; Park, C.-K.; Kim, Y.; Chung, G. Ion Channels Involved in Tooth Pain. *Int. J. Mol. Sci.* **2019**, *20*, 2266. [CrossRef] [PubMed]
18. Shteinikov, V.Y.; Potapieva, N.N.; E Gmiro, V.; Tikhonov, D.B. Hydrophobic Amines and Their Guanidine Analogues Modulate Activation and Desensitization of ASIC3. *Int. J. Mol. Sci.* **2019**, *20*, 1713. [CrossRef] [PubMed]

International Journal of
*Molecular Sciences*

MDPI

*Review*

# Emerging Perspectives on Pain Management by Modulation of TRP Channels and ANO1

**Yasunori Takayama [1,*,†], Sandra Derouiche [2,*,†], Kenta Maruyama [3,*,†] and Makoto Tominaga [2,*,†]**

1 Department of Physiology, Showa University School of Medicine, 1-5-8 Hatanodai, Shinagawa, Tokyo 142-8555, Japan
2 Thermal Biology group, Exploratory Research Center on Life and Living Systems, National Institutes for Natural Sciences, 5-1 Aza-higashiyama, Myodaiji, Okazaki, Aichi 444-8787, Japan
3 National Institute for Physiological Sciences, National Institutes for Natural Sciences, 5-1 Aza-higashiyama, Myodaiji, Okazaki, Aichi 444-8787, Japan
* Correspondence: ytakayama@med.showa-u.ac.jp (Y.T.); derosand@nips.ac.jp (S.D.); maruken@nips.ac.jp (K.M.); tominaga@nips.ac.jp (M.T.)
† These authors contributed equally.

Received: 28 May 2019; Accepted: 9 July 2019; Published: 11 July 2019

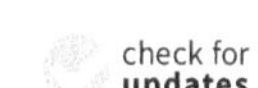

**Abstract:** Receptor-type ion channels are critical for detection of noxious stimuli in primary sensory neurons. Transient receptor potential (TRP) channels mediate pain sensations and promote a variety of neuronal signals that elicit secondary neural functions (such as calcitonin gene-related peptide [CGRP] secretion), which are important for physiological functions throughout the body. In this review, we focus on the involvement of TRP channels in sensing acute pain, inflammatory pain, headache, migraine, pain due to fungal infections, and osteo-inflammation. Furthermore, action potentials mediated via interactions between TRP channels and the chloride channel, anoctamin 1 (ANO1), can also generate strong pain sensations in primary sensory neurons. Thus, we also discuss mechanisms that enhance neuronal excitation and are dependent on ANO1, and consider modulation of pain sensation from the perspective of both cation and anion dynamics.

**Keywords:** TRPA1; TRPV1; TRPM3; ANO1; acute pain; inflammatory pain; migraine; Candidiasis

## 1. Introduction

The ability to sense elements of the natural environment (including temperature, pH, pressure, light, and noxious compounds) is critical for survival. Detection and response to environmental agents and stimuli are frequently mediated by receptor-type plasma membrane proteins, particularly ion channels that show versatile function in a range of organisms from prokaryotes to eukaryotes. Relative to G-protein coupled receptors (GPCRs), ion channels can directly impact neural excitation by both sensing natural stimuli and converting these signals into electrical changes to affect the polarization state of the plasma membrane.

In this review, we focus on several transient receptor potential (TRP) channels that are specifically activated by natural compounds and largely localize to primary sensory neurons. There are three types of nerves in primary sensory neurons, including Aβ-, Aδ-, and C-fibers. Aβ-fibers are myelinated afferent nerves that respond to innocuous mechanical stimuli. Aδ-fibers are also myelinated nerves, but alternatively this nervous pathway responds to rapid noxious stimuli. C-fibers are nonmyelinated nerves involved in slow pain [1]. The TRP channel superfamily comprises six subfamilies: TRP vanilloid (TRPV), canonical (TRPC), mucolipin (TRPML), polycystin (TRPP), ankyrin (TRPA), and melastatin (TRPM). Several TRP channels are expressed in small-size dorsal root ganglion (DRG) and trigeminal ganglion (TG) neurons (C fibers and Aδ fibers) [2]. While TRPV1 and TRPA1 are considered to be the major receptors of this superfamily involved in nociception [3]. In particular, TRPV1 and sensitized TRPA1 are activated by heat and cold, respectively, and as such are important for detection of noxious temperature changes. Recently, TRPM3 involvement in heat sensation was also reported in mice [4].

The calcium-activated chloride channel, anoctamin 1 (ANO1, also known as TMEM16A) [5–7], was recently reported to be directly activated in DRG neurons by extremely rapid temperature changes that reach noxious ranges [8,9]. ANO1 can also be activated immediately downstream of Gq protein-coupled receptors (GqPCRs), including the bradykinin receptor, as evidenced by direct interaction of ANO1 with inositol trisphosphate ($IP_3$) receptors on endoplasmic reticulum membranes [10]. Chloride channels typically function in neuronal suppression in the central nervous system, in part because intracellular chloride concentrations are maintained at low levels by the potassium–chloride co-transporter type 2 (KCC2). However, in DRG neurons, KCC2 expression is either absent or very low, whereas expression of the sodium–potassium–chloride co-transporter type 1 (which is an important molecule in the chloride intake pathway) is high [11]. Thus, chloride efflux through ANO1 activation is a key pathway for generation of neuronal excitation in many primary sensory neurons.

Here, we summarize the physiological significance of TRP and ANO1 channels. First, we describe current understanding of representative ion channels, namely TRPA1, TRPV1, and ANO1 (Part 1). Second, we discuss the multiple functions of TRP channels and ANO1 (Part 2). Finally, we propose the significance of those functions in clinical situations, including headache, migraine, and fungus infection (Parts 3 and 4).

## 2. Basic Understanding of Ion Channels in Primary Sensory Neurons

### 2.1. TRPA1

#### 2.1.1. TRPA1 Activation by Natural Ligands

TRPA1 is activated by many natural ligands such as allyl isothiocyanate (AITC), tetrahydrocannabinol, cinnamaldehyde, allicin, diallyl sulfide, carvacrol, eugenol, gingerol, methyl salicylate, capsiate, thymol, propofol, 1,4-cineole, oleocanthal, and carbon dioxide, and by membrane extension and intracellular alkalization [12–26]. Moreover, TRPA1 is activated by calcium [27]. The magnitude of TRPA1 currents gradually increases during application of a given agonist, as observed for the lagging peak current induced after application of -eudesmol from hops to HEK293 cells expressing human TRPA1 [28]. Although the precise mechanism of TRPA1 activation remains unclear, covalent protein modification is involved. Carbons of AITC and *N*-methyl maleimide covalently bind to cysteine in the N-terminus of TRPA1 to enhance channel activation, whereas C-terminal lysine and arginine are important for AITC-mediated activation [29–31]. Menthol also has agonistic effects on human and mouse TRPA1, although the effects are bimodal [32,33]. The agonistic and antagonistic effects on mouse TRPA1 involves serine 876 and threonine 877 in the transmembrane (TM)-5 region [33]. Interestingly, G878 is also important for TRPA1-mediated cold sensitivity in rodents [34]. Rodent TRPA1 can be activated by cold stimulation and is involved in cold hyperalgesia after application of complete Freund's adjuvant (CFA) [35,36]. Although human TRPA1 does not show a cold response, it nonetheless responds to cold at approximately 18 °C if oxidization with dehydroxylation at proline 394 occurs [37].

#### 2.1.2. TRPA1 in Pathological Conditions

TRPA1 activation induces hyperalgesia during inflammation because inflammatory factors (such as bradykinin released by tissue injury) activate and sensitize TRPA1 in DRG neurons. In this pathway, protein kinase A (PKA) and phospholipase C (PLC) are important for TRPA1 sensitization [38]. Adenosine triphosphate (ATP) is another important inducer in inflammatory pain. Pain sensations are enhanced via a similar pathway through activation of purinergic P2Y receptors expressed in DRG neurons. Moreover, P2X receptors are involved in neuropathic pain via phospholipase A2 (PLA2) signaling [39], which activates protein kinase C (PKC), and in turn sensitizes TRPA1 [40].

TRPA1 expression increases after application of nerve growth factor (NGF) and is inhibited by the p38 mitogen-activated protein kinase (MAPK) inhibitor, SB203580. In DRG neurons, NGF released from inflamed tissue phosphorylates p38, which subsequently enhances TRPA1 expression [36]. Thus, the NGF–p38 MAPK–TRPA1 axis is one of the pathways that exacerbates TRPA1-mediated pain sensation in DRG neurons. For example, gastric distension-induced visceral pain relies on activation of both TRPA1 and p38 [41]. TRPA1 localization can be modified by pathological stimulation. TRPA1 localization to the plasma membrane is enhanced in forskolin-treated DRG neurons [42]. Further, in mice, full-length TRPA1 positively translocates to the plasma membrane by co-expression of a TRPA1 splicing variant [43]. There are two TRPA1 splicing variants: TRPA1a is the full-length protein whereas TRPA1b lacks exon 20, which encodes part of TM2 and the intracellular domain between TM2 and TM3. TRPA1b has no ion channel activity but instead enhances TRPA1a translocation to the plasma membrane. One-day of CFA treatment or partial sciatic nerve ligation (PSL) causes inflammatory and neuropathic pain, respectively. In both cases, TRPA1a expression levels increase transiently. Interestingly, TRPA1b expression levels significantly increase while TRPA1a expression reduces to basal levels at five days after CFA treatment or PSL. As such, up-regulation of TRPA1a translocation via TRPA1b overexpression causes a continuous pathological condition.

As with CFA, lipopolysaccharide (LPS) is also often used to induce the inflammatory condition. LPS can activate Toll-like receptor (TLR)-4, and cause subsequent release of multiple cytokines from immune cells [44]. The cytokine, tumor necrosis factor-alpha (TNF-$\alpha$), enhances AITC-induced calcium increases in nodose and jugular ganglion neurons from rats [45]. However, a recent report suggested that LPS-induced calcium increases in nodose ganglion neurons from mice do not depend on TLR4, even although the responses are reduced in TRPA1 knockout mice [46]. Altogether, these results suggest that LPS directly activates TRPA1. Further, LPS increases single channel activity in TRPA1-expressing CHO cells. Ultimately, this novel relationship between bacteria and primary sensory nerves suggests that TRPA1 antagonists could be valuable for reducing pain induced by bacterial infections.

#### 2.1.3. TRPA1 Activation by Reactive Oxygen Species and Hypoxia

In addition to thermal stimuli and environmental agents, TRPA1 is activated by reactive oxygen species (ROS) such as hydrogen peroxide ($H_2O_2$) [47–49]. Responses to certain pathological conditions involving increased ROS synthesis (such as dysesthesia in ischemia and reperfusion of blood flow) are dependent on TRPA1 activity in mice [50]. Pain-related behavior due to dysesthesia is reduced by activation of prolyl hydroxylase (PHD)-2 involving hydroxylation at proline. Under normoxic conditions, TRPA1 steady status activity is maintained by PHD-mediated hydroxylation of proline 394, but under hypoxic conditions hydroxylation is inhibited and $H_2O_2$-induced TRPA1 activity is enhanced [50]. In contrast, high concentrations of oxygen also activate TRPA1 by directly modifying TRPA1 cysteines [51]. Collectively, these two functions allow TRPA1 to act as an oxygen sensor under both hypoxic and hyperoxic conditions.

Side effects of the anti-cancer agent oxaliplatin include induction of various dysesthesias, including peripheral nerve disorder and cold hyperalgesia. These dysesthesias are associated with enhanced TRPA1 expression in DRG and PHD-induced modification of TRPA1 [52–54]. Moreover, the oxaliplatin degradation product, oxalate, inhibits PHD and subsequently TRPA1 dehydroxylation, and also promotes cold hypersensitivity upon activation of TRPA1 in response to ROS production by mitochondria [37,55]. Mechanical allodynia associated with oxaliplatin treatment can be inhibited by the TRPA1 antagonist, ADM_09 [56]. Together, these results clearly indicate the importance of the relationship between TRPA1 and the PHD cascade, and also that TRPA1 could be targeted as part of treatment for dysesthesia induced by ischemia and hypoxia, as well as drug-induced cold and mechanical hyperalgesia.

### 2.1.4. pH Sensing by TRPA1

The relationship between oxygen and pH is physiologically important because hypoxic conditions caused by ischemia induce intracellular acidification. Neuronal death may occur upon low levels of oxygen and glucose, as well as excessive release of glutamic acid from astrocytes, which induces a fatal calcium influx in neurons. Intracellular pH of astrocytes is drastically reduced by lactic acid production due to anaerobic respiration in response to hypoxia. Subsequent acidification induces glutamic acid exocytosis led to brain damage [57]. Importantly, TRPA1 expressed in astrocytes may be activated by acidification [58–60]. In addition, activation of TRPA1 expressed in oligodendrocytes can damage myelin [61]. In contrast, intracellular alkalization also affects TRPA1 activity [26]. TRPA1 is activated at approximately pH 8.0, and the alkalization-induced pain-related behavior is significantly reduced in TRPA1 deficient mice. Consequently, the pH dependency of TRPA1 may be beneficial target for the treatment of central nervous system diseases, not only pain.

### 2.1.5. Neural Networks Involving TRPA1-Mediated Pain Sensation

A-fiber and C-fiber primary sensory nerves govern fast and slow responses to pain, respectively. Aδ-fibers (mid-sized DRG neurons) innervate lamina I and V, whereas C-fibers (small-sized DRG neurons) innervate lamina I and II of the dorsal horn of the spinal cord [1]. C-fibers also contain peptidergic and nonpeptidergic neurons. Peptidergic neurons contain substance P and CGRP, with both peptides released upon neural excitation. TRPA1-positive neurons are immunoreactive for CGRP in DRG neurons [15]. In healthy mice, these CGRP-positive neurons enhance heat sensation and suppress cold sensation [62]. These findings suggest that TRPA1 in CGRP-positive DRG neurons contributes less significantly to noxious cold sensation.

Neural transmission in the spinal cord can modify pain perception. Substantia gelatinosa (SG) neurons in lamina II are important targets for investigation of how pain sensations are transmitted from the periphery to the central nervous system. Initial understanding on in vivo SG neuronal responses to peripheral stimulation is that SG neuronal activity mediated through non-*N*-methyl-*D*-aspartate (non-NMDA) receptors is enhanced by mechanical stimuli, such as pinch and air flow, but not thermal changes [63]. However, excitatory postsynaptic currents enhanced by capsaicin treatment are detected in approximately 80% of SG neurons in slice patch-clamp recordings [64]. Interestingly, there is no neuronal response to AITC alone (Figure 1).

Since AITC responses depend on both NMDA and non-NMDA receptors [65], TRPA1-mediated pain signals are likely integrated with TRPV1-mediated pain signals in lamina II of the spinal cord. Importantly, there are three types of DRG neurons: those that express both TRPA1 and TRPV1, TRPA1 alone, or TRPV1 alone. Meanwhile, one study demonstrated that spinal TRPA1 activation by intrathecal administration of the acetaminophen metabolite, N-acetyl-p-benzoquinone imine, enhanced anti-nociception in the spinal cord of mice [66]. Therefore, components of TRPA1-mediated neural systems may participate in pain reduction, while nociception by TRPA1 activation can function in central termini of DRG neurons. For instance, TRPA1 activation by hepoxilin causes mechanical allodynia in rats, whereas pinch-evoked SG neuronal excitation is reduced by increases in inhibitory postsynaptic currents mediated by TRPA1 activation in vivo [67,68]. Thus, consideration of TRPA1 activation in the central nervous system may also be important for investigating pain mechanisms.

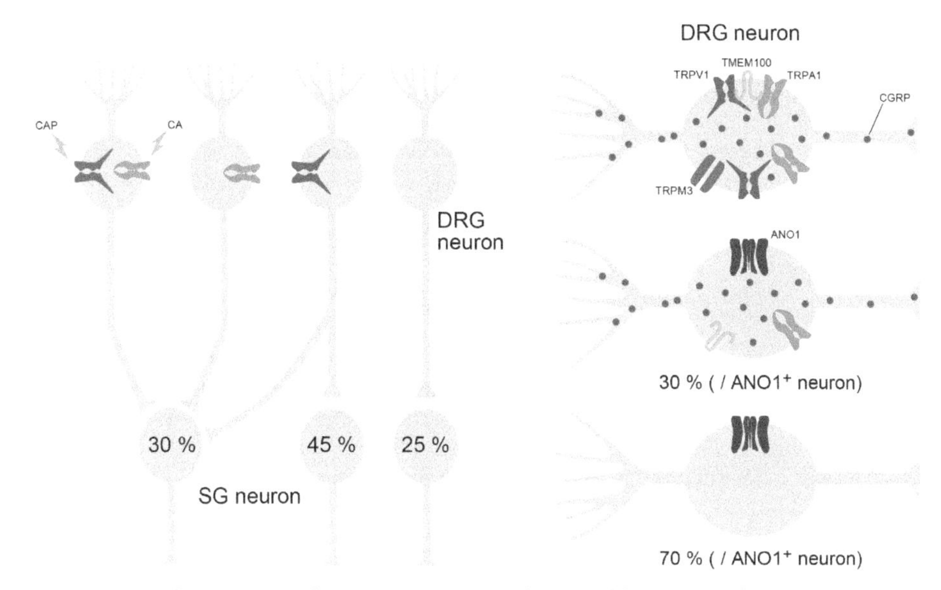


**Figure 1.** Nociceptor populations in substantia gelatinosa (SG) neurons of lamina II, which receive nociceptive inputs from dorsal root ganglion (DRG) neurons, no neurons respond only to TRPA1-associated stimuli. Approximately 30% of SG neurons are double-positive to capsaicin (CAP) and cinnamaldehyde (CA), 45% of SG neurons response to only CAP, and 25% of neurons show no effect to either CAP or CA. There are calcitonin gene-related peptide (CGRP)-positive and -negative neurons in peripheral sensory nerves. Most TRPV1–TMEM100–TRPA1 complexes and TRPV1–TRPA1–TRPM3 trios are expressed in CGRP-positive neurons. Anoctamin 1 (ANO1) is also expressed in CGRP-positive neurons, however approximately 70% of ANO1-expressing neurons are CGRP-negative.

#### 2.1.6. TRPA1 Activation by microRNA in the Central Nervous System

Although TRPA1 is expressed in both brain and spinal cord cells, whether it is activated via direct or indirect pathways is unclear. In the central nervous system, only one obvious possibility exists for direct activation of TRPA1 by an endogenous ligand, namely microRNA (miRNA). Among the miRNAs present in cerebrospinal fluid, increased levels of let-7b are particularly associated with the incidence of Alzheimer's disease. Meanwhile, astrocytes can exacerbate symptoms associated with amyloid-induced TRPA1 activation [69,70]. Let-7b activates TLR7, which is followed by cytotoxicity and direct activation of TRPA1 [71]. Importantly, let-7b release is enhanced by formalin application to DRG, while let-7b injection induces both nociceptive behavior and mechanical allodynia, which are reduced in TRPA1- and TLR7-deficient mice. These results demonstrate a relationship between let-7b and TRPA1 as a possible molecular mechanism of inflammatory symptoms in central nervous system diseases.

### 2.2. *TRPV1*

Among TRP channels expressed in primary sensory neurons, TRPV1 is well-known for its role in pain [3]. TRPV1 is mainly expressed in small DRG and TG neurons and is activated by capsaicin, capsiate, camphor, allicin, 2-aminoethoxydiphenyl borate, anandamide, N-arachidonoyl dopamine, resiniferatoxin, nitrogen oxide, low pH, noxious heat, hypertonicity, and the double-knot toxin in tarantula venom [14,72–81]. Expression levels of TRPV1 are enhanced by NGF receptor activation in DRG neurons, while TRPV1 is transported to peripheral termini [82]. TRPV1 activation is enhanced upon phosphorylation by PKA and PKC via A-kinase anchor protein in DRG neurons activated by GPCRs [83–86]. Due to the lowered thermal threshold of phosphorylated TRPV1 at body temperature, allodynia can be caused by inflammatory pathways depending on GqPCR activation. Many inflammatory factors, including prostaglandin E2, adenosine, ATP, bradykinin, protease, and NGF, are released following tissue injury, with protein kinases activated downstream of each GPCR [1]. In addition, the ionotropic ATP receptor, P2X, may also be involved in modulating activity of protein kinases that target TRP channels, as evidenced by activation of cytosolic PLA2 by P2X3 and P2X2/3 receptors during neuropathic pain [39]. Activated PLA2 can in turn activate PKC to promote phosphorylation of TRP [40]. Taken together, these findings suggest that TRP channel phosphorylation may be caused by both P2X and metabotropic P2Y receptor activation in primary sensory neurons.

### 2.3. *Anoctamin 1*

ANO1 is a calcium-activated chloride channel [5–7]. Although the ANO family includes ten subtypes, only ANO1 and ANO2 exhibit marked activity as calcium-activated chloride channels, and conductance of ANO1 is larger than ANO2. The crystal structure of fungal ANO has recently been determined at high resolution [87]. Furthermore, the dimer structure of mouse ANO1 (which contains ten TM regions in one subunit) has been clarified by cryo-electron microscopy [88,89]. Accordingly, the calcium binding site was shown to be encompassed by TM6 to TM8. Interestingly, each ANO1 subunit has one pore region surrounded by TM3 to TM8. Structural analysis showed that the dynamic movement of TM6 may be critical for calcium-mediated channel opening.

Although ANO1 can be activated by global increases in intracellular calcium via activation of voltage-gated calcium channels, ANO1 activation through GqPCR is also likely to be important due to direct interactions between ANO1 and $IP_3R$ in endoplasmic reticulum calcium stores [10,90]. Therefore, ANO1 is possibly involved in nociception induced by inflammatory factors such as bradykinin [91]. ANO1 is also activated by noxious heat in DRG neurons and induces a burning pain sensation [8]. These characteristics may explain why chloride channel activity evokes neural excitation in primary sensory neurons, in that higher intracellular chloride concentrations can be maintained as the equilibrium potential in these cells is more positive than the resting potential in DRG neurons [11].

## 3. Collaboration of Ion Channels

Although each ion channel, including TRP channels, independently work as detectors of nociceptive stimuli, some ion channels make physical or functional complexes that are critically involved in pain sensation. In this part, we summarize ion channel interactions and nociceptor populations according to recent reports (Figure 1).

### 3.1. *TRPV1–TMEM100–TRPA1 Interaction*

TRPV1 is co-expressed with TRPA1 in DRG neurons. Since TRPA1 activity is enhanced by intracellular calcium, it had been thought that calcium influx through TRPV1 activation could affect TRPA1 function. However, the TRPV1 entity reduces the probability of TRPA1 ion channel opening accelerated by mustard oil [92]. It appears that TRPA1-associated pain is normally reduced by TRPV1 expression, which may be prevented by TMEM100 [92]. TMEM100 is a small membrane protein, and its expression pattern highly overlaps with CGRP. Interestingly, TRPA1 almost co-localizes with TRPV1, TMEM100, and CGRP in DRG neurons (Figure 1). Together, TRPV1, TRPA1, and TMEM100 form a complex, and the interaction between TRPV1 and TRPA1 is suppressed by interposition of TMEM100. Furthermore, a mutant peptide of TMEM100 (T100-Mut) can permeate the plasma membrane and disturb correct binding of TMEM100, thereby inhibiting TRPA1-associated pain-related behavior. This may provide a novel strategy for reducing pain sensation.

### 3.2. *TRPV1–ANO1 Interaction*

Approximately 80% of TRPV1-positive DRG neurons also express ANO1 by immunostaining [8,93]. To examine the function of this co-expression, we investigated whether these ion channels interact in a setting of acute pain induced by capsaicin. We found that capsaicin-induced currents in isolated DRG neurons were suppressed by the ANO1 inhibitor, T16Ainh-A01 [94]. Although capsaicin-induced action potentials were also inhibited by T16Ainh-A01, inhibition was observed in response to second application of capsaicin (10 min after first capsaicin application). Some DRG neurons may not have exhibited second action potentials, yet there were DRG neurons that did show action potentials in the second application [93]. Another study showed that TRPV1 is desensitized by calmodulin binding. Moreover, TRPV1 function spontaneously and fully recovered after one hour, while desensitization was inhibited by TRPV1 phosphorylation [85]. Accordingly, the second response in some neurons is thought to depend on random phosphorylation levels in each neuron. Nonetheless, T16Ainh-A01 almost completely inhibits these second action potentials. These results suggest that chloride efflux elicited by ANO1 activation may accelerate depolarization to induce secondary action potentials, and that ANO1 inhibition may be effective at reducing pain sensation. In fact, capsaicin-induced pain-related behavior in mice is inhibited by T16Ainh-A01 [93]. Taken together, these findings indicate that the TRPV1–ANO1 interaction is critical for sensation of noxious stimuli.

ANO1 is also co-expressed with TRPV1 in TG neurons and is functionally involved in heat sensation [95,96]. In addition, TRPV1 and ANO1 expression levels are enhanced by estrogen in female rats [97]. Based on these observations, the TRPV1–ANO1 interaction may be a crucial target for pain therapies. According to a previous report, approximately 70% of ANO1-positive neurons do not colocalize with CGRP [8]. Although contribution of the TRPV1–ANO1 interaction in CGRP release is unknown, the TRPV1–ANO1 interaction may encompass the alternative side of the TRPV1–TMEM100–TRPA1 interaction system (Figure 1).

### 3.3. *Triple Conjugation of TRPV1, TRPA1, and TRPM3*

TRPM3 is a heat sensitive TRP channel that functionally couples with TRPV1 and TRPA1 [98]. Although TRPM3-deficient mice show a delayed tail flick at 57 °C, the effect of TRPM3 alone on heat sensation is unclear because tail flick behavior induced at 57 °C in TRPM3/TRPA1 double-deficient mice is no different to wild-type mice [4]. However, triple conjugation of TRPV1, TRPA1, and TRPM3 is important for detecting the noxious heat environment [4]. Withdrawal latency of TRPM3-deficient mice in the hot-plate test (50 °C) is the same as in wild-type mice [98]. Interestingly, this behavior disappears in TRPV1/TRPA1/TRPM3 triple-deficient mice, while the other responses to nociceptive stimuli are normal. Furthermore, wild-type and triple-deficient mice show a similar distribution on a gradient temperature plate (from 5 to 50 °C). In addition, CGRP-expressing DRG neurons are involved in heat sensation [62], and CGRP release from skin preparations is enhanced by the TRPM3 agonist, CIM0216, which is the same as for capsaicin treatment [99]. These findings indicate that the likely multiple function of TRPV1, TRPA1, and TRPM3 in peptidergic DRG neurons is to escape from a noxious heat environment.

## 4. Headache and Migraine

Primary headaches (i.e., those that are not associated with another disorder) are one of the most common causes of disability worldwide. The various types of headache include migraine, tension, and trigeminal autonomic cephalalgia [100]. Migraine is a multifactorial and incapacitating neurovascular disorder characterized by recurrent attacks of severe, unilateral, and throbbing headache, which can be aggravated by routine physical activity and can last from several hours to several days [100]. Migraine attacks often involve not only head pain, but also several premonitory and postdromal symptoms that occur before the headache initiates and persist after the headache ends, respectively. These symptoms are diverse and may include hypersensitivity to light, sound, smell, fatigue, neck stiffness, yawning, mood change, nausea, vomiting, cutaneous allodynia, and transient visual disturbances termed aura [101–104]. Migraine attacks can be triggered by many internal and external stimuli such as stress, hormonal fluctuations, sleep disturbances, skipping meals, weather changes, and ingestion of alcoholic beverage or certain types of food [105,106]. This multifactorial origin, as well as the variety of symptoms, have complicated identification of the underlying mechanisms of migraines. Although the events that trigger migraines remain unknown, the pain phase of migraine headaches is thought to involve activation and sensitization of primary afferent nociceptors that innervate the dural and meningeal vasculature. Indeed, the trigeminovascular system (TGVS) is a key component in pain initiation and transmission in migraine. Specifically, perivascular TG nerve endings are known to release CGRP, which induces vasodilation of cranial blood vessels and degranulation of meningeal mast cells, leading to neurogenic inflammation [107,108]. Interestingly, recent evidence indicated that a variety of ion channels, including TRP channels, make important contributions to migraine physiopathology (Figure 2 and Table 1). Several recent reviews discussed involvement of TRP channels in migraine, confirming significant interest in these channels as molecular targets for treatment of migraine [109–111]. In the following sections, we will discuss recent findings regarding TRP channels in migraine.

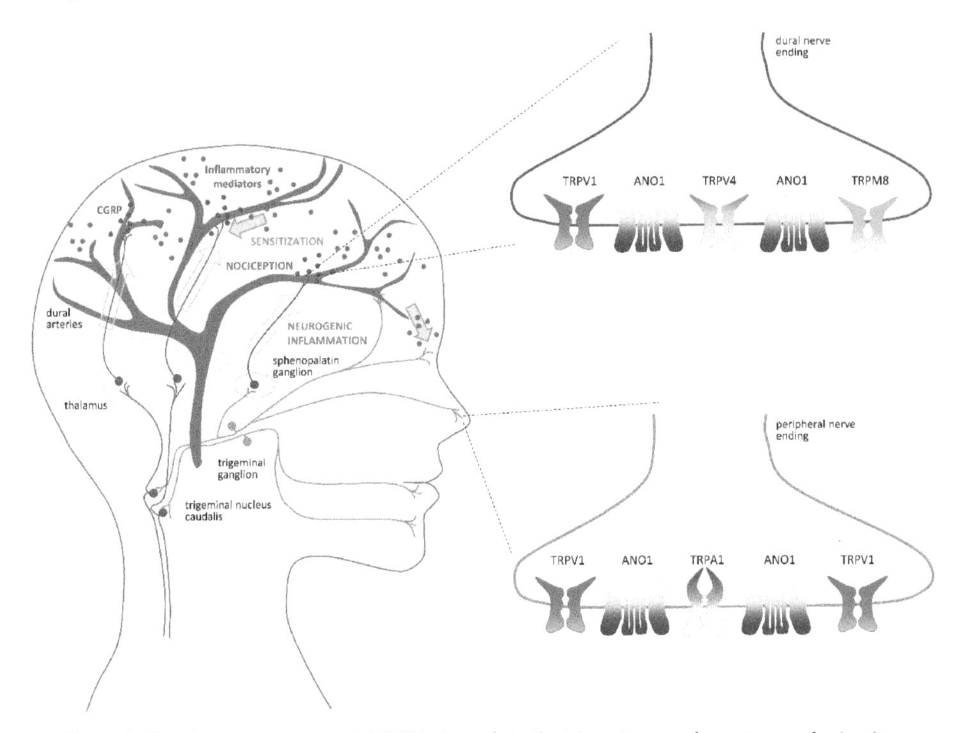


**Figure 2.** Transient receptor potential (TRP) channels in the trigeminovascular system and migraine development. Migraine triggering events remain unknown and may be a central or peripheral process. The precise order of events contributing to migraine pain is also debated. It is generally admitted that activation and sensitization of primary afferent nociceptors that innervate the dural and meningeal vasculature trigger both calcitonin gene-related peptide (CGRP)-induced vasodilatation and neurogenic inflammation. Pain signals pass through the trigeminal nucleus caudalis, which relays signals to higher order neurons in the thalamus and cortex (green arrows). A trigemino–parasympathetic or trigeminal autonomic reflex arc passes through the sphenopalatine ganglion and is responsible for migraine pain by mediating neurogenic inflammation (blue arrow). Central and peripheral sensitization (pink arrows) may contribute to maintenance of pain signals and predispose to future migraine attacks. Transient receptor potential (TRP) vanilloid 1 (TRPV1) and anoctamin 1 (ANO1) might be involved in initiation, nociception, and sensitization processes of migraine. TRP ankyrin 1 (TRPA1) might be more relevant in the initiation phase. There are few studies available on TRPV4 and TRP melastatin 8 (TRPM8), yet these receptors might be important in pain signal transmission or neurogenic inflammation.

**Table 1.** Roles of transient receptor potential (TRP) channels in migraine and their potential as therapeutic targets. RNS: reactive nitrogen species, ROS: reactive oxygen species, TRPA1: transient receptor potential ankyrin 1, TRPM8: transient receptor potential melastatin 8, TRPV1: transient receptor potential vanilloid 1, TRPV4: transient receptor potential vanilloid 4.

| | Action | Potential Endogenous Modulators | Antagonists with Potential Clinical Use |
|---|---|---|---|
| TRPV1 | CGRP release and vasodilatation [112]<br>Sensitization of peripheral and central neurons (during and between migraine attacks) [113–117] | Alcohol [118]<br>Cortical spreading depression (lowered pH in cortical neurons) [119]<br>Inflammatory mediators [113–115]<br>Estrogens [97]<br>Obesity-related mechanisms [120] | JNJ-38893777 [121]<br>JNJ-17203212 [121] |
| TRPA1 | CGRP release and vasodilatation [122] | Volatile irritants [123]<br>ROS and RNS [123] | Compound 16-8 [124] |
| TRPV4 | Cutaneous allodynia [125]<br>Nociception [126,127] | Mechanical force [126]<br>Formalin (directly or undirectly) [127] | Compound 16-8 [124] |
| TRPM8 | Genetic predisposition [128–132]<br>Anti- and pronociception (depending on context) [133] | Unknown | Unknown |

### 4.1. *TRPV1 as a Crucial Migraine Initiator*

The first hint that TRPV1 is involved in migraine was demonstration of its co-expression with CGRP in rat TG neurons [134,135] and mouse dural afferent neurons [136], suggesting a crucial role for TRPV1 in migraine. Moreover, in rat dura mater, application of capsaicin is accompanied by vasodilatation mediated by CGRP release from sensory afferent nerves [112]. Another substantial link between migraine and TRPV1 was the demonstration that a frequent migraine trigger, ethanol, induced neurogenic vasodilation via TRPV1 activation and subsequent CGRP release in the TGVS of guinea pigs [118]. Although the triggering event that actually initiates a migraine attack remains elusive, Meents et al., proposed that the premonitory aura exhibited by some migraineurs promotes endogenous activation of TRPV1 [137]. Aura arises from a phenomenon called cortical spreading depression (CSD), a short-lasting depolarization of cortical neurons that is known to increase extracellular concentrations of $H^+$, which can ultimately activate TRPV1 and induce CGRP release [119]. However, not all migraineurs experience aura, indicating that other endogenous mechanisms likely contribute to TRPV1 activation within the dura. Hence, although TRPV1 is currently recognized as a key player in migraine initiation, it is also likely involved in other phases and characteristics of migraine.

Sensitization of peripheral and central trigeminovascular neurons is usually observed following migraine attack onset. Peripheral sensitization mediates the throbbing perception of a headache, whereas sensitization of second-order neurons from the spinal trigeminal nucleus mediates cephalic allodynia and muscle tenderness [138]. Sensitization of TG neurons may contribute to direct sensitization of TRPV1, either by its increased activity or translocation to the cell membrane, or by increased protein production. Interestingly, the cerebrospinal fluid of chronic migraineurs (i.e., patients who have more than 15 migraine attacks per month) exhibit elevated levels of inflammatory mediators, including NGF [139]. Bradykinin and prostaglandin E2 are inflammatory mediators also released during neurogenic inflammation, and are commonly used in an animal model of headache to induce a chronic state of trigeminal hypersensitivity [113–115]. As already noted, NGF can trigger TRPV1 translocation to the plasma membrane, while bradykinin and prostaglandin E2 can orchestrate TRPV1 phosphorylation, which lowers its activation threshold. Moreover, TRPV1 expression is up-regulated in nerve fibers that innervate arteries in the scalp of chronic migraine patients [116]. Therefore, neurogenic inflammation that occurs during migraine attacks likely contributes to sensitization by modulating TRPV1 channel activity and expression.

Schwedt and colleagues proposed an interesting idea, namely that a state of persistent sensitization is maintained in migraineurs that enable more ready firing of TGVS [117]. They hypothesized that a cyclical process of migraine headaches causes interictal sensitization that contributes to predisposition to future migraine attacks. In their study, episodic and chronic migraineurs displayed enhanced sensitivity to thermal stimulation (decreased heat and cold pain threshold and tolerance) during the interictal period, compared with non-migraine controls. Such sensitization can partly be attributed to TRPV1 and may also explain why migraine patients do not tolerate ambient temperature changes [140]. In the same manner, a recent study showed that migraine patients exhibit enhanced extracephalic capsaicin-induced pain sensation during interictal periods, supporting the contribution of TRPV1 to interictal sensitization [141].

Another characteristic of migraines is their difference in prevalence and perception between men and women. Women are three-times more likely to suffer from migraine than men, and women experience more frequent, longer-lasting, and more intense migraine attacks than men [142]. Higher estrogen levels may in part be responsible for these differences. Interestingly, estrogen was recently shown to increase pain sensation by up-regulating expression of TRPV1 and ANO1 in TG neurons from female rats [97]. This could explain why women exhibit a higher susceptibility to migraine, and suggests that potential interaction of TRPV1 and ANO1 in TG neurons may be involved in migraine initiation.

Obesity has also been linked to migraine prevalence since it increases the risk of developing a migraine. Further persons with obesity suffer from more frequent and severe headache attacks [143]. A study involving mice fed a high-fat diet showed that facial intradermal injection of lower capsaicin doses are needed to induce photophobic behavior in obese mice compared to non-obese control mice [120]. Also, cell size distribution among TRPV1-positive cultured TG neurons from obese mice shifted towards larger cell diameters compared with control mice, and a higher capsaicin-induced calcium influx was observed in these neurons. In another mouse model that induced obesity through feeding of a high-fat and high-sucrose diet, dural application of capsaicin induced enhanced vasodilatory and vasoconstrictor responses compared with control animals. Basal and capsaicin-induced CGRP release from meningeal afferents was also increased [144]. These findings may explain why diet-induced obesity is associated with TGVS sensitization, which might occur via TRPV1 modulation, although the precise molecular mechanism is unclear.

### 4.2. *TRPA1 as an Intriguing Migraine Contributor*

TRPA1 is also co-expressed in CGRP-positive nociceptors [136,145] and has attracted significant attention in the context of migraine pathophysiology due to its sensitivity to numerous exogenous and endogenous compounds. Indeed, environmental irritants such as cigarette smoke or formaldehyde, as well as ROS and reactive nitrogen species (RNS), can activate TRPA1 (for review see [123]). A link with migraine can subsequently be readily deduced from the ability of these compounds to generate headache. Appropriately, TRPA1 activation in trigeminal nerve endings located in the nasal mucosa are suspected to trigger headache when irritants are inhaled. Indeed, intranasal administration of irritant compounds in rats can induce CGRP release via TRPA1 activation and increase cerebral blood flow [122]. Similarly, intranasal administration of umbellulone, the volatile active compound from *Umbellaria californica*, known as the "headache tree", evokes TRPA1-mediated and CGRP-dependent neurogenic meningeal vasodilation in mice [146,147]. In contrast, compounds known for their anti-headache properties were shown to desensitize TRPA1. For example, stimulation of rat TG neurons with parthenolide, a compound extracted from the feverfew herb (*Tanacetum parthenium* L.), which has been used for centuries to reduce pain, fever, and headaches [148]), induced potent and prolonged desensitization of TRPA1 channels, which rendered peptidergic neurons unresponsive to any stimulus and unable to release CGRP [149]. Similar properties were observed for isopetasin, a major constituent of extracts from butterbur, a plant known to have anti-migraine effects. Isopetasin visibly desensitized TRPA1 in patch-clamp experiments with rat TG neurons, while it also inhibited nociception and neurogenic dural vasodilatation mediated by TRPA1 in vivo [150].

Another important migraine trigger is ROS. Several studies reported increased oxidative stress in migraine patients both during headache attacks and in the interictal period (the period between migraine attacks) [151–153]. As already noted, ROS are potent TRPA1 activators, and in a recent study were shown to mediate the CSD responsible for aura [154]. In that study, exogenous $H_2O_2$ activated TRPA1 expressed in cortical neurons in mice brain slices, raising their susceptibility to CSD. Conversely, endogenous ROS produced upon CSD development [155] activated TRPA1 expression in TG neurons and mediated CGRP production, leading to a positive feedback loop that regulates cortical susceptibility to CSD. Based on these findings, it was proposed that reducing ROS production together with blockade of neuronal TRPA1 could help prevent stress-triggered migraine.

RNS can also act as TRPA1 agonists [79], and have been linked to headaches and migraine development. Indeed, an increase in endogenous nitric oxide (NO) production is observed during migraine attacks [156]. Eberhardt and colleagues reported that nitroxyl, generated by a redox reaction between NO and hydrogen sulfide can trigger TRPA1 activation in the TGVS, leading to CGRP release in the cranial dura mater of rats [145]. This pathway ultimately resulted in vasodilation and increased meningeal blood flow, and could also account for the headache phase of a migraine attack. Similarly, the well-known headache inducer, glyceryl trinitrate, targets TRPA1 in TG neurons to generate periorbital oxidative stress and mechanical allodynia [157].

### 4.3. *TRPM8 as a Familial Migraine Instigator*

TRPM8 is found on both Aδ and C fiber afferents, and is important for the activation of peripheral sensory neurons by cold temperature. It is activated at non-noxious cold temperatures (< 26 °C) and by compounds that produce a cooling sensation such as menthol or eucalyptol [158,159]. While its role as a cold sensor has been firmly established, it is not the case regarding its role in pain sensation. It is still under debate whether TRPM8 reduces or exacerbates pain sensation, and the most recent view on the matter is that TRPM8-expressing afferent fibers have the ability to both produce and alleviate pain, and the outcome will be determined by context (see for review [133,160]). As such, TRPM8 has begun to gather attention in the migraine field. A genetic predisposition to migraine is well-recognized: migraineurs presenting a hereditary component account for 42% of patients with migraine, as shown in studies on families and twins [161,162]. Migraine is genetically complex because many genetic variants with small effects and environmental factors can confer migraine susceptibility [163]. However, several genome-wide association studies from different cohorts identified single nucleotide polymorphisms (SNPs) in the gene encoding TRPM8, suggesting an important role for this TRP channel in migraine pathophysiology [128–132]. Several of these variants are located in regions involved in transcriptional regulation and may therefore impact upon TRPM8 expression levels. Moreover, in calcium imaging experiments, some TRPM8 SNP variants heterologously expressed in HEK293 cells showed alterations of channel functionality [164]. Based on these results, TRPM8 variants identified in migraine patients likely contribute to migraine pathology. In adult mice, TRPM8 is also expressed in dural trigeminal nerve endings, albeit rather sparsely [136,165]. Age-dependent decreases in TRPM8 expression in TG neurons appears to play a role in pathways that are differentially regulated with age, in that both the density and number of branches of TRPM8-expressing fibers are comparable to CGRP-expressing fibers in postnatal mouse dura. Specifically, both are reduced by half in adult mouse dura [165]. However, the functional consequence of this reduction remains unclear.

Although TRPM8 is a well-established cold transducer, limited temperature fluctuations in the skull suggest that this activity is less important in dural tissue. Thus, endogenous TRPM8 activators within the dura are unknown. Similarly, whether TRPM8 activation within the dura has a pro- or anti-nociception effect is unclear. The most recent studies yielded opposite results. Ren and colleagues observed that dural application of menthol resulted in inhibition of nocifensive behavior in a mouse migraine model induced by inflammatory mediators, suggesting an anti-nociceptive effect of TRPM8 [165]. In contrast, dural application of icilin produced migraine-like behavior in mice, such as cutaneous facial and hind paw allodynia. Pretreatment with the TRPM8 antagonist, AMG1161, attenuated these behaviors [133]. The contrary results obtained in these two studies may be due to the model used: when activated alone, dural TRPM8 appears to have a pro-nociceptive effect, but when activated together with inflammatory mediators, TRPM8 has an anti-nociceptive effect. Ultimately, TRPM8 activation may act as a migraine initiator in the first instance, but have another role during the neurogenic inflammation phase. Moreover, as suggested by Dussor and Cao, these different outcomes might also reflect how TRPM8-expressing fibers project to central neurons as well as the context dependence of TRPM8 activation [166]. More studies are needed to fully understand the role of TRPM8 in dural afferents and migraine pathophysiology.

### 4.4. *TRPV4 as an Indirect Migraine Modulator*

TRP Vanilloid 4 (TRPV4) is a widely distributed cationic channel that participates in the transduction of both physical (osmotic, mechanical, and heat) and chemical (endogenous, plant-derived, and synthetic ligands) stimuli (see for review [126]). As a mechanosensitive channel, TRPV4 has attracted increasing interest in the context of migraine. Indeed, headaches can be influenced by changes in intracranial pressure. Recently, TRPV4 was shown to be expressed in dural afferents, and its activation in the dura of freely-moving rats could produce migraine-like behavior such as cutaneous allodynia [125]. Although dural afferents are known to be mechanically sensitive, whether TRPV4 activation that contributes to migraine is due to mechanical stimulation or another endogenous mechanism remains to be elucidated. Another study showed that both TRPV4 and TRPA1 can be activated by the irritant formalin in the TGVS, and result in downstream MEK–ERK pathway activation and pain behavior in mice [127]. However, whether formalin directly or indirectly activates TRPV4 is unknown. Nevertheless, taken together, these findings suggest that TRPV4 could also be a promising target for agents that provide relief from pain that originates in the trigeminal system.

### 4.5. *TRP Channel Modulators for Acute Treatment of Migraine Attacks*

As an initial line of investigation, TRPV1 agonists were considered as potential analgesics to treat headaches. Intranasal applications of civamide and capsaicin were reported to alleviate headache pain during migraine attack [167,168]. However, in most patients these agents caused severe side effects, such as nasal burning and lacrimation, and thus impeded their clinical use for the treatment of acute migraine. Instead, TRP channels antagonists show more promise as a novel approach to prevent or treat acute migraine attacks. However, an initial clinical randomized trial conducted in 2009 showed that TRPV1 channel blockers failed to treat migraine attacks. In this study, the compound SB-705498 did not relieve headache pain for up to 24 h post-dose [169]. Although this outcome does not exclude a contribution of TRPV1 to migraine pathology, it indicates that selectively targeting TRPV1 alone is not sufficient for acute treatment of migraine attacks. Another explanation for the lack of success with TRPV1 channel blockers is that SB-705498 is largely ineffective in humans. Several other clinical trials using this compound showed no or poor efficacy for treating different conditions. Notably, SB-705498 did not relieve itching arising from histamine-induced pruritus, prevent coughing in refractory chronic cough, or alleviate symptoms elicited by cold, dry air in non-allergic rhinitis [170–172], despite documented involvement of TRPV1 in these disorders. To date, no other clinical trials using TRP channel antagonists have been performed for migraine, but numerous in vivo studies show their potential for the development of new therapeutic strategies. Indeed, in a recent study, two TRPV1 antagonists, JNJ-38893777 and JNJ-17203212, reduced or even completely abolished capsaicin-induced CGRP release from TG neurons in two different animal models of migraine [121]. These compounds, used alone or together with other blockers of important molecular players, could be promising pain relief medicines.

Interestingly, a new molecule, Compound 16-8, which specifically co-targets TRPV4 and TRPA1, was developed based on the TRPV4 antagonist, GSK205 [124]. Compound 16-8 was reported to inhibit both channels at sub-micromolar potency and also abolish formalin-induced trigeminal pain in an in vivo model. This suggests that dual inhibitors may be more effective in treating pain elicited by several molecular players, such as pain that occurs in headaches and migraine induced by irritant compounds. To date, no clinical study has focused on TRPV4. This may be due to limited research concerning TRPV4, and the fact that dual inhibition strategies have not yet been considered for the treatment of migraine. Moreover, although the contribution of ANO1 to pain mechanisms in TG neurons is not fully elucidated [96] and interactions between TRP channels and ANO1 await investigation in the TGVS, we contend that ANO1 in TG neurons likely behaves similarly to that seen in DRG neurons. Thus, simultaneous blockage of TRP and ANO1 channels has potential to provide strong pain relief from headache.

Although TRPM8 variants are associated with migraine susceptibility, whether therapeutic strategies that target this channel should be agonists or antagonists, is unclear. As such, additional information about the role of TRPM8 in migraine development is needed before new therapeutics that focus on this channel can be pursued.

## 5. Infection and Immunity

Pain sensation is a negative stress for animals, and CGRP release from nociceptors exacerbates symptoms. Conversely, we recently clarified that CGRP release dependent on TRPV1 and TRPA1 activity in DRG neurons is involved in bone protection during fungus infection. This phenomenon is supported by several physiological mechanisms, including ATP release from keratinocytes, neural excitation of sodium channel 1.8 (Nav1.8)-positive DRG neurons, and CGRP-dependent suppression of osteoclasts activated via TNF-. Thus, in this part, we comprehensively describe the physiological and pathological systems involved in cutaneous infection to bone inflammation (Figure 3).

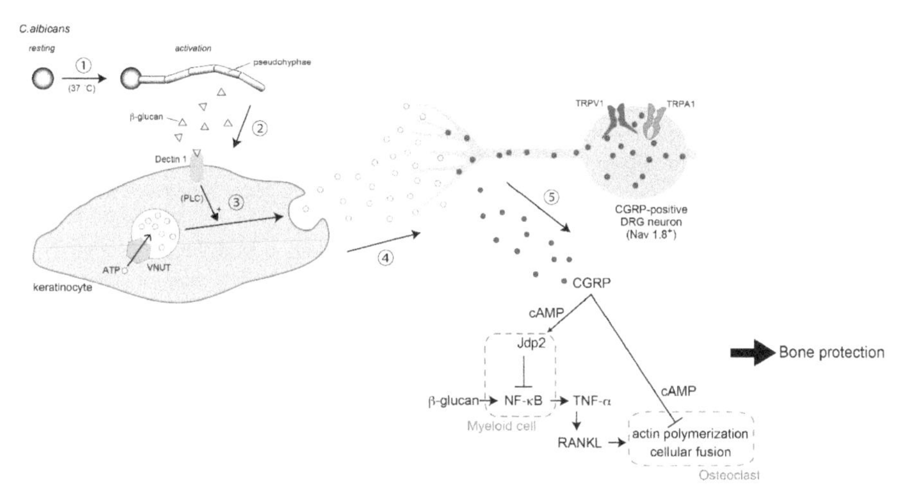


**Figure 3.** Bone protection system from *Candida* infection. Bone disruption followed by inflammation is worsened by overactive osteoclasts. Nociceptors suppress osteoclast development through calcitonin gene-related peptide (CGRP) release. The process involves: (1) *Candida albicans* is activated in an optimal environment (e.g., body temperature); (2) β-glucan released from pseudohyphae bind to its receptor, dectin-1, on the plasma membrane of keratinocytes; (3) ATP release from keratinocytes is enhanced through the phospholipase C (PLC) pathway; (4) neuronal excitation in voltage-gated sodium channel 1.8 (Nav1.8)-positive dorsal root ganglion (DRG) neurons; and (5) CGRP release from DRG neurons. Jun dimerization protein 2 (Jdp2) is activated by CGRP through a cAMP cascade in myeloid cells. In turn, tumor necrosis factor-alpha (TNF-α release (which accelerates osteoclast development) is suppressed. TNF-α-dependent inflammation is induced by the direct effect of β-glucan on myeloid cells. Furthermore, the CGRP–cAMP axis in osteoclasts also inhibits over-development. Thus, these pathways from skin to bone induce bone protection and inhibit bone inflammation during fungus infection.

## 5.1. Nociception by C. albicans

*Candida albicans* infections can cause skin or vulvar pain. Breast candidiasis is characterized by severe pain around the nipple [173]. *C. albicans* in the vagina causes itching and mechanical allodynia [174]. *C. albicans* can also enter skeletal tissue and induce painful bone infection [175]. Although *C. albicans* has algesic activity, the mechanisms by which this fungus triggers pain remains completely unknown.

TLR4 expressed on myeloid cells are involved in recognition of fungal mannan and cytokine production upon MyD88 and TIR-domain-containing adapter-inducing interferon-beta (TRIF) activation. The fungal cell wall contains β-glucan and mannan on the intracellular and extracellular face, respectively [176]. Surface exposure of β-glucan is sensed by dectin-1 [177]. Activated dectin-1 assembles as a multimeric complex and induces signaling via an ITAM-like motif, promoting formation of CARD-9–Bcl-10–Malt-1 trimers (CBM trimer) and activation of the NLRP3–ASC–ICE complex (NLRP3 inflammasome). CBM trimers and NLRP3 inflammasome activation are both required to induce cytokine production [178].

Candidalysin was recently discovered and is the first fungal cytolytic peptide [179]. This peptide may also contribute to the pathogenesis of fungal invasion. Pain induced by fungal infections is thought to be caused by inflammation, but a recent study suggested that both *Staphylococcus aureus*-derived N-formulated peptides and α-hemolysin can directly stimulate nociceptors [180]. Therefore, nociceptors may be able to sense pathogens, but the underlying molecular mechanisms behind fungal nociception remain unclear.

In the colonization phase, *C. albicans* forms yeast-like structures that are harmless because colony growth is suppressed by host immunity and the natural antagonistic effects of microbial flora. When the yeast form of *C. albicans* attaches to the skin of an immunocompromised host, budding growth is immediately induced and the soluble β-glucan form is secreted. Notably, β-glucan-induced allodynia is relatively severe compared with that induced by mannan and other pathological components such as Candidalysin. Furthermore, dectin-1-deficient mice are completely unresponsive to *C. albicans* or β-glucan-induced pain in a MyD88/TRIF/inflammasome-independent manner. Moreover, we discovered that *C. albicans* induces acute pain by stimulating Nav1.8-positive nociceptors in primary sensory neurons via the dectin-1-mediated PLCγ2–TRPV1/TRPA1 axis. β-Glucan also induces allodynia, which is dependent on dectin-1-mediated ATP secretion from keratinocytes. Notably, keratinocyte-derived extracellular ATP stimulates sensory neurons via P2X receptors. We also found that mice deficient in the ATP transporter, vesicular nucleotide transporter (VNUT), are unresponsive to β-glucan-induced allodynia, while the VNUT inhibitor clodronate has potent prophylactic potential to target fungal nociceptive symptoms. Together, these findings suggest that ATP- or VNUT-targeted therapies such as clodronate treatment may be a promising therapeutic option for treating pain or allodynia associated with fungal infections [181].

## 5.2. Secondary Symptoms Following Nociception

Nociceptor innervation is seen in skin and bone. Although the function of nociceptors in the osteo–immune system is unclear, ion channels in the DRG may be responsible for sensing noxious stimuli [182]. During inflammation, pro-algesic cytokines derived from immune cells gradually evoke allodynia, leading to production of neuropeptides such as CGRP [183], which in turn causes vasodilatation, impaired insulin release, and enhanced Th17 cell function [184–186]. Meanwhile, depletion of TRPV1-positive neurons or CGRP deficiency can lead to osteoporosis [187,188]. Thus, nociceptors may modulate osteo–immune system activity, but how they influence pathogen-induced inflammation and bone destruction in a physiological context remains unknown.

To investigate these questions, we injected LPS or β-glucan into the hind paw of *Nav1.8CreRosa26DTA* mice, a nociceptor-deficient line. Notably, LPS-induced osteo-inflammation was unaffected, suggesting that nociceptors do not affect TLR-induced osteo-inflammation. In contrast, *Nav1.8CreRosa26DTA* mice injected with β-glucan exhibit severe skin inflammation and bone destruction, indicating that nociceptors are negative regulators of fungal osteo-inflammation. Similar to *Nav1.8CreRosa26DTA* mice, TRPV1/TRPA1 double-deficient mice exhibit severe osteo-inflammation in response to β-glucan, and this phenotype was rescued by CGRP administration. Notably, β-glucan injection into the hind paws of both *Nav1.8CreRosa26DTA* and TRPV1/TRPA1 double-deficient mice abolished serum CGRP, indicating that TRP channels acting as nociceptors are required for CGRP induction. To address how CGRP inhibits osteo-inflammation, we assessed the effects of CGRP on osteoclast formation and cytokine production. Intriguingly, we discovered that nerve-derived CGRP inhibits osteoclast actin polymerization via cAMP induction, leading to impaired osteoclast multinucleation. We also found that the CGRP-induced transcriptional repressor, Jun dimerization protein 2, selectively blocks dectin-1-mediated pro-inflammatory cytokine production in myeloid cells via direct inhibition of p65. These unexpected roles for β-glucan-stimulated nociceptors suggests the existence of novel sensocrine pathways that may play a role in fungal osteo-inflammation [181].

## 6. Conclusions

Studies from the last two decades show the importance of TRP channels in pain sensations caused by noxious temperatures and many chemicals. In particular, TRPA1 is activated in many conditions and its activity evokes an extremely uncomfortable sensation. Therefore, TRPA1 may be a crucial target for pain treatment, although TRPV1 contribution to more specific nociception cannot be disregarded. While some pharmaceutical companies are already focused on the development of TRPA1 and TRPV1 antagonists [189], ANO1 inhibition may also be effective in treating pain because the role of ANO1 is akin to an amplifier, and its suppression does not generate a painless condition. Namely, a level of nociception that is sufficient enough to sense damage for survival can be maintained in an ANO1-blocked state, but not with shutdown of the detectors, i.e., TRP channels.

Sensory systems do not only detect noxious stimuli but also participate in the establishment of inflammation or chronic pain. Neural excitation induces CGRP release from nociceptor termini, inducing inflammation that can ultimately lead to sensitization of nociceptors. Therefore, inhibition of CGRP release by suppression of TRP channel activity is expected to provide relief for intractable pain, headache, and migraine. However, this strategy may result in dangerous secondary effects in some diseases, including fungus infection. Bone is often disrupted in *Candida* infection, a situation induced by osteoclast activity. It remains unclear why osteoclast activity is up-regulated during infection, but in this case, CGRP release from nociceptors becomes beneficial by suppressing two pathways: TNF-α release from myeloid cells and overactivation of osteoclasts. Thus, inflammation induced by CGRP is not detrimental in some pathological conditions, yet targeting it for pain relief might not always be the best strategy.

In fact, pain sensation negatively controls our physiological conditions, including our emotions. Recent reports indicate that complete abolishment of pain induces a severe pathological condition to our body. Therefore, we believe that the important point of pain management is to decrease pain to a tolerable level, but one that is sufficient enough to maintain natural protection against tissue damage. To develop this strategy, there is a need for the discovery of new TRP channels and ANO1 inhibitors that could be used concomitantly and adjusted depending on each patient's condition.

**Author Contributions:** Writing—Original Draft Preparation (Parts 1 and 2), Y.T.; Writing—Original Draft Preparation (Parts 3 and 4), S.D. and K.M., respectively; Writing—Review & Editing, M.T.

**Funding:** This research was funded by Japan Society for the Prmotion of Science (JP17K15793 to Y.T.), Takeda Science Foundation (to Y.T.), and Japan Society for the Prmotion of Science (JP18H02970, JP19K22712 to K.M.).

**Acknowledgments:** This paper was supported by the Showa University School of Medicine. We thank Rachel James, from Edanz Group (www.edanzediting.com/ac) for editing a draft of this manuscript.

**Conflicts of Interest:** The authors declare no conflict of interest.

## References


1. Basbaum, A.I.; Bautista, D.M.; Scherrer, G.; Julius, D. Cellular and molecular mechanisms of pain. *Cell* **2009**, *139*, 267–284. [CrossRef] [PubMed]
2. Vandewauw, I.; Owsianik, G.; Voets, T. Systematic and quantitative mRNA expression analysis of TRP channel genes at the single trigeminal and dorsal root ganglion level in mouse. *BMC Neurosci.* **2013**, *14*, 21. [CrossRef] [PubMed]
3. Julius, D. TRP channels and pain. *Annu. Rev. Cell Dev. Biol.* **2013**, *29*, 355–384. [CrossRef] [PubMed]
4. Vandewauw, I.; De Clercq, K.; Mulier, M.; Held, K.; Pinto, S.; Van Ranst, N.; Segal, A.; Voet, T.; Vennekens, R.; Zimmermann, K.; et al. A TRP channel trio mediates acute noxious heat sensing. *Nature* **2018**, *555*, 662–666. [CrossRef] [PubMed]
5. Caputo, A.; Caci, E.; Ferrera, L.; Pedemonte, N.; Barsanti, C.; Sondo, E.; Pfeffer, U.; Ravazzolo, R.; Zegarra-Moran, O.; Galietta, L.J. TMEM16A, a membrane protein associated with calcium-dependent chloride channel activity. *Science* **2008**, *322*, 590–594. [CrossRef] [PubMed]
6. Yang, Y.D.; Cho, H.; Koo, J.Y.; Tak, M.H.; Cho, Y.; Shim, W.S.; Park, S.P.; Lee, J.; Lee, B.; Kim, B.M.; et al. TMEM16A confers receptor-activated calcium-dependent chloride conductance. *Nature* **2008**, *455*, 1210–1215. [CrossRef] [PubMed]
7. Schroeder, B.C.; Cheng, T.; Jan, Y.N.; Jan, L.Y. Expression cloning of TMEM16A as a calcium-activated chloride channel subunit. *Cell* **2008**, *134*, 1019–1029. [CrossRef] [PubMed]
8. Cho, H.; Yang, Y.D.; Lee, J.; Lee, B.; Kim, T.; Jang, Y.; Back, S.K.; Na, H.S.; Harfe, B.D.; Wang, F.; et al. The calcium-activated chloride channel anoctamin 1 acts as a heat sensor in nociceptive neurons. *Nat. Neurosci.* **2012**, *15*, 1015–1021. [CrossRef] [PubMed]
9. Oh, U.; Jung, J. Cellular functions of TMEM16/anoctamin. *Pflug. Arch.* **2016**, *468*, 443–453. [CrossRef]
10. Jin, X.; Shah, S.; Liu, Y.; Zhang, H.; Lees, M.; Fu, Z.; Lippiat, J.D.; Beech, D.J.; Sivaprasadarao, A.; Baldwin, S.A.; et al. Activation of the Cl- channel ANO1 by localized calcium signals in nociceptive sensory neurons requires coupling with the IP3 receptor. *Sci. Signal* **2013**, *6*, ra73. [CrossRef]
11. Mao, S.; Garzon-Muvdi, T.; Di Fulvio, M.; Chen, Y.; Delpire, E.; Alvarez, F.J.; Alvarez-Leefmans, F.J. Molecular and functional expression of cation-chloride cotransporters in dorsal root ganglion neurons during postnatal maturation. *J. Neurophysiol.* **2012**, *108*, 834–852. [CrossRef] [PubMed]
12. Jordt, S.E.; Bautista, D.M.; Chuang, H.H.; McKemy, D.D.; Zygmunt, P.M.; Hogestatt, E.D.; Meng, I.D.; Julius, D. Mustard oils and cannabinoids excite sensory nerve fibres through the TRP channel ANKTM1. *Nature* **2004**, *427*, 260–265. [CrossRef] [PubMed]
13. Bandell, M.; Story, G.M.; Hwang, S.W.; Viswanath, V.; Eid, S.R.; Petrus, M.J.; Earley, T.J.; Patapoutian, A. Noxious cold ion channel TRPA1 is activated by pungent compounds and bradykinin. *Neuron* **2004**, *41*, 849–857. [CrossRef]
14. Macpherson, L.J.; Geierstanger, B.H.; Viswanath, V.; Bandell, M.; Eid, S.R.; Hwang, S.; Patapoutian, A. The pungency of garlic: activation of TRPA1 and TRPV1 in response to allicin. *Curr. Biol.* **2005**, *15*, 929–934. [CrossRef] [PubMed]
15. Bautista, D.M.; Movahed, P.; Hinman, A.; Axelsson, H.E.; Sterner, O.; Hogestatt, E.D.; Julius, D.; Jordt, S.E.; Zygmunt, P.M. Pungent products from garlic activate the sensory ion channel TRPA1. *Proc. Natl. Acad. Sci. USA* **2005**, *102*, 12248–12252. [CrossRef] [PubMed]
16. Xu, H.; Delling, M.; Jun, J.C.; Clapham, D.E. Oregano, thyme and clove-derived flavors and skin sensitizers activate specific TRP channels. *Nat. Neurosci.* **2006**, *9*, 628–635. [CrossRef] [PubMed]
17. Koizumi, K.; Iwasaki, Y.; Narukawa, M.; Iitsuka, Y.; Fukao, T.; Seki, T.; Ariga, T.; Watanabe, T. Diallyl sulfides in garlic activate both TRPA1 and TRPV1. *Biochem. Biophys. Res. Commun.* **2009**, *382*, 545–548. [CrossRef]
18. Shintaku, K.; Uchida, K.; Suzuki, Y.; Zhou, Y.; Fushiki, T.; Watanabe, T.; Yazawa, S.; Tominaga, M. Activation of transient receptor potential A1 by a non-pungent capsaicin-like compound, capsiate. *Br. J. Pharmacol.* **2012**, *165*, 1476–1486. [CrossRef]
19. Lee, S.P.; Buber, M.T.; Yang, Q.; Cerne, R.; Cortes, R.Y.; Sprous, D.G.; Bryant, R.W. Thymol and related alkyl phenols activate the hTRPA1 channel. *Br. J. Pharmacol.* **2008**, *153*, 1739–1749. [CrossRef]
20. Fischer, M.J.; Leffler, A.; Niedermirtl, F.; Kistner, K.; Eberhardt, M.; Reeh, P.W.; Nau, C. The general anesthetic propofol excites nociceptors by activating TRPV1 and TRPA1 rather than GABAA receptors. *J. Biol. Chem.* **2010**, *285*, 34781–34792. [CrossRef]

21. Nishimoto, R.; Kashio, M.; Tominaga, M. Propofol-induced pain sensation involves multiple mechanisms in sensory neurons. *Pflug. Arch.* **2015**, *467*, 2011–2020. [CrossRef] [PubMed]
22. Takaishi, M.; Fujita, F.; Uchida, K.; Yamamoto, S.; Sawada Shimizu, M.; Hatai Uotsu, C.; Shimizu, M.; Tominaga, M. 1,8-cineole, a TRPM8 agonist, is a novel natural antagonist of human TRPA1. *Mol. Pain* **2012**, *8*, 86. [CrossRef] [PubMed]
23. Peyrot des Gachons, C.; Uchida, K.; Bryant, B.; Shima, A.; Sperry, J.B.; Dankulich-Nagrudny, L.; Tominaga, M.; Smith, A.B., 3rd; Beauchamp, G.K.; Breslin, P.A. Unusual pungency from extra-virgin olive oil is attributable to restricted spatial expression of the receptor of oleocanthal. *J. Neurosci.* **2011**, *31*, 999–1009. [CrossRef] [PubMed]
24. Wang, Y.Y.; Chang, R.B.; Liman, E.R. TRPA1 is a component of the nociceptive response to CO2. *J. Neurosci.* **2010**, *30*, 12958–12963. [CrossRef] [PubMed]
25. Fujita, F.; Uchida, K.; Takayama, Y.; Suzuki, Y.; Takaishi, M.; Tominaga, M. Hypotonicity-induced cell swelling activates TRPA1. *J. Physiol. Sci.* **2018**, *68*, 431–440. [CrossRef] [PubMed]
26. Fujita, F.; Uchida, K.; Moriyama, T.; Shima, A.; Shibasaki, K.; Inada, H.; Sokabe, T.; Tominaga, M. Intracellular alkalization causes pain sensation through activation of TRPA1 in mice. *J. Clin. Investing* **2008**, *118*, 4049–4057. [CrossRef] [PubMed]
27. Zurborg, S.; Yurgionas, B.; Jira, J.A.; Caspani, O.; Heppenstall, P.A. Direct activation of the ion channel TRPA1 by Ca2+. *Nat. Neurosci* **2007**, *10*, 277–279. [CrossRef] [PubMed]
28. Ohara, K.; Fukuda, T.; Okada, H.; Kitao, S.; Ishida, Y.; Kato, K.; Takahashi, C.; Katayama, M.; Uchida, K.; Tominaga, M. Identification of significant amino acids in multiple transmembrane domains of human transient receptor potential ankyrin 1 (TRPA1) for activation by eudesmol, an oxygenized sesquiterpene in hop essential oil. *J. Biol. Chem.* **2015**, *290*, 3161–3171. [CrossRef]
29. Hinman, A.; Chuang, H.H.; Bautista, D.M.; Julius, D. TRP channel activation by reversible covalent modification. *Proc. Natl. Acad. Sci. USA* **2006**, *103*, 19564–19568. [CrossRef]
30. Macpherson, L.J.; Dubin, A.E.; Evans, M.J.; Marr, F.; Schultz, P.G.; Cravatt, B.F.; Patapoutian, A. Noxious compounds activate TRPA1 ion channels through covalent modification of cysteines. *Nature* **2007**, *445*, 541–545. [CrossRef]
31. Samad, A.; Sura, L.; Benedikt, J.; Ettrich, R.; Minofar, B.; Teisinger, J.; Vlachova, V. The C-terminal basic residues contribute to the chemical- and voltage-dependent activation of TRPA1. *J. Biochem.* **2011**, *433*, 197–204. [CrossRef] [PubMed]
32. Karashima, Y.; Damann, N.; Prenen, J.; Talavera, K.; Segal, A.; Voets, T.; Nilius, B. Bimodal action of menthol on the transient receptor potential channel TRPA1. *J. Neurosci.* **2007**, *27*, 9874–9884. [CrossRef] [PubMed]
33. Xiao, B.; Dubin, A.E.; Bursulaya, B.; Viswanath, V.; Jegla, T.J.; Patapoutian, A. Identification of transmembrane domain 5 as a critical molecular determinant of menthol sensitivity in mammalian TRPA1 channels. *J. Neurosci.* **2008**, *28*, 9640–9651. [CrossRef] [PubMed]
34. Chen, J.; Kang, D.; Xu, J.; Lake, M.; Hogan, J.O.; Sun, C.; Walter, K.; Yao, B.; Kim, D. Species differences and molecular determinant of TRPA1 cold sensitivity. *Nat. Commun.* **2013**, *4*, 2501. [CrossRef] [PubMed]
35. Story, G.M.; Peier, A.M.; Reeve, A.J.; Eid, S.R.; Mosbacher, J.; Hricik, T.R.; Earley, T.J.; Hergarden, A.C.; Andersson, D.A.; Hwang, S.W.; et al. ANKTM1, a TRP-like channel expressed in nociceptive neurons, is activated by cold temperatures. *Cell* **2003**, *112*, 819–829. [CrossRef]
36. Obata, K.; Katsura, H.; Mizushima, T.; Yamanaka, H.; Kobayashi, K.; Dai, Y.; Fukuoka, T.; Tokunaga, A.; Tominaga, M.; Noguchi, K. TRPA1 induced in sensory neurons contributes to cold hyperalgesia after inflammation and nerve injury. *J. Clin. Investing* **2005**, *115*, 2393–2401. [CrossRef]
37. Miyake, T.; Nakamura, S.; Zhao, M.; So, K.; Inoue, K.; Numata, T.; Takahashi, N.; Shirakawa, H.; Mori, Y.; Nakagawa, T.; et al. Cold sensitivity of TRPA1 is unveiled by the prolyl hydroxylation blockade-induced sensitization to ROS. *Nat. Commun.* **2016**, *7*, 12840. [CrossRef]
38. Wang, S.; Dai, Y.; Fukuoka, T.; Yamanaka, H.; Kobayashi, K.; Obata, K.; Cui, X.; Tominaga, M.; Noguchi, K. Phospholipase C and protein kinase A mediate bradykinin sensitization of TRPA1: a molecular mechanism of inflammatory pain. *Brain* **2008**, *131*, 1241–1251. [CrossRef]
39. Tsuda, M.; Hasegawa, S.; Inoue, K. P2X receptors-mediated cytosolic phospholipase A2 activation in primary afferent sensory neurons contributes to neuropathic pain. *J. Neurochem.* **2007**, *103*, 1408–1416. [CrossRef]
40. Nishizuka, Y. Intracellular signaling by hydrolysis of phospholipids and activation of protein kinase C. *Science* **1992**, *258*, 607–614. [CrossRef]

41. Kondo, T.; Sakurai, J.; Miwa, H.; Noguchi, K. Activation of p38 MAPK through transient receptor potential A1 in a rat model of gastric distension-induced visceral pain. *Neuroreport* **2013**, *24*, 68–72. [CrossRef] [PubMed]
42. Schmidt, M.; Dubin, A.E.; Petrus, M.J.; Earley, T.J.; Patapoutian, A. Nociceptive signals induce trafficking of TRPA1 to the plasma membrane. *Neuron* **2009**, *64*, 498–509. [CrossRef] [PubMed]
43. Zhou, Y.; Suzuki, Y.; Uchida, K.; Tominaga, M. Identification of a splice variant of mouse TRPA1 that regulates TRPA1 activity. *Nat. Commun.* **2013**, *4*, 2399. [CrossRef] [PubMed]
44. Palsson-McDermott, E.M.; O'Neill, L.A. Signal transduction by the lipopolysaccharide receptor, Toll-like receptor-4. *Immunology* **2004**, *113*, 153–162. [CrossRef] [PubMed]
45. Hu, Y.; Gu, Q.; Lin, R.L.; Kryscio, R.; Lee, L.Y. Calcium transient evoked by TRPV1 activators is enhanced by tumor necrosis factor-{alpha} in rat pulmonary sensory neurons. *Am. J. Physiol. Lung Cell Mol. Physiol.* **2010**, *299*, L483–L492. [CrossRef] [PubMed]
46. Meseguer, V.; Alpizar, Y.A.; Luis, E.; Tajada, S.; Denlinger, B.; Fajardo, O.; Manenschijn, J.A.; Fernandez-Pena, C.; Talavera, A.; Kichko, T.; et al. TRPA1 channels mediate acute neurogenic inflammation and pain produced by bacterial endotoxins. *Nat. Commun.* **2014**, *5*, 3125. [CrossRef] [PubMed]
47. Andersson, D.A.; Gentry, C.; Moss, S.; Bevan, S. Transient receptor potential A1 is a sensory receptor for multiple products of oxidative stress. *J. Neurosci.* **2008**, *28*, 2485–2494. [CrossRef] [PubMed]
48. Bessac, B.F.; Sivula, M.; von Hehn, C.A.; Escalera, J.; Cohn, L.; Jordt, S.E. TRPA1 is a major oxidant sensor in murine airway sensory neurons. *J. Clin. Investing* **2008**, *118*, 1899–1910. [CrossRef] [PubMed]
49. Sawada, Y.; Hosokawa, H.; Matsumura, K.; Kobayashi, S. Activation of transient receptor potential ankyrin 1 by hydrogen peroxide. *Eur. J. Neurosci.* **2008**, *27*, 1131–1142. [CrossRef] [PubMed]
50. So, K.; Tei, Y.; Zhao, M.; Miyake, T.; Hiyama, H.; Shirakawa, H.; Imai, S.; Mori, Y.; Nakagawa, T.; Matsubara, K.; et al. Hypoxia-induced sensitisation of TRPA1 in painful dysesthesia evoked by transient hindlimb ischemia/reperfusion in mice. *Sci. Rep.* **2016**, *6*, 23261. [CrossRef]
51. Takahashi, N.; Kuwaki, T.; Kiyonaka, S.; Numata, T.; Kozai, D.; Mizuno, Y.; Yamamoto, S.; Naito, S.; Knevels, E.; Carmeliet, P.; et al. TRPA1 underlies a sensing mechanism for O2. *Nat. Chem. Biol.* **2011**, *7*, 701–711. [CrossRef] [PubMed]
52. Anand, U.; Otto, W.R.; Anand, P. Sensitization of capsaicin and icilin responses in oxaliplatin treated adult rat DRG neurons. *Mol. Pain* **2010**, *6*, 82. [CrossRef] [PubMed]
53. Nassini, R.; Gees, M.; Harrison, S.; De Siena, G.; Materazzi, S.; Moretto, N.; Failli, P.; Preti, D.; Marchetti, N.; Cavazzini, A.; et al. Oxaliplatin elicits mechanical and cold allodynia in rodents via TRPA1 receptor stimulation. *Pain* **2011**, *152*, 1621–1631. [CrossRef] [PubMed]
54. Yamamoto, K.; Chiba, N.; Chiba, T.; Kambe, T.; Abe, K.; Kawakami, K.; Utsunomiya, I.; Taguchi, K. Transient receptor potential ankyrin 1 that is induced in dorsal root ganglion neurons contributes to acute cold hypersensitivity after oxaliplatin administration. *Mol. Pain* **2015**, *11*, 69. [CrossRef] [PubMed]
55. Bailey, S.R.; Mitra, S.; Flavahan, S.; Flavahan, N.A. Reactive oxygen species from smooth muscle mitochondria initiate cold-induced constriction of cutaneous arteries. *Am. J. Physiol. Heart Circ. Physiol.* **2005**, *289*, H243–H250. [CrossRef] [PubMed]
56. Nativi, C.; Gualdani, R.; Dragoni, E.; Di Cesare Mannelli, L.; Sostegni, S.; Norcini, M.; Gabrielli, G.; la Marca, G.; Richichi, B.; Francesconi, O.; et al. A TRPA1 antagonist reverts oxaliplatin-induced neuropathic pain. *Sci. Rep.* **2013**, *3*, 2005. [CrossRef] [PubMed]
57. Beppu, K.; Sasaki, T.; Tanaka, K.F.; Yamanaka, A.; Fukazawa, Y.; Shigemoto, R.; Matsui, K. Optogenetic countering of glial acidosis suppresses glial glutamate release and ischemic brain damage. *Neuron* **2014**, *81*, 314–320. [CrossRef]
58. Shigetomi, E.; Tong, X.; Kwan, K.Y.; Corey, D.P.; Khakh, B.S. TRPA1 channels regulate astrocyte resting calcium and inhibitory synapse efficacy through GAT-3. *Nat. Neurosci.* **2011**, *15*, 70–80. [CrossRef] [PubMed]
59. Takahashi, N.; Mizuno, Y.; Kozai, D.; Yamamoto, S.; Kiyonaka, S.; Shibata, T.; Uchida, K.; Mori, Y. Molecular characterization of TRPA1 channel activation by cysteine-reactive inflammatory mediators. *Channels (Austin)* **2008**, *2*, 287–298. [CrossRef] [PubMed]
60. De la Roche, J.; Eberhardt, M.J.; Klinger, A.B.; Stanslowsky, N.; Wegner, F.; Koppert, W.; Reeh, P.W.; Lampert, A.; Fischer, M.J.; Leffler, A. The molecular basis for species-specific activation of human TRPA1 protein by protons involves poorly conserved residues within transmembrane domains 5 and 6. *J. Biol. Chem.* **2013**, *288*, 20280–20292. [CrossRef] [PubMed]

61. Hamilton, N.B.; Kolodziejczyk, K.; Kougioumtzidou, E.; Attwell, D. Proton-gated Ca(2+)-permeable TRP channels damage myelin in conditions mimicking ischaemia. *Nature* **2016**, *529*, 523–527. [CrossRef] [PubMed]
62. McCoy, E.S.; Taylor-Blake, B.; Street, S.E.; Pribisko, A.L.; Zheng, J.; Zylka, M.J. Peptidergic CGRPalpha primary sensory neurons encode heat and itch and tonically suppress sensitivity to cold. *Neuron* **2013**, *78*, 138–151. [CrossRef] [PubMed]
63. Furue, H.; Narikawa, K.; Kumamoto, E.; Yoshimura, M. Responsiveness of rat substantia gelatinosa neurones to mechanical but not thermal stimuli revealed by in vivo patch-clamp recording. *J. Physiol.* **1999**, *521*, 529–535. [CrossRef] [PubMed]
64. Uta, D.; Furue, H.; Pickering, A.E.; Rashid, M.H.; Mizuguchi-Takase, H.; Katafuchi, T.; Imoto, K.; Yoshimura, M. TRPA1-expressing primary afferents synapse with a morphologically identified subclass of substantia gelatinosa neurons in the adult rat spinal cord. *Eur. J. Neurosci.* **2010**, *31*, 1960–1973. [CrossRef]
65. Kosugi, M.; Nakatsuka, T.; Fujita, T.; Kuroda, Y.; Kumamoto, E. Activation of TRPA1 channel facilitates excitatory synaptic transmission in substantia gelatinosa neurons of the adult rat spinal cord. *J. Neurosci.* **2007**, *27*, 4443–4451. [CrossRef] [PubMed]
66. Andersson, D.A.; Gentry, C.; Alenmyr, L.; Killander, D.; Lewis, S.E.; Andersson, A.; Bucher, B.; Galzi, J.L.; Sterner, O.; Bevan, S.; et al. TRPA1 mediates spinal antinociception induced by acetaminophen and the cannabinoid Delta(9)-tetrahydrocannabiorcol. *Nat. Commun.* **2011**, *2*, 551. [CrossRef] [PubMed]
67. Gregus, A.M.; Doolen, S.; Dumlao, D.S.; Buczynski, M.W.; Takasusuki, T.; Fitzsimmons, B.L.; Hua, X.Y.; Taylor, B.K.; Dennis, E.A.; Yaksh, T.L. Spinal 12-lipoxygenase-derived hepoxilin A3 contributes to inflammatory hyperalgesia via activation of TRPV1 and TRPA1 receptors. *Proc. Natl. Acad. Sci. USA* **2012**, *109*, 6721–6726. [CrossRef]
68. Yamanaka, M.; Taniguchi, W.; Nishio, N.; Hashizume, H.; Yamada, H.; Yoshida, M.; Nakatsuka, T. In vivo patch-clamp analysis of the antinociceptive actions of TRPA1 activation in the spinal dorsal horn. *Mol. Pain* **2015**, *11*, 20. [CrossRef]
69. Lehmann, S.M.; Kruger, C.; Park, B.; Derkow, K.; Rosenberger, K.; Baumgart, J.; Trimbuch, T.; Eom, G.; Hinz, M.; Kaul, D.; et al. An unconventional role for miRNA: let-7 activates Toll-like receptor 7 and causes neurodegeneration. *Nat. Neurosci.* **2012**, *15*, 827–835. [CrossRef]
70. Lee, K.I.; Lee, H.T.; Lin, H.C.; Tsay, H.J.; Tsai, F.C.; Shyue, S.K.; Lee, T.S. Role of transient receptor potential ankyrin 1 channels in Alzheimer's disease. *J. Neuroinflammation* **2016**, *13*, 92. [CrossRef]
71. Park, C.K.; Xu, Z.Z.; Berta, T.; Han, Q.; Chen, G.; Liu, X.J.; Ji, R.R. Extracellular microRNAs activate nociceptor neurons to elicit pain via TLR7 and TRPA1. *Neuron* **2014**, *82*, 47–54. [CrossRef] [PubMed]
72. Caterina, M.J.; Schumacher, M.A.; Tominaga, M.; Rosen, T.A.; Levine, J.D.; Julius, D. The capsaicin receptor: a heat-activated ion channel in the pain pathway. *Nature* **1997**, *389*, 816–824. [CrossRef] [PubMed]
73. Tominaga, M.; Caterina, M.J.; Malmberg, A.B.; Rosen, T.A.; Gilbert, H.; Skinner, K.; Raumann, B.E.; Basbaum, A.I.; Julius, D. The cloned capsaicin receptor integrates multiple pain-producing stimuli. *Neuron* **1998**, *21*, 531–543. [CrossRef]
74. Iida, T.; Moriyama, T.; Kobata, K.; Morita, A.; Murayama, N.; Hashizume, S.; Fushiki, T.; Yazawa, S.; Watanabe, T.; Tominaga, M. TRPV1 activation and induction of nociceptive response by a non-pungent capsaicin-like compound, capsiate. *Neuropharmacology* **2003**, *44*, 958–967. [CrossRef]
75. Xu, H.; Blair, N.T.; Clapham, D.E. Camphor activates and strongly desensitizes the transient receptor potential vanilloid subtype 1 channel in a vanilloid-independent mechanism. *J. Neurosci.* **2005**, *25*, 8924–8937. [CrossRef] [PubMed]
76. Gu, Q.; Lin, R.L.; Hu, H.Z.; Zhu, M.X.; Lee, L.Y. 2-aminoethoxydiphenyl borate stimulates pulmonary C neurons via the activation of TRPV channels. *Am. J. Physiol. Lung Cell Mol. Physiol.* **2005**, *288*, L932–L941. [CrossRef] [PubMed]
77. De Petrocellis, L.; Bisogno, T.; Maccarrone, M.; Davis, J.B.; Finazzi-Agro, A.; Di Marzo, V. The activity of anandamide at vanilloid VR1 receptors requires facilitated transport across the cell membrane and is limited by intracellular metabolism. *J. Biol. Chem.* **2001**, *276*, 12856–12863. [CrossRef] [PubMed]
78. Huang, S.M.; Bisogno, T.; Trevisani, M.; Al-Hayani, A.; De Petrocellis, L.; Fezza, F.; Tognetto, M.; Petros, T.J.; Krey, J.F.; Chu, C.J.; et al. An endogenous capsaicin-like substance with high potency at recombinant and native vanilloid VR1 receptors. *Proc. Natl. Acad. Sci. USA* **2002**, *99*, 8400–8405. [CrossRef]
79. Miyamoto, T.; Dubin, A.E.; Petrus, M.J.; Patapoutian, A. TRPV1 and TRPA1 mediate peripheral nitric oxide-induced nociception in mice. *PLoS ONE* **2009**, *4*, e7596. [CrossRef]

80. Nishihara, E.; Hiyama, T.Y.; Noda, M. Osmosensitivity of transient receptor potential vanilloid 1 is synergistically enhanced by distinct activating stimuli such as temperature and protons. *PLoS ONE* **2011**, *6*, e22246. [CrossRef]
81. Bohlen, C.J.; Priel, A.; Zhou, S.; King, D.; Siemens, J.; Julius, D. A bivalent tarantula toxin activates the capsaicin receptor, TRPV1, by targeting the outer pore domain. *Cell* **2010**, *141*, 834–845. [CrossRef] [PubMed]
82. Ji, R.R.; Samad, T.A.; Jin, S.X.; Schmoll, R.; Woolf, C.J. p38 MAPK activation by NGF in primary sensory neurons after inflammation increases TRPV1 levels and maintains heat hyperalgesia. *Neuron* **2002**, *36*, 57–68. [CrossRef]
83. Premkumar, L.S.; Ahern, G.P. Induction of vanilloid receptor channel activity by protein kinase C. *Nature* **2000**, *408*, 985–990. [CrossRef] [PubMed]
84. Tominaga, M.; Wada, M.; Masu, M. Potentiation of capsaicin receptor activity by metabotropic ATP receptors as a possible mechanism for ATP-evoked pain and hyperalgesia. *Proc. Natl. Acad. Sci. USA* **2001**, *98*, 6951–6956. [CrossRef] [PubMed]
85. Schnizler, K.; Shutov, L.P.; Van Kanegan, M.J.; Merrill, M.A.; Nichols, B.; McKnight, G.S.; Strack, S.; Hell, J.W.; Usachev, Y.M. Protein kinase A anchoring via AKAP150 is essential for TRPV1 modulation by forskolin and prostaglandin E2 in mouse sensory neurons. *J. Neurosci.* **2008**, *28*, 4904–4917. [CrossRef] [PubMed]
86. Zhang, X.; Li, L.; McNaughton, P.A. Proinflammatory mediators modulate the heat-activated ion channel TRPV1 via the scaffolding protein AKAP79/150. *Neuron* **2008**, *59*, 450–461. [CrossRef] [PubMed]
87. Brunner, J.D.; Lim, N.K.; Schenck, S.; Duerst, A.; Dutzler, R. X-ray structure of a calcium-activated TMEM16 lipid scramblase. *Nature* **2014**, *516*, 207–212. [CrossRef] [PubMed]
88. Dang, S.; Feng, S.; Tien, J.; Peters, C.J.; Bulkley, D.; Lolicato, M.; Zhao, J.; Zuberbuhler, K.; Ye, W.; Qi, L.; et al. Cryo-EM structures of the TMEM16A calcium-activated chloride channel. *Nature* **2017**, *552*, 426–429. [CrossRef]
89. Paulino, C.; Kalienkova, V.; Lam, A.K.M.; Neldner, Y.; Dutzler, R. Activation mechanism of the calcium-activated chloride channel TMEM16A revealed by cryo-EM. *Nature* **2017**, *552*, 421–425. [CrossRef]
90. Jin, X.; Shah, S.; Du, X.; Zhang, H.; Gamper, N. Activation of Ca(2+) -activated Cl(-) channel ANO1 by localized Ca(2+) signals. *J. Physiol.* **2016**, *594*, 19–30. [CrossRef]
91. Lee, B.; Cho, H.; Jung, J.; Yang, Y.D.; Yang, D.J.; Oh, U. Anoctamin 1 contributes to inflammatory and nerve-injury induced hypersensitivity. *Mol. Pain* **2014**, *10*, 5. [CrossRef] [PubMed]
92. Weng, H.J.; Patel, K.N.; Jeske, N.A.; Bierbower, S.M.; Zou, W.; Tiwari, V.; Zheng, Q.; Tang, Z.; Mo, G.C.; Wang, Y.; et al. Tmem100 Is a Regulator of TRPA1-TRPV1 Complex and Contributes to Persistent Pain. *Neuron* **2015**, *85*, 833–846. [CrossRef] [PubMed]
93. Takayama, Y.; Uta, D.; Furue, H.; Tominaga, M. Pain-enhancing mechanism through interaction between TRPV1 and anoctamin 1 in sensory neurons. *Proc. Natl. Acad. Sci. USA* **2015**, *112*, 5213–5218. [CrossRef] [PubMed]
94. Namkung, W.; Phuan, P.W.; Verkman, A.S. TMEM16A inhibitors reveal TMEM16A as a minor component of calcium-activated chloride channel conductance in airway and intestinal epithelial cells. *J. Biol. Chem.* **2011**, *286*, 2365–2374. [CrossRef] [PubMed]
95. Kanazawa, T.; Matsumoto, S. Expression of transient receptor potential vanilloid 1 and anoctamin 1 in rat trigeminal ganglion neurons innervating the tongue. *Brain Res. Bull.* **2014**, *106*, 17–20. [CrossRef] [PubMed]
96. Suzuki, A.; Shinoda, M.; Honda, K.; Shirakawa, T.; Iwata, K. Regulation of transient receptor potential vanilloid 1 expression in trigeminal ganglion neurons via methyl-CpG binding protein 2 signaling contributes tongue heat sensitivity and inflammatory hyperalgesia in mice. *Mol. Pain* **2016**, *12*, 1744806916633206. [CrossRef]
97. Yamagata, K.; Sugimura, M.; Yoshida, M.; Sekine, S.; Kawano, A.; Oyamaguchi, A.; Maegawa, H.; Niwa, H. Estrogens Exacerbate Nociceptive Pain via Up-Regulation of TRPV1 and ANO1 in Trigeminal Primary Neurons of Female Rats. *Endocrinology* **2016**, *157*, 4309–4317. [CrossRef]
98. Vriens, J.; Owsianik, G.; Hofmann, T.; Philipp, S.E.; Stab, J.; Chen, X.; Benoit, M.; Xue, F.; Janssens, A.; Kerselaers, S.; et al. TRPM3 is a nociceptor channel involved in the detection of noxious heat. *Neuron* **2011**, *70*, 482–494. [CrossRef]
99. Held, K.; Kichko, T.; De Clercq, K.; Klaassen, H.; Van Bree, R.; Vanherck, J.C.; Marchand, A.; Reeh, P.W.; Chaltin, P.; Voets, T.; et al. Activation of TRPM3 by a potent synthetic ligand reveals a role in peptide release. *Proc. Natl. Acad. Sci. USA* **2015**, *112*, E1363–E1372. [CrossRef]

100. Headache Classification Committee of the International Headache Society (IHS) The International Classification of Headache Disorders, 3rd ed. *Cephalalgia* **2018**, *38*, 1–211. [CrossRef]
101. Laurell, K.; Artto, V.; Bendtsen, L.; Hagen, K.; Haggstrom, J.; Linde, M.; Soderstrom, L.; Tronvik, E.; Wessman, M.; Zwart, J.A.; et al. Premonitory symptoms in migraine: A cross-sectional study in 2714 persons. *Cephalalgia* **2016**, *36*, 951–959. [CrossRef] [PubMed]
102. Karsan, N.; Goadsby, P.J. Biological insights from the premonitory symptoms of migraine. *Nat. Rev. Neurol.* **2018**, *14*, 699–710. [CrossRef] [PubMed]
103. Kelman, L. The postdrome of the acute migraine attack. *Cephalalgia* **2006**, *26*, 214–220. [CrossRef] [PubMed]
104. Giffin, N.J.; Lipton, R.B.; Silberstein, S.D.; Olesen, J.; Goadsby, P.J. The migraine postdrome: An electronic diary study. *Neurology* **2016**, *87*, 309–313. [CrossRef] [PubMed]
105. Fukui, P.T.; Goncalves, T.R.; Strabelli, C.G.; Lucchino, N.M.; Matos, F.C.; Santos, J.P.; Zukerman, E.; Zukerman-Guendler, V.; Mercante, J.P.; Masruha, M.R.; et al. Trigger factors in migraine patients. *Arq. Neuropsiquiatr.* **2008**, *66*, 494–499. [CrossRef]
106. Mollaoglu, M. Trigger factors in migraine patients. *J. Health Psychol.* **2013**, *18*, 984–994. [CrossRef]
107. Lassen, L.H.; Haderslev, P.A.; Jacobsen, V.B.; Iversen, H.K.; Sperling, B.; Olesen, J. CGRP may play a causative role in migraine. *Cephalalgia* **2002**, *22*, 54–61. [CrossRef]
108. Olesen, J.; Diener, H.C.; Husstedt, I.W.; Goadsby, P.J.; Hall, D.; Meier, U.; Pollentier, S.; Lesko, L.M. BIBN 4096 BS Clinical Proof of Concept Study Group. Calcitonin gene-related peptide receptor antagonist BIBN 4096 BS for the acute treatment of migraine. *N. Engl. J. Med.* **2004**, *350*, 1104–1110. [CrossRef]
109. Benemei, S.; De Cesaris, F.; Fusi, C.; Rossi, E.; Lupi, C.; Geppetti, P. TRPA1 and other TRP channels in migraine. *J. Headache Pain* **2013**, *14*, 71. [CrossRef]
110. Dussor, G.; Yan, J.; Xie, J.Y.; Ossipov, M.H.; Dodick, D.W.; Porreca, F. Targeting TRP channels for novel migraine therapeutics. *ACS Chem. Neurosci.* **2014**, *5*, 1085–1096. [CrossRef]
111. Benemei, S.; Dussor, G. TRP Channels and Migraine: Recent Developments and New Therapeutic Opportunities. *Pharmaceuticals (Basel)* **2019**, *12*, 54. [CrossRef] [PubMed]
112. Dux, M.; Santha, P.; Jancso, G. Capsaicin-sensitive neurogenic sensory vasodilatation in the dura mater of the rat. *J. Physiol.* **2003**, *552*, 859–867. [CrossRef] [PubMed]
113. Oshinsky, M.L.; Gomonchareonsiri, S. Episodic dural stimulation in awake rats: a model for recurrent headache. *Headache* **2007**, *47*, 1026–1036. [CrossRef]
114. Becerra, L.; Bishop, J.; Barmettler, G.; Kainz, V.; Burstein, R.; Borsook, D. Brain network alterations in the inflammatory soup animal model of migraine. *Brain Res.* **2017**, *1660*, 36–46. [CrossRef] [PubMed]
115. Chen, N.; Su, W.; Cui, S.H.; Guo, J.; Duan, J.C.; Li, H.X.; He, L. A novel large animal model of recurrent migraine established by repeated administration of inflammatory soup into the dura mater of the rhesus monkey. *Neural. Regen. Res.* **2019**, *14*, 100–106. [PubMed]
116. Del Fiacco, M.; Quartu, M.; Boi, M.; Serra, M.P.; Melis, T.; Boccaletti, R.; Shevel, E.; Cianchetti, C. TRPV1, CGRP and SP in scalp arteries of patients suffering from chronic migraine. *J. Neurol. Neurosurg. Psychiatry* **2015**, *86*, 393–397. [CrossRef]
117. Schwedt, T.J.; Krauss, M.J.; Frey, K.; Gereau, R.W.t. Episodic and chronic migraineurs are hypersensitive to thermal stimuli between migraine attacks. *Cephalalgia* **2011**, *31*, 6–12. [CrossRef]
118. Nicoletti, P.; Trevisani, M.; Manconi, M.; Gatti, R.; De Siena, G.; Zagli, G.; Benemei, S.; Capone, J.A.; Geppetti, P.; Pini, L.A. Ethanol causes neurogenic vasodilation by TRPV1 activation and CGRP release in the trigeminovascular system of the guinea pig. *Cephalalgia* **2008**, *28*, 9–17. [CrossRef] [PubMed]
119. Lauritzen, M. Pathophysiology of the migraine aura. The spreading depression theory. *Brain* **1994**, *117*, 199–210. [CrossRef]
120. Rossi, H.L.; Broadhurst, K.A.; Luu, A.S.; Lara, O.; Kothari, S.D.; Mohapatra, D.P.; Recober, A. Abnormal trigeminal sensory processing in obese mice. *Pain* **2016**, *157*, 235–246. [CrossRef]
121. Meents, J.E.; Hoffmann, J.; Chaplan, S.R.; Neeb, L.; Schuh-Hofer, S.; Wickenden, A.; Reuter, U. Two TRPV1 receptor antagonists are effective in two different experimental models of migraine. *J. Headache Pain* **2015**, *16*, 57. [CrossRef] [PubMed]
122. Kunkler, P.E.; Ballard, C.J.; Oxford, G.S.; Hurley, J.H. TRPA1 receptors mediate environmental irritant-induced meningeal vasodilatation. *Pain* **2011**, *152*, 38–44. [CrossRef] [PubMed]
123. Benemei, S.; Fusi, C.; Trevisan, G.; Geppetti, P. The TRPA1 channel in migraine mechanism and treatment. *Br. J. Pharmacol.* **2014**, *171*, 2552–2567. [CrossRef] [PubMed]

124. Kanju, P.; Chen, Y.; Lee, W.; Yeo, M.; Lee, S.H.; Romac, J.; Shahid, R.; Fan, P.; Gooden, D.M.; Simon, S.A.; et al. Small molecule dual-inhibitors of TRPV4 and TRPA1 for attenuation of inflammation and pain. *Sci. Rep.* **2016**, *6*, 26894. [CrossRef] [PubMed]
125. Wei, X.; Edelmayer, R.M.; Yan, J.; Dussor, G. Activation of TRPV4 on dural afferents produces headache-related behavior in a preclinical rat model. *Cephalalgia* **2011**, *31*, 1595–1600. [CrossRef] [PubMed]
126. Nilius, B.; Flockerzi, V. Mammalian transient receptor potential (TRP) cation channels. Preface. *Handb. Exp. Pharmacol.* **2014**, *223*, v–vi. [PubMed]
127. Chen, Y.; Kanju, P.; Fang, Q.; Lee, S.H.; Parekh, P.K.; Lee, W.; Moore, C.; Brenner, D.; Gereau, R.W.T.; Wang, F.; et al. TRPV4 is necessary for trigeminal irritant pain and functions as a cellular formalin receptor. *Pain* **2014**, *155*, 2662–2672. [CrossRef] [PubMed]
128. Chasman, D.I.; Schurks, M.; Anttila, V.; de Vries, B.; Schminke, U.; Launer, L.J.; Terwindt, G.M.; van den Maagdenberg, A.M.; Fendrich, K.; Volzke, H.; et al. Genome-wide association study reveals three susceptibility loci for common migraine in the general population. *Nat. Genet.* **2011**, *43*, 695–698. [CrossRef]
129. Freilinger, T.; Anttila, V.; de Vries, B.; Malik, R.; Kallela, M.; Terwindt, G.M.; Pozo-Rosich, P.; Winsvold, B.; Nyholt, D.R.; van Oosterhout, W.P.; et al. Genome-wide association analysis identifies susceptibility loci for migraine without aura. *Nat. Genet.* **2012**, *44*, 777–782. [CrossRef]
130. Esserlind, A.L.; Christensen, A.F.; Le, H.; Kirchmann, M.; Hauge, A.W.; Toyserkani, N.M.; Hansen, T.; Grarup, N.; Werge, T.; Steinberg, S.; et al. Replication and meta-analysis of common variants identifies a genome-wide significant locus in migraine. *Eur. J. Neurol.* **2013**, *20*, 765–772. [CrossRef]
131. Sintas, C.; Fernandez-Morales, J.; Vila-Pueyo, M.; Narberhaus, B.; Arenas, C.; Pozo-Rosich, P.; Macaya, A.; Cormand, B. Replication study of previous migraine genome-wide association study findings in a Spanish sample of migraine with aura. *Cephalalgia* **2015**, *35*, 776–782. [CrossRef] [PubMed]
132. Anttila, V.; Wessman, M.; Kallela, M.; Palotie, A. Genetics of migraine. *Handb. Clin. Neurol.* **2018**, *148*, 493–503. [PubMed]
133. Burgos-Vega, C.C.; Ahn, D.D.; Bischoff, C.; Wang, W.; Horne, D.; Wang, J.; Gavva, N.; Dussor, G. Meningeal transient receptor potential channel M8 activation causes cutaneous facial and hindpaw allodynia in a preclinical rodent model of headache. *Cephalalgia* **2016**, *36*, 185–193. [CrossRef] [PubMed]
134. Ichikawa, H.; Sugimoto, T. VR1-immunoreactive primary sensory neurons in the rat trigeminal ganglion. *Brain Res.* **2001**, *890*, 184–188. [CrossRef]
135. Bae, Y.C.; Oh, J.M.; Hwang, S.J.; Shigenaga, Y.; Valtschanoff, J.G. Expression of vanilloid receptor TRPV1 in the rat trigeminal sensory nuclei. *J. Comp. Neurol.* **2004**, *478*, 62–71. [CrossRef] [PubMed]
136. Huang, D.; Li, S.; Dhaka, A.; Story, G.M.; Cao, Y.Q. Expression of the transient receptor potential channels TRPV1, TRPA1 and TRPM8 in mouse trigeminal primary afferent neurons innervating the dura. *Mol. Pain* **2012**, *8*, 66. [CrossRef] [PubMed]
137. Meents, J.E.; Neeb, L.; Reuter, U. TRPV1 in migraine pathophysiology. *Trends. Mol. Med.* **2010**, *16*, 153–159. [CrossRef]
138. Noseda, R.; Burstein, R. Migraine pathophysiology: anatomy of the trigeminovascular pathway and associated neurological symptoms, CSD, sensitization and modulation of pain. *Pain* **2013**, *154* (Suppl. 1), S44–S53. [CrossRef]
139. Sarchielli, P.; Mancini, M.L.; Floridi, A.; Coppola, F.; Rossi, C.; Nardi, K.; Acciarresi, M.; Pini, L.A.; Calabresi, P. Increased levels of neurotrophins are not specific for chronic migraine: Evidence from primary fibromyalgia syndrome. *J. Pain* **2007**, *8*, 737–745. [CrossRef]
140. Yang, A.C.; Fuh, J.L.; Huang, N.E.; Shia, B.C.; Wang, S.J. Patients with migraine are right about their perception of temperature as a trigger: time series analysis of headache diary data. *J. Headache Pain* **2015**, *16*, 533. [CrossRef]
141. You, D.S.; Haney, R.; Albu, S.; Meagher, M.W. Generalized Pain Sensitization and Endogenous Oxytocin in Individuals With Symptoms of Migraine: A Cross-Sectional Study. *Headache* **2018**, *58*, 62–77. [CrossRef] [PubMed]
142. Celentano, D.D.; Linet, M.S.; Stewart, W.F. Gender differences in the experience of headache. *Soc. Sci. Med.* **1990**, *30*, 1289–1295. [CrossRef]
143. Chai, N.C.; Scher, A.I.; Moghekar, A.; Bond, D.S.; Peterlin, B.L. Obesity and headache: part I–a systematic review of the epidemiology of obesity and headache. *Headache* **2014**, *54*, 219–234. [CrossRef] [PubMed]

144. Marics, B.; Peitl, B.; Pazmandi, K.; Bacsi, A.; Nemeth, J.; Oszlacs, O.; Jancso, G.; Dux, M. Diet-Induced Obesity Enhances TRPV1-Mediated Neurovascular Reactions in the Dura Mater. *Headache* **2017**, *57*, 441–454. [CrossRef] [PubMed]
145. Eberhardt, M.; Dux, M.; Namer, B.; Miljkovic, J.; Cordasic, N.; Will, C.; Kichko, T.I.; de la Roche, J.; Fischer, M.; Suarez, S.A.; et al. H2S and NO cooperatively regulate vascular tone by activating a neuroendocrine HNO-TRPA1-CGRP signalling pathway. *Nat. Commun.* **2014**, *5*, 4381. [CrossRef] [PubMed]
146. Benemei, S.; Appendino, G.; Geppetti, P. Pleasant natural scent with unpleasant effects: cluster headache-like attacks triggered by Umbellularia californica. *Cephalalgia* **2010**, *30*, 744–746. [CrossRef] [PubMed]
147. Nassini, R.; Materazzi, S.; Vriens, J.; Prenen, J.; Benemei, S.; De Siena, G.; la Marca, G.; Andre, E.; Preti, D.; Avonto, C.; et al. The 'headache tree' via umbellulone and TRPA1 activates the trigeminovascular system. *Brain* **2012**, *135*, 376–390. [CrossRef] [PubMed]
148. Levin, M. Herbal treatment of headache. *Headache* **2012**, *52*, 76–80. [CrossRef] [PubMed]
149. Materazzi, S.; Benemei, S.; Fusi, C.; Gualdani, R.; De Siena, G.; Vastani, N.; Andersson, D.A.; Trevisan, G.; Moncelli, M.R.; Wei, X.; et al. Parthenolide inhibits nociception and neurogenic vasodilatation in the trigeminovascular system by targeting the TRPA1 channel. *Pain* **2013**, *154*, 2750–2758. [CrossRef] [PubMed]
150. Benemei, S.; De Logu, F.; Li Puma, S.; Marone, I.M.; Coppi, E.; Ugolini, F.; Liedtke, W.; Pollastro, F.; Appendino, G.; Geppetti, P.; et al. The anti-migraine component of butterbur extracts, isopetasin, desensitizes peptidergic nociceptors by acting on TRPA1 cation channel. *Br. J. Pharmacol.* **2017**, *174*, 2897–2911. [CrossRef]
151. Tuncel, D.; Tolun, F.I.; Gokce, M.; Imrek, S.; Ekerbicer, H. Oxidative stress in migraine with and without aura. *Biol. Trace Elem. Res.* **2008**, *126*, 92–97. [CrossRef] [PubMed]
152. Bockowski, L.; Sobaniec, W.; Kulak, W.; Smigielska-Kuzia, J. Serum and intraerythrocyte antioxidant enzymes and lipid peroxides in children with migraine. *Pharmacol. Rep.* **2008**, *60*, 542–548. [PubMed]
153. Bernecker, C.; Ragginer, C.; Fauler, G.; Horejsi, R.; Moller, R.; Zelzer, S.; Lechner, A.; Wallner-Blazek, M.; Weiss, S.; Fazekas, F.; et al. Oxidative stress is associated with migraine and migraine-related metabolic risk in females. *Eur. J. Neurol.* **2011**, *18*, 1233–1239. [CrossRef] [PubMed]
154. Jiang, L.; Ma, D.; Grubb, B.D.; Wang, M. ROS/TRPA1/CGRP signaling mediates cortical spreading depression. *J. Headache Pain* **2019**, *20*, 25. [CrossRef] [PubMed]
155. Shatillo, A.; Koroleva, K.; Giniatullina, R.; Naumenko, N.; Slastnikova, A.A.; Aliev, R.R.; Bart, G.; Atalay, M.; Gu, C.; Khazipov, R.; et al. Cortical spreading depression induces oxidative stress in the trigeminal nociceptive system. *Neuroscience* **2013**, *253*, 341–349. [CrossRef] [PubMed]
156. Sarchielli, P.; Alberti, A.; Codini, M.; Floridi, A.; Gallai, V. Nitric oxide metabolites, prostaglandins and trigeminal vasoactive peptides in internal jugular vein blood during spontaneous migraine attacks. *Cephalalgia* **2000**, *20*, 907–918. [CrossRef] [PubMed]
157. Marone, I.M.; De Logu, F.; Nassini, R.; De Carvalho Goncalves, M.; Benemei, S.; Ferreira, J.; Jain, P.; Li Puma, S.; Bunnett, N.W.; Geppetti, P.; et al. TRPA1/NOX in the soma of trigeminal ganglion neurons mediates migraine-related pain of glyceryl trinitrate in mice. *Brain* **2018**, *141*, 2312–2328. [CrossRef] [PubMed]
158. McKemy, D.D.; Neuhausser, W.M.; Julius, D. Identification of a cold receptor reveals a general role for TRP channels in thermosensation. *Nature* **2002**, *416*, 52–58. [CrossRef] [PubMed]
159. Peier, A.M.; Moqrich, A.; Hergarden, A.C.; Reeve, A.J.; Andersson, D.A.; Story, G.M.; Earley, T.J.; Dragoni, I.; McIntyre, P.; Bevan, S.; et al. A TRP channel that senses cold stimuli and menthol. *Cell* **2002**, *108*, 705–715. [CrossRef]
160. Weyer, A.D.; Lehto, S.G. Development of TRPM8 Antagonists to Treat Chronic Pain and Migraine. *Pharmaceuticals (Basel)* **2017**, *10*, 37. [CrossRef]
161. Stewart, W.F.; Bigal, M.E.; Kolodner, K.; Dowson, A.; Liberman, J.N.; Lipton, R.B. Familial risk of migraine: variation by proband age at onset and headache severity. *Neurology* **2006**, *66*, 344–348. [CrossRef] [PubMed]
162. Mulder, E.J.; Van Baal, C.; Gaist, D.; Kallela, M.; Kaprio, J.; Svensson, D.A.; Nyholt, D.R.; Martin, N.G.; MacGregor, A.J.; Cherkas, L.F.; et al. Genetic and environmental influences on migraine: a twin study across six countries. *Twin. Res.* **2003**, *6*, 422–431. [CrossRef] [PubMed]
163. De Boer, I.; van den Maagdenberg, A.; Terwindt, G.M. Advance in genetics of migraine. *Curr. Opin. Neurol.* **2019**, *32*, 413–421. [CrossRef] [PubMed]
164. Morgan, K.; Sadofsky, L.R.; Morice, A.H. Genetic variants affecting human TRPA1 or TRPM8 structure can be classified in vitro as 'well expressed', 'poorly expressed' or 'salvageable'. *Biosci. Rep.* **2015**, *35*, e00255. [CrossRef] [PubMed]

165. Ren, L.; Dhaka, A.; Cao, Y.Q. Function and postnatal changes of dural afferent fibers expressing TRPM8 channels. *Mol. Pain* **2015**, *11*, 37. [CrossRef] [PubMed]
166. Dussor, G.; Cao, Y.Q. TRPM8 and Migraine. *Headache* **2016**, *56*, 1406–1417. [CrossRef] [PubMed]
167. Diamond, S.; Freitag, F.; Phillips, S.B.; Bernstein, J.E.; Saper, J.R. Intranasal civamide for the acute treatment of migraine headache. *Cephalalgia* **2000**, *20*, 597–602. [CrossRef]
168. Fusco, B.M.; Barzoi, G.; Agro, F. Repeated intranasal capsaicin applications to treat chronic migraine. *Br. J. Anaesth.* **2003**, *90*, 812. [CrossRef] [PubMed]
169. Chizh, B.; Palmer, J.; Lai, R.; Guillard, F.; Bullman, J.; Baines, A.; Napolitano, A.; Appleby, J. 702 A randomised, two-period cross-over study to investigate the efficacy of the trpv1 antagonist sb-705498 in acute migraine. *Eur. J. Pain* **2009**, *13*, S202a–S202. [CrossRef]
170. Gibson, R.A.; Robertson, J.; Mistry, H.; McCallum, S.; Fernando, D.; Wyres, M.; Yosipovitch, G. A randomised trial evaluating the effects of the TRPV1 antagonist SB705498 on pruritus induced by histamine, and cowhage challenge in healthy volunteers. *PLoS ONE* **2014**, *9*, e100610. [CrossRef]
171. Khalid, S.; Murdoch, R.; Newlands, A.; Smart, K.; Kelsall, A.; Holt, K.; Dockry, R.; Woodcock, A.; Smith, J.A. Transient receptor potential vanilloid 1 (TRPV1) antagonism in patients with refractory chronic cough: A double-blind randomized controlled trial. *J. Allergy Clin. Immunol.* **2014**, *134*, 56–62. [CrossRef] [PubMed]
172. Murdoch, R.D.; Bareille, P.; Denyer, J.; Newlands, A.; Bentley, J.; Smart, K.; Yarnall, K.; Patel, D. TRPV1 inhibition does not prevent cold dry air-elicited symptoms in non-allergic rhinitis. *Int. J. Clin. Pharmacol. Ther.* **2014**, *52*, 267–276. [CrossRef] [PubMed]
173. Amir, L.H.; Donath, S.M.; Garland, S.M.; Tabrizi, S.N.; Bennett, C.M.; Cullinane, M.; Payne, M.S. Does Candida and/or Staphylococcus play a role in nipple and breast pain in lactation? A cohort study in Melbourne, Australia. *BMJ Open* **2013**, *3*, e002351. [CrossRef] [PubMed]
174. Farmer, M.A.; Taylor, A.M.; Bailey, A.L.; Tuttle, A.H.; MacIntyre, L.C.; Milagrosa, Z.E.; Crissman, H.P.; Bennett, G.J.; Ribeiro-da-Silva, A.; Binik, Y.M.; et al. Repeated vulvovaginal fungal infections cause persistent pain in a mouse model of vulvodynia. *Sci. Transl. Med.* **2011**, *3*, 101ra91. [CrossRef] [PubMed]
175. Gamaletsou, M.N.; Kontoyiannis, D.P.; Sipsas, N.V.; Moriyama, B.; Alexander, E.; Roilides, E.; Brause, B.; Walsh, T.J. Candida osteomyelitis: analysis of 207 pediatric and adult cases (1970–2011). *Clin. Infect. Dis.* **2012**, *55*, 1338–1351. [CrossRef]
176. Gow, N.A.; van de Veerdonk, F.L.; Brown, A.J.; Netea, M.G. Candida albicans morphogenesis and host defence: discriminating invasion from colonization. *Nat. Rev. Microbiol.* **2011**, *10*, 112–122. [CrossRef] [PubMed]
177. Saijo, S.; Fujikado, N.; Furuta, T.; Chung, S.H.; Kotaki, H.; Seki, K.; Sudo, K.; Akira, S.; Adachi, Y.; Ohno, N.; et al. Dectin-1 is required for host defense against Pneumocystis carinii but not against Candida albicans. *Nat. Immunol.* **2007**, *8*, 39–46. [CrossRef]
178. Underhill, D.M.; Iliev, I.D. The mycobiota: interactions between commensal fungi and the host immune system. *Nat. Rev. Immunol.* **2014**, *14*, 405–416. [CrossRef]
179. Moyes, D.L.; Wilson, D.; Richardson, J.P.; Mogavero, S.; Tang, S.X.; Wernecke, J.; Hofs, S.; Gratacap, R.L.; Robbins, J.; Runglall, M.; et al. Candidalysin is a fungal peptide toxin critical for mucosal infection. *Nature* **2016**, *532*, 64–68. [CrossRef]
180. Chiu, I.M.; Heesters, B.A.; Ghasemlou, N.; Von Hehn, C.A.; Zhao, F.; Tran, J.; Wainger, B.; Strominger, A.; Muralidharan, S.; Horswill, A.R.; et al. Bacteria activate sensory neurons that modulate pain and inflammation. *Nature* **2013**, *501*, 52–57. [CrossRef]
181. Maruyama, K.; Takayama, Y.; Sugisawa, E.; Yamanoi, Y.; Yokawa, T.; Kondo, T.; Ishibashi, K.I.; Sahoo, B.R.; Takemura, N.; Mori, Y.; et al. The ATP Transporter VNUT Mediates Induction of Dectin-1-Triggered Candida Nociception. *iScience* **2018**, *6*, 306–318. [CrossRef] [PubMed]
182. Woolf, C.J.; Costigan, M. Transcriptional and posttranslational plasticity and the generation of inflammatory pain. *Proc. Natl. Acad. Sci. USA* **1999**, *96*, 7723–7730. [CrossRef] [PubMed]
183. Rosenfeld, M.G.; Mermod, J.J.; Amara, S.G.; Swanson, L.W.; Sawchenko, P.E.; Rivier, J.; Vale, W.W.; Evans, R.M. Production of a novel neuropeptide encoded by the calcitonin gene via tissue-specific RNA processing. *Nature* **1983**, *304*, 129–135. [CrossRef] [PubMed]
184. Brain, S.D.; Williams, T.J.; Tippins, J.R.; Morris, H.R.; MacIntyre, I. Calcitonin gene-related peptide is a potent vasodilator. *Nature* **1985**, *313*, 54–56. [CrossRef] [PubMed]

185. Pettersson, M.; Ahren, B.; Bottcher, G.; Sundler, F. Calcitonin gene-related peptide: Occurrence in pancreatic islets in the mouse and the rat and inhibition of insulin secretion in the mouse. *Endocrinology* **1986**, *119*, 865–869. [CrossRef] [PubMed]
186. Mikami, N.; Watanabe, K.; Hashimoto, N.; Miyagi, Y.; Sueda, K.; Fukada, S.; Yamamoto, H.; Tsujikawa, K. Calcitonin gene-related peptide enhances experimental autoimmune encephalomyelitis by promoting Th17-cell functions. *Int. Immunol.* **2012**, *24*, 681–691. [CrossRef] [PubMed]
187. Ding, Y.; Arai, M.; Kondo, H.; Togari, A. Effects of capsaicin-induced sensory denervation on bone metabolism in adult rats. *Bone* **2010**, *46*, 1591–1596. [CrossRef] [PubMed]
188. Huebner, A.K.; Schinke, T.; Priemel, M.; Schilling, S.; Schilling, A.F.; Emeson, R.B.; Rueger, J.M.; Amling, M. Calcitonin deficiency in mice progressively results in high bone turnover. *J. Bone Miner. Res.* **2006**, *21*, 1924–1934. [CrossRef]
189. Kaneko, Y.; Szallasi, A. Transient receptor potential (TRP) channels: A clinical perspective. *Br. J. Pharmacol.* **2014**, *171*, 2474–2507. [CrossRef]

International Journal of
*Molecular Sciences*

*Article*

# Expression and Function of Transient Receptor Potential Ankyrin 1 Ion Channels in the Caudal Nucleus of the Solitary Tract

**Lin Feng [1], Victor V. Uteshev [1,2] and Louis S. Premkumar [1,*]**

1 Department of Pharmacology, Southern Illinois University School of Medicine, Springfield, IL 62702, USA; lfeng@siumed.edu (L.F.); Victor.Uteshev@unthsc.edu (V.V.U.)
2 Department of Pharmacology and Neuroscience, University of North Texas Health Science Center, Fort Worth, TX 76107, USA
* Correspondence: lpremkumar@siumed.edu; Tel.: +217-545-2179; Fax: +217-545-0145

Received: 18 March 2019; Accepted: 17 April 2019; Published: 26 April 2019

**Abstract:** The nucleus of the solitary tract (NTS) receives visceral information via the solitary tract (ST) that comprises the sensory components of the cranial nerves VII, IX and X. The Transient Receptor Potential Ankyrin 1 (TRPA1) ion channels are non-selective cation channels that are expressed primarily in pain-related sensory neurons and nerve fibers. Thus, TRPA1 expressed in the primary sensory afferents may modulate the function of second order NTS neurons. This hypothesis was tested and confirmed in the present study using acute brainstem slices and caudal NTS neurons by RT-PCR, immunostaining and patch-clamp electrophysiology. The expression of TRPA1 was detected in presynaptic locations, but not the somata of caudal NTS neurons that did not express TRPA1 mRNA or proteins. Moreover, caudal NTS neurons did not show somatodendritic responsiveness to TRPA1 agonists, while TRPA1 immunostaining was detected only in the afferent fibers. Electrophysiological recordings detected activation of presynaptic TRPA1 in glutamatergic terminals synapsing on caudal NTS neurons evidenced by the enhanced glutamatergic synaptic neurotransmission in the presence of TRPA1 agonists. The requirement of TRPA1 for modulation of spontaneous synaptic activity was confirmed using TRPA1 knockout mice where TRPA1 agonists failed to alter synaptic efficacy. Thus, this study provides the first evidence of the TRPA1-dependent modulation of the primary afferent inputs to the caudal NTS. These results suggest that the second order caudal NTS neurons act as a TRPA1-dependent interface for visceral noxious-innocuous integration at the level of the caudal brainstem.

**Keywords:** ion channel; transient receptor potential ankyrin 1 (TRPA1); synaptic transmission; nucleus tractus solitarius (NTS)

---

## 1. Introduction

The brainstem nucleus of the solitary tract (NTS) is the key integrating relay in the central processing of sensory information from the thoracic and most subdiaphragmatic viscera [1–3]. The solitary tract (ST) is a bundle of sensory nerve fibers that extends longitudinally and bilaterally through the brainstem medulla. It comprises the sensory components of the cranial nerves VII, IX and X and relays information from both nociceptors and innocuous sensory receptors of the visceral organs and other tissues to the NTS. The ST relays information to the NTS from sensory receptors of the visceral organs and other tissues [4–9]. The NTS is a highly heterogeneous population of neurons, where seemingly indistinguishable neighboring neurons could participate in very different autonomic (e.g., gastrointestinal and cardiorespiratory reflexes) and nociceptive functions [8,10–14].

While glutamate is the major excitatory neurotransmitter in the brainstem, synaptic transmission at the level of the NTS can be modulated via the activation of multiple types of presynaptic ligand-gated ion channels such as transient receptor potential (TRP) channels [15–20]. The TRP ankyrin 1 (TRPA1) ion channels are nonselective cation channels highly permeable to $Ca^{2+}$ ions. Results from animal models of visceral pain suggest that activation of TRPA1 is critical for transmission of visceral pain and may be implicated in visceral pain sensation in patients with colitis, gastric distention and inflammatory bowel disease [21–27].

TRPA1 may modulate neuronal and synaptic activity via diverse pathways because thermal, chemical and mechanical stimuli have been shown to activate TRPA1 in various animal models [28–34]. However, while TRPA1 are expressed predominantly in sensory neurons (trigeminal, superior cervical, nodose and dorsal root ganglia neurons), but not in second order neurons [30,35–37] and thus, not expected to be expressed in the NTS. Activation of TRPA1 in nerve terminals that synapse onto NTS neurons may have prominent effects on neuronal function and synaptic transmission within the NTS. This hypothesis is tested in the present study using acute brainstem slices of caudal NTS neurons. Our findings suggest that the second order caudal NTS neurons act as a TRPA1-dependent interface for visceral noxious-innocuous integration at the level of the caudal brainstem.

## 2. Results

### 2.1. Presynaptic Expression of TRPA1 in the Caudal NTS

The existing literature indicates that sensory TRP channels are predominantly expressed in peripheral neurons and their expression in central neurons is limited [38,39]. To determine whether this rule applies to TRPA1 in the NTS, we used RT-PCR and immunohistochemistry. The TRPA1 mRNA was not detected in caudal NTS neurons while its presence was clearly detected in DRG neurons that served as a positive control (Figure 1A). Immunofluorescent staining then revealed that TRPA1 in the caudal NTS are selectively expressed only in nerve fibers i.e., pre-synaptically (Figure 1B,C), which is consistent with the expression of TRPA1 in the solitary tract fibers, but not the somatic expression in the second order caudal NTS neurons. However, these results alone do not rule out the expression of TRPA1 in non-solitary tract terminals and will be further confirmed in conjunction with our electrophysiological data (see Figures 2–4).

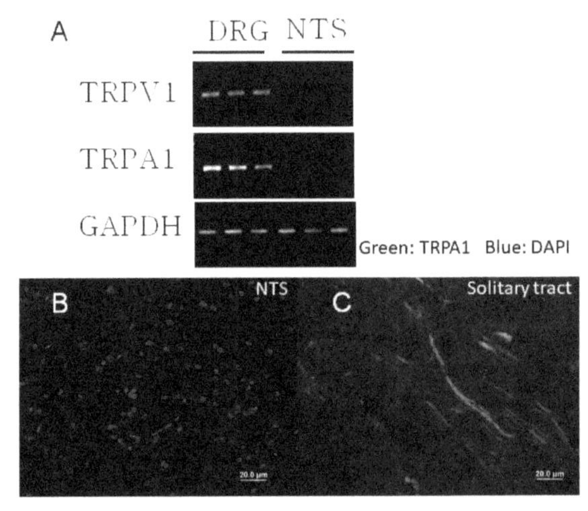


**Figure 1.** The lack of expression of TRPA1 in the nucleus of the solitary tract (NTS). (**A**) Expression of TRPA1 mRNA in sensory neurons but not in the NTS. (**B**,**C**) Immunostaining of TRPA1 (green), DAPI (blue) in the NTS. The TRPA1 staining was detected abundantly on the solitary tract, but not within the NTS. DRG: dorsal root ganglion.

## 2.2. Modulation of Synaptic Transmission by AITC in the Caudal NTS

To determine the functional characteristics of TRPA1 expressed in pre-synaptic terminals that synapse onto caudal NTS neurons, we used acute horizontal brainstem slices and patch-clamp electrophysiology. In voltage-clamp experiments, visualized caudal NTS neurons were held at −60 mV and synaptic currents were recorded at various experimental conditions. The expression of functional TRPA1 was confirmed using TRPA1 agonist, allyl isothiocyanate (i.e., AITC). Focal pressure puffs of AITC applied to recording caudal NTS neurons robustly increased the frequency of spontaneous excitatory synaptic currents (sEPSCs) (data not shown). To determine the effects of AITC on miniature excitatory synaptic activity (mEPSCs), patch-clamp recordings were conducted in the presence of 1 μM tetrodotoxin (TTX) to inhibit voltage-gated $Na^+$ channels and prevent action potential-dependent synaptic events. In these experiments, application of AITC (200 μM) significantly increased the frequency of mEPSCs in a pulse duration-/concentration-dependent manner (Figure 2).

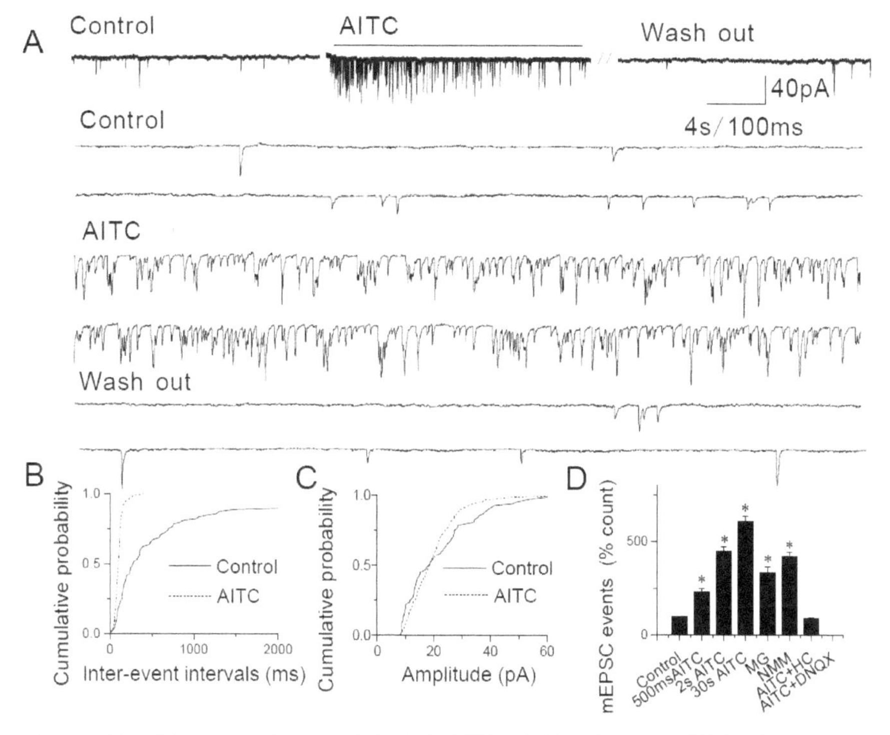


**Figure 2.** Modulation of synaptic transmission in the NTS by allyl isothiocyanate (AITC). (**A**) Application of AITC (200 μM) increases the frequency of miniature excitatory synaptic activity (mEPSCs) in a reversible manner. The synaptic events are shown in higher time resolution below. (**B**) Cumulative probability plot showing decreased inter-event intervals representing increased frequency of mEPSCs ($p < 0.0001$, KS test). (**C**) The increase in frequency is not accompanied by a change in the amplitude. (**D**) Summary graph showing AITC-mediated increases in the frequency of mEPSCs in a dose-dependent manner (* $p < 0.05$). Furthermore, the increase in AITC -induced synaptic events are blocked by 300 μM HC030031 ($n = 5$, * $p < 0.05$). The asterisk (*) represents $p < 0.05$ as compared to control.

The increase in mEPSC frequency is expressed as a percentage of control (i.e., no drugs applied): 500 ms, 230.25 ± 20.41% ($n = 8$, $p < 0.05$): 2 s, 450.63 ± 22.74% ($n = 13$, $p < 0.05$); 30 s, 610.79 ± 30.81% ($n = 9$, $p < 0.05$) (Figure 2A,B,D). The means mEPSC amplitude did not significantly change (control, 19.53 ± 2.53 pA, $n = 5$; after AITC, 21.2 ± 3.09 pA, $n = 12$) (Figure 2C). The increase in the mEPSC frequency, but not the amplitude suggests that AITC acts only pre-synaptically, which is consistent with the pre-synaptic expression of TRPA1 (Figure 1). These potentiating effects of AITC were observed in ~40% ($n$ = 143/358) of tested caudal NTS neurons. These results suggest that only neurons that received primary sensory afferent input responded by increasing the frequency of mEPSCs following AITC application.

mEPSCs were completely blocked by 6,7-Dinitroquinoxaline-2,3-dione (i.e., DNQX; 16 µM), a selective antagonist of AMPA receptors (Figure 2D). The involvement of TRPA1 was confirmed using HC030031 (i.e., HC), a TRPA1 selective antagonist. HC (300 µM) abolished the effects of AITC on mEPSC frequency without affecting the background synaptic activity (Figure 2D).

Similar results were obtained in experiments where two other TRPA1 agonists (i.e., N-methylmaleimide (i.e., NMM), an oxidizing agent that forms a covalent bond with TRPA1 and methylglyoxal (i.e., MG), a reactive molecule and an endogenous TRPA1 agonist, produced during hyperglycemia) were used. Pressure puffs of NMM or MG increased the frequency, but not amplitude of mEPSCs in caudal NTS neurons (Figures 3 and 4): 100 µM NMM (421.61 ± 24.09%, $n = 8$, $p < 0.05$; Figure 3B,D) and 50 µM MG (334.76 ± 30.62% $n = 7$, $p < 0.05$; Figure 4B,D).

By contrast, the somatodendritic responsiveness between NTS neurons to AITC, NMM and MG has not been detected in this study. Together with the molecular biological results (Figure 1), these data (Figures 2–4) support a strictly pre-synaptic expression of TRPA1 either in the primary afferent (solitary) terminals and/or pre-synaptic glutamatergic terminals of second- or higher-order NTS neurons.

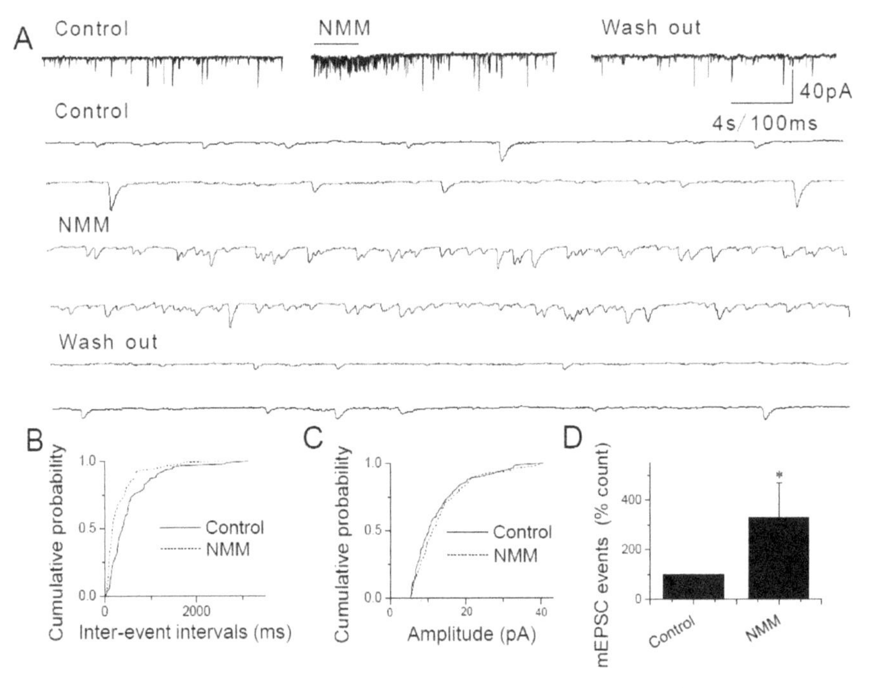


**Figure 3.** Modulation of synaptic transmission in the NTS by N-methylmaleimide (NMM). (**A**) Application of NMM (100 µM) increases the frequency of mEPSCs in a reversible manner. The synaptic events are shown in higher time resolution below. (**B**) Cumulative probability plot showing decreased inter-event intervals representing increased frequency of mEPSCs ($p < 0.0001$, KS test). (**C**) The increase in frequency is not accompanied by a change in the amplitude. (**D**) Summary graphs showing that the NMM-mediated increase in mEPSCs ($n = 8$, * $p < 0.05$). The asterisk (*) represents $p < 0.05$ as compared to control.

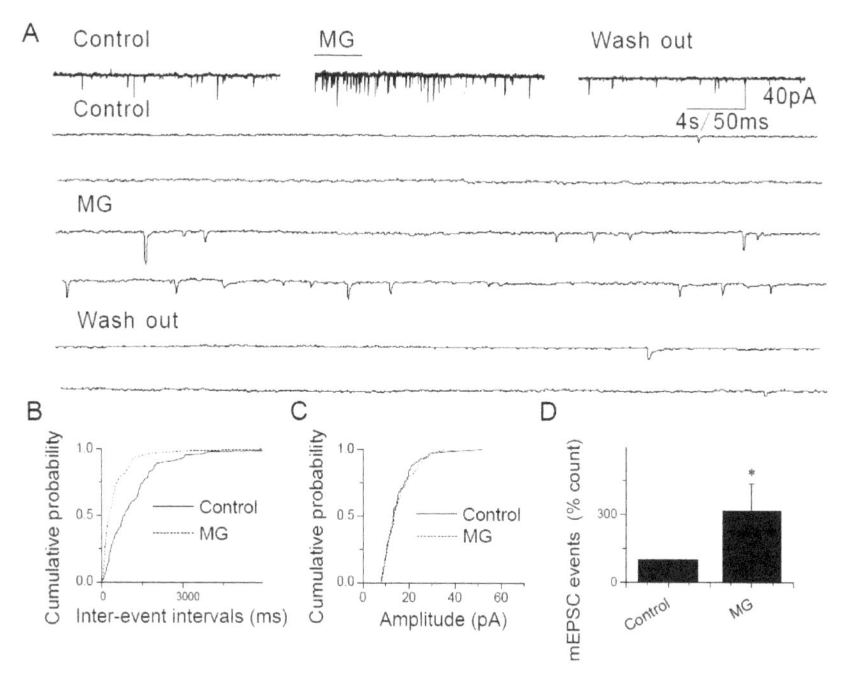


**Figure 4.** Modulation of synaptic transmission in the NTS by methylglyoxal (MG). (**A**) Application of MG (50 μM) increases the frequency of mEPSCs in a reversible manner. The synaptic events are shown in higher time resolution below. (**B**) Cumulative probability plot showing decreased interevent intervals representing increased frequency of mEPSCs ($p < 0.0001$, KS test). (**C**) The increase in frequency is not accompanied by a change in the amplitude. (**D**) Summary graphs showing that MG-mediated increase in mEPSCs ($n = 7$, * $p < 0.05$). The asterisk (*) represents $p < 0.05$ as compared to control.

### 2.3. Changes in mEPSCs in Response to Continuous and Repeated Application of AITC

TRPA1 agonists AITC, NMM and MG have been shown to activate the channel by covalent modification of cysteine and lysine residues [34,40]. Although covalent modification is expected to be an irreversible process, within the time course of electrophysiological experiments, NMM and AITC activate TRPA1 in a reversible manner [34,41]. We found that brief (2–10 s) puffs of AITC (200 μM), NMM (100 μM) or MG (50 μM) induced responses that were readily reversible (Figures 2–4) and a continuous application of AITC decreased the frequency of mEPSCs over time (Figure 2A).

In a separate experiment, changes in mEPSC frequency were analyzed with continuous application of AITC (Figure 5A,B). The AITC-mediated facilitation of mEPSCs showed a gradual decrease with time (Figure 5B). A persistent depolarization of presynaptic terminal and/or $Ca^{2+}$-dependent decrease in TRPA1 activity and/or desensitization of TRPA1 may be responsible for the observed run-down of synaptic activity.

To determine whether AITC causes tachyphylaxis, AITC (2 s duration) was repeatedly applied to the recorded caudal NTS neurons every 10 s separated by a washout. Successive applications of AITC (200 μM) gradually decreased the mEPSC frequency (first application, 657.53 ± 135.2%, $n = 5$; second application, 487.64 ± 122.57%, $n = 5$; third application, 374.85 ± 73.50% of control, $n = 6$) (Figure 5A,C,D). These data support that TRPA1-mediated transmitter release is decreased with repeated application of AITC. Thus, in all experiments with multiple applications of agonists, at least 3 min of washout was given after each agonist application to avoid desensitization.

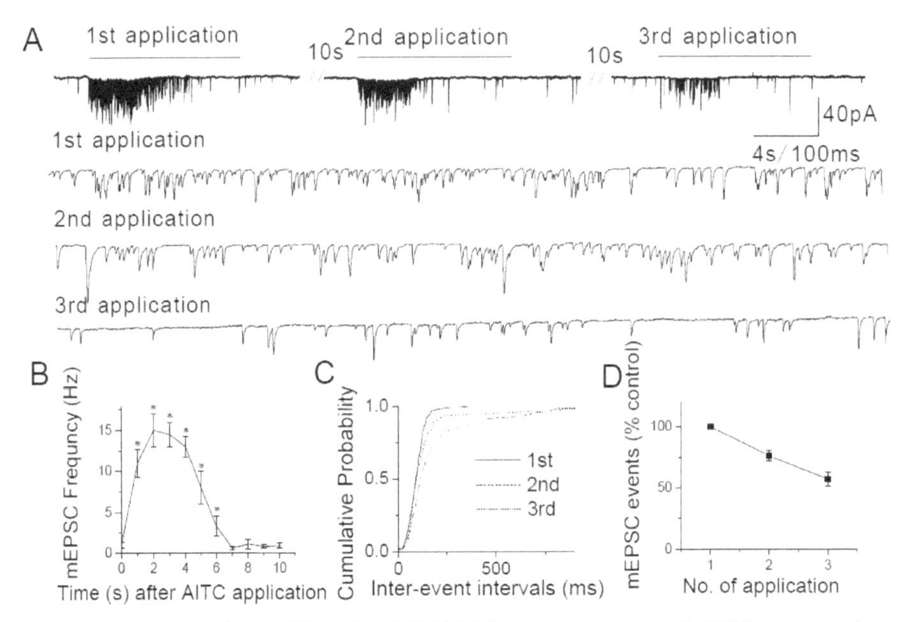


**Figure 5.** The tachyphylaxis of TRPA1. (**A**) AITC (200 μM) causes enhancement of mEPSCs progressively decreases with repeated AITC application. Synaptic currents are shown in a higher time resolution below. (**B**) Summary graph representing a progressive decrease in the number of mEPSC events immediately following AITC application. (**C**) Cumulative probability plots showing decreased mEPSC frequency as indicated by a progressive increase in the inter-event intervals with each subsequent AITC application. (**D**) Summary graph representing a progressive decline in mEPSC frequency with repeated AITC application. A significant decrease in frequency is observed only between the first and third AITC application (first application, $n = 5$; third application, $n = 5$, $p < 0.05$) (first application, 357.53 ± 135.2%, $n = 5$; second application, 271.64 ± 122.57%, $n = 5$; third application, 203.85 ± 73.50% of control, $n = 6$). The asterisk (*) represents $p < 0.05$ as compared to control.

## 2.4. The Lack of Sensitization of Presynaptic TRPA1 by PDBu in the Caudal NTS

Application of phorbol-12,13-dibutyrate (PDBu, 1–5 μM, a PKC activator) alone induced a dose-dependent increase in the frequency, but not amplitude of mEPSCs in caudal NTS neurons (Figure 6A,B,D). The effects of PDBu on mEPSC frequency were significant: 1 μM, 117.84 ± 19.66%, $n = 6$; 3 μM, 195.76 ± 22.15%, $n = 5$, $p < 0.01$; 5 μM, 295.13 ± 28.57%, $n = 6$, $p < 0.01$ (Figure 6D), whereas the mEPSC amplitudes remained unaltered (Figure 6C), supporting a presynaptic mechanism of action of PDBu [42–45].

Following application of PDBu (3 μM) to the ACSF, AITC increased the mEPSC frequency, but not amplitude (AITC alone, 298.51 ± 21.86%; AITC + PDBu, 358.69 ± 42.58% ($n = 9$, $p > 0.05$)) (Figure 7A,B,D). The effects of PDBu and AITC on mEPSC frequency appear to be additive, not synergistic suggesting that PDBu and AITC employ two different and independent mechanisms of action in presyanptic terminals of caudal NTS neurons.

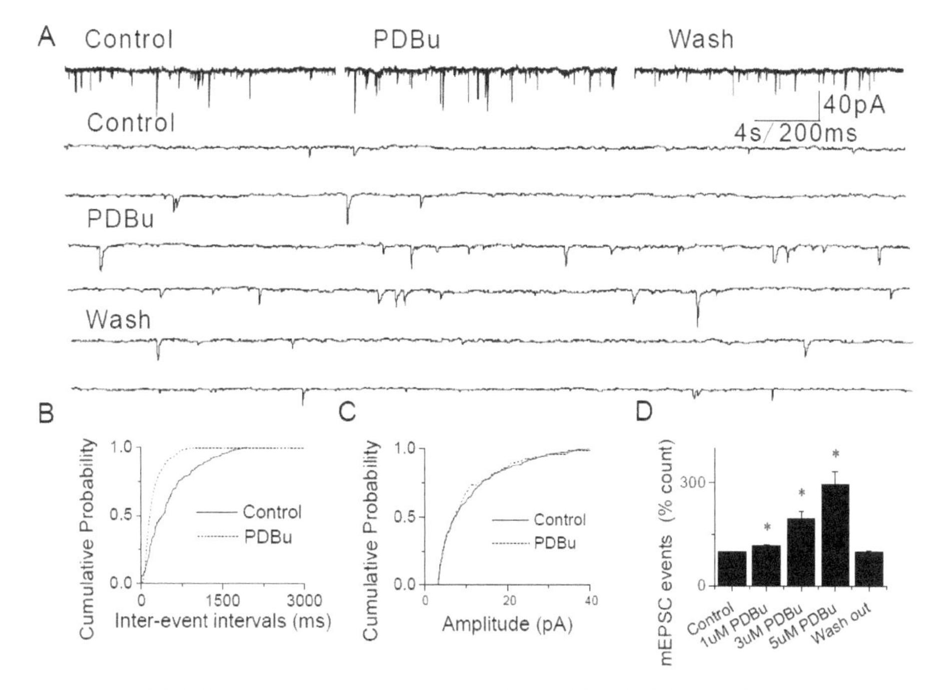


**Figure 6.** Modulation of synaptic transmission in the NTS by phorbol-12,13-dibutyrate (PDBu). (**A**) PDBu increases the frequency of mEPSCs in a reversible manner. Synaptic currents are shown in higher time resolution below. (**B**,**C**) Cumulative probability graphs showing enhanced frequency of mEPSCs ($p < 0.001$,KS test) in response to PDBu without change in their amplitudes. (**D**) Summary graphs showing that PDBu-mediated increase in mEPSCs (1 μM, 117.84 ± 19.66%, $n = 6$; 3 μM, 195.76 ± 30.15%, $n = 5$, * $p < 0.01$; 5 μM, 295.13 ± 38.57%, $n = 6$, * $p < 0.01$). The asterisk (*) represents $p < 0.05$ as compared to control.

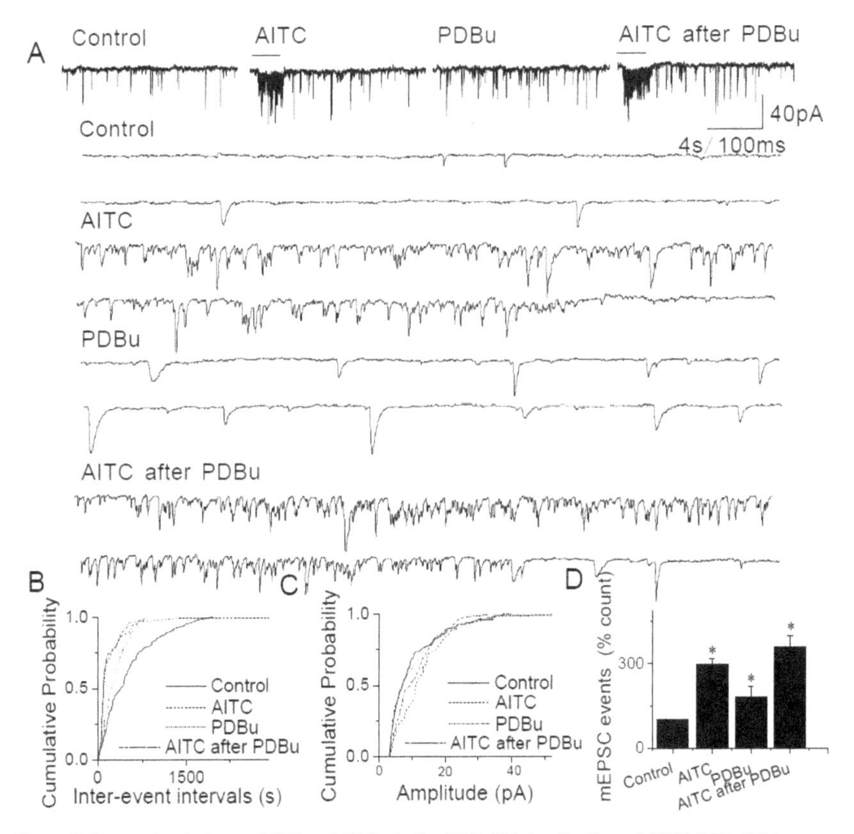


**Figure 7.** Interaction between AITC and PDBu in the NTS. (**A**) Application of AITC (200 μM) increases the frequency of mEPSCs and this action is enhanced by incubation in 3 μM PDBu. Synaptic events are shown at a higher time resolution below. (**B**) The cumulative probability plots show an AITC-mediated increase in the frequency of synaptic events. These effects are further significantly enhanced by PDBu. (**C**) The increase in frequency of events is not accompanied by a change in the amplitude. (**D**) Summary graphs showing AITC-induced an increase in mEPSC frequency ($n$ = 9) and potentiation by PDBu ($n$ = 8). The asterisk (*) represents $p < 0.05$ as compared to control.

### *2.5. The Lack of Modulatory Effects of AITC on Inhibitory Synapses in the Caudal NTS*

In experiments where AMPA and NMDA receptors were blocked with DNQX (16 μM) and 2-amino-5-phosphonovaleric acid (APV, 20 μM), respectively, added to ACSF, pressure application of AITC (500 μM) failed to alter either the frequency or the amplitude of mIPSCs (AITC, 97.86 ± 8.64%, $n$ = 10, Figure 8). Therefore, AITC does not affect the release of inhibitory synaptic neurotransmitters.

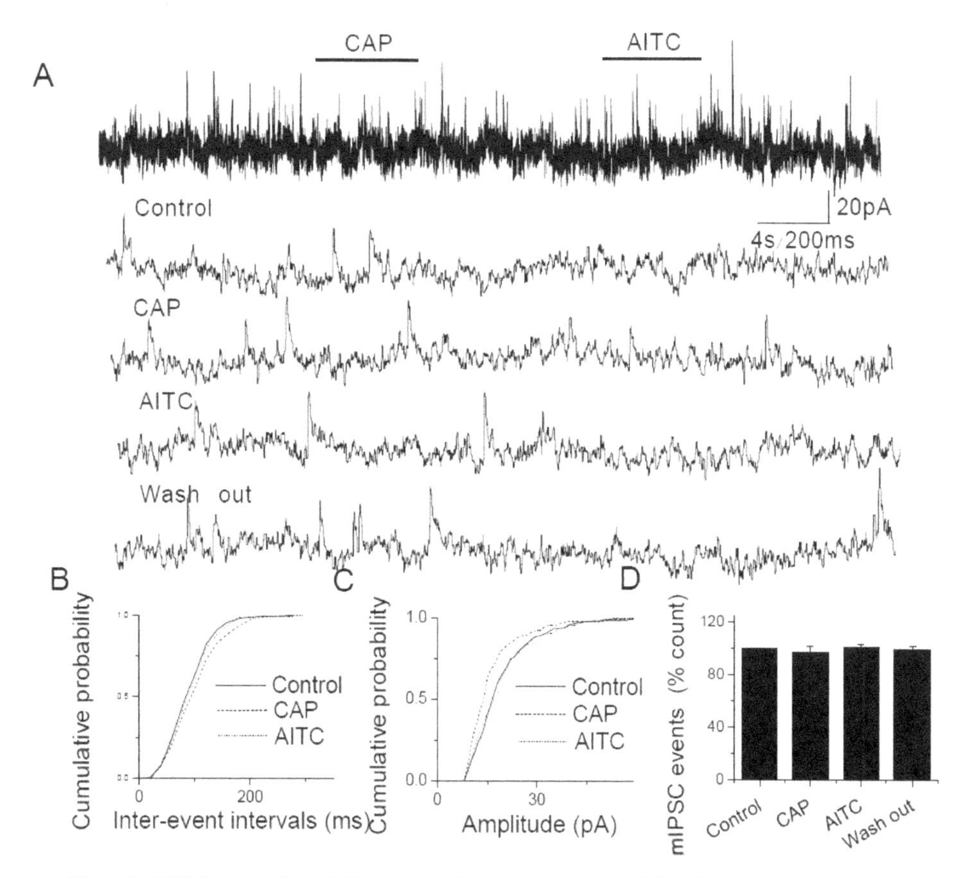


**Figure 8.** AITC does not alter inhibitory synaptic transmission in the NTS. (**A**) Administration of AITC (500 µM) or capsaicin (i.e., CAP; 200 nM) did not have any effect on the frequency of mIPSCs (top trace). Bottom four traces: the same synaptic events are shown at a higher time resolution. (**B**,**C**) The cumulative probability plots show no significant changes ($p < 0.0001$, KS test) in the distributions of inter-event intervals (**B**) and amplitudes (**C**) of mIPSCs. (**D**) A summary graph illustrating a lack of effects of AITC on the mIPSC frequency ($n = 10$).

### *2.6. Modulation of Evoked EPSCs by AITC in the Caudal NTS*

AITC significantly modulated evoked EPSCs generated by electrical stimulation of the solitary tract (50–400 µs, pulse duration; 100–500 µA, pulse intensity; 60 s inter-pulse interval). The average amplitude of EPSCs was 131.86 pA ± 41.29 ($n = 14$) (ranged between 40.00 pA to 396.19 pA) and AITC (100 µM) failed to alter the EPSC amplitude or kinetics ($p > 0.05$; $n = 14$, Figure 9A,C). In experiments where a paired pulse protocol was employed, AITC significantly depressed the paired-pulse ratio (PPR) in a concentration-dependent manner (control, 0.6877 ± 0.163, $n = 12$; 100 µM, 0.770 ± 0.060, $n = 10$; 200 µM, 0.7 ± 0.086, $n = 4$; 500 µM, 0.517 ± 0.051, $n = 7$, $p < 0.05$; 1 mM, 0.117 ± 0.179, $n = 8$, $p < 0.05$; Figure 9B,D). It is interesting to note that stimulation of TRPV1 significantly inhibited the evoked responses as reported by Peters et al., 2010 [46]. The differential modulation of synaptic transmission by TRPA1 and TRPV1 is intriguing. These results are consistent with the expression of TRPA1 in the primary afferent solitary tract terminals in the caudal NTS.

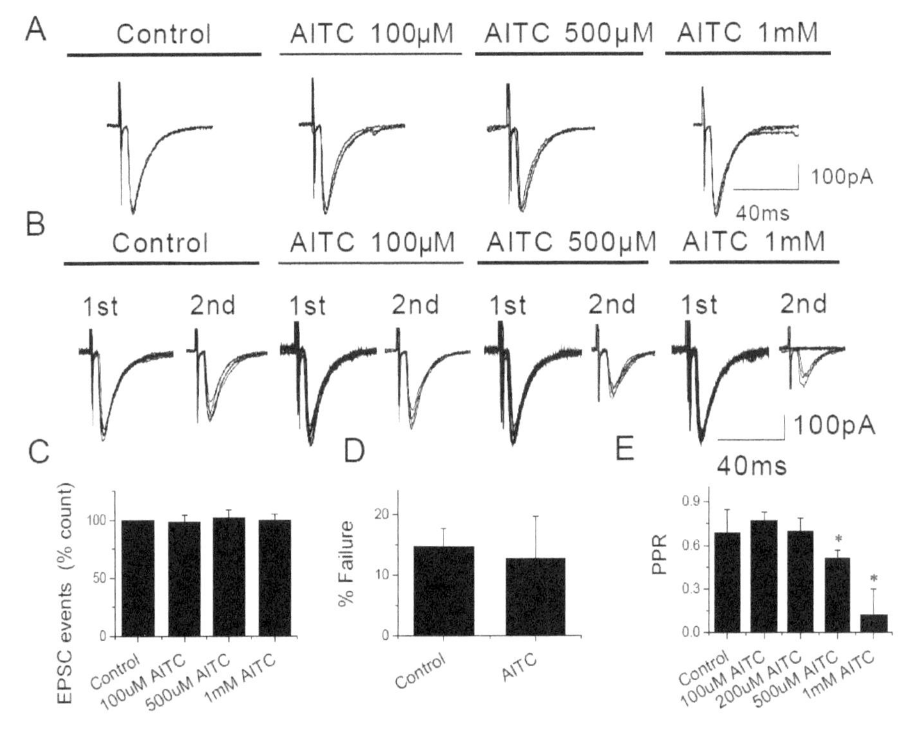


**Figure 9.** AITC-induced effects on solitary tract-evoked synaptic responses in horizontal brainstem slices. (**A**,**B**) Representative traces of solitary tract-stimulated EPSCs (ST-EPSCs) showing seven overlapping sweeps. (**A**,**C**) AITC does not have any significant effect on the amplitude of the ST-EPSCs ($n = 14$). (**B**,**E**) AITC depressed the paired pulse ratio (PPR, EPSC2/EPSC1) in a concentration dependent manner. (**D**) AITC does not show any significant effect on the failure rate (% failures = (number of failures/total stimulations) × 100%) in NTS neurons ($n = 11$). The asterisk (*) represents $p < 0.05$ as compared to control.

*2.7. Data from TRPA1 Knockout Mice Support the Modulatory Role of TRPA1 in the Caudal NTS*

TRPA1 knockout mice were used to further elucidate the modulatory role of presynaptic TRPA1 in the caudal NTS. In experiments using brainstem slices from TRPA1 knockout mice, AITC (200 μM) was ineffective in increasing mEPSC frequency (as a percent of control): 105.21 ± 12.23% ($n = 15$, Figure 10A,B,D; vs. effects of AITC in caudal NTS neurons obtained from wild-type mice: 450.63 ± 22.74% ($n = 13$, $p < 0.05$) Figure 2A,B,D). In the same experiments, TRPV1 agonist, capsaicin (100 nM) significantly increased the frequency of mEPSCs (309.88 ± 42.06%, $n = 7$, Figure 10A,B,D). These experiments further confirmed that the observed effects of AITC were mediated by TRPA1.

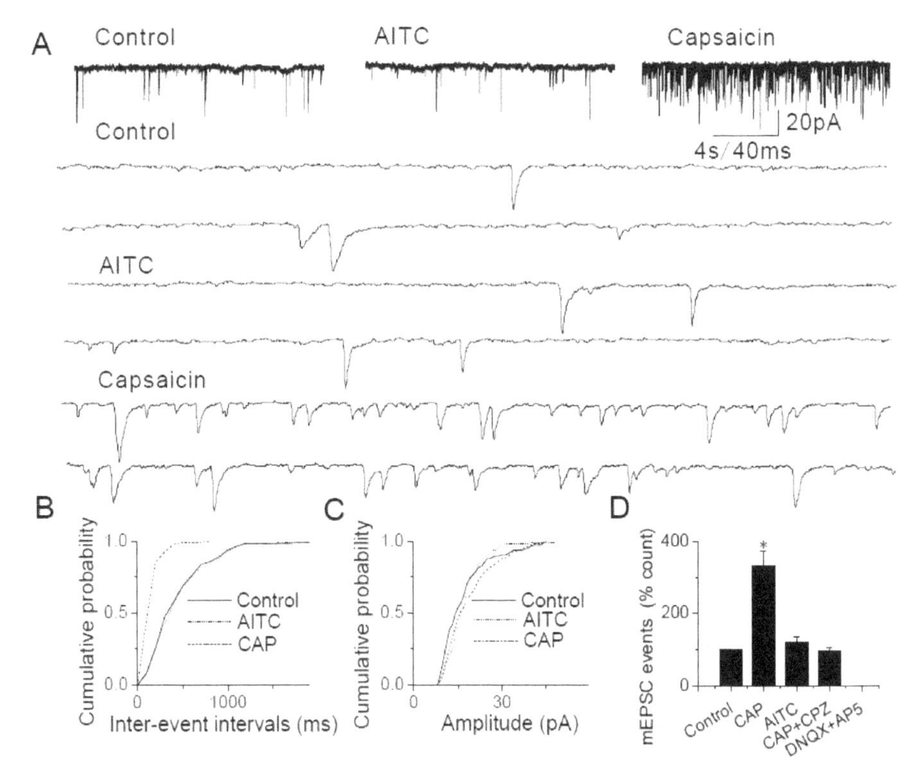


**Figure 10.** AITC does not induce enhancement of synaptic transmission in the TRPA1 knockout mice. (**A**) Application of AITC (200 μM) does not increases the frequency of mEPSC in the TRPA1 knockout mice (105.21 ± 12.23%, $n = 15$, $p > 0.05$); whereas, capsaicin increase the frequency of mEPSC significantly (309 ± 42%, $n = 7$, $p < 0.05$); The synaptic events are shown in higher time resolution below. (**B**) Cumulative probability plot showing decreased interevent intervals representing increased frequency of mEPSCs mediated by capsaicin (i.e., CAP; $p < 0.0001$, KS test), but no change of interevent intervals mediated by AITC. (**C**) There was no change in the amplitude in all three groups. (**D**) Summary graph showing only capsaicin-mediated increases but no AITC mediated changes in the frequency of mEPSCs (* $p < 0.05$). The asterisk (*) represents $p < 0.05$ as compared to control.

## 3. Discussion

In this study, the molecular biological and electrophysiological techniques were used to demonstrate TRPA1-mediated modulation of synaptic transmission at the NTS and the expression of TRPA1 in caudal NTS is strictly presynaptic. Together with the existing literature, these results support the exclusive expression of TRPA1 in the first order sensory neurons and indicate that the expression of TRPA1 in the caudal NTS is restricted to the primary afferent solitary tract terminals [30,35,36,47]. Activation of presynaptic TRPA1 by selective agonists was detected by its facilitating and modulatory effects on spontaneous and evoked glutamatergic synaptic transmission, respectively, recorded electrophysiologically in acute horizontal brainstem slices; while the specificity of TRPA1 in triggering synaptic glutamate release was confirmed in TRPA1 knockout mice. The effects of TRPA1 activation on glutamatergic synaptic transmission in the caudal NTS are consistent with those observed previously in other central sensory nuclei: the substantia gelatinosa and the caudal spinal trigeminal nucleus [30,48–53]. Agonists of TRPA1 (AITC) and TRPV1 (capsaicin; i.e., CAP) caused paired pulse depression. However, AITC did not depress evoked EPSC amplitude, while capsaicin did depress the evoked EPSC amplitude as described previously [45]. This observation is curious because

TRPA1 and TRPV1 channels are often co-expressed in the same subset of neurons [30] and their effects on synaptic neurotransmitter release are expected to be comparable. Differences in the sensitivity of presynaptic terminals to AITC and capsaicin, as well as the exact channel distribution within presynaptic terminals may be responsible for the apparent dichotomy of TRPA1- and TRPV1-mediated evoked presynaptic responses, respectively.

The functional role of TRPA1 in sensory signaling is suggested by the expression of these channels in subsets of primary sensory neurons in the trigeminal, jugular, geniculate, nodose and dorsal root ganglia [43,54–56]. The TRPA1 channels act as primary sensors for thermal (cold), mechanical and inflammatory noxious stimuli generated by environmental and endogenous agents [30–32,43,57–62]. Because TRPA1 channels are critical for transmission of visceral and inflammatory pain [21–27], the expression of functional presynaptic TRPA1 in primary afferents within the caudal NTS suggests a potential for TRPA1-dependent central component of nociception and its sensitization by TRPA1 agents at the level of the caudal brainstem.

Visceral pain is a common symptom of functional gastrointestinal disorder such as irritable bowel syndrome, ulcerative colitis and dyspepsia. As visceral structures are highly sensitive to distention, ischemia and inflammation, the features of chronic visceral pain are inflammatory and mechanical hyperalgesia and allodynia. Adopting strategies to reduce inflammatory and mechanosensory transduction may be particularly useful in relieving visceral pain [63–66]. The role of TRPA1 in gastrointestinal inflammatory disorders is becoming increasingly important as TRPA1 up-regulation has been confirmed in several disease model systems. Thus, TRPA1 may represent a useful target in treatments of chronic visceral pain. In fact, several TRPA1 antagonists have already entered clinical trials including one in phase II clinical trial [66].

A potential role of protein kinase C (PKC) in modulating TRPA1-mediated synaptic transmission was examined using PDBu, a PKC activator. The PKC-mediated phosphorylation plays an important role in modulation of synaptic neurotransmission [40,43–45,67,68]. For example, PDBu substantially increased the amplitude of TRPV1-mediated currents but had no effect on TRPA1-mediated currents in dorsal root ganglia neurons [41,69]. The direct PDBu-mediated facilitation of the frequency of synaptic events was observed in all neurons tested in this study, even in those cases where AITC failed to facilitate synaptic release. However, the effects of AITC and PDBu were simply additive (i.e., not synergistic) suggesting that AITC and PDBu likely employ independent pathways and thus, the TRPA1-dependent machinery is unlikely to involve PKC-mediated phosphorylation. This finding may reflect the nature of agonists used. AITC and NMM activate TRPA1 by covalent modification of cysteine residues. As a result, the affinity of these ligands for TRPA1 binding site(s) may not be altered by phosphorylation. In that event, the TRPA1 activity may still be susceptible to modulation by non-covalent modifying agonists and physical inputs, such as thermal (cold) and mechanical stimuli.

Taken together, the results of this study provide the first evidence of the TRPA1-dependent modulation of the primary afferent inputs to the caudal NTS and suggest that second order caudal NTS neurons serve as a TRPA1-dependent interface for visceral noxious-innocuous integration at the level of the caudal brainstem. Thus, TRPA1 may represent a useful target in treatments of chronic visceral and inflammatory pain.

## 4. Materials and Methods

### 4.1. Animals

Young adult male Sprague-Dawley rats (P30-50) and TRPA1 knockout mice (Jackson Laboratories, Bar Harbor, ME, USA) were used in accordance with the Guide for the Care and Use of Laboratory Animals (NIH 865-23, Bethesda, MD, USA), and was approved by the Animal Care and Use Committee of Southern Illinois University (A3209-01; Approval date: 7/25/2013).

### 4.2. Preparation of Brainstem Slices

Brains were removed and placed in an ice-cold oxygenated solution of the following composition (in mM): sucrose 250, KCl 1.5, $NaH_2PO_4$ 2.23, $MgCl_2$ 5, $CaCl_2$ 0.5, $NaHCO_3$ 26, glucose 10 (pH 7.4) and bubbled with carbogen (95% $O_2$ and 5% $CO_2$). The brainstem was isolated and transferred to the cutting chamber of Vibratome-1000+ slicer (Leica Microsystems, Wetzlar, Germany) where horizontal brainstem slices (250–300 μm thickness) containing the NTS were prepared. The slices were then transferred to a storage chamber and incubated at 30 °C for ~60 min in an oxygenated artificial cerebrospinal fluid (ACSF) of the following composition (in mM): NaCl 125, KCl 1.5, $NaH_2PO_4$ 2.23, $MgCl_2$ 1, $CaCl_2$ 2, $NaHCO_3$ 26, glucose 10 (pH 7.4). Slices were then stored in the identical oxygenated ACSF at 24 °C for up to 8 h.

### 4.3. Electrophysiological Patch-Clamp Recordings

For patch-clamp recordings, brainstem slices were placed in the recording chamber and perfused with oxygenated ACSF at a rate of 1 mL/min using a 2232 Microperpex S peristaltic pump (LK.B, Upsalla, Sweden). Slices were secured using a nylon mesh attached to platinum ring insert. Whole-cell recordings were conducted at room temperature. The patch electrode solution contained (in mM): K-gluconate 140, NaCl 1, $MgCl_2$ 2, Mg-ATP 2, Na-GTP 0.3, HEPES 10, KOH 0.42 (pH 7.4). Membrane voltages were not corrected for the liquid junction potential which was calculated using software available through pCamp: VLJ = 16.2 mV. The electrophysiological data were recorded using MultiClamp-700B patch-clamp amplifier (Molecular Devices, Sunyvale, CA, USA). The seal resistance was >2 GΩ. The access resistance was <30 MΩ and was not compensated. Patches with access resistances >30 MΩ were corrected by applying additional negative suction or discarded. Input resistance and series resistance were measured every five minutes and cells showing greater that 20% change in series resistance were not included in analysis. Data were sampled at 10–20 kHz and filtered at 5 kHz. The final drug concentration in the bath was calculated based on the known concentration of the stock solution and adjustable rates of pumps [70]. A picospritzer (Parker Hannifin Instrumentation, Cleveland, OH, USA) was used for agonist applications via pipettes (4–7 MΩ) identical to those used for patch clamp recordings. Agonist application was standardized by positioning the tip of the application pipette 15 μm from the recorded neuron. This distance was calibrated and marked on the TV monitor and used for visualization of neurons while patching. Off-line data analysis was done with the program Clampfit 9 (Molecular Devices, Sunnyvale, CA, USA). To obtain evoked EPSCs, a Grass Stimulator (S88) with stimulus isolation unit with constant current output (Grass Instrument, Quincy, MA, USA) was used to stimulate a concentric bipolar electrode (Rhodes Medical Instruments, Tujunga, CA, USA) placed on the solitary tract. The following parameters of electrical stimulation were used: stimulus duration, 50–400 μs; stimulus intensity,100–500 μA; inter-stimulus interval, 60 s.

### 4.4. Immunohistochemistry and Peptide Absorption Studies

Rats were anesthetized by intraperitoneal (i.p.) injections of ketamine (85 mg/kg) and xylazine (10 mg/kg). The anesthetized animals were perfused transcardially with 4% paraformaldehyde. The brainstem was harvested and stored in the phosphate buffer saline (PBS, pH7.4) containing 30% sucrose for at least 24 h. Then the tissue was frozen in powdered dry ice and stored at −80 °C. Serial horizontal sections were cut at 20 μm using a Leica CM1850 cryostat at −18 °C. Selected sections were thawed and mounted onto Superfrost/Plus slides. The sections were rinsed in PBS and then blocked in 10% normal donkey serum in PBS for 60 min. The sections were incubated with rabbit anti-TRPA1 antibody (1:100, Santa Cruz, Dallas, TX, USA) overnight at 4 °C and FITC donkey anti-rabbit IgG (1:100, Jackson Immunoresearch Laboratories Inc., West Grove, PA, USA). Images were captured by a fluorescence microscope.

### 4.5. Total RNA Extraction and RT-PCR

Total RNA was extracted by Trizol reagent (Invitrogen Co., Carlsbad, CA, USA) from nucleus of the solitary tract (NTS) and dorsal root ganglion (DRG). cDNAs were prepared by reverse transcription. PCR was performed using a standard approach and PCR green master mix (Promega Corporation, Madison, WI, USA). The PCR products were electrophoresed in 1.5% agarose gel with ethidium bromide in Tris/Borate/EDTA buffer. The gel was scanned using Versa Doc imaging system (Bio-Rad, Hercules, CA, USA) and the blot band density was quantified by Quantity One (Bio-Rad, Hercules, CA, USA).

### 4.6. Data Analysis

All data are shown as means ± SEM. Significance is tested using unpaired Student's *t*-test, and the data were considered significant at $p < 0.05$. For analysis of synaptic currents, Kolmogorov-Smirnov (KS) test was used to compare the cumulative probability plots for inter-event intervals and amplitude between various treatment groups. Data are represented as means ± SE and expressed as percentage of control, which is scaled to 100%.

The spontaneous/miniature postsynaptic currents (s/mPSCs) were analyzed off-line using MiniAnalysis 6.0.3 (Synaptosoft Inc., Fort Lee, NJ, USA) and the threshold for event detection (usually 10 pA) was at least three times baseline noise levels.

All the chemicals used in this study were obtained from Sigma (St. Louis, MO, USA).

**Author Contributions:** Conceptualization, L.S.P. and L.F.; methodology, V.V.U. and L.F.; formal analysis, L.F. and V.V.U; investigation, L.F.; resources, L.S.P.; data curation, L.F.; writing—original draft preparation, L.F. and V.V.U., L.S.P.; writing—review and editing, L.F., V.V.U. and L.S.P.; project administration, L.S.P.; funding acquisition, L.S.P.

**Funding:** This work was supported with a grant from National Institutes of Health (DA028017 to L.S.P).

**Conflicts of Interest:** The authors declare no conflict of interest. The funders had no role in the design of the study; in the collection, analyses, or interpretation of data; in the writing of the manuscript, or in the decision to publish the results.

## References

1. Ionescu, D.A.; Lugoji, G.; Radula, D. The nucleus of the solitary tract: A review of its anatomy and functions, with emphasis on its role in a putative central-control of brain-capillaries permeability. *Neurol. Psychiatr.* **1986**, *24*, 69–85.
2. McAllen, R.M.; Spyer, K.M. The location of cardiac vagal preganglionic motoneurones in the medulla of the cat. *J. Physiol.* **1976**, *258*, 187–204. [CrossRef]
3. Miura, M.; Reis, D.J. Termination and secondary projections of carotid sinus nerve in the cat brain stem. *Am. J. Physiol.* **1969**, *217*, 142–153. [CrossRef] [PubMed]
4. Beckstead, R.M.; Norgren, R. An autoradiographic examination of the central distribution of the trigeminal, facial, glossopharyngeal, and vagal nerves in the monkey. *J. Comp. Neurol.* **1979**, *184*, 455–472. [CrossRef] [PubMed]
5. Gamboa-Esteves, F.O.; Tavares, I.; Almeida, A.; Batten, T.F.; McWilliam, P.N.; Lima, D. Projection sites of superficial and deep spinal dorsal horn cells in the nucleus tractus solitarii of the rat. *Brain Res.* **2001**, *921*, 195–205. [CrossRef]
6. Hamilton, R.B.; Norgren, R. Central projections of gustatory nerves in the rat. *J. Comp. Neurol.* **1984**, *222*, 560–577. [CrossRef]
7. Kalia, M.; Sullivan, J.M. Brainstem projections of sensory and motor components of the vagus nerve in the rat. *J. Comp. Neurol.* **1982**, *211*, 248–265. [CrossRef]
8. Menetrey, D.; Basbaum, A.I. Spinal and trigeminal projections to the nucleus of the solitary tract: A possible substrate for somatovisceral and viscerovisceral reflex activation. *J. Comp. Neurol.* **1987**, *255*, 439–450. [CrossRef]
9. Travers, S.P.; Pfaffmann, C.; Norgren, R. Convergence of lingual and palatal gustatory neural activity in the nucleus of the solitary tract. *Brain Res.* **1986**, *365*, 305–320. [CrossRef]

10. Boscan, P.; Pickering, A.E.; Paton, J.F. The nucleus of the solitary tract: An integrating station for nociceptive and cardiorespiratory afferents. *Exp. Physiol.* **2002**, *87*, 259–266. [CrossRef]
11. Feng, L.; Uteshev, V.V. Projection target-specific action of nicotine in the caudal nucleus of the solitary tract. *J. Neurosci. Res.* **2014**, *92*, 1560–1572. [CrossRef]
12. Ricardo, J.A.; Koh, E.T. Anatomical evidence of direct projections from the nucleus of the solitary tract to the hypothalamus, amygdala, and other forebrain structures in the rat. *Brain Res.* **1978**, *153*, 1–26. [CrossRef]
13. Spyer, K.M.; Donoghue, S.; Felder, R.B.; Jordan, D. Processing of afferent inputs in cardiovascular control. *Clin. Exp. Hypertens. A* **1984**, *6*, 173–184. [CrossRef]
14. Wiertelak, E.P.; Roemer, B.; Maier, S.F.; Watkins, L.R. Comparison of the effects of nucleus tractus solitarius and ventral medial medulla lesions on illness-induced and subcutaneous formalin-induced hyperalgesias. *Brain Res.* **1997**, *748*, 143–150. [CrossRef]
15. Caterina, M.J.; Schumacher, M.A.; Tominaga, M.; Rosen, T.A.; Levine, J.D.; Julius, D. The capsaicin receptor: A heat-activated ion channel in the pain pathway. *Nature* **1997**, *389*, 816–824. [CrossRef]
16. Andresen, M.C.; Hofmann, M.E.; Fawley, J.A. The unsilent majority-TRPV1 drives "spontaneous" transmission of unmyelinated primary afferents within cardiorespiratory NTS. *Am. J. Physiol. Regul. Integr. Comp. Physiol.* **2012**, *303*, R1207–R1216. [CrossRef]
17. Bailey, T.W.; Jin, Y.H.; Doyle, M.W.; Andresen, M.C. Vanilloid-sensitive afferents activate neurons with prominent A-type potassium currents in nucleus tractus solitarius. *J. Neurosci.* **2002**, *22*, 8230–8237. [CrossRef]
18. Doyle, M.W.; Bailey, T.W.; Jin, Y.H.; Andresen, M.C. Vanilloid receptors presynaptically modulate cranial visceral afferent synaptic transmission in nucleus tractus solitarius. *J. Neurosci.* **2002**, *22*, 8222–8229. [CrossRef]
19. Sun, B.; Bang, S.I.; Jin, Y.H. Transient receptor potential A1 increase glutamate release on brain stem neurons. *Neuroreport* **2009**, *20*, 1002–1006. [CrossRef]
20. Smith, D.V.; Uteshev, V.V. Heterogeneity of nicotinic acetylcholine receptor expression in the caudal nucleus of the solitary tract. *Neuropharmacology* **2008**, *54*, 445–453. [CrossRef]
21. Cattaruzza, F.; Spreadbury, I.; Miranda-Morales, M.; Grady, E.F.; Vanner, S.; Bunnett, N.W. Transient receptor potential ankyrin-1 has a major role in mediating visceral pain in mice. *Am. J. Physiol. Gastrointest. Liver Physiol.* **2010**, *298*, G81–G91. [CrossRef]
22. Kimball, E.S.; Prouty, S.P.; Pavlick, K.P.; Wallace, N.H.; Schneider, C.R.; Hornby, P.J. Stimulation of neuronal receptors, neuropeptides and cytokines during experimental oil of mustard colitis. *Neurogastroenterol. Motil.* **2007**, *19*, 390–400. [CrossRef]
23. Vermeulen, W.; De Man, J.G.; De Schepper, H.U.; Bult, H.; Moreels, T.G.; Pelckmans, P.A.; De Winter, B.Y. Role of TRPV1 and TRPA1 in visceral hypersensitivity to colorectal distension during experimental colitis in rats. *Eur. J. Pharmacol.* **2013**, *698*, 404–412. [CrossRef]
24. Yang, J.; Li, Y.; Zuo, X.; Zhen, Y.; Yu, Y.; Gao, L. Transient receptor potential ankyrin-1 participates in visceral hyperalgesia following experimental colitis. *Neurosci. Lett.* **2008**, *440*, 237–241. [CrossRef]
25. Engel, M.A.; Leffler, A.; Niedermirtl, F.; Babes, A.; Zimmermann, K.; Filipović, M.R.; Izydorczyk, I.; Eberhardt, M.; Kichko, T.I.; Mueller-Tribbensee, S.M.; et al. TRPA1 and substance P mediate colitis in mice. *Gastroenterology* **2011**, *141*, 1346–1358. [CrossRef]
26. Kondo, T.; Sakurai, J.; Miwa, H.; Noguchi, K. Activation of p38 MAPK through transient receptor potential A1 in a rat model of gastric distension-induced visceral pain. *Neuroreport* **2013**, *24*, 68–72. [CrossRef]
27. Laird, J.M.; Martinez-Caro, L.; Garcia-Nicas, E.; Cervero, F. A new model of visceral pain and referred hyperalgesia in the mouse. *Pain* **2001**, *92*, 335–342. [CrossRef]
28. Kwan, K.Y.; Glazer, J.M.; Corey, D.P.; Rice, F.L.; Stucky, C.L. TRPA1 modulates mechanotransduction in cutaneous sensory neurons. *J. Neurosci.* **2009**, *29*, 4808–4819. [CrossRef]
29. Sotomayor, M.; Corey, D.P.; Schulten, K. In search of the hair-cell gating spring elastic properties of ankyrin and cadherin repeats. *Structure* **2005**, *13*, 669–682. [CrossRef]
30. Story, G.M.; Peier, A.M.; Reeve, A.J.; Eid, S.R.; Mosbacher, J.; Hricik, T.R.; Earley, T.J.; Hergarden, A.C.; Andersson, D.A.; Hwang, S.W.; et al. ANKTM1, a TRP-like channel expressed in nociceptive neurons, is activated by cold temperatures. *Cell* **2003**, *112*, 819–829. [CrossRef]
31. Jordt, S.E.; Bautista, D.M.; Chuang, H.H.; McKemy, D.D.; Zygmunt, P.M.; Högestätt, E.D.; Meng, I.D.; Julius, D. Mustard oils and cannabinoids excite sensory nerve fibres through the TRP channel ANKTM1. *Nature* **2004**, *427*, 260–265. [CrossRef] [PubMed]

32. Zurborg, S.; Yurgionas, B.; Jira, J.A.; Caspani, O.; Heppenstall, P.A. Direct activation of the ion channel TRPA1 by $Ca^{2+}$. *Nat. Neurosci.* **2007**, *10*, 277–279. [CrossRef] [PubMed]
33. Bandell, M.; Story, G.M.; Hwang, S.W.; Viswanath, V.; Eid, S.R.; Petrus, M.J.; Earley, T.J.; Patapoutian, A. Noxious cold ion channel TRPA1 is activated by pungent compounds and bradykinin. *Neuron* **2004**, *41*, 849–857. [CrossRef]
34. Hinman, A.; Chuang, H.H.; Bautista, D.M.; Julius, D. TRP channel activation by reversible covalent modification. *Proc. Natl. Acad. Sci. USA* **2006**, *103*, 19564–19568. [CrossRef] [PubMed]
35. Bautista, D.M.; Pellegrino, M.; Tsunozaki, M. TRPA1: A gatekeeper for inflammation. *Annu. Rev. Physiol.* **2013**, *75*, 181–200. [CrossRef]
36. Hondoh, A.; Ishida, Y.; Ugawa, S.; Ueda, T.; Shibata, Y. Distinct expression of cold receptors (TRPM8 and TRPA1) in the rat nodose-petrosal ganglion complex. *Brain Res.* **2010**, *1319*, 60–69. [CrossRef]
37. Katsura, H.; Tsuzuki, K.; Noguchi, K.; Sakagami, M. Differential expression of capsaicin-, menthol-, and mustard oil-sensitive receptors in naive rat geniculate ganglion neurons. *Chem. Senses* **2006**, *31*, 681–688. [CrossRef]
38. Cavanaugh, D.J.; Chesler, A.T.; Jackson, A.C.; Sigal, Y.M.; Yamanaka, H.; Grant, R.; O'Donnell, D.; Nicoll, R.A.; Shah, N.M.; Julius, D.; et al. Trpv1 reporter mice reveal highly restricted brain distribution and functional expression in arteriolar smooth muscle cells. *J. Neurosci.* **2011**, *31*, 5067–5177. [CrossRef]
39. Mishra, S.K.; Tisel, S.M.; Orestes, P.; Bhangoo, S.K.; Hoon, M.A. TRPV1-lineage neurons are required for thermal sensation. *EMBO J.* **2011**, *30*, 582–593. [CrossRef]
40. Macpherson, L.J.; Dubin, A.E.; Evans, M.J.; Marr, F.; Schultz, P.G.; Cravatt, B.F.; Patapoutian, A. Noxious compounds activate TRPA1 ion channels through covalent modification of cysteines. *Nature* **2007**, *445*, 541–545. [CrossRef]
41. Raisinghani, M.; Zhong, L.; Jeffry, J.A.; Bishnoi, M.; Pabbidi, R.M.; Pimentel, F.; Cao, D.S.; Evans, M.S.; Premkumar, L.S. Activation characteristics of transient receptor potential ankyrin 1 and its role in nociception. *Am. J. Physiol. Cell Physiol.* **2011**, *301*, C587–C600. [CrossRef]
42. Brose, N.; Rosenmund, C. Move over protein kinase C, you've got company: Alternative cellular effectors of diacylglycerol and phorbol esters. *J. Cell Sci.* **2002**, *115*, 4399–4411. [CrossRef]
43. Cao, D.S.; Yu, S.Q.; Premkumar, L.S. Modulation of transient receptor potential Vanilloid 4-mediated membrane currents and synaptic transmission by protein kinase C. *Mol. Pain* **2009**, *5*, 5. [CrossRef]
44. Lou, X.; Korogod, N.; Brose, N.; Schneggenburger, R. Phorbol esters modulate spontaneous and $Ca^{2+}$-evoked transmitter release via acting on both Munc13 and protein kinase C. *J. Neurosci.* **2008**, *28*, 8257–8267. [CrossRef] [PubMed]
45. Peters, J.H.; McDougall, S.J.; Fawley, J.A.; Smith, S.M.; Andresen, M.C. Primary afferent activation of thermosensitive TRPV1 triggers asynchronous glutamate release at central neurons. *Neuron* **2010**, *65*, 657–669. [CrossRef] [PubMed]
46. Zanotto, K.L.; Iodi Carstens, M.; Carstens, E. Cross-desensitization of responses of rat trigeminal subnucleus caudalis neurons to cinnamaldehyde and menthol. *Neurosci. Lett.* **2008**, *430*, 29–33. [CrossRef] [PubMed]
47. Kim, Y.S.; Son, J.Y.; Kim, T.H.; Paik, S.K.; Dai, Y.; Noguchi, K.; Ahn, D.K.; Bae, Y.C. Expression of transient receptor potential ankyrin 1 (TRPA1) in the rat trigeminal sensory afferents and spinal dorsal horn. *J. Comp. Neurol.* **2010**, *518*, 687–698. [CrossRef] [PubMed]
48. Kobayashi, K.; Fukuoka, T.; Obata, K.; Yamanaka, H.; Dai, Y.; Tokunaga, A.; Noguchi, K. Distinct expression of TRPM8, TRPA1, and TRPV1 mRNAs in rat primary afferent neurons with adelta/c-fibers and colocalization with trk receptors. *J. Comp. Neurol.* **2005**, *493*, 596–606. [CrossRef] [PubMed]
49. Kosugi, M.; Nakatsuka, T.; Fujita, T.; Kuroda, Y.; Kumamoto, E. Activation of TRPA1 channel facilitates excitatory synaptic transmission in substantia gelatinosa neurons of the adult rat spinal cord. *J. Neurosci.* **2007**, *27*, 4443–4451. [CrossRef] [PubMed]
50. Yokoyama, T.; Ohbuchi, T.; Saito, T.; Sudo, Y.; Fujihara, H.; Minami, K.; Nagatomo, T.; Uezono, Y.; Ueta, Y. Allyl isothiocyanates and cinnamaldehyde potentiate miniature excitatory postsynaptic inputs in the supraoptic nucleus in rats. *Eur. J. Pharmacol.* **2011**, *655*, 31–37. [CrossRef]
51. Sikand, P.; Premkumar, L.S. Potentiation of glutamatergic synaptic transmission by protein kinase C-mediated sensitization of TRPV1 at the first sensory synapse. *J. Physiol.* **2007**, *581*, 631–647. [CrossRef] [PubMed]

52. Jeffry, J.A.; Yu, S.Q.; Sikand, P.; Parihar, A.; Evans, M.S.; Premkumar, L.S. Selective targeting of TRPV1 expressing sensory nerve terminals in the spinal cord for long lasting analgesia. *PLoS ONE* **2009**, *4*, e7021. [CrossRef]
53. Grace, M.S.; Belvisi, M.G. TRPA1 receptors in cough. *Pulm. Pharmacol. Ther.* **2011**, *24*, 286–288. [CrossRef]
54. Ramsey, I.S.; Delling, M.; Clapham, D.E. An introduction to TRP channels. *Annu. Rev. Physiol.* **2006**, *68*, 619–647. [CrossRef]
55. Sadofsky, L.R.; Boa, A.N.; Maher, S.A.; Birrell, M.A.; Belvisi, M.G.; Morice, A.H. TRPA1 is activated by direct addition of cysteine residues to the N-hydroxysuccinyl esters of acrylic and cinnamic acids. *Pharmacol. Res.* **2011**, *63*, 30–36. [CrossRef]
56. Andersson, D.A.; Gentry, C.; Moss, S.; Bevan, S. Transient receptor potential A1 is a sensory receptor for multiple products of oxidative stress. *J. Neurosci.* **2008**, *28*, 2485–2494. [CrossRef]
57. Bessac, B.F.; Sivula, M.; von Hehn, C.A.; Escalera, J.; Cohn, L.; Jordt, S.E. TRPA1 is a major oxidant sensor in murine airway sensory neurons. *J. Clin. Investig.* **2008**, *118*, 1899–1910. [CrossRef] [PubMed]
58. Obata, K.; Katsura, H.; Mizushima, T.; Yamanaka, H.; Kobayashi, K.; Dai, Y.; Fukuoka, T.; Tokunaga, A.; Tominaga, M.; Noguchi, K. TRPA1 induced in sensory neurons contributes to cold hyperalgesia after inflammation and nerve injury. *J. Clin. Investig.* **2005**, *115*, 2393–2401. [CrossRef] [PubMed]
59. Kwan, K.Y.; Allchorne, A.J.; Vollrath, M.A.; Christensen, A.P.; Zhang, D.S.; Woolf, C.J.; Corey, D.P. TRPA1 contributes to cold, mechanical, and chemical nociception but is not essential for hair-cell transduction. *Neuron* **2006**, *50*, 277–289. [CrossRef] [PubMed]
60. Kerstein, P.C.; del Camino, D.; Moran, M.M.; Stucky, C.L. Pharmacological blockade of TRPA1 inhibits mechanical firing in nociceptors. *Mol. Pain* **2009**, *5*, 19. [CrossRef]
61. Bautista, D.M.; Jordt, S.E.; Nikai, T.; Tsuruda, P.R.; Read, A.J.; Poblete, J.; Yamoah, E.N.; Basbaum, A.I.; Julius, D. TRPA1 mediates the inflammatory actions of environmental irritants and proalgesic agents. *Cell* **2006**, *124*, 1269–1282. [CrossRef]
62. Christianson, J.A.; Bielefeldt, K.; Altier, C.; Cenac, N.; Davis, B.M.; Gebhart, G.F.; High, K.W.; Kollarik, M.; Randich, A.; Undem, B.; et al. Development, plasticity and modulation of visceral afferents. *Brain Res. Rev.* **2009**, *60*, 171–186. [CrossRef]
63. Blackshaw, L.A.; Brierley, S.M.; Hughes, P.A. TRP channels: New targets for visceral pain. *Gut* **2010**, *59*, 126–135. [CrossRef] [PubMed]
64. Poole, D.P.; Pelayo, J.C.; Cattaruzza, F.; Kuo, Y.M.; Gai, G.; Chiu, J.V.; Bron, R.; Furness, J.B.; Grady, E.F.; Bunnett, N.W. Transient receptor potential ankyrin 1 is expressed by inhibitory motoneurons of the mouse intestine. *Gastroenterology* **2011**, *141*, 565–575.e4. [CrossRef] [PubMed]
65. Premkumar, L.S.; Abooj, M. TRP channels and analgesia. *Life Sci.* **2013**, *92*, 415–424. [CrossRef] [PubMed]
66. Moran, M.M.; McAlexander, M.A.; Biro, T.; Szallasi, A. Transient receptor potential channels as therapeutic targets. *Nat. Rev. Drug Discov.* **2011**, *10*, 601–620. [PubMed]
67. Hori, Y.; Endo, K.; Takahashi, T. Long-lasting synaptic facilitation induced by serotonin in superficial dorsal horn neurones of the rat spinal cord. *J. Physiol.* **1996**, *492*, 867–876. [CrossRef]
68. Malenka, R.C.; Madison, D.V.; Nicoll, R.A. Potentiation of synaptic transmission in the hippocampus by phorbol esters. *Nature* **1986**, *321*, 175–177. [CrossRef]
69. Premkumar, L.S.; Ahern, G.P. Induction of vanilloid receptor channel activity by protein kinase C. *Nature* **2000**, *408*, 985–990. [CrossRef] [PubMed]
70. Kalappa, B.I.; Feng, L.; Kem, W.R.; Gusev, A.G.; Uteshev, V.V. Mechanisms of facilitation of synaptic glutamate release by nicotinic agonists in the nucleus of the solitary tract. *Am. J. Physiol. Cell Physiol.* **2011**, *301*, C347–C361. [CrossRef] [PubMed]

International Journal of *Molecular Sciences*

MDPI

*Review*

# Noncanonical Ion Channel Behaviour in Pain

**Cosmin I. Ciotu [1,†], Christoforos Tsantoulas [2,†], Jannis Meents [3,†], Angelika Lampert [3], Stephen B. McMahon [2], Andreas Ludwig [4] and Michael J.M. Fischer [1,*]**

1 Center for Physiology and Pharmacology, Medical University of Vienna, 1090 Vienna, Austria
2 Wolfson Centre for Age-Related Diseases, King's College London, London SE1 1UR, UK
3 Institute of Physiology, University Hospital RWTH Aachen, 52074 Aachen, Germany
4 Institute of Experimental and Clinical Pharmacology and Toxicology, Friedrich-Alexander-Universität Erlangen-Nürnberg, 91054 Erlangen, Germany
* Correspondence: michael.jm.fischer@meduniwien.ac.at; Tel.: +43-1-40160-31410
† These authors contributed equally to this work.

Received: 7 August 2019; Accepted: 12 September 2019; Published: 15 September 2019

check for updates

**Abstract:** Ion channels contribute fundamental properties to cell membranes. Although highly diverse in conductivity, structure, location, and function, many of them can be regulated by common mechanisms, such as voltage or (de-)phosphorylation. Primarily considering ion channels involved in the nociceptive system, this review covers more novel and less known features. Accordingly, we outline noncanonical operation of voltage-gated sodium, potassium, transient receptor potential (TRP), and hyperpolarization-activated cyclic nucleotide (HCN)-gated channels. Noncanonical features discussed include properties as a memory for prior voltage and chemical exposure, alternative ion conduction pathways, cluster formation, and silent subunits. Complementary to this main focus, the intention is also to transfer knowledge between fields, which become inevitably more separate due to their size.

**Keywords:** pharmacology; drug development; sodium channel; potassium channel; TRP channel; HCN channel

## 1. Overview

The main aim of this review is to illustrate unexpected behaviour of ion channels, which might cross-pollinate advances between fields. To at least partly fulfil this aim, we restricted coverage to the pain field and the subsections are written by authors with a focus on the respective ion channel families. We consider as canonical any feature of an ion channel pore-forming protein, which can allow the flow of ions across membranes [1]. Although not universal, common features include regulation of the permeation (gating) by ligands and voltage; preference or selectivity for some ions over others; interaction with other cytoplasmic or membrane proteins; trafficking between the plasma membrane and reserve pools; heteromerisation of the channels; modulation by intracellular cascades e.g., by a change of phosphorylation state; and a change of expression levels, e.g., in inflammatory conditions. Less common and more on the line between canonical and noncanonical are features such as interaction with phospholipids or accessory subunits.

## 2. Sodium Channels

Voltage-gated sodium channels (Navs) are responsible for the generation of action potentials in most excitable cells, such as neurons and muscle cells. Ten different isoforms have been described in mammals (Nav1.1-1.9 and Nax), which vary in tissue expression and electrophysiological properties [2,3]. Generally, Navs are highly voltage sensitive and open in response to small membrane depolarisations. They are selective for the conduction of sodium ions, thus amplifying membrane depolarisation and initiating action potentials. For the study and treatment of pain, the subtypes Nav1.7, Nav1.8, and Nav1.9, mostly expressed in peripheral sensory neurons, have received large interest over recent years considering that mutations in these channel isoforms can lead to a variety of pain syndromes in patients [4,5]. Well-described canonical features of Nav channel activity comprise voltage-dependent gating and fast inactivation during membrane depolarisation as well as channel deactivation upon cell membrane repolarisation [2,6,7]. Of note in this respect are the recently published 3-D crystal structures of different Nav isoforms that have shed a new light on these well-known functions [8–11]. There are, however, a number of rather unexpected and less well-understood channel functions that will be discussed in the following. Some of these have already been reviewed in a similar context by Barbosa and Cummins [12].

In addition to fast inactivation, which occurs within milliseconds after channel opening, Navs can also undergo slow inactivation, a process that takes place on a time scale of seconds to minutes. Under experimental conditions, this process can be observed during prolonged depolarisations (e.g., 30–60 s). Physiologically, slow inactivation is believed to take place during high-frequency firing, also serving to modulate it. Slow inactivation in Navs has been known for several decades [13,14]. Slow inactivation is different depending on Nav channel isoform, and the exact molecular determinants for slow inactivation are difficult to pinpoint as many positions and residues have been described that seem to affect slow inactivation. Generally, the process of slow inactivation in Navs bears similarities to the C-type inactivation of potassium channels and involves the channel pore [15]. Especially a ring of four negatively charged amino acids directly above the selectivity filter (E409, E764, D1248, and D1539 in hNav1.4 [9]) seems to be involved in this process [16]. However, many other residues both inside and outside the channel pore have been implicated in slow inactivation (reviewed in References [13,14]). With regards to nociception, slow inactivation has been found to be modulated by different mutations in Nav1.7 that cause the chronic pain syndrome erythromelalgia [17–29] (reviewed in References [12,30]). Slow inactivation in peripheral Navs seems to be enhanced by cold temperatures, with the exception of Nav1.8, which inactivates cold-independently and thus mediates cold nociception in mice [31].

Slow inactivation can be regarded as a sort of negative hysteresis, i.e., Nav channels "remember" a previous prolonged or high-frequency stimulation and, as a result, remain inactive. In other ion channels, such as transient receptor potential and potassium channels, different forms of hysteresis have been described, including an *increase* in conduction or sensitivity upon prolonged or repeated stimulation (see below and in Reference [32]. However, our group has failed to show changes in voltage dependence of activation of Nav1.7 after a series of depolarizing pre-pulses, thus questioning the role of positive hysteresis in Nav1.7 [33].

According to the canonical view, Nav channels inactivate milliseconds after opening and remain impassive to sodium flux until the cell membrane has repolarised and the channel has returned to its resting state. However, an unusual Nav current, termed resurgent current, has been described to occur during membrane repolarisation. Since their first description in the late '90s in cerebellar Purkinje neurons [34], resurgent currents have become highly investigated and are believed to modulate high-frequency action-potential firing in different types of neurons, including nociceptors [35]. The molecular process of resurgent currents consists of an open channel block by a positively charged intracellular blocking particle, which occludes the channel pore before fast inactivation occurs. During membrane repolarisation, this particle is released from the channel pore due to its positive charge, thus leading to a very brief inward sodium current before the channel inactivates [35,36]. The most likely candidate for the open channel blocking particle is the C-terminal end of the Navβ4 subunit [35,37]. Peripheral sensory neurons express fast and slow resurgent currents, mainly mediated by Nav1.6, Nav1.2, and potentially Nav1.8 [38–40]. Nav1.7 has also been shown to produce resurgent currents of small amplitudes. However, mutations in Nav1.7 that cause the chronic pain phenotype paroxysmal extreme pain disorder (PEPD) enhance resurgent currents in this Nav subtype, whereas mutations leading to erythromelalgia do not [29,41]. Interestingly, there seems to be a direct correlation between resurgent current generation and slow inactivation in Nav1.6 and Nav1.7: enhanced slow inactivation impairs resurgent currents and vice-versa [29].

Local anaesthetics can be a useful tool for quick and localized pain treatment. These drugs have been shown to bind inside the central cavity of the channel pore [42–44]. Recent findings in prokaryotic and mammalian Nav channels have substantiated earlier reports, which suggested that entry of local anaesthetics into the central cavity can be mediated via the lipid phase of the cell membrane [45]. The recently published 3-D crystal structures as well as earlier models show side fenestrations of the channel pore, which are large enough to be permeated by small molecules, such as local anaesthetics [9,10,46–49]. This may have important implications for the future development of Nav channel blocking compounds.

Several naturally occurring mutations in Nav1.2, Nav1.4, and Nav1.5 have been reported to conduct so-called gating pore (or omega) currents. These currents originate from mutations of gating charge residues in the S4 voltage sensor, leading to an alternative ion permeation pathway across the membrane [50–54] (reviewed in Reference [55]). Whereas such gating pore currents have not yet, to our knowledge, been investigated in nociceptive Nav channel isoforms, it might still be worthwhile to check for such currents, especially in Nav1.6–Nav1.9, as leak currents through these channels would almost certainly affect nociceptor excitability and pain perception.

## 3. Potassium Channels

Potassium channels are the most populous, diverse, and widely distributed ion channel superfamily. Once regarded as "innocent bystanders" that could nevertheless be pharmacologically exploited to counteract neuronal hyperexcitability rising from maladaptive activity of other ion channels, potassium channels are increasingly viewed as key players that can directly promote pain pathogenesis [56]. Indeed, an ever-growing number of studies report causative links between reduced function of specific potassium channel subunits and development of neuronal hyperexcitability and pain sensation [57]. Furthermore, in not electrically excitable cells, potassium channels participate in several neurophysiological processes that are independent of ion conduction, such as proliferation, migration, and exocytosis, and the mechanisms governing these noncanonical functions may also be of relevance to pain syndromes [58–60].

The best studied group, voltage-gated potassium channels (Kv), comprises 40 members which assemble as homo- or hetero-tetramers and mediate a hyperpolarising $K^+$ efflux that limits neuronal excitability by opposing action-potential generation. Membrane depolarisation triggers Kv opening via movement of the voltage sensor, which is coupled through a 15aa helical S4-S5 linker to the channel pore [61]. Recent work, however, in Drosophila's Shaker potassium channels (closely related to human Kv1 channels) identified an additional electromechanical coupling between residues of transmembrane domains S4 and S5 which facilitate movement of the helices in a "rack-and-pinion" fashion [62]. This noncanonical mechanism is a good candidate to explain pore opening in channels like hERG (Kv11.1), which contain a short S4–S5 linker, expendable for voltage-gating [63,64].

The traditional view of Kv opening exclusively gated by voltage was challenged by Hao et al., who demonstrated that Kv1.1 is a bona fide mechanoreceptor in sensory neurons [65]. In thorough experiments, it was shown that a variety of mechanical manipulations such as piezo-electrically driven force, membrane stretching, and hypoosmotic shock directly activate Kv1.1 channels to mediate a mechanosusceptive current, dubbed $I_{Kmech}$. Mechanistically, generation of $I_{Kmech}$ results from a change in the voltage dependence of the open probability, favouring the open conformation of the channel. Traditional mechanotransducers use a mechanical sensor linked to cytoskeletal elements to convert membrane tension energy into conformation changes. In contrast, Kv1.1 mechanoactivation may depend on inherent properties of the voltage sensor because mechanosensitivity is retained in excised patches of DRG neurons [65]. The authors postulated that the local membrane distortion induced by applied forces alters the energetic stability of the voltage-sensing machinery by physical movement of charges within the channel. Whatever the precise mechanism, Kv1.1 activation by mechanical stimulation can—because it reduces neuronal excitability—tune sensory neuron excitability by opposing excitatory influences of mechanosensitive cation channels. The net outcome of this process depends on the exact ionic channel complement of the neuron; in C-high threshold mechanoreceptors which mediate slowly adapting mechanosensitive cation currents [66], $I_{Kmech}$ opposes depolarisation and increases mechanical thresholds. In contrast, in Aβ mechanoreceptors which encode rapidly adapting mechanosensitive currents, $I_{Kmech}$ is not engaged sufficiently to influence firing thresholds but can nevertheless regulate firing rates. This elaborate control of mechanosensitivity by a Kv channel is a novel mechanism of mechanosensation, and inhibition of this pathway due to injury or inflammation could promote mechanically induced pain. Consistent with this, blocking Kv1.1 activity in mice either genetically or pharmacologically triggers mechanical hypersensitivity, without affecting heat pain responses [65].

The closely related member Kv1.2 also stands out because its function is subject to epigenetic silencing by G9a (histone-lysine N-methyltransferase 2) [67]. Neuropathic injury induces the Myeloid Zinc Finger 1 transcription factor, which in turn upregulates a long noncoding antisense RNA which attenuates Kv1.2 expression and activity, leading to hyperexcitability and pain sensitivity in rodents. Blocking induction of the antisense RNA spares Kv1.2 expression and is protective against pain [68]. This and other emerging pathways regulating Kv-dependent excitability via epigenetic modifications [69] might constitute a dynamic mechanism which shapes neuronal activity in development and disease [70,71]; the applicability of this theme in pain pathology remains to be further established but could critically inform gene therapy approaches in the near future.

Kv2.1 is another interesting channel as it features unique subcellular localisation, regulation by silent subunits, and nonconducting functions. Being a high-threshold channel with characteristically slow kinetics, Kv2.1 becomes particularly important during prolonged stimulation, like that encountered in central neurons during seizures [72,73], or in peripheral nociceptors during spontaneous firing. Accordingly, inhibiting Kv2.1 currents in DRG neurons allows higher firing rates during sustained input [74], while Kv2.1 knockout in the CNS results in neuronal hyperexcitability reminiscent of epilepsy [72]. It therefore appears that Kv2.1 acts as a resistor that filters elevated neuronal firing and is compromised in syndromes linked to neuronal hyperexcitability, including chronic pain. Kv2.1 is downregulated in damaged sensory neurons thus promoting hyperexcitability [74], but as Kv2.1 levels are not completely abolished, this may be exploitable for pharmacological enhancement with Kv openers. Constituent reduction of Kv2.1 activity can also occur via mutations in the *KNCB1* gene; missense variants located within the pore domain result in loss of $K^+$ selectivity and generation of a depolarizing inward sodium current at negative voltages [75] or even loss of voltage dependence, causing Kv2.1 to remain tonically open [73]. It will be interesting to investigate whether similar Kv2.1 mutations are linked to human pain channelopathies.

Kv2.1 is robustly regulated by members of the Kv5, Kv6, Kv8, and Kv9 families, which comprise the so-called "silent subunits" (KvS). These enigmatic proteins are incapable of conducting currents on their own but can form functional tetramers with Kv2.1, substantially altering the biophysical properties of the channel [76]. For instance, association of Kv2.1 with any of Kv5.1, Kv6.1, Kv9.1, or Kv9.3 hyperpolarises the voltage dependence of inactivation in pyramidal neurons, while Kv2.1/Kv5.1 exhibits accelerated rates of (open-state) inactivation and slower closing rates upon repolarization (deactivation) [77]. Mechanistically, these modulatory effects can be mediated by direct changes in the gating mechanism or indirectly by promoting $Ca^{2+}$/calmodulin-dependent dephosphorylation [77,78]. It is becoming increasingly evident that heteromerization of Kv2.1 with different KvS endows neurons with functional diversity that is often essential for normal physiology. For example, mammalian photoreceptors depend on Kv2.1/Kv8.2 channels to mediate transient hyperpolarizing overshoots of the membrane potential [79,80] and Kv8.2 mutations cause a cone dystrophy disorder [81]. Kv6.1, Kv8.1, and Kv8.2 have also been implicated in hyperexcitability of hippocampal neurons relevant to epilepsy [82–84].

KvS have also been implicated in chronic pain. Kv9.1 co-localises with Kv2.1 in myelinated sensory neurons that become hyperexcitable following nerve damage. Injury-induced Kv9.1 downregulation decreases Kv2.1 activity and enhances excitability, including spontaneous and evoked firing, and triggers pain hypersensitivity in rodents [85]. Consistent with this, deletion of the KCNS1 gene encoding Kv9.1 in mice results in basal and neuropathic pain sensitivity [86]. Kv2.1, but not Kv9.1, is also expressed in small nociceptors [74], but the composition of the native tetramers is not known. Since Kv2.1 conduction is sculpted by the modulatory influence of silent subunits, it is plausible that different Kv2.1/KvS combinations and stoichiometry can fine-tune excitability in distinct classes of sensory neurons. For example, Kv9.3 hyperpolarises the voltage dependence of Kv2 inactivation more substantially than Kv9.1 [78,87,88] and, even for a given KvS, the physiological impact is predicted to depend on whether firing is limited by the inactivation of inward currents [88]. The importance of Kv2.1 modulation by Kv9.1 in nociception is further underscored by the identification of two SNPs in the human Kv9.1 gene, which predisposes to the development of chronic pain [89]. Similarly, mutations in Kv6.4 may promote excitability of trigeminal neurons during migraine attacks [90] as well as pain during labor [91]. The role of KvS in pain is still poorly understood, and it may even include regulation of noncanonical Kv2.1 functions such as channel clustering and protein trafficking, as discussed below. Untangling this pathway could provide unique opportunities for pain treatments, which may prove advantageous compared to targeting the ubiquitously expressed Kv2.1 subunit.

A prime example of noncanonical Kv function is the formation of large clusters by Kv2.1 channels which localise to the neuronal membrane of the soma, proximal dendrites, and axon initial segment of CNS neurons [92]. In contrast to the active, diffused form of the channel, these micrometer-sized clusters are found to be primarily nonconductive [93,94] and are dispersed in response to neuronal activity, glutamate-induced excitotoxicity, hypoxia, or second messengers [95,96]. Regulation of cluster formation was originally thought to fine-tune neuronal excitability by dynamic control of the active vs. inactive forms but recent evidence hints towards a nonconducting role for Kv2 clusters. Thus, Kv2.1 and Kv2.2 clusters play a structural role in the formation of plasma membrane–endoplasmic reticulum junctions which serve as trafficking hubs for recruitment of several proteins (e.g., voltage-activated $Ca^{2+}$ channels, VAMPs, AKAPs, kinases, and syntaxin), important for many neurophysiological processes like trafficking, neurotransmitter release, $Ca^{2+}$ homeostasis, and burst firing [97–101]. Such a role is consistent with the identification of three Kv2.1 mutations which cause neurodevelopmental disorders despite the fact that they do not alter Kv2.1 conductance per se [102]. While Kv2 cluster formation has not been confirmed in peripheral sensory neurons, it is plausible that similar mechanisms operate in the pain-signaling pathways and that disruption of normal Kv2 clustering due to inflammation or injury affects nociceptive excitability.

Two P(ore) domain potassium channels (K2Ps) are known for facilitating a passive and rapid $K^+$ flow at a range of membrane potentials. This leak (also called background) outward conductance stabilises resting membrane potential, assists repolarisation, and even enables AP generation in the absence of classical Kv channels [103]. Surprisingly, however, additional voltage-dependent activation has been documented in some K2P channels despite the absence of a canonical voltage-sensing domain with a positively charged S4-helix for gating by depolarisation [104,105]. Instead, voltage sensitivity appears to derive from movement of three or four ions into the high electric field of the selectivity filter [106], which then acts to "gate" the movement of $K^+$ ions. Thus, in contrast to classical Kv channels where the properties of voltage sensing, activation, and inactivation can be mapped to distinct regions of the channel, K2Ps carry out these functions by employing different structural states of the selectivity filter. Moreover, many stimuli relevant to physiological functions such as $PIP_2$, acidosis, and membrane stretch can switch off this voltage activation [107], limiting K2Ps to leak conductance by locking them open. Altogether, K2P channels are increasingly recognised as important modulators of polymodal pain perception. The best studied members TRAAK, TREK1, and TREK2 are mechano- and thermosensitive, albeit at different temperature ranges (TRAAK and TREK1, noxious temperatures;

TREK2, moderate temperatures) [108–112]. Accordingly, deletion of these channels affects mechanical, heat, and oxaliplatin-induced cold sensitivity [108,109,112,113].

When covering atypical potassium channel function, a special mention should be made on inwardly rectifying potassium channels (Kir) which are mainly found in supporting cells such as glia. Kir channels are unique in that, at depolarised potentials, they preferentially mediate movement of $K^+$ ions towards the inside of the cell in contrast to other potassium channels. The resulting inward currents help maintain resting membrane potentials and are therefore important in a number of physiological processes such as microglial activation during inflammation [114]. In addition, the buffering activity of Kir in nonneuronal cells prevents extracellular $K^+$ accumulation, which would cause action potential "short-circuiting" and could detrimentally impact neuronal excitability [115,116]. Several members of the seven Kir subfamilies (Kir1–Kir7) have been specifically implicated in pain modulation. Kir4.1 channels expressed in satellite glia cells of the trigeminal ganglion appear to be important for facial pain. Silencing Kir4.1 expression in rats to mimic the effect of nerve injury or inflammation induces hyperexcitability and facial pain behaviours [117,118], while Kir4.1 knockout mice exhibit depolarised membrane potentials and inhibition of $K^+$ uptake [119,120]. Members of the Kir3 family (also known as G protein-regulated inward rectifiers $K^+$ channels, GIRK) are crucial mediators of spinal analgesia because their coupling to G proteins underlies analgesia conferred by endogenous and exogenous opioids [121,122]. Consistent with this, variations in the gene encoding GIRK2 are associated with pain phenotypes, as well as analgesic responses in humans [121–123]. Besides their established role in the CNS, GIRK2 expressed in sensory neurons may also contribute to peripheral antinociception induced by opioids [124].

Together, there is considerable interest in developing novel forms of analgesia by modulating $K^+$ channel function, either correcting a primary pathology underpinning a pain state or nonspecifically reducing neuronal excitability. Much of the early interest focused on Kv channel openers, but more recent data suggests that drugs activating K2Ps may prove useful for a variety of pain symptoms [125]. Recently, a somewhat counterintuitive observation was that opening of ATP-sensitive potassium channels induces migraine attacks in migraineurs using a randomized, double-blind, placebo-controlled, crossover design [126]. The mechanism is not clear but might involve neuronal or nonneuronal processes, given the widespread distribution of these channels. It is worth noting that other $K^+$ channels have a much more restricted distribution to nociceptive neurons.

## 4. TRP Channels

Transient receptor potential (TRP) channels form a group of 28 ion channels (27 in humans) organized into 7 families TRPC (canonical), TRPM (melastatin), TRPV (vanilloid), TRPA (ankyrin), TRPML (mucolipin), TRPP (polycystic), and TRPN (no mechanoreceptor potential C) [127,128]. Members of this family continue to be among the most studied in the ion channel field, in particular for pain as their relevance for certain pathophysiological conditions and peripheral sensory perception prompts them as targets for therapeutic modulation [129–131]. Canonical features of this family would be weak voltage-dependence [132], conductance for cations including divalent cations [133], and frequently, channel-specific rectification [134]. Below, some unexpected channel behavior is discussed.

TRPM3 has been found to generate an unexpected conductance when a combination of agonists was applied, namely pregnenolone sulfate and clotrimazole (or sole application of the agonist CIM0216) [135,136]. This ion permeation pathway, that allows inward rectification driven by $Na^+$, has been likened to the omega pore in classical voltage-gated cation channels; however, the latter has been uncovered in disease-inducing mutations, whereas in the case of TRMP3, it exists in the wild-type channel. Moreover, in the case of clinically relevant clotrimazole plasma levels, it is feasible that, at 37 °C, circulating levels of pregnenolone sulfate can open this alternative pore [135]. Further analysis of the voltage-sensing domain of TRPM3 by means of site-directed mutagenesis revealed a critical role of several amino acids in the voltage-sensing domain for the formation of this alternative ion-permeation pathway [136]. In other channel families, including potassium, sodium, and proton channels, channel mutations have also been shown to cause noncanonical pores, as discussed above [51,137,138]. Ultimately, such unexpected behaviour can contribute to a larger goal, understanding the gating and overall ion channel molecular mechanics.

The majority of TRP channels is outwardly rectifying, despite some exceptions such as TRPML1/2/3 [139] and TRPV5/6 [140]. In case the rectification is not dependent on divalent cations, causing asymmetry by an open channel block [141,142], this is an intrinsic property of the channel. This property can be changed by a single helix-breaking amino acid, as shown for the TRPML3-A419P mutation [143].

Increasing response to continuous agonist exposure: TRPA1 stands out by being most sensitive to modification by electrophilic molecules, a feature which critically involves cysteine residues on the N terminus of the channel [144–147]. Irrespective of the mode of activation, a dilation of the pore has been described for TRPA1, a feature attributed only to a few ion channels, including TRPV1 [148]. This pore dilation has a time constant below 10 s [149,150]. It should be mentioned that an alternative explanation to dilations of the pore has been proposed [151] (summarized in Reference [152]).

However, a slow but several-fold increase in conductance upon continuous agonist exposure with a much longer time course has been demonstrated [153]. These current increases can be better studied in the absence of calcium, as calcium influx causes a calcium-dependent desensitisation, and both mechanisms seem to balance each other. This allows continuous activation through TRPA1, where other channels show extensive tachyphylaxis or desensitisation. The mechanism is not PKA- or PKC-dependent. The topic has been further investigated using the noncovalent agonist carvacrol [154]. In contrast to the slow covalent action of TRPA1 by allyl isothiocyanate, the time constant for activation by carvacrol was 3.1 s, which allowed for tracking the current faster. A similar agonist-induced current increase was detected, and the current observed after a previous exposure is picked up almost invariable to the period between stimuli. The time constant of agonist-induced sensitisation was 130 s, which is well above all other described processes. Agonist exposure is required for this effect, as it could not be reproduced by opening the channel using voltage stimulation. Agonist-induced sensitisation occurred between covalent and non-covalent agonists, indicating a modification which is common to all agonists but upstream or independent of voltage-induced gating. However, a current through the channel was not required, as the exposure time-dependent current increase progresses when flux is inhibited by the additional presence of an antagonist. Similar to allyl isothiocyanate, a desensitisation was observed for saturating concentrations of carvacrol; the reason for this remaining unclear. The agonist-induced sensitisation was assumed to bring TRPA1 into a hypothesized state, which has a far left-shifted voltage dependence [154]. Inhibition of ATP-dependent mechanisms and membrane trafficking also did not affect the observation. TRPA1 has been investigated using long exposures mainly due to the slow onset required for the covalent agonists. This is not required for other channels; therefore, such protocols might simply not have been tested so far. It should be mentioned that a shift in concentration-dependent binding with prolonged agonist exposure has been reported in other receptors [155]; an "imprinting" by a lasting conformational change was hypothesized. For TRPA1, the change in receptor binding after prolonged exposure has not been investigated.

## 5. HCN Channels

Hyperpolarization-activated cyclic nucleotide-gated (HCN) channels comprise a small family with four members, HCN1–HCN4 [156–160]. The channels are related to CNG and Kv channels [161–163] but are distinguished by several unique features. HCN channels are controlled by the membrane potential; however, in contrast to most other voltage-gated ion channels, hyperpolarization but not depolarization opens them. Second, the channels contain the GYG motif in the pore region, which constitutes the potassium selectivity filter in potassium channels. Nevertheless, HCN channels are nonselective cation channels conducting both sodium and potassium ions (selectivity for $K^+/Na^+$ ~4:1). Under physiological conditions, activation of the channels leads to influx of sodium ions, resulting in depolarization. Third, cyclic nucleotides, particularly cAMP, stimulate the channels by accelerating their activation kinetics and by shifting the activation curve in the positive direction. However, cAMP is not required for channel opening. HCN channels contain a cyclic nucleotide-binding domain (CNBD) in the carboxyterminus. Truncation experiments have shown that the cyclic nucleotide-binding domain inhibits gating in the cAMP-unbound state, whereas cAMP binding relieves this inhibition [164].

The recent cryo-EM structure of HCN1 [165] together with the crystal structure of a cyclic nucleotide-binding domain [166] yielded important insights into the peculiar characteristic outlined above. As compared to potassium channels, the outer half of the selectivity filter in HCNs is enlarged and two of the potential four potassium binding sites are lacking. This results in a loss of the kinetic selectivity for potassium present in potassium channels, where four binding sites in the permeation pathway are present. Second, the closed pore is stabilized by the voltage sensor (S4 segment) and other domains in the depolarized state. It is proposed that the hyperpolarization-induced downward movement of S4 disrupts these interactions, leading to spontaneous pore opening. Third, binding of cAMP induces small conformational changes, leading to a rotation towards opening of the inner gate. However, HCN1 is barely modulated by cyclic nucleotides, so it remains to be seen if this structural mechanism also operates in HCN2 and HCN4, which are strongly modulated by cAMP.

The individual isoforms possess characteristic properties, which have been investigated in heterologous expression systems and were confirmed by using knockout mice of each isoform [167–171]. Beneath the different sensitivities toward cyclic nucleotides, the isoforms strongly differ in the rate of channel opening. HCN1 is the fastest activating isoform, HCN2 and HCN3 possess an intermediate activation kinetic, and HCN4 is the slowest HCN channel with an activation time constant up to several seconds.

In principle, activation of HCN channels leads to depolarisation and promotes AP generation. Since neuronal hyperexcitability and spontaneous AP generation of nociceptors contribute to the generation of pathological pain, HCN channels and, in particular, HCN2 may be involved in the sensitization of nociceptors in chronic pain conditions. In line with this assumption, an enhancement of the current flowing through these channels ($I_h$) has been directly shown in different models of neuropathic [172–174] and inflammatory [175] pain. The increase in $I_h$ has been attributed to an upregulation of HCN transcript and/or protein [113,175–179], upregulation of the potential auxiliary subunit MiRP1 [174], increased intracellular cAMP levels [180], and PKA-dependent phosphorylation of HCN2 [181]. Nociceptor-specific deletion of HCN2 by using a Nav1.8-Cre transgene to delete the floxed HCN2 exons directly demonstrated the important role of this channel in pathological pain conditions [179,182]. In two different models of neuropathic pain, HCN2 emerged as a key regulator since its deletion strongly reduced [183] and even abolished [182] the increase in nociceptive sensitivity. Moreover, in diabetic mice, deletion or block of HCN2 prevented the mechanical allodynia following diabetic neuropathy [180]. In inflammatory pain, the importance of HCN2 was also shown, but the extent differed between the inflammatory compound (carrageenan, PGE2, 8-bromo-cAMP, zymosan A, and CFA) and behavioural test (mechanical and heat hypersensitivity) used [179,182]. It is proposed that HCN2 channels determine nociceptor hypersensitivity if the inflammatory signal transduction pathways result in an increase of cAMP, which may directly modulate channel activity via binding to the CNBD [184] or indirectly via activation of PKA and phosphorylation of HCN2 or associated proteins [183].

However, in spite of these promising findings in murine models, a recent human phase 2 study did not find any effect of ivabradine on capsaicin-induced hyperalgesia and pain in healthy volunteers [185]. Ivabradine caused a significant heart-rate reduction, indicating that the dose was sufficient to block HCN4 and HCN1 channels in the sinoatrial node. These results suggest that it might be necessary to develop HCN2-selective substances (which do not cross the blood–brain barrier [167]) to serve as analgesics. Beyond that, it is still possible that ivabradine is effective in other human pain models distinct from the neurogenic inflammation induced by TRPV1 activation.

## 6. Conclusion

Unexpected properties of several ion channels with importance for the pain field were discussed (Table 1). We hope that the selective and non-comprehensive choices help to transfer knowledge within the field. Considering the possibility that such findings in other channels might explain otherwise not understood issues and facilitate scientific progress.

**Table 1.** Overview for the mentioned noncanonical ion channel features: canonical features not discussed in the review are considered ligand- and voltage-gating, selective permeability, interaction with other proteins, fractional presence at the plasma membrane, heteromerisation, modulation by intracellular cascades, and a change of expression.

| Noncanonical Property | Ion Channel and References |
|---|---|
| slow inactivation | **Nav** [9,12–33] |
| resurgent current during membrane repolarisation | **Nav (1.2, 1.5, 1.6, 1.7, 1.8)** [29,34–41] |
| side fenestrations of the channel pore | **Nav** [9,10,46–49] |
| alternative ion permeation pathway across the membrane | **Nav (1.2, 1.4, 1.5)** [50–55]<br>**TRPM3** [135–138] |
| electromechanical coupling of transmembrane domain S4 and S5 residues, facilitating "rack-and-pinion" movements | **Kv** [62–64] |
| mechanically induced current | **Kv (1.1)** [65,66] |
| susceptibility to epigenetic silencing | **Kv (1.2)** [67–71] |
| action potential frequency filtering by silent subunits | **Kv (2.1)–KvS (members of the Kv5, 6, 8, 9 families)** [72–88] |
| formation of channel clusters | **Kv (2.1, 2.2)** [92–101] |
| voltage-dependent activation in the absence of a canonical voltage-sensing domain | **K2P** [104–107] |
| inward currents at depolarizing potentials | **Kir** [114–120]<br>**TRPML1-3** [139,141–143]<br>**TRPV5/6** [140] |
| pore dilation | **TRPA1** [149,150,152]<br>**TRPV1** [148] |
| sensitization due to continuous agonist exposure | **TRPA1** [32,147,149–154] |
| open probability increases upon hyperpolarization | **HCN 1–4** [156–163] |
| nonselective cation channel despite a typical potassium selectivity filter motif | **HCN 1–4** [156–163] |

**Author Contributions:** C.I.C., C.T., J.M., A.L. (Angelika Lampert), S.B.M., A.L. (Andreas Ludwig), and M.J.M.F. wrote the manuscript.

**Funding:** This research was supported by the Austrian Science Fund (FWF): P 32534.

**Conflicts of Interest:** The authors declare no conflict of interest.

## Abbreviations

| | |
|---|---|
| AKAP | A-kinase anchor protein |
| HCN | Hyperpolarization-activated cyclic nucleotide-gated channel |
| Kv | voltage-gated potassium channel |
| KvS | Silent potassium channel subunit |
| K2P | Two Pore domain potassium channel |
| Kir | Kir inwardly rectifying potassium channel |
| Nav | Voltage-gated sodium channel |
| TRP | Transient receptor potential channels |
| VAMP | Vesicle-associated membrane protein |

## References


1. Hille, B. *Ion Channels of Excitable Membranes*, 3rd ed.; Sinauer Associates: Sunderland, MA, USA, 2001.
2. Catterall, W.A. From Ionic Currents to Molecular Mechanisms: The Structure and Function of Voltage-Gated Sodium Channels. *Neuron* **2000**, *26*, 13–25. [CrossRef]
3. Catterall, W.A. International Union of Pharmacology. XLVII. Nomenclature and Structure-Function Relationships of Voltage-Gated Sodium Channels. *Pharmacol. Rev.* **2005**, *57*, 397–409. [CrossRef] [PubMed]
4. Lampert, A.; O'Reilly, A.O.; Reeh, P.; Leffler, A. Sodium channelopathies and pain. *Pflugers Arch.* **2010**, *460*, 249–263. [CrossRef] [PubMed]
5. Emery, E.C.; Luiz, A.P.; Wood, J.N. Nav1.7 and other voltage-gated sodium channels as drug targets for pain relief. *Expert Opin Ther Targets* **2016**, *20*, 975–983. [CrossRef] [PubMed]
6. Ahern, C.A.; Payandeh, J.; Bosmans, F.; Chanda, B. The hitchhiker's guide to the voltage-gated sodium channel galaxy. *J. Gen. Physiol.* **2016**, *147*, 1–24. [CrossRef] [PubMed]
7. Peters, C.H.; Ruben, P.C. Introduction to Sodium Channels. In *Voltage Gated Sodium Channels*; Ruben, P.C., Ed.; Handbook of Experimental Pharmacology; Springer: Berlin/Heidelberg, Germany, 2014; pp. 1–6.
8. Yan, Z.; Zhou, Q.; Wang, L.; Wu, J.; Zhao, Y.; Huang, G.; Peng, W.; Shen, H.; Lei, J.; Yan, N. Structure of the Nav1.4-β1 Complex from Electric Eel. *Cell* **2017**, *170*, 470–482. [CrossRef] [PubMed]
9. Pan, X.; Li, Z.; Zhou, Q.; Shen, H.; Wu, K.; Huang, X.; Chen, J.; Zhang, J.; Zhu, X.; Lei, J.; et al. Structure of the human voltage-gated sodium channel Nav1.4 in complex with β1. *Science* **2018**, *362*, eaau2486. [CrossRef]
10. Shen, H.; Liu, D.; Wu, K.; Lei, J.; Yan, N. Structures of human Nav1.7 channel in complex with auxiliary subunits and animal toxins. *Science* **2019**, *363*, 1303–1308. [CrossRef]
11. Clairfeuille, T.; Cloake, A.; Infield, D.T.; Llongueras, J.P.; Arthur, C.P.; Li, Z.R.; Jian, Y.; Martin-Eauclaire, M.-F.; Bougis, P.E.; Ciferri, C.; et al. Structural basis of α-scorpion toxin action on Nav channels. *Science* **2019**, *363*, eaav8573. [CrossRef]
12. Barbosa, C.; Cummins, T.R. Unusual Voltage-Gated Sodium Currents as Targets for Pain. In *Current Topics in Membranes*; French, R.J., Noskov, S.Y., Eds.; Academic Press: Cambridge, MA, USA, 2016; Volume 78, pp. 599–638.
13. Ulbricht, W. Sodium Channel Inactivation: Molecular Determinants and Modulation. *Physiol. Rev.* **2005**, *85*, 1271–1301. [CrossRef]
14. Silva, J. Slow Inactivation of Na+ Channels. In *Voltage Gated Sodium Channels*; Ruben, P.C., Ed.; Handbook of Experimental Pharmacology; Springer: Berlin/Heidelberg, Germany, 2014; pp. 33–49.
15. Chatterjee, S.; Vyas, R.; Chalamalasetti, S.V.; Sahu, I.D.; Clatot, J.; Wan, X.; Lorigan, G.A.; Deschênes, I.; Chakrapani, S. The voltage-gated sodium channel pore exhibits conformational flexibility during slow inactivation. *J. Gen. Physiol.* **2018**, *150*, 1333–1347. [CrossRef] [PubMed]
16. Xiong, W.; Li, R.A.; Tian, Y.; Tomaselli, G.F. Molecular Motions of the Outer Ring of Charge of the Sodium Channel: Do They Couple to Slow Inactivation? *J. Gen. Physiol.* **2003**, *122*, 323–332. [CrossRef] [PubMed]
17. Cummins, T.R.; Dib-Hajj, S.D.; Waxman, S.G. Electrophysiological Properties of Mutant Nav1.7 Sodium Channels in a Painful Inherited Neuropathy. *J. Neurosci.* **2004**, *24*, 8232–8236. [CrossRef] [PubMed]
18. Dib-Hajj, S.D.; Rush, A.M.; Cummins, T.R.; Hisama, F.M.; Novella, S.; Tyrrell, L.; Marshall, L.; Waxman, S.G. Gain-of-function mutation in Nav1.7 in familial erythromelalgia induces bursting of sensory neurons. *Brain* **2005**, *128*, 1847–1854. [CrossRef] [PubMed]
19. Choi, J.-S.; Dib-Hajj, S.D.; Waxman, S.G. Inherited erythermalgia: Limb pain from an S4 charge-neutral Na channelopathy. *Neurology* **2006**, *67*, 1563–1567. [CrossRef] [PubMed]
20. Harty, T.P.; Dib-Hajj, S.D.; Tyrrell, L.; Blackman, R.; Hisama, F.M.; Rose, J.B.; Waxman, S.G. Nav1.7 Mutant A863P in Erythromelalgia: Effects of Altered Activation and Steady-State Inactivation on Excitability of Nociceptive Dorsal Root Ganglion Neurons. *J. Neurosci.* **2006**, *26*, 12566–12575. [CrossRef]
21. Lampert, A.; Dib-Hajj, S.D.; Tyrrell, L.; Waxman, S.G. Size Matters: Erythromelalgia Mutation S241T in Nav1.7 Alters Channel Gating. *J. Biol. Chem.* **2006**, *281*, 36029–36035. [CrossRef]
22. Sheets, P.L.; Jackson, J.O.; Waxman, S.G.; Dib-Hajj, S.D.; Cummins, T.R. A Nav1.7 channel mutation associated with hereditary erythromelalgia contributes to neuronal hyperexcitability and displays reduced lidocaine sensitivity. *J. Physiol.* **2007**, *581*, 1019–1031. [CrossRef]

23. Cheng, X.; Dib-Hajj, S.D.; Tyrrell, L.; Waxman, S.G. Mutation I136V alters electrophysiological properties of the NaV1.7 channel in a family with onset of erythromelalgia in the second decade. *Mol. Pain* **2008**, *4*, 1. [CrossRef]
24. Han, C.; Dib-Hajj, S.D.; Lin, Z.; Li, Y.; Eastman, E.M.; Tyrrell, L.; Cao, X.; Yang, Y.; Waxman, S.G. Early- and late-onset inherited erythromelalgia: Genotype-phenotype correlation. *Brain* **2009**, *132*, 1711–1722. [CrossRef]
25. Ahn, H.-S.; Dib-Hajj, S.D.; Cox, J.J.; Tyrrell, L.; Elmslie, F.V.; Clarke, A.A.; Drenth, J.P.H.; Woods, C.G.; Waxman, S.G. A new Nav1.7 sodium channel mutation I234T in a child with severe pain. *Eur. J. Pain* **2010**, *14*, 944–950. [CrossRef] [PubMed]
26. Estacion, M.; Choi, J.S.; Eastman, E.M.; Lin, Z.; Li, Y.; Tyrrell, L.; Yang, Y.; Dib-Hajj, S.D.; Waxman, S.G. Can robots patch-clamp as well as humans? Characterization of a novel sodium channel mutation. *J. Physiol.* **2010**, *588*, 1915–1927. [CrossRef] [PubMed]
27. Cheng, X.; Dib-Hajj, S.D.; Tyrrell, L.; te Morsche, R.H.; Drenth, J.P.H.; Waxman, S.G. Deletion mutation of sodium channel NaV1.7 in inherited erythromelalgia: Enhanced slow inactivation modulates dorsal root ganglion neuron hyperexcitability. *Brain* **2011**, *134*, 1972–1986. [CrossRef] [PubMed]
28. Estacion, M.; Yang, Y.; Dib-Hajj, S.D.; Tyrrell, L.; Lin, Z.; Yang, Y.; Waxman, S.G. A new Nav1.7 mutation in an erythromelalgia patient. *Biochem. Biophys. Res. Commun.* **2013**, *432*, 99–104. [CrossRef] [PubMed]
29. Hampl, M.; Eberhardt, E.; O'Reilly, A.O.; Lampert, A. Sodium channel slow inactivation interferes with open channel block. *Sci. Rep.* **2016**, *6*, 25974. [CrossRef]
30. Lampert, A.; Eberhardt, M.; Waxman, S.G. Altered Sodium Channel Gating as Molecular Basis for Pain: Contribution of Activation, Inactivation, and Resurgent Currents. In *Voltage Gated Sodium Channels*; Ruben, P.C., Ed.; Handbook of Experimental Pharmacology; Springer: Berlin/Heidelberg, Germany, 2014; pp. 91–110.
31. Zimmermann, K.; Leffler, A.; Babes, A.; Cendan, C.M.; Carr, R.W.; Kobayashi, J.; Nau, C.; Wood, J.N.; Reeh, P.W. Sensory neuron sodium channel $Na_V1.8$ is essential for pain at low temperatures. *Nature* **2007**, *447*, 856–859. [CrossRef] [PubMed]
32. Villalba-Galea, C.A. Hysteresis in voltage-gated channels. *Channels (Austin)* **2016**, *11*, 140–155. [CrossRef]
33. Meents, J.E.; Bressan, E.; Sontag, S.; Foerster, A.; Hautvast, P.; Rösseler, C.; Hampl, M.; Schüler, H.; Goetzke, R.; Le, T.K.C.; et al. The role of Nav1.7 in human nociceptors: Insights from human induced pluripotent stem cell–derived sensory neurons of erythromelalgia patients. *Pain.* **2019**, *160*, 1327. [CrossRef]
34. Raman, I.M.; Bean, B.P. Resurgent Sodium Current and Action Potential Formation in Dissociated Cerebellar Purkinje Neurons. *J. Neurosci.* **1997**, *17*, 4517–4526. [CrossRef]
35. Lewis, A.H.; Raman, I.M. Resurgent current of voltage-gated Na+ channels. *J. Physiol.* **2014**, *592*, 4825–4838. [CrossRef]
36. Meents, J.E.; Lampert, A. Studying Sodium Channel Gating in Heterologous Expression Systems. In *Advanced Patch-Clamp Analysis for Neuroscientists*; Korngreen, A., Ed.; Neuromethods; Springer: New York, NY, USA, 2016; pp. 37–65.
37. Grieco, T.M.; Malhotra, J.D.; Chen, C.; Isom, L.L.; Raman, I.M. Open-channel block by the cytoplasmic tail of sodium channel beta4 as a mechanism for resurgent sodium current. *Neuron* **2005**, *45*, 233–244. [CrossRef]
38. Rush, A.M.; Dib-Hajj, S.D.; Waxman, S.G. Electrophysiological properties of two axonal sodium channels, Nav1.2 and Nav1.6, expressed in mouse spinal sensory neurones. *J. Physiol.* **2005**, *564*, 803–815. [CrossRef]
39. Cummins, T.R.; Dib-Hajj, S.D.; Herzog, R.I.; Waxman, S.G. Nav1.6 channels generate resurgent sodium currents in spinal sensory neurons. *FEBS Lett.* **2005**, *579*, 2166–2170. [CrossRef]
40. Tan, Z.-Y.; Piekarz, A.D.; Priest, B.T.; Knopp, K.L.; Krajewski, J.L.; McDermott, J.S.; Nisenbaum, E.S.; Cummins, T.R. Tetrodotoxin-Resistant Sodium Channels in Sensory Neurons Generate Slow Resurgent Currents That Are Enhanced by Inflammatory Mediators. *J. Neurosci.* **2014**, *34*, 7190–7197. [CrossRef]
41. Theile, J.W.; Jarecki, B.W.; Piekarz, A.D.; Cummins, T.R. Nav1.7 mutations associated with paroxysmal extreme pain disorder, but not erythromelalgia, enhance Navβ4 peptide-mediated resurgent sodium currents. *J. Physiol.* **2011**, *589*, 597–608. [CrossRef]
42. Ragsdale, D.S.; McPhee, J.C.; Scheuer, T.; Catterall, W.A. Molecular determinants of state-dependent block of Na+ channels by local anesthetics. *Science* **1994**, *265*, 1724–1728. [CrossRef]

43. Yarov-Yarovoy, V.; Brown, J.; Sharp, E.M.; Clare, J.J.; Scheuer, T.; Catterall, W.A. Molecular Determinants of Voltage-dependent Gating and Binding of Pore-blocking Drugs in Transmembrane Segment IIIS6 of the Na+ Channel α Subunit. *J. Biol. Chem.* **2001**, *276*, 20–27. [CrossRef]
44. Ahern, C.A.; Eastwood, A.L.; Dougherty, D.A.; Horn, R. Electrostatic contributions of aromatic residues in the local anesthetic receptor of voltage-gated sodium channels. *Circ. Res.* **2008**, *102*, 86–94. [CrossRef]
45. Hille Bertil Local anesthetics: Hydrophilic and hydrophobic pathways for the drug-receptor reaction. | JGP. Available online: http://jgp.rupress.org/content/69/4/497.long (accessed on 3 July 2019).
46. O'Reilly, A.O.; Eberhardt, E.; Weidner, C.; Alzheimer, C.; Wallace, B.A.; Lampert, A. Bisphenol A Binds to the Local Anesthetic Receptor Site to Block the Human Cardiac Sodium Channel. *PLoS ONE* **2012**, *7*, e41667. [CrossRef]
47. Kaczmarski, J.A.; Corry, B. Investigating the size and dynamics of voltage-gated sodium channel fenestrations. *Channels (Austin)* **2014**, *8*, 264–277. [CrossRef]
48. Smith, N.E.; Corry, B. Mutant bacterial sodium channels as models for local anesthetic block of eukaryotic proteins. *Channels (Austin)* **2016**, *10*, 225–237. [CrossRef]
49. El-Din, T.M.G.; Lenaeus, M.J.; Zheng, N.; Catterall, W.A. Fenestrations control resting-state block of a voltage-gated sodium channel. *PNAS* **2018**, *115*, 13111–13116. [CrossRef]
50. Sokolov, S.; Scheuer, T.; Catterall, W.A. Ion Permeation through a Voltage- Sensitive Gating Pore in Brain Sodium Channels Having Voltage Sensor Mutations. *Neuron* **2005**, *47*, 183–189. [CrossRef]
51. Sokolov, S.; Scheuer, T.; Catterall, W.A. Gating pore current in an inherited ion channelopathy. *Nature* **2007**, *446*, 76–78. [CrossRef]
52. Sokolov, S.; Scheuer, T.; Catterall, W.A. Depolarization-activated gating pore current conducted by mutant sodium channels in potassium-sensitive normokalemic periodic paralysis. *PNAS* **2008**, *105*, 19980–19985. [CrossRef]
53. Gosselin-Badaroudine, P.; Delemotte, L.; Moreau, A.; Klein, M.L.; Chahine, M. Gating pore currents and the resting state of Nav1.4 voltage sensor domains. *PNAS* **2012**, *109*, 19250–19255. [CrossRef]
54. Moreau, A.; Gosselin-Badaroudine, P.; Mercier, A.; Burger, B.; Keller, D.I.; Chahine, M. A leaky voltage sensor domain of cardiac sodium channels causes arrhythmias associated with dilated cardiomyopathy. *Sci. Rep.* **2018**, *8*, 13804. [CrossRef]
55. Moreau, A.; Gosselin-Badaroudine, P.; Chahine, M. Gating pore currents, a new pathological mechanism underlying cardiac arrhythmias associated with dilated cardiomyopathy. *Channels (Austin)* **2015**, *9*, 139–144. [CrossRef]
56. Tsantoulas, C.; McMahon, S.B. Opening paths to novel analgesics: The role of potassium channels in chronic pain. *Trends Neurosci.* **2014**, *37*, 146–158. [CrossRef]
57. Tsantoulas, C. Emerging potassium channel targets for the treatment of pain. *Curr. Opin. Support. Palliat. Care* **2015**, *9*, 147–154. [CrossRef]
58. Pillozzi, S.; Brizzi, M.F.; Bernabei, P.A.; Bartolozzi, B.; Caporale, R.; Basile, V.; Boddi, V.; Pegoraro, L.; Becchetti, A.; Arcangeli, A. VEGFR-1 (FLT-1), β1 integrin, and hERG K+ channel for a macromolecular signaling complex in acute myeloid leukemia: Role in cell migration and clinical outcome. *Blood* **2007**, *110*, 1238–1250. [CrossRef]
59. Feinshreiber, L.; Singer-Lahat, D.; Ashery, U.; Lotan, I. Voltage-gated Potassium Channel as a Facilitator of Exocytosis. *Ann. N. Y. Acad. Sci.* **2009**, *1152*, 87–92. [CrossRef]
60. Jang, S.H.; Choi, C.; Hong, S.-G.; Yarishkin, O.V.; Bae, Y.M.; Kim, J.G.; O'Grady, S.M.; Yoon, K.-A.; Kang, K.-S.; Ryu, P.D.; et al. Silencing of Kv4.1 potassium channels inhibits cell proliferation of tumorigenic human mammary epithelial cells. *Biochem. Biophys. Res. Commun.* **2009**, *384*, 180–186. [CrossRef]
61. MacKinnon, R. Potassium channels. *FEBS Lett.* **2003**, *555*, 62–65. [CrossRef]
62. Fernández-Mariño, A.I.; Harpole, T.J.; Oelstrom, K.; Delemotte, L.; Chanda, B. Gating interaction maps reveal a noncanonical electromechanical coupling mode in the Shaker K+ channel. *Nat. Struct. Mol. Biol.* **2018**, *25*, 320–326. [CrossRef]
63. Lörinczi, É.; Gómez-Posada, J.C.; de la Peña, P.; Tomczak, A.P.; Fernández-Trillo, J.; Leipscher, U.; Stühmer, W.; Barros, F.; Pardo, L.A. Voltage-dependent gating of KCNH potassium channels lacking a covalent link between voltage-sensing and pore domains. *Nat. Commun.* **2015**, *6*, 6672. [CrossRef]
64. Wang, W.; MacKinnon, R. Cryo-EM Structure of the Open Human Ether-à-go-go -Related K + Channel hERG. *Cell* **2017**, *169*, 422–430. [CrossRef]

65. Hao, J.; Padilla, F.; Dandonneau, M.; Lavebratt, C.; Lesage, F.; Noël, J.; Delmas, P. Kv1.1 Channels Act as Mechanical Brake in the Senses of Touch and Pain. *Neuron* **2013**, *77*, 899–914. [CrossRef]
66. Hao, J.; Delmas, P. Multiple Desensitization Mechanisms of Mechanotransducer Channels Shape Firing of Mechanosensory Neurons. *J. Neurosci.* **2010**, *30*, 13384–13395. [CrossRef]
67. Liang, L.; Gu, X.; Zhao, J.-Y.; Wu, S.; Miao, X.; Xiao, J.; Mo, K.; Zhang, J.; Lutz, B.M.; Bekker, A.; et al. G9a participates in nerve injury-induced Kcna2 downregulation in primary sensory neurons. *Sci. Rep.* **2016**, *6*, 37704. [CrossRef]
68. Zhao, X.; Tang, Z.; Zhang, H.; Atianjoh, F.E.; Zhao, J.-Y.; Liang, L.; Wang, W.; Guan, X.; Kao, S.-C.; Tiwari, V.; et al. A long noncoding RNA contributes to neuropathic pain by silencing Kcna2 in primary afferent neurons. *Nat. Neurosci.* **2013**, *16*, 1024–1031. [CrossRef]
69. Laumet, G.; Garriga, J.; Chen, S.-R.; Zhang, Y.; Li, D.-P.; Smith, T.M.; Dong, Y.; Jelinek, J.; Cesaroni, M.; Issa, J.-P.; et al. G9a is essential for epigenetic silencing of K+ channel genes in acute-to-chronic pain transition. *Nat. Neurosci.* **2015**, *18*, 1746–1755. [CrossRef]
70. Briggs, J.A.; Wolvetang, E.J.; Mattick, J.S.; Rinn, J.L.; Barry, G. Mechanisms of Long Non-coding RNAs in Mammalian Nervous System Development, Plasticity, Disease, and Evolution. *Neuron* **2015**, *88*, 861–877. [CrossRef]
71. Barry, G.; Briggs, J.A.; Hwang, D.W.; Nayler, S.P.; Fortuna, P.R.J.; Jonkhout, N.; Dachet, F.; Maag, J.L.V.; Mestdagh, P.; Singh, E.M.; et al. The long non-coding RNA NEAT1 is responsive to neuronal activity and is associated with hyperexcitability states. *Sci. Rep.* **2017**, *7*, 40127. [CrossRef]
72. Speca, D.J.; Ogata, G.; Mandikian, D.; Bishop, H.I.; Wiler, S.W.; Eum, K.; Wenzel, H.J.; Doisy, E.T.; Matt, L.; Campi, K.L.; et al. Deletion of the Kv2.1 delayed rectifier potassium channel leads to neuronal and behavioral hyperexcitability: Kv2.1 deletion and hyperexcitability. *Genes Brain Behav.* **2014**, *13*, 394–408. [CrossRef]
73. Torkamani, A.; Bersell, K.; Jorge, B.S.; Bjork, R.L.; Friedman, J.R.; Bloss, C.S.; Cohen, J.; Gupta, S.; Naidu, S.; Vanoye, C.G.; et al. De novo *KCNB1* mutations in epileptic encephalopathy: *KCNB1* Mutations. *Ann. Neurol.* **2014**, *76*, 529–540. [CrossRef]
74. Tsantoulas, C.; Zhu, L.; Yip, P.; Grist, J.; Michael, G.J.; McMahon, S.B. Kv2 dysfunction after peripheral axotomy enhances sensory neuron responsiveness to sustained input. *Exp. Neurol.* **2014**, *251*, 115–126. [CrossRef]
75. Thiffault, I.; Speca, D.J.; Austin, D.C.; Cobb, M.M.; Eum, K.S.; Safina, N.P.; Grote, L.; Farrow, E.G.; Miller, N.; Soden, S.; et al. A novel epileptic encephalopathy mutation in *KCNB1* disrupts Kv2.1 ion selectivity, expression, and localization. *J. Gen. Physiol.* **2015**, *146*, 399–410. [CrossRef]
76. Bocksteins, E.; Snyders, D.J. Electrically Silent Kv Subunits: Their Molecular and Functional Characteristics. *Physiology* **2012**, *27*, 73–84. [CrossRef]
77. Kramer, J.W.; Post, M.A.; Brown, A.M.; Kirsch, G.E. Modulation of potassium channel gating by coexpression of Kv2.1 with regulatory Kv5.1 or Kv6.1 α-subunits. *Am. J. Physiol. Cell Physiol.* **1998**, *274*, C1501–C1510. [CrossRef]
78. Salinas, M.; Duprat, F.; Heurteaux, C.; Hugnot, J.P.; Lazdunski, M. New modulatory alpha subunits for mammalian Shab K+ channels. *J. Biol. Chem.* **1997**, *272*, 24371–24379. [CrossRef]
79. Wu, H.; Cowing, J.A.; Michaelides, M.; Wilkie, S.E.; Jeffery, G.; Jenkins, S.A.; Mester, V.; Bird, A.C.; Robson, A.G.; Holder, G.E.; et al. Mutations in the gene KCNV2 encoding a voltage-gated potassium channel subunit cause "cone dystrophy with supernormal rod electroretinogram" in humans. *Am. J. Hum. Genet.* **2006**, *79*, 574–579. [CrossRef]
80. Czirják, G.; Tóth, Z.E.; Enyedi, P. Characterization of the Heteromeric Potassium Channel Formed by Kv2.1 and the Retinal Subunit Kv8.2 in Xenopus Oocytes. *J. Neurophysiol.* **2007**, *98*, 1213–1222. [CrossRef]
81. Stockman, A.; Henning, G.B.; Michaelides, M.; Moore, A.T.; Webster, A.R.; Cammack, J.; Ripamonti, C. Cone Dystrophy With "Supernormal" Rod ERG: Psychophysical Testing Shows Comparable Rod and Cone Temporal Sensitivity Losses With No Gain in Rod Function. *Invest. Ophthalmol. Vis. Sci.* **2014**, *55*, 832–840. [CrossRef]
82. Sano, A.; Mikami, M.; Nakamura, M.; Ueno, S.-I.; Tanabe, H.; Kaneko, S. Positional candidate approach for the gene responsible for benign adult familial myoclonic epilepsy. *Epilepsia* **2002**, *43*, 26–31. [CrossRef]
83. Bergren, S.K.; Rutter, E.D.; Kearney, J.A. Fine Mapping of an Epilepsy Modifier Gene on Mouse Chromosome 19. *Mamm. Genome* **2009**, *20*, 359–366. [CrossRef]

84. Jorge, B.S.; Campbell, C.M.; Miller, A.R.; Rutter, E.D.; Gurnett, C.A.; Vanoye, C.G.; George, A.L.; Kearney, J.A. Voltage-gated potassium channel KCNV2 (Kv8.2) contributes to epilepsy susceptibility. *Proc. Natl. Acad. Sci. USA* **2011**, *108*, 5443–5448. [CrossRef]
85. Tsantoulas, C.; Zhu, L.; Shaifta, Y.; Grist, J.; Ward, J.P.T.; Raouf, R.; Michael, G.J.; McMahon, S.B. Sensory Neuron Downregulation of the Kv9.1 Potassium Channel Subunit Mediates Neuropathic Pain following Nerve Injury. *J. Neurosci.* **2012**, *32*, 17502–17513. [CrossRef]
86. Tsantoulas, C.; Denk, F.; Signore, M.; Nassar, M.A.; Futai, K.; McMahon, S.B. Mice lacking KCNS1 in peripheral neurons show increased basal and neuropathic pain sensitivity. *Pain* **2018**, *159*, 1641–1651. [CrossRef]
87. Patel, A.J.; Lazdunski, M.; Honoré, E. Kv2.1/Kv9.3, a novel ATP-dependent delayed-rectifier K+ channel in oxygen-sensitive pulmonary artery myocytes. *EMBO J.* **1997**, *16*, 6615–6625. [CrossRef]
88. Richardson, F.C.; Kaczmarek, L.K. Modification of delayed rectifier potassium currents by the Kv9.1 potassium channel subunit. *Hear. Res.* **2000**, *147*, 21–30. [CrossRef]
89. Costigan, M.; Belfer, I.; Griffin, R.S.; Dai, F.; Barrett, L.B.; Coppola, G.; Wu, T.; Kiselycznyk, C.; Poddar, M.; Lu, Y.; et al. Multiple chronic pain states are associated with a common amino acid-changing allele in KCNS1. *Brain* **2010**, *133*, 2519–2527. [CrossRef]
90. Lafrenière, R.G.; Rouleau, G.A. Identification of novel genes involved in migraine. *Headache* **2012**, *52*, 107–110. [CrossRef]
91. Lee, M.C.; Nahorski, M.S.; Hockley, J.R.F.; Lu, V.B.; Stouffer, K.; Fletcher, E.; Ison, G.; Brown, C.; Wheeler, D.; Ernfors, P.; et al. Human labor pain is influenced by the voltage-gated potassium channel KV6.4 subunit. *Biorxiv* **2018**. [CrossRef]
92. Sarmiere, P.D.; Weigle, C.M.; Tamkun, M.M. The Kv2.1 K+ channel targets to the axon initial segment of hippocampal and cortical neurons in culture and in situ. *BMC Neurosci.* **2008**, *9*, 112. [CrossRef]
93. O'Connell, K.M.S.; Loftus, R.; Tamkun, M.M. Localization-dependent activity of the Kv2.1 delayed-rectifier K+ channel. *Proc. Natl. Acad. Sci. USA* **2010**, *107*, 12351–12356. [CrossRef]
94. Fox, P.D.; Loftus, R.J.; Tamkun, M.M. Regulation of Kv2.1 K(+) conductance by cell surface channel density. *J. Neurosci.* **2013**, *33*, 1259–1270. [CrossRef]
95. Misonou, H.; Trimmer, J.S. Determinants of voltage-gated potassium channel surface expression and localization in Mammalian neurons. *Crit. Rev. Biochem. Mol. Biol.* **2004**, *39*, 125–145. [CrossRef]
96. Romer, S.H.; Deardorff, A.S.; Fyffe, R.E.W. Activity-dependent redistribution of Kv2.1 ion channels on rat spinal motoneurons. *Physiol. Rep.* **2016**, *4*, e13039. [CrossRef]
97. Antonucci, D.E.; Lim, S.T.; Vassanelli, S.; Trimmer, J.S. Dynamic localization and clustering of dendritic Kv2.1 voltage-dependent potassium channels in developing hippocampal neurons. *Neuroscience* **2001**, *108*, 69–81. [CrossRef]
98. King, A.N.; Manning, C.F.; Trimmer, J.S. A unique ion channel clustering domain on the axon initial segment of mammalian neurons. *J. Comp. Neurol.* **2014**, *522*, 2594–2608. [CrossRef] [PubMed]
99. Fox, P.D.; Haberkorn, C.J.; Akin, E.J.; Seel, P.J.; Krapf, D.; Tamkun, M.M. Induction of stable ER–plasma-membrane junctions by Kv2.1 potassium channels. *J. Cell Sci.* **2015**, *128*, 2096–2105. [CrossRef] [PubMed]
100. Johnson, B.; Leek, A.N.; Solé, L.; Maverick, E.E.; Levine, T.P.; Tamkun, M.M. Kv2 potassium channels form endoplasmic reticulum/plasma membrane junctions via interaction with VAPA and VAPB. *PNAS* **2018**, *115*, E7331–E7340. [CrossRef] [PubMed]
101. Kirmiz, M.; Vierra, N.C.; Palacio, S.; Trimmer, J.S. Identification of VAPA and VAPB as Kv2 Channel-Interacting Proteins Defining Endoplasmic Reticulum–Plasma Membrane Junctions in Mammalian Brain Neurons. *J. Neurosci.* **2018**, *38*, 7562–7584. [CrossRef] [PubMed]
102. De Kovel, C.G.F.; Syrbe, S.; Brilstra, E.H.; Verbeek, N.; Kerr, B.; Dubbs, H.; Bayat, A.; Desai, S.; Naidu, S.; Srivastava, S.; et al. Neurodevelopmental Disorders Caused by De Novo Variants in KCNB1 Genotypes and Phenotypes. *JAMA Neurol.* **2017**, *74*, 1228–1236. [CrossRef] [PubMed]
103. MacKenzie, G.; Franks, N.P.; Brickley, S.G. Two-pore domain potassium channels enable action potential generation in the absence of voltage-gated potassium channels. *Pflugers Arch.* **2015**, *467*, 989–999. [CrossRef]
104. Bockenhauer, D.; Zilberberg, N.; Goldstein, S.a.N. KCNK2: Reversible conversion of a hippocampal potassium leak into a voltage-dependent channel. *Nat. Neurosci.* **2001**, *4*, 486. [CrossRef]

105. Brickley, S.G.; Revilla, V.; Cull-Candy, S.G.; Wisden, W.; Farrant, M. Adaptive regulation of neuronal excitability by a voltage- independent potassium conductance. *Nature* **2001**, *409*, 88. [CrossRef]
106. Schewe, M.; Nematian-Ardestani, E.; Sun, H.; Musinszki, M.; Cordeiro, S.; Bucci, G.; de Groot, B.L.; Tucker, S.J.; Rapedius, M.; Baukrowitz, T. A Non-canonical Voltage-Sensing Mechanism Controls Gating in K2P K(+) Channels. *Cell* **2016**, *164*, 937–949. [CrossRef]
107. Chemin, J.; Patel, A.J.; Duprat, F.; Lauritzen, I.; Lazdunski, M.; Honoré, E. A phospholipid sensor controls mechanogating of the K+ channel TREK-1. *EMBO J.* **2005**, *24*, 44–53. [CrossRef]
108. Alloui, A.; Zimmermann, K.; Mamet, J.; Duprat, F.; Noël, J.; Chemin, J.; Guy, N.; Blondeau, N.; Voilley, N.; Rubat-Coudert, C.; et al. TREK-1, a K+ channel involved in polymodal pain perception. *EMBO J.* **2006**, *25*, 2368–2376. [CrossRef] [PubMed]
109. Noël, J.; Zimmermann, K.; Busserolles, J.; Deval, E.; Alloui, A.; Diochot, S.; Guy, N.; Borsotto, M.; Reeh, P.; Eschalier, A.; et al. The mechano-activated K+ channels TRAAK and TREK-1 control both warm and cold perception. *EMBO J.* **2009**, *28*, 1308–1318. [CrossRef] [PubMed]
110. Brohawn, S.G.; Su, Z.; MacKinnon, R. Mechanosensitivity is mediated directly by the lipid membrane in TRAAK and TREK1 K+ channels. *PNAS* **2014**, *111*, 3614–3619. [CrossRef] [PubMed]
111. Brohawn, S.G.; Campbell, E.B.; MacKinnon, R. Physical mechanism for gating and mechanosensitivity of the human TRAAK K+ channel. *Nature* **2014**, *516*, 126–130. [CrossRef] [PubMed]
112. Pereira, V.; Busserolles, J.; Christin, M.; Devilliers, M.; Poupon, L.; Legha, W.; Alloui, A.; Aissouni, Y.; Bourinet, E.; Lesage, F.; et al. Role of the TREK2 potassium channel in cold and warm thermosensation and in pain perception. *Pain* **2014**, *155*, 2534–2544. [CrossRef] [PubMed]
113. Descoeur, J.; Pereira, V.; Pizzoccaro, A.; Francois, A.; Ling, B.; Maffre, V.; Couette, B.; Busserolles, J.; Courteix, C.; Noel, J.; et al. Oxaliplatin-induced cold hypersensitivity is due to remodelling of ion channel expression in nociceptors. *EMBO Mol. Med.* **2011**, *3*, 266–278. [CrossRef] [PubMed]
114. Franchini, L.; Levi, G.; Visentin, S. Inwardly rectifying K+ channels influence Ca2+ entry due to nucleotide receptor activation in microglia. *Cell Calcium* **2004**, *35*, 449–459. [CrossRef] [PubMed]
115. Janigro, D.; Gasparini, S.; D'Ambrosio, R.; Ii, G.M.; DiFrancesco, D. Reduction of K+ Uptake in Glia Prevents Long-Term Depression Maintenance and Causes Epileptiform Activity. *J. Neurosci.* **1997**, *17*, 2813–2824. [CrossRef] [PubMed]
116. D'Ambrosio, R.; Gordon, D.S.; Winn, H.R. Differential role of KIR channel and Na(+)/K(+)-pump in the regulation of extracellular K(+) in rat hippocampus. *J. Neurophysiol.* **2002**, *87*, 87–102. [CrossRef]
117. Vit, J.-P.; Ohara, P.T.; Bhargava, A.; Kelley, K.; Jasmin, L. Silencing the Kir4.1 potassium channel subunit in satellite glial cells of the rat trigeminal ganglion results in pain-like behavior in the absence of nerve injury. *J. Neurosci.* **2008**, *28*, 4161–4171. [CrossRef]
118. Takeda, M.; Tsuboi, Y.; Kitagawa, J.; Nakagawa, K.; Iwata, K.; Matsumoto, S. Potassium channels as a potential therapeutic target for trigeminal neuropathic and inflammatory pain. *Mol. Pain* **2011**, *7*, 5. [CrossRef] [PubMed]
119. Djukic, B.; Casper, K.B.; Philpot, B.D.; Chin, L.-S.; McCarthy, K.D. Conditional knock-out of Kir4.1 leads to glial membrane depolarization, inhibition of potassium and glutamate uptake, and enhanced short-term synaptic potentiation. *J. Neurosci.* **2007**, *27*, 11354–11365. [CrossRef] [PubMed]
120. Tang, X.; Hang, D.; Sand, A.; Kofuji, P. Variable loss of Kir4.1 channel function in SeSAME syndrome mutations. *Biochem. Biophys. Res. Commun.* **2010**, *399*, 537–541. [CrossRef] [PubMed]
121. Nishizawa, D.; Nagashima, M.; Katoh, R.; Satoh, Y.; Tagami, M.; Kasai, S.; Ogai, Y.; Han, W.; Hasegawa, J.; Shimoyama, N.; et al. Association between KCNJ6 (GIRK2) gene polymorphisms and postoperative analgesic requirements after major abdominal surgery. *PLoS ONE* **2009**, *4*, e7060. [CrossRef] [PubMed]
122. Nishizawa, D.; Fukuda, K.; Kasai, S.; Ogai, Y.; Hasegawa, J.; Sato, N.; Yamada, H.; Tanioka, F.; Sugimura, H.; Hayashida, M.; et al. Association between KCNJ6 (GIRK2) gene polymorphism rs2835859 and post-operative analgesia, pain sensitivity, and nicotine dependence. *J. Pharmacol. Sci.* **2014**, *126*, 253–263. [CrossRef] [PubMed]
123. Bruehl, S.; Denton, J.S.; Lonergan, D.; Koran, M.E.; Chont, M.; Sobey, C.; Fernando, S.; Bush, W.S.; Mishra, P.; Thornton-Wells, T.A. Associations between KCNJ6 (GIRK2) gene polymorphisms and pain-related phenotypes. *Pain* **2013**, *154*, 2853–2859. [CrossRef] [PubMed]

124. Nockemann, D.; Rouault, M.; Labuz, D.; Hublitz, P.; McKnelly, K.; Reis, F.C.; Stein, C.; Heppenstall, P.A. The K(+) channel GIRK2 is both necessary and sufficient for peripheral opioid-mediated analgesia. *EMBO Mol. Med.* **2013**, *5*, 1263–1277. [CrossRef] [PubMed]
125. Lolicato, M.; Arrigoni, C.; Mori, T.; Sekioka, Y.; Bryant, C.; Clark, K.A.; Minor, D.L. K2P2.1 (TREK-1)-activator complexes reveal a cryptic selectivity filter binding site. *Nature* **2017**, *547*, 364–368. [CrossRef] [PubMed]
126. Al-Karagholi, M.A.-M.; Hansen, J.M.; Severinsen, J.; Jansen-Olesen, I.; Ashina, M. The KATP channel in migraine pathophysiology: A novel therapeutic target for migraine. *J. Headache Pain* **2017**, *18*. [CrossRef]
127. Nilius, B.; Prenen, J.; Owsianik, G. Irritating channels: The case of TRPA1. *J. Physiol. (Lond.)* **2011**, *589*, 1543–1549. [CrossRef]
128. Walker, R.G.; Willingham, A.T.; Zuker, C.S. A Drosophila mechanosensory transduction channel. *Science* **2000**, *287*, 2229–2234. [CrossRef] [PubMed]
129. Dai, Y. TRPs and pain. *Semin. Immunopathol.* **2016**, *38*, 277–291. [CrossRef] [PubMed]
130. Julius, D. TRP Channels and Pain. *Annu. Rev. Cell Dev. Biol.* **2013**, *29*, 355–384. [CrossRef] [PubMed]
131. González-Ramírez, R.; Chen, Y.; Liedtke, W.B.; Morales-Lázaro, S.L. TRP Channels and Pain. In *Neurobiology of TRP Channels*; Emir, T.L.R., Ed.; Frontiers in Neuroscience; CRC Press/Taylor & Francis: Boca Raton, FL, USA, 2017.
132. Nilius, B.; Talavera, K.; Owsianik, G.; Prenen, J.; Droogmans, G.; Voets, T. Gating of TRP channels: A voltage connection?: Voltage dependence of TRP channels. *J. Physiol.* **2005**, *567*, 35–44. [CrossRef] [PubMed]
133. Bouron, A.; Kiselyov, K.; Oberwinkler, J. Permeation, regulation and control of expression of TRP channels by trace metal ions. *Pflügers Arch.* **2015**, *467*, 1143–1164. [CrossRef] [PubMed]
134. Lev, S.; Minke, B. Constitutive Activity of TRP Channels. In *Methods in Enzymology*; Elsevier: Amsterdam, Netherlands, 2010; Volume 484, pp. 591–612.
135. Vriens, J.; Held, K.; Janssens, A.; Tóth, B.I.; Kerselaers, S.; Nilius, B.; Vennekens, R.; Voets, T. Opening of an alternative ion permeation pathway in a nociceptor TRP channel. *Nat. Chem. Biol.* **2014**, *10*, 188–195. [CrossRef]
136. Held, K.; Gruss, F.; Aloi, V.D.; Janssens, A.; Ulens, C.; Voets, T.; Vriens, J. Mutations in the voltage-sensing domain affect the alternative ion permeation pathway in the TRPM3 channel. *J. Physiol. (Lond.)* **2018**, *596*, 2413–2432. [CrossRef]
137. Starace, D.M.; Bezanilla, F. Histidine scanning mutagenesis of basic residues of the S4 segment of the shaker k+ channel. *J. Gen. Physiol.* **2001**, *117*, 469–490. [CrossRef]
138. Tombola, F.; Ulbrich, M.H.; Isacoff, E.Y. The Voltage-Gated Proton Channel Hv1 Has Two Pores, Each Controlled by One Voltage Sensor. *Neuron* **2008**, *58*, 546–556. [CrossRef]
139. Xu, X.-Z.S.; Li, H.-S.; Guggino, W.B.; Montell, C. Coassembly of TRP and TRPL Produces a Distinct Store-Operated Conductance. *Cell* **1997**, *89*, 1155–1164. [CrossRef]
140. Owsianik, G.; Talavera, K.; Voets, T.; Nilius, B. Permeation and Selectivity of Trp Channels. *Annu. Rev. Physiol.* **2006**, *68*, 685–717. [CrossRef] [PubMed]
141. Nadler, M.J.S.; Hermosura, M.C.; Inabe, K.; Perraud, A.-L.; Zhu, Q.; Stokes, A.J.; Kurosaki, T.; Kinet, J.-P.; Penner, R.; Scharenberg, A.M.; et al. LTRPC7 is a Mg·ATP-regulated divalent cation channel required for cell viability. *Nature* **2001**, *411*, 590–595. [CrossRef] [PubMed]
142. Parnas, M.; Katz, B.; Minke, B. Open channel block by Ca2+ underlies the voltage dependence of drosophila TRPL channel. *J. Gen. Physiol.* **2007**, *129*, 17–28. [CrossRef] [PubMed]
143. Grimm, C.; Cuajungco, M.P.; van Aken, A.F.J.; Schnee, M.; Jörs, S.; Kros, C.J.; Ricci, A.J.; Heller, S. A helix-breaking mutation in TRPML3 leads to constitutive activity underlying deafness in the varitint-waddler mouse. *Proc. Natl. Acad. Sci. USA* **2007**, *104*, 19583–19588. [CrossRef] [PubMed]
144. Hinman, A.; Chuang, H.; Bautista, D.M.; Julius, D. TRP channel activation by reversible covalent modification. *Proc. Natl. Acad. Sci. USA* **2006**, *103*, 19564–19568. [CrossRef] [PubMed]
145. Macpherson, L.J.; Dubin, A.E.; Evans, M.J.; Marr, F.; Schultz, P.G.; Cravatt, B.F.; Patapoutian, A. Noxious compounds activate TRPA1 ion channels through covalent modification of cysteines. *Nature* **2007**, *445*, 541–545. [CrossRef] [PubMed]
146. Takahashi, N.; Kuwaki, T.; Kiyonaka, S.; Numata, T.; Kozai, D.; Mizuno, Y.; Yamamoto, S.; Naito, S.; Knevels, E.; Carmeliet, P.; et al. TRPA1 underlies a sensing mechanism for $O_2$. *Nat. Chem. Biol.* **2011**, *7*, 701–711. [CrossRef]

147. Bahia, P.K.; Parks, T.A.; Stanford, K.R.; Mitchell, D.A.; Varma, S.; Stevens, S.M.; Taylor-Clark, T.E. The exceptionally high reactivity of Cys 621 is critical for electrophilic activation of the sensory nerve ion channel TRPA1. *J. Gen. Physiol.* **2016**, *147*, 451–465. [CrossRef]
148. Chung, M.-K.; Güler, A.D.; Caterina, M.J. TRPV1 shows dynamic ionic selectivity during agonist stimulation. *Nat. Neurosci.* **2008**, *11*, 555–564. [CrossRef]
149. Chen, J.; Kim, D.; Bianchi, B.R.; Cavanaugh, E.J.; Faltynek, C.R.; Kym, P.R.; Reilly, R.M. Pore dilation occurs in TRPA1 but not in TRPM8 channels. *Mol. Pain* **2009**, *5*, 3. [CrossRef]
150. Banke, T.G. The dilated TRPA1 channel pore state is blocked by amiloride and analogues. *Brain Res.* **2011**, *1381*, 21–30. [CrossRef] [PubMed]
151. Li, M.; Toombes, G.E.S.; Silberberg, S.D.; Swartz, K.J. Physical basis of apparent pore dilation of ATP-activated P2X receptor channels. *Nat. Neurosci.* **2015**, *18*, 1577–1583. [CrossRef] [PubMed]
152. Bean, B.P. Pore dilation reconsidered. *Nat. Neurosci.* **2015**, *18*, 1534–1535. [CrossRef] [PubMed]
153. Raisinghani, M.; Zhong, L.; Jeffry, J.A.; Bishnoi, M.; Pabbidi, R.M.; Pimentel, F.; Cao, D.-S.; Steven Evans, M.; Premkumar, L.S. Activation characteristics of transient receptor potential ankyrin 1 and its role in nociception. *Am. J. Physiol. Cell Physiol.* **2011**, *301*, C587–C600. [CrossRef] [PubMed]
154. Meents, J.E.; Fischer, M.J.M.; McNaughton, P.A. Agonist-induced sensitisation of the irritant receptor ion channel TRPA1. *J. Physiol. (Lond.)* **2016**, *594*, 6643–6660. [CrossRef] [PubMed]
155. Birdsong, W.T.; Arttamangkul, S.; Clark, M.J.; Cheng, K.; Rice, K.C.; Traynor, J.R.; Williams, J.T. Increased agonist affinity at the μ-opioid receptor induced by prolonged agonist exposure. *J. Neurosci.* **2013**, *33*, 4118–4127. [CrossRef] [PubMed]
156. Gauss, R.; Seifert, R.; Kaupp, U.B. Molecular identification of a hyperpolarization-activated channel in sea urchin sperm. *Nature* **1998**, *393*, 583. [CrossRef]
157. Ludwig, A.; Zong, X.; Jeglitsch, M.; Hofmann, F.; Biel, M. A family of hyperpolarization-activated mammalian cation channels. *Nature* **1998**, *393*, 587. [CrossRef]
158. Santoro, B.; Liu, D.T.; Yao, H.; Bartsch, D.; Kandel, E.R.; Siegelbaum, S.A.; Tibbs, G.R. Identification of a Gene Encoding a Hyperpolarization-Activated Pacemaker Channel of Brain. *Cell* **1998**, *93*, 717–729. [CrossRef]
159. Ludwig, A.; Zong, X.; Stieber, J.; Hullin, R.; Hofmann, F.; Biel, M. Two pacemaker channels from human heart with profoundly different activation kinetics. *EMBO J.* **1999**, *18*, 2323–2329. [CrossRef]
160. Seifert, R.; Scholten, A.; Gauss, R.; Mincheva, A.; Lichter, P.; Kaupp, U.B. Molecular characterization of a slowly gating human hyperpolarization-activated channel predominantly expressed in thalamus, heart, and testis. *Proc. Natl. Acad. Sci. USA* **1999**, *96*, 9391–9396. [CrossRef] [PubMed]
161. Robinson, R.B.; Siegelbaum, S.A. Hyperpolarization-Activated Cation Currents: From Molecules to Physiological Function. *Annu. Rev. Physiol.* **2003**, *65*, 453–480. [CrossRef] [PubMed]
162. Craven, K.B.; Zagotta, W.N. CNG and HCN channels: Two peas, one pod. *Annu. Rev. Physiol.* **2006**, *68*, 375–401. [CrossRef] [PubMed]
163. Biel, M.; Wahl-Schott, C.; Michalakis, S.; Zong, X. Hyperpolarization-activated cation channels: From genes to function. *Physiol. Rev.* **2009**, *89*, 847–885. [CrossRef] [PubMed]
164. Wainger, B.J.; DeGennaro, M.; Santoro, B.; Siegelbaum, S.A.; Tibbs, G.R. Molecular mechanism of cAMP modulation of HCN pacemaker channels. *Nature* **2001**, *411*, 805. [CrossRef]
165. Lee, C.-H.; MacKinnon, R. Structures of the Human HCN1 Hyperpolarization-Activated Channel. *Cell* **2017**, *168*, 111–120.e11. [CrossRef]
166. Zagotta, W.N.; Olivier, N.B.; Black, K.D.; Young, E.C.; Olson, R.; Gouaux, E. Structural basis for modulation and agonist specificity of HCN pacemaker channels. *Nature* **2003**, *425*, 200. [CrossRef]
167. Ludwig, A.; Budde, T.; Stieber, J.; Moosmang, S.; Wahl, C.; Holthoff, K.; Langebartels, A.; Wotjak, C.; Munsch, T.; Zong, X.; et al. Absence epilepsy and sinus dysrhythmia in mice lacking the pacemaker channel HCN2. *EMBO J.* **2003**, *22*, 216–224. [CrossRef]
168. Nolan, M.F.; Malleret, G.; Lee, K.H.; Gibbs, E.; Dudman, J.T.; Santoro, B.; Yin, D.; Thompson, R.F.; Siegelbaum, S.A.; Kandel, E.R.; et al. The hyperpolarization-activated HCN1 channel is important for motor learning and neuronal integration by cerebellar Purkinje cells. *Cell* **2003**, *115*, 551–564. [CrossRef]
169. Herrmann, S.; Stieber, J.; Stöckl, G.; Hofmann, F.; Ludwig, A. HCN4 provides a "depolarization reserve" and is not required for heart rate acceleration in mice. *EMBO J.* **2007**, *26*, 4423–4432. [CrossRef]

170. Fenske, S.; Mader, R.; Scharr, A.; Paparizos, C.; Cao-Ehlker, X.; Michalakis, S.; Shaltiel, L.; Weidinger, M.; Stieber, J.; Feil, S.; et al. HCN3 Contributes to the Ventricular Action Potential Waveform in the Murine Heart. *Circ. Res.* **2011**, *109*, 1015–1023. [CrossRef] [PubMed]
171. Zobeiri, M.; Chaudhary, R.; Blaich, A.; Rottmann, M.; Herrmann, S.; Meuth, P.; Bista, P.; Kanyshkova, T.; Lüttjohann, A.; Narayanan, V.; et al. The Hyperpolarization-Activated HCN4 Channel is Important for Proper Maintenance of Oscillatory Activity in the Thalamocortical System. *Cereb. Cortex* **2019**, *29*, 2291–2304. [CrossRef] [PubMed]
172. Chaplan, S.R.; Guo, H.-Q.; Lee, D.H.; Luo, L.; Liu, C.; Kuei, C.; Velumian, A.A.; Butler, M.P.; Brown, S.M.; Dubin, A.E. Neuronal Hyperpolarization-Activated Pacemaker Channels Drive Neuropathic Pain. *J. Neurosci.* **2003**, *23*, 1169–1178. [CrossRef] [PubMed]
173. Yao, H.; Donnelly, D.F.; Ma, C.; LaMotte, R.H. Upregulation of the Hyperpolarization-Activated Cation Current after Chronic Compression of the Dorsal Root Ganglion. *J. Neurosci.* **2003**, *23*, 2069–2074. [CrossRef] [PubMed]
174. Resta, F.; Micheli, L.; Laurino, A.; Spinelli, V.; Mello, T.; Sartiani, L.; Di Cesare Mannelli, L.; Cerbai, E.; Ghelardini, C.; Romanelli, M.N.; et al. Selective HCN1 block as a strategy to control oxaliplatin-induced neuropathy. *Neuropharmacology* **2018**, *131*, 403–413. [CrossRef] [PubMed]
175. Weng, X.; Smith, T.; Sathish, J.; Djouhri, L. Chronic inflammatory pain is associated with increased excitability and hyperpolarization-activated current (Ih) in C- but not Aδ-nociceptors. *Pain.* **2012**, *153*, 900. [CrossRef]
176. Jiang, Y.-Q.; Xing, G.-G.; Wang, S.-L.; Tu, H.-Y.; Chi, Y.-N.; Li, J.; Liu, F.-Y.; Han, J.-S.; Wan, Y. Axonal accumulation of hyperpolarization-activated cyclic nucleotide-gated cation channels contributes to mechanical allodynia after peripheral nerve injury in rat. *Pain.* **2008**, *137*, 495–506. [CrossRef]
177. Papp, I.; Holló, K.; Antal, M. Plasticity of hyperpolarization-activated and cyclic nucleotid-gated cation channel subunit 2 expression in the spinal dorsal horn in inflammatory pain. *Eur. J. Neurosci.* **2010**, *32*, 1193–1201. [CrossRef]
178. Acosta, C.; McMullan, S.; Djouhri, L.; Gao, L.; Watkins, R.; Berry, C.; Dempsey, K.; Lawson, S.N. HCN1 and HCN2 in Rat DRG neurons: Levels in nociceptors and non-nociceptors, NT3-dependence and influence of CFA-induced skin inflammation on HCN2 and NT3 expression. *PLoS ONE* **2012**, *7*, e50442. [CrossRef]
179. Schnorr, S.; Eberhardt, M.; Kistner, K.; Rajab, H.; Käer, J.; Hess, A.; Reeh, P.; Ludwig, A.; Herrmann, S. HCN2 channels account for mechanical (but not heat) hyperalgesia during long-standing inflammation. *Pain.* **2014**, *155*, 1079. [CrossRef]
180. Tsantoulas, C.; Laínez, S.; Wong, S.; Mehta, I.; Vilar, B.; McNaughton, P.A. Hyperpolarization-activated cyclic nucleotide-gated 2 (HCN2) ion channels drive pain in mouse models of diabetic neuropathy. *Sci. Transl. Med.* **2017**, *9*, eaam6072. [CrossRef] [PubMed]
181. Herrmann, S.; Rajab, H.; Christ, I.; Schirdewahn, C.; Höfler, D.; Fischer, M.J.M.; Bruno, A.; Fenske, S.; Gruner, C.; Kramer, F.; et al. Protein kinase A regulates inflammatory pain sensitization by modulating HCN2 channel activity in nociceptive sensory neurons. *Pain.* **2017**, *158*, 2012. [CrossRef] [PubMed]
182. Emery, E.C.; Young, G.T.; Berrocoso, E.M.; Chen, L.; McNaughton, P.A. HCN2 ion channels play a central role in inflammatory and neuropathic pain. *Science* **2011**, *333*, 1462–1466. [CrossRef] [PubMed]
183. Herrmann, S.; Schnorr, S.; Ludwig, A. HCN channels–modulators of cardiac and neuronal excitability. *Int. J. Mol. Sci.* **2015**, *16*, 1429–1447. [CrossRef] [PubMed]
184. Emery, E.C.; Young, G.T.; McNaughton, P.A. HCN2 ion channels: An emerging role as the pacemakers of pain. *Trends Pharmacol. Sci.* **2012**, *33*, 456–463. [CrossRef] [PubMed]
185. Lee, M.C.; Bond, S.; Wheeler, D.; Scholtes, I.; Armstrong, G.; McNaughton, P.; Menon, D. A randomised, double blind, placebo-controlled crossover trial of the influence of the HCN channel blocker ivabradine in a healthy volunteer pain model: An enriched population trial. *Pain* **2019**. [CrossRef] [PubMed]

International Journal of
*Molecular Sciences*

*Article*

# Expression and Activity of TRPA1 and TRPV1 in the Intervertebral Disc: Association with Inflammation and Matrix Remodeling

**Takuya Kameda [1,2], Joel Zvick [1], Miriam Vuk [1], Aleksandra Sadowska [1], Wai Kit Tam [3], Victor Y. Leung [3], Kata Bölcskei [4,5], Zsuzsanna Helyes [4,5], Lee Ann Applegate [6], Oliver N. Hausmann [7], Juergen Klasen [8], Olga Krupkova [1,*,†] and Karin Wuertz-Kozak [1,9,10,†]**

1 Institute for Biomechanics, ETH Zurich, Hoenggerbergring 64, 8093 Zurich, Switzerland; sa57384@cd6.so-net.ne.jp (T.K.); joel.zvick@hest.ethz.ch (J.Z.); vukm@student.ethz.ch (M.V.); aleksandra.sadowska@hest.ethz.ch (A.S.); kwuertz@ethz.ch (K.W.-K.)
2 Department of Orthopaedic Surgery, Fukushima Medical University, 1 Hikarigaoka, Fukushima City, Fukushima 960-1295, Japan
3 Department of Orthopaedics and Traumatology, The University of Hong Kong, 21 Sassoon Road, Pokfulam, Hong Kong SAR, China; tamwk1@hku.hk (W.K.T.); vicleung@hku.hk (V.Y.L.)
4 Department of Pharmacology and Pharmacotherapy, University of Pécs, Szigeti út 12., H-7624 Pécs, Hungary; kata.bolcskei@aok.pte.hu (K.B.); zsuzsanna.helyes@aok.pte.hu (Z.H.)
5 János Szentágothai Research Centre, University of Pécs, Ifjúság útja 20., H-7624 Pécs, Hungary
6 Department of Musculoskeletal Medicine, Unit of Regenerative Therapy (UTR), University Hospital Lausanne, EPCR/02 Chemin des Croisettes 22, 1066 Epalinges, Switzerland; Lee.Laurent-Applegate@chuv.ch
7 Neuro- and Spine Center, St. Anna Hospital, Sankt-Anna-Strasse 32, 6006 Luzern, Switzerland; ohausmann@hin.ch
8 Clinic Prodorso, Walchestrasse 15, 8006 Zurich, Switzerland; info@prodorso.ch
9 Schön Clinic Munich Harlaching, Spine Center, Academic Teaching Hospital and Spine Research Institute of the Paracelsus Medical University Salzburg (AU), Harlachinger Str. 51, 81547 Munich, Germany
10 Department of Health Sciences, University of Potsdam, Am Neuen Palais 10, 14469 Potsdam, Germany
* Correspondence: okrupkova@ethz.ch; Tel.: +41-44-633-2901
† The authors contributed equally to this work.

Received: 7 March 2019; Accepted: 30 March 2019; Published: 10 April 2019

check for updates

**Abstract:** Transient receptor potential (TRP) channels have emerged as potential sensors and transducers of inflammatory pain. The aims of this study were to investigate (1) the expression of TRP channels in intervertebral disc (IVD) cells in normal and inflammatory conditions and (2) the function of Transient receptor potential ankyrin 1 (TRPA1) and Transient receptor potential vanilloid 1 (TRPV1) in IVD inflammation and matrix homeostasis. RT-qPCR was used to analyze human fetal, healthy, and degenerated IVD tissues for the gene expression of TRPA1 and TRPV1. The primary IVD cell cultures were stimulated with either interleukin-1 beta (IL-1β) or tumor necrosis factor alpha (TNF-α) alone or in combination with TRPA1/V1 agonist allyl isothiocyanate (AITC, 3 and 10 μM), followed by analysis of calcium flux and the expression of inflammation mediators (RT-qPCR/ELISA) and matrix constituents (RT-qPCR). The matrix structure and composition in caudal motion segments from TRPA1 and TRPV1 wild-type (WT) and knock-out (KO) mice was visualized by FAST staining. Gene expression of other TRP channels (A1, C1, C3, C6, V1, V2, V4, V6, M2, M7, M8) was also tested in cytokine-treated cells. TRPA1 was expressed in fetal IVD cells, 20% of degenerated IVDs, but not in healthy mature IVDs. TRPA1 expression was not detectable in untreated cells and it increased upon cytokine treatment, while TRPV1 was expressed and concomitantly reduced. In inflamed IVD cells, 10 μM AITC activated calcium flux, induced gene expression of IL-8, and reduced disintegrin and metalloproteinase with thrombospondin motifs 5 (ADAMTS5) and collagen 1A1, possibly via upregulated TRPA1. TRPA1 KO in mice was associated with signs of degeneration in the nucleus pulposus and the vertebral growth plate, whereas TRPV1 KO did not show profound changes. Cytokine treatment also affected the gene expression of TRPV2 (increase), TRPV4 (increase), and TRPC6 (decrease). TRPA1 might be expressed in developing IVD, downregulated during its maturation, and upregulated again in degenerative disc disease, participating in matrix homeostasis. However, follow-up studies with larger sample sizes are needed to fully elucidate the role of TRPA1 and other TRP channels in degenerative disc disease.

**Keywords:** low back pain; TRP channels; pro-inflammatory cytokines; aggrecanases; collagen; TRPA1; TRPV1; TRPV2; TRPV4; TRPC6

---

## 1. Introduction

Low back pain (LBP) is the leading cause of disability, activity limitation and lost productivity throughout the world today, with approximately 80% of all people suffering from back pain at least once in their life [1]. Degenerative disc disease (DDD), which is a progressive multifactorial disorder of the intervertebral disc (IVD), is an important factor that is involved in the development of LBP [1]. DDD is characterized by the release of inflammatory and catabolic mediators, including interleukin-1 beta (IL-1β), tumor necrosis factor alpha (TNF-α), prostaglandins, and proteases, which further promote the degradation of extracellular matrix (ECM) [2,3]. Pro-inflammatory cytokines IL-1β and TNF-α directly act as nociceptive triggers, but also activate the expression of other potentially nociceptive molecules including neuropeptides, interleukin-6 (IL-6), and interleukin-8 (IL-8) [4,5]. Molecular mechanisms involved in transduction and modulation of IL-1β and TNF-α signaling in DDD are not yet well-understood [6–8].

Signals that are provided by pro-inflammatory cytokines can be mediated via membrane channels [9,10]. Recently, transient receptor potential (TRP) channels have emerged as putative receptors for inflammation-associated molecules, positive/negative regulators of inflammation, and transducers of inflammatory pain [11–13]. TRP channels are cation selective transmembrane receptors with diverse physiological functions. Six families of mammalian TRP channels have been identified, classifying TRP channels according to their sequence homology and topological differences: TRPA (ankyrin), TRPC (canonical), TRPM (melastatin), TRPV (vanillin), TRPP (polycystin), and TRPML (mucolipin). Apart from TRPA, every subfamily has several members [14,15]. The dysregulation of TRP channels is implicated in many pathologies, including cardiovascular diseases, muscular dystrophies, and hyperalgesia [14,16]. Interestingly, the expression and activity of certain TRP channels is altered in painful joints and IVDs [11,12]. For example, the expression of TRPV4 in human IVDs was found to be elevated in regions of aggrecan depletion [17], while the gene expression of TRPC6 was associated with the severity of disc degeneration, increased expression of IL-6, and cell senescence [18,19]. TRPA1 and TRPV1 are involved in inflammation/nociception in sensory neurons and non-neuronal tissues [20–22]. TRPV1 is a non-selective calcium channel, the expression and activity of which increases after inflammatory stimulation in dorsal root ganglions (DRGs), possibly causing chronic hyperalgesia. TRPV1 also expressed in chondrocytes [23]. TRPA1 is a calcium permeable non-selective cation channel that is also widely expressed in sensory neurons and in non-neuronal cells, including chondrocytes [11]. TRPA1 is involved in various sensory and homeostatic functions, depending on the cell type [12,24]. TRPA1 and TRPV1 were shown to complement each other's activities [21,25]. As numerous publications have linked TRPA1 and TRPV1 with inflammatory pain [21], therapeutic inhibition or the activation of TRPA1 and/or TRPV1 channels may be beneficial in the treatment of DDD. Therefore, the aims of this study were to investigate the (1) expression of TRP channels in IVD cells in normal and inflammatory conditions and (2) the function of TRPA1 and TRPV1 in disc inflammation and matrix homeostasis.

## 2. Results

### 2.1. Gene Expression of TRPA1 in Human IVD Tissue

Gene expression of TRPA1 was tested in human non-degenerated ($n$ = 4) and degenerated ($n$ = 20) IVD tissue in relation to disc region, disease type, pain score (for degenerated discs only), grade, and age. In the degenerated tissue, TRPA1 was found expressed in 20% of tested donors (four out of 20). Although only expressed in a subset of degenerated donors, TRPA1 was found in both annulus fibrosus (AF) and nucleus pulposus (NP), in an age range of 39–76 years, at pain scores 2 (= intense) and 3 (= disabling) and at Pfirrmann grades 2 and 3 (Table 1). TRPA1 was only expressed in one non-degenerated NP sample (in one out of four donors: male, 17 years old, grade 1, $2^{-dCt} = 4.57 \times 10^{-5}$). Interestingly, TRPA1 was found expressed in cells isolated from human fetal IVD tissue ($n$ = 4; $2^{-dCt} = 0.0463 \pm 0.04742$). It is known that TRPA1 can associate with TRPV1, thereby regulating its intrinsic properties independently of intracellular calcium. Table 1 summarizes the expression of TRPV1. A manuscript that provides details on the expression of TRPV1 in IVD tissue is currently in revision [26].

**Table 1.** Gene expression of TRPA1 and TRPV1 in human intervertebral disc (IVD) tissue. Some donor tissues were divided into nucleus pulposus (NP) and annulus fibrosus (AF), resulting in more AF and NP samples ($n$ = 21 in region, pain score, grade) than a total number of donors ($n$ = 20).

| Degenerated Lumbar IVDs | $n$ | $2^{-dCt}$ (Mean ± SD) | Region | Pain score | Grade | Age (Mean ± SD) |
|---|---|---|---|---|---|---|
| **TRPA1** | in 4 out of 20 | 0.0006 ± 0.001 | 2 in AF<br>2 in NP | 2 pain score 2;<br>2 pain score 3 | 2 grade 2;<br>2 grade 3 | 60 ± 15.6<br>max: 76;<br>min: 39 |
| **TRPV1** [26] | 19 out of 20 | 0.0047 ± 0.0024 | 10 in NP<br>9 in AF<br>2 in mix | 6 in pain score 1;<br>11 in pain score 2;<br>4 in pain score 3 | 4 grade 2;<br>8 grade 3;<br>6 grade 4;<br>3 grade 5 | 54 ± 15<br>max: 80;<br>min: 31 |

### 2.2. Gene Expression of TRPA1 and TRPV1 in Human IVD Cells Treated with Pro-Inflammatory Cytokines

TRPA1 and TRPV1 can be involved in modulating inflammation in both neuronal and non-neuronal cells [25]. Therefore, changes in the gene expression of TRPA1 and TRPV1 were tested in IVD cells stimulated with pro-inflammatory cytokines IL-1β and TNF-α (both 5 and 10 ng/mL) ($n$ = 5) (Table 2). TRPA1 was under the detection limit in most untreated IVD cells, while its gene expression tended to increase with IL-1β treatment ($p$ = 0.07 for IL-1 5 ng/mL) and it significantly increased with TNF-α (Figure 1A). TNF-α, but not IL-1β, significantly reduced gene expression of TRPV1 (Figure 1B). The induction of an inflammatory-catabolic shift upon cytokine treatment was demonstrated by an increase of IL-6, IL-8, cyclooxygenase-2 (COX-2), nerve growth factor (NGF), matrix metalloproteinase 1 (MMP1), matrix metalloproteinase 3 (MMP3), a disintegrin and metalloproteinase with thrombospondin motifs 4 (ADAMTS4), a disintegrin and metalloproteinase with thrombospondin motifs 5 (ADAMTS5), and a reduction in COL2A1. Tissue inhibitor of matrix metalloproteinase 1 (TIMP1), tissue inhibitor of matrix metalloproteinase 2 (TIMP2), Aggrecan and COL1A1 were unchanged (Supplementary Figure S1). In IVD cells seeded in three-dimensional (3D) alginate beads and treated with IL-1β (5 ng/mL, 15 days, $n$ = 4–10), gene expression of TRPA1 significantly increased on day 1, 8, and 15 (Figure 1C), while the gene expression of TRPV1 remained unchanged (Figure 1D). Immunostaining confirmed TRPA1 protein induction upon IL-1β treatment (5 ng/mL) ($n$ = 3) (Figure 1E). Cell viability in alginate beads was monitored by Calcein/Ethidium homodimer staining and an average of 87% of living cells per treatment and time point was found (Supplementary Figure S2).

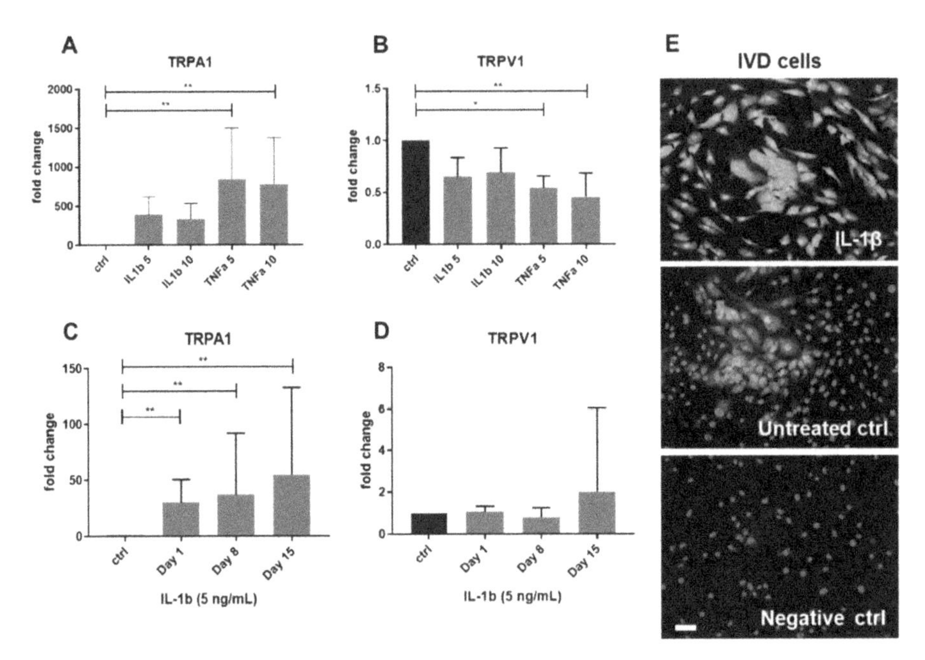


**Figure 1.** Gene expression of TRPA1 and TRPV1 in IVD cells treated with 5 and 10 ng/mL interleukin-1 beta (IL-1β) or tumor necrosis factor alpha (TNF-α). Gene expression of (**A**) TRPA1 and (**B**) TRPV1 in two-dimensional (2D) culture (Graph shows $2^{-ddCt}$ (mean ± SD, $n$ = 5). Gene expression of (**C**) TRPA1 and (**D**) TRPV1 in IVD cells cultured in three-dimensional (3D) alginate beads and treated with IL-1β for 15 days. Graph shows $2^{-ddCt}$ (mean ± SD, $n$ = 4–10). Asterisks indicate statistical significance (* $p < 0.05$, ** $p < 0.01$, Kruskal–Wallis test and Dunn's multiple comparison test). (**E**) Protein expression of TRPA1 in IVD cells treated with IL-1β and untreated (DAPI = blue, TRPA1 = green). Negative control images show cells without secondary antibody. Scale bar is 50 μm.

**Table 2.** Gene expression of TRPA1 and TRPV1 in untreated and treated human IVD cells (2D). $n$ = 5 donors. Values show fold change relative to the untreated cells. In case of expression under the detection limit, $dC_t$ was set at 40 cycles.

| Treatment | Untreated | IL-1β 5 ng/mL | IL-1β 10 ng/mL | TNF-α 5 ng/mL | TNF-α 10 ng/mL |
|---|---|---|---|---|---|
| **TRPA1** | No expression | 385.18 ± 233.06 | 333.42 ± 199.80 | 842.06 ± 659.98 | 780.23 ± 600.17 |
| **TRPV1** | Expression | 0.64 ± 0.18 | 0.69 ± 0.23 | 0.53 ± 0.11 | 0.45 ± 0.22 |

### *2.3. Gene Expression of Other TRP Channels in Human IVD Cells Treated with Pro-Inflammatory Cytokines*

Changes in the gene expression of other TRP channels, namely TRPC1, TRPC3, TRPC6, TRPV2, TRPV4, TRPV6, TRPM2, TRPM7, and TRPM8, were also tested in IVD cells upon cytokine stimulation (IL-1β and TNF-α, both 5 and 10 ng/mL) ($n$ = 5). IL-1β significantly induced gene expression of TRPV4 and reduced TRPC6. TNF-α significantly activated the gene expression of TRPV2. The expression of other TRP channels showed no clear association with a pro-inflammatory treatment in the tested experimental settings (Figure 2A–G). Gene expression of TRPM2 and TRPM8 was under the detection limit in all of the treatment groups.

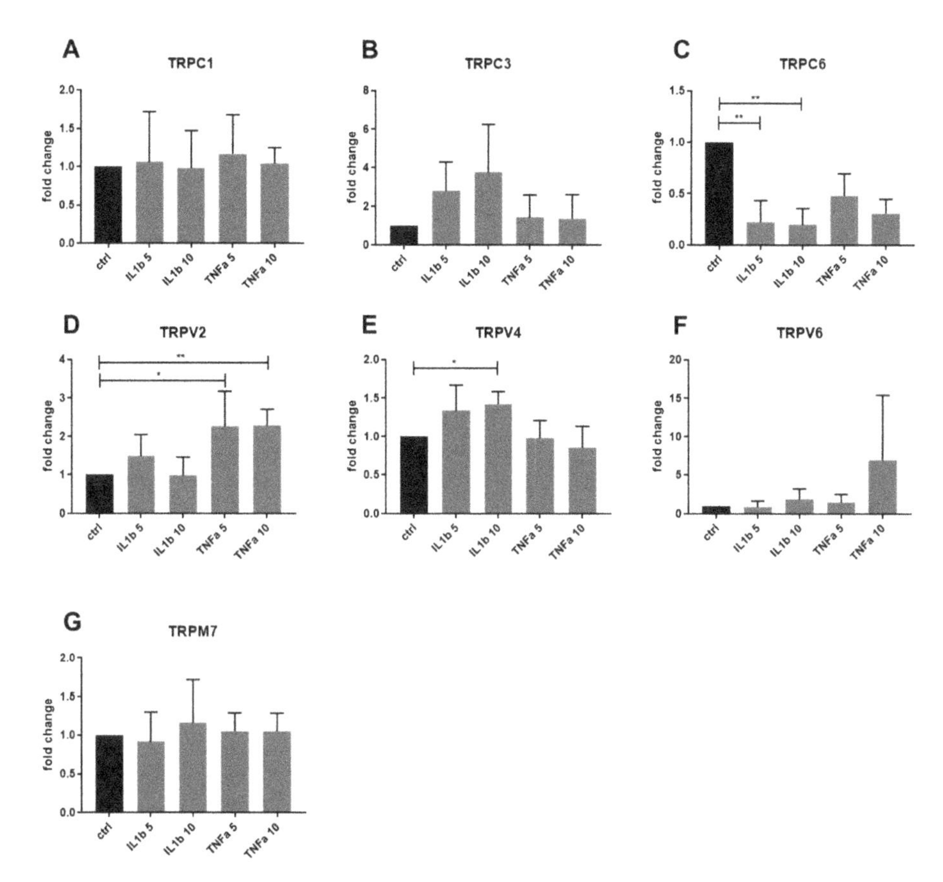


**Figure 2.** Gene expression of TRP channels in IVD cells treated with 5 and 10 ng/mL IL-1β or TNF-α. Gene expression of (**A**) TRPC1, (**B**) TRPC3, (**C**) TRPC6, (**D**) TRPV2, (**E**) TRPV4, (**F**) TRPV6, and (**G**) TRPM7. Graphs show $2^{-ddCt}$ (mean ± SD, $n = 5$). Asterisks indicate statistical significance (* $p < 0.05$, ** $p < 0.01$, Kruskal–Wallis test and Dunn's multiple comparison test).

## 2.4. Functional Analysis of TRPA1 and TRPV1 in IVD Cells

Allyl isothiocyanate (AITC, 3 and 10 μM), which is an agonist for TRPA1 and TRPV1, was used to test the involvement of these channels in (1) inflammation responses and (2) matrix homeostasis within the IVD compartment. The activity of AITC was tested by Calcium flux assay ($n = 3$). AITC was applied in untreated IVD cells (expressing TRPV1) as well as in cells that were treated with pro-inflammatory cytokines (expressing TRPA1 and TRPV1). AITC did not induce calcium flux in the untreated IVD cells. AITC significantly induced calcium flux in both IL-1β and TNF-α-treated cells, suggesting the involvement of TRPA1 (Figure 3). Calcium flux in the TNF-α treated cells was significantly higher than in IL-1β-treated cells, likely due to the overall higher induction of TRPA1 expression by TNF-α. Therefore, it is possible that TRPA1 (not TRPV1) can be functionally involved in the calcium-induced responses of inflamed IVD cells. To verify the function of TRPA1/TRPV1, AITC was used to study gene and protein expression of inflammation and catabolic mediators in untreated as well as cytokine-treated IVD cells.

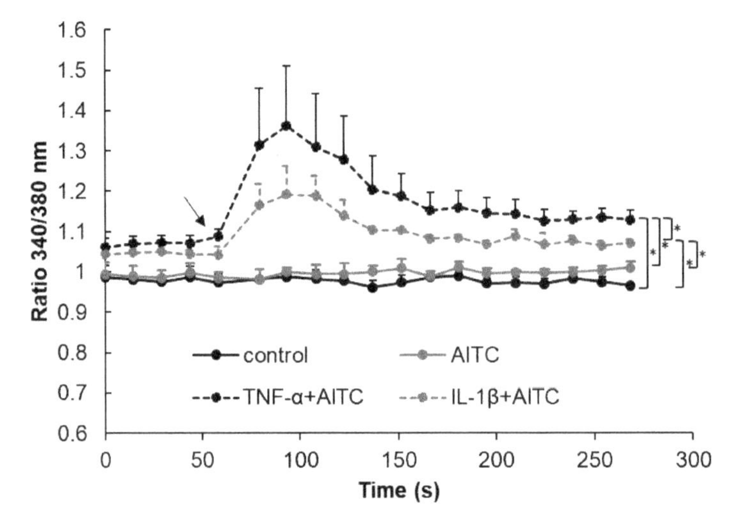


**Figure 3.** Calcium flux in IVD cells untreated (control) and treated with 100 μM Allyl isothiocyanate (AITC) with and without 10 ng/mL IL-1β or 10 ng/mL TNF-α. Graph shows calcium flux as 340/380 signal ratio (mean ± SD, *n* = 3). Asterisks indicate statistical significance (* $p < 0.05$, Kruskal–Wallis test and Dunn's multiple comparison test). The arrow indicates AITC application.

Cytokine untreated IVD cells (expressing TRPV1) were used to test the downstream effects of TRPV1 activation (*n* = 3–4). In unstimulated IVD cells, the gene expression of inflammation/pain mediators (IL-6, IL-8, NGF, COX-2) (Figure 4A–D) and most of the ECM remodeling enzymes (ADAMTS4, ADAMTS5, MMP3, TIMP1, TIMP2) was unchanged by AITC (Figure 4E–J). IL-6 and IL-8 concentration in culture media was close to the lower detection limit (~zero) (not shown). MMP1 was significantly induced by 10μM AITC (Figure 4H). The gene expression of COL1A1, COL2A1, and aggrecan in AITC-treated cells was not significantly different from the control (Figure 4K–M). Due to the fact that cytokine-untreated cells did not express TRPA1, the observed changes in MMP1 expression were not TRPA1-dependent. MMP1 upregulation by AITC may be an unrelated non-specific effect (as AITC did not induce calcium flux in untreated IVD cells).

IVD cells treated with pro-inflammatory cytokines (expressing TRPA1 and TRPV1) were used to test the downstream effects of TRPA1 activation (*n* = 3–4). In IL-1β-treated cells, AITC did not influence gene expression of inflammation mediators IL-6 and IL-8 (Figure 5A,D) and pain mediators NGF and COX-2 (Figure 5C,F). Interestingly, the protein release of IL-8 was significantly reduced by 3 μM AITC (Figure 5E). 10 μM AITC significantly reduced the gene expression of ADAMTS5 (Figure 5H), while MMP1 was induced (Figure 5J). Gene expression of other ECM remodeling enzymes (ADAMTS4, MMP3, TIMP1, TIMP2) and ECM genes (Figure 5M–O) in AITC + IL-1β-treated cells was not different from the IL-1β-only controls.

10 μM AITC significantly induced the gene and protein expression of IL-8 (Figure 6D,E) in TNF-α-treated cells (expressing TRPA1, reduced TRPV1). AITC did not influence the expression of IL-6 (Figure 6A,B), NGF, and COX-2 (Figure 6C,F). The gene expression of ADAMTS5 was significantly reduced (Figure 6H), while MMP1 was induced by 10 μM AITC in TNF-α-treated cells (Figure 6J). Gene expression of other ECM remodeling enzymes (MMP3, ADAMTS4, TIMP1, and TIMP2) was not different from TNF-α-treated control. Gene expression of COL1A1 was significantly reduced (Figure 6M), while the other tested ECM proteins (COL2A1 and aggrecan) were not influenced by AITC in TNF-α-treated cells. AITC did not affect the gene expression of TRPA1 itself (Supplementary Figure S3).

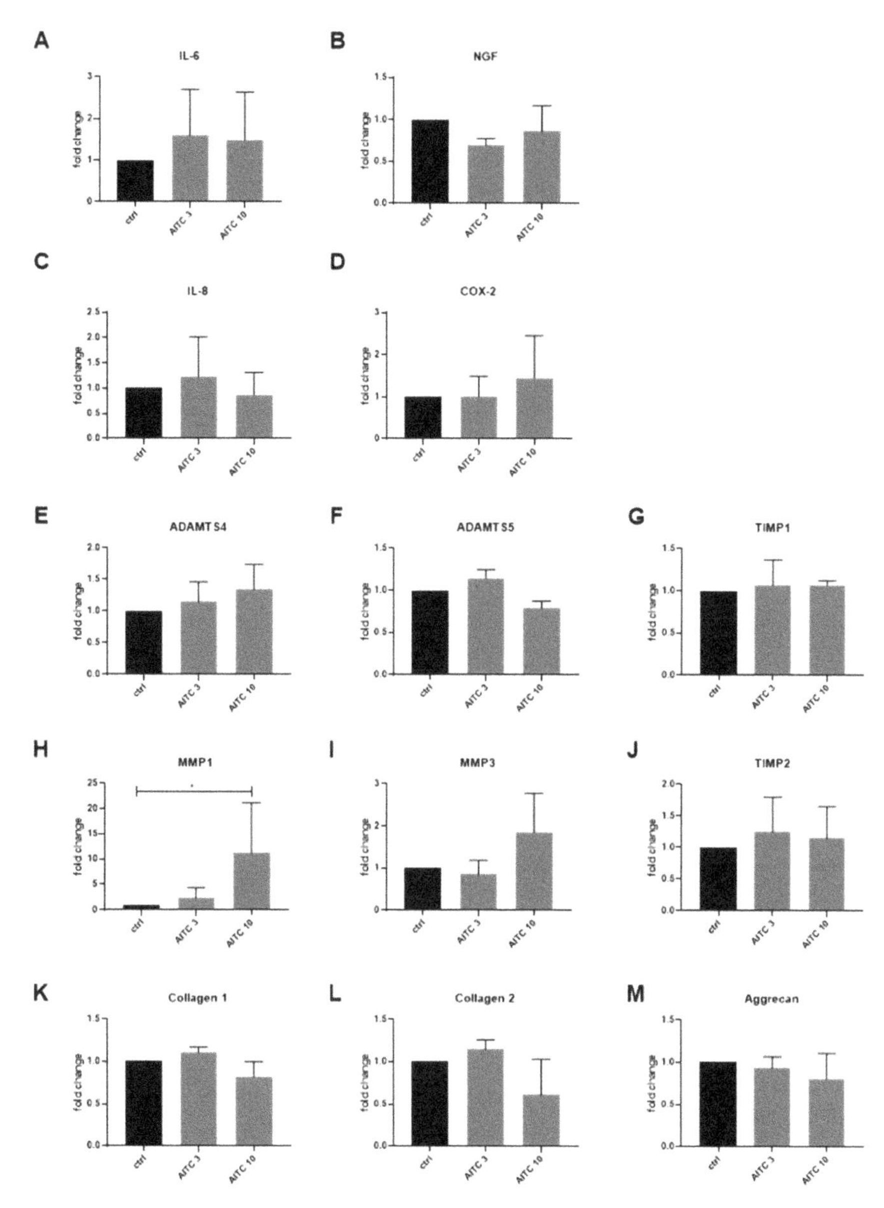


**Figure 4.** The effects of TRPA1 agonist allyl isothiocyanate (AITC) on the gene expression of inflammation markers and extracellular matrix (ECM) molecules in IVD cells without cytokine pre-treatment. Gene expression of (**A**) interleukin-6 (IL-6), (**B**) nerve growth factor (NGF), (**C**) interleukin 8 (IL-8), (**D**) cyclooxygenase-2 (COX-2), (**E**) ADAMTS4, (**F**) ADAMTS5, (**G**) tissue inhibitor of matrix metalloproteinase 1 (TIMP1), (**H**) matrix metalloproteinase 1 (MMP1), (**I**) matrix metalloproteinase 3 (MMP3), (**J**) tissue inhibitor of matrix metalloproteinase 2 (TIMP2), (**K**) COL1A1, (**L**) COL2A1, and (**M**) Aggrecan in IVD cells treated with 3 and 10 μM AITC. Graphs show gene expression and protein release calculated relative control ($2^{-ddCt}$, mean ± SD, $n$ = 3). Asterisks indicate statistical significance (* $p < 0.05$, Kruskal–Wallis test and Dunn's multiple comparison test).

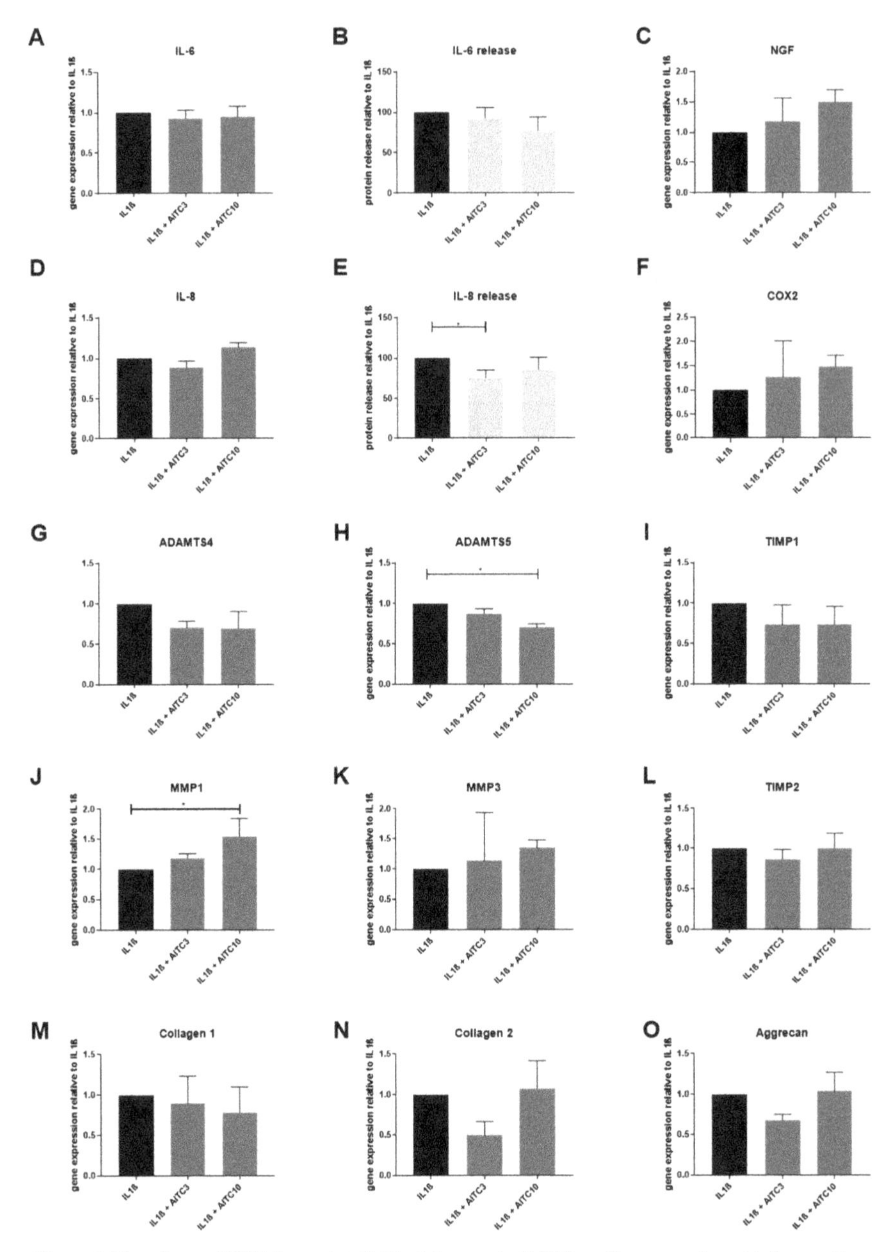


**Figure 5.** The effects of TRPA1 agonist allyl isothiocyanate (AITC) on the expression of inflammation markers and ECM molecules in IL-1β-treated cells. Gene expression of (**A**) IL-6 and (**D**) IL-8 in IVD cells treated with 10 ng/mL IL-1β ± 3 and 10 μM AITC. Protein release of (**B**) IL-6 and (**E**) IL-8 in IVD cells that were treated with 10 ng/mL IL-1β ± 3 and 10 μM AITC. Gene expression of (**C**) NGF, (**F**) COX-2, (**G**) ADAMTS4, (**H**) ADAMTS5, (**J**) MMP1, (**K**) MMP3, (**I**) TIMP1 and (**L**) TIMP2, (**M**) COL1A1, (**N**) COL2A1, and (**O**) Aggrecan in IVD cells that were treated with 10 ng/mL IL-1β ± 3 or 10 μM AITC. Graphs show gene expression and protein release calculated relative to IL-1β treatment (mean ± SD, $n$ = 3–4). Asterisks indicate statistical significance (* $p < 0.05$, Kruskal–Wallis test and Dunn's multiple comparison test).

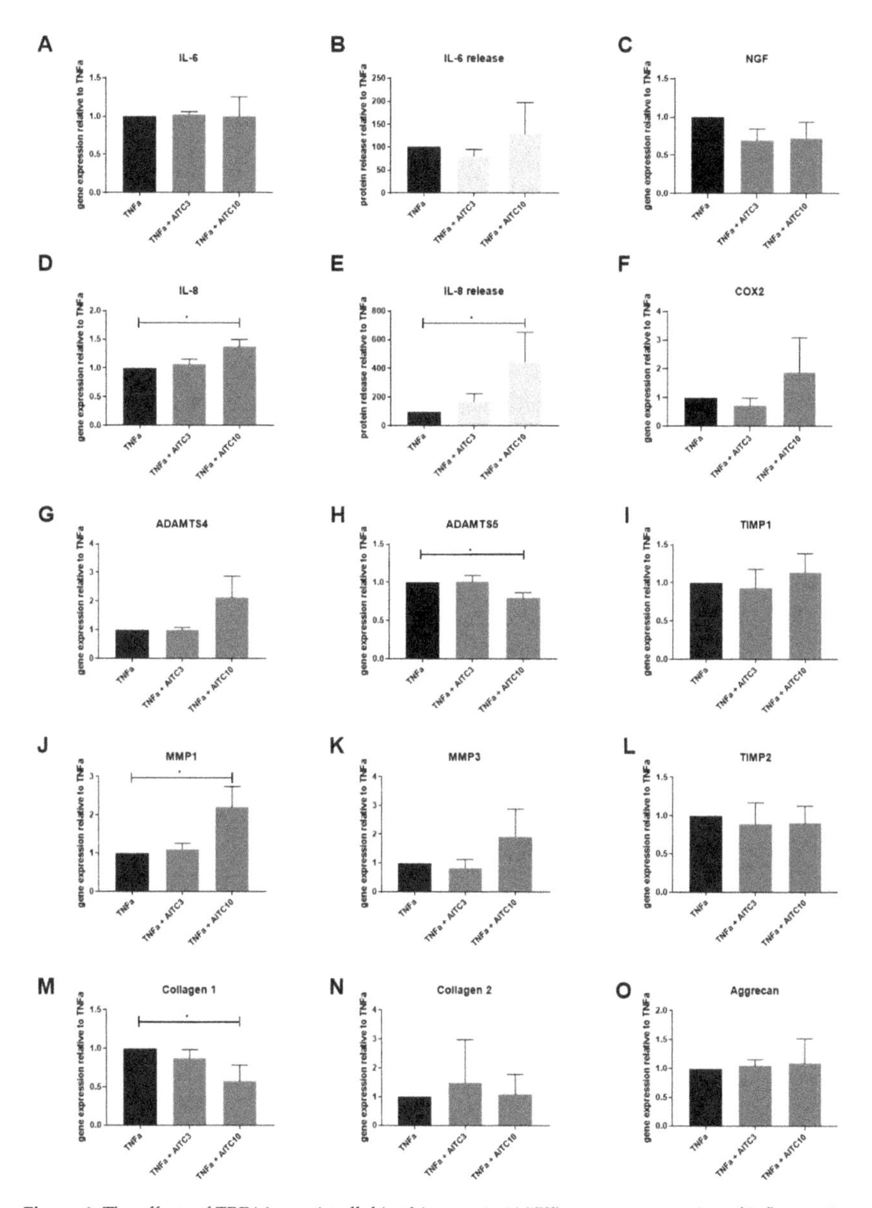


**Figure 6.** The effects of TRPA1 agonist allyl isothiocyanate (AITC) on gene expression of inflammation markers and ECM molecules in TNF-α-treated cells. Gene expression of (**A**) IL-6 and (**D**) IL-8 in IVD cells treated with 10 ng/mL TNF-α ± 3 or 10 μM AITC. Protein release of (**B**) IL-6 and (**E**) IL-8 in IVD cells treated with 10 ng/mL TNF-α ± 3 or 10 μM AITC. Gene expression of (**C**) NGF, (**F**) COX-2, (**G**) ADAMTS4, (**H**) ADAMTS5, (**I**) TIMP1, (**J**) MMP1, (**K**) MMP3, and (**L**) TIMP2, (**M**) COL1A1, (**N**) COL2A1, and (**O**) Aggrecan in IVD cells treated with 10 ng/mL TNF-α ± 3 and 10 μM AITC. Graphs show gene expression and protein release calculated relative to TNF-α treatment (mean ± SD, $n$ = 3-4). Asterisks indicate statistical significance (* $p < 0.05$, Kruskal–Wallis test and Dunn's multiple comparison test).

### 2.5. Motion Segments of TRPA1 and TRPV1-Deficient Mice

As our in vitro study showed a possible involvement of TRPA1/TRPV1 in IVD metabolism, we also focused at their effects in vivo. The possible involvement of TRPA1 and TRPV1 in ECM homeostasis was studied by comparing the tail motion segments of young (two months old) and mature (seven months old) TRPA1 wild type (WT) and knock-out (KO) mice ($n$ = 5 in each group) as well as young (four months old) and mature (seven months old) TRPV1 KO mice ($n$ = 5 in each group). Anatomically, IVD structures from TRPA1 KO and TRPV1 KO mouse were intact with a distinctive central NP tissue, surrounded by lamella fibers of annulus fibrosus (AF) and sandwiched with cartilaginous endplates. However, FAST staining revealed a depletion of sulfated glycoproteins (Alcian blue) in the NP and a reduction of glycosaminoglycan (GAGs) (Safranin O) in the outer AF and vertebral growth plates of matured TRPA1 KO mice, when compared with TRPA1 WT matured controls. No discernible changes of GAG contents were detected in the young TRPA1 KO mouse IVD (Figure 7). On the contrary, no significant changes in the GAG contents were evidenced in the NP and vertebral growth plates of TRPV1 KO mice (Figure 8). The data suggested a functional importance of TRPA1 in GAG production during IVD maturation.

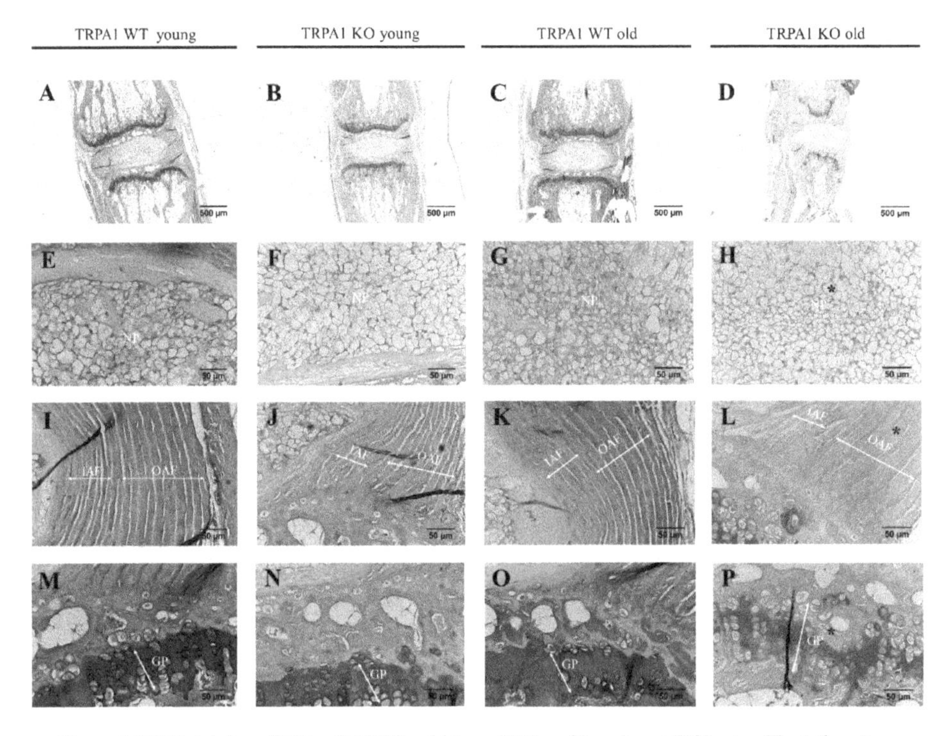


**Figure 7.** FAST staining of IVDs of TRPA1 wild-type (WT) and knock-out (KO) mice. The tail motion segments of TRPA1 young WT (**A**,**E**,**I**,**M**), TRPA1 young KO (**B**,**F**,**J**,**N**), TRPA1 old WT (**C**,**G**,**K**,**O**), and TRPA1 old KO (**D**,**H**,**L**,**P**) mice. The nucleus pulposus: NP (**E**–**H**); inner annulus fibrosus: IAF and outer annulus fibrosus: OAF (**I**–**L**); vertebral growth plate: GP (**M**–**P**) are also shown in higher magnification. Asterisks (*) indicate depletion of glycosaminoglycan deposition in IVD. Scale bars indicate 500 µm in upper panel (**A**–**D**), but 50 µm in lower panels (**E**–**P**).

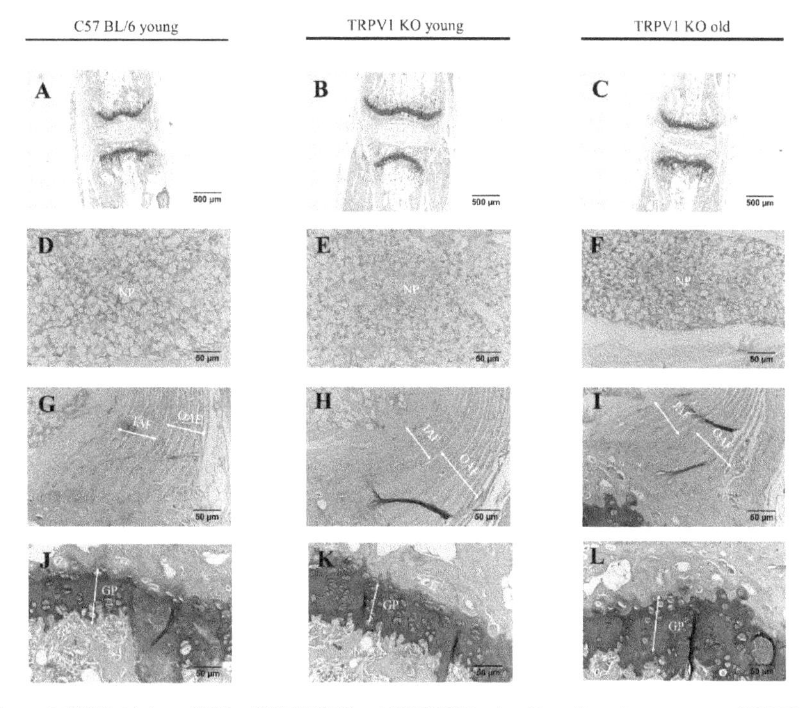


**Figure 8.** FAST staining of IVDs of TRPA1 WT and TRPV1 KO mice. The tail motion segments of C57 BL/6 young WT (**A**,**D**,**G**,**J**), TRPV1 young KO (**B**,**E**,**H**,**K**), and TRPV1 old KO (**C**,**F**,**I**,**L**) mice. The nucleus pulposus: NP (**D**–**F**); inner annulus fibrosus: IAF and outer annulus fibrosus: OAF (**G**–**I**); and, vertebral growth plate: GP (**J**–**L**) are also shown in higher magnification. Scale bars indicate 500 µm in upper panel (**A**–**C**), but 50 µm in lower panels (**D**–**L**).

## 3. Discussion

Several TRP channels are expressed in joints and IVDs, but their potential biological function and therapeutic relevance are not fully understood. The first aim of this study was to investigate the expression of TRP channels in IVD cells in normal and inflammatory conditions, as inflammation is one of the major hallmarks of DDD. We showed that IL-1β significantly induced gene expression of TRPA1 and TRPV4 and reduced TRPC6. TNF-α significantly increased the gene expression of TRPA1 and TRPV2, while reducing TRPV1. It was previously reported that TRPA1 and TRPV1 are commonly expressed in sensory neurons that can innervate joints and IVDs as well as in chondrocytes, where they are associated with degenerative changes [24,27,28].

We found that gene expression of TRPA1 is undetectable in mature healthy human IVDs and untreated cultured IVD cells. Interestingly, TRPA1 was expressed in 20% of degenerated IVDs, possibly due to the presence of pro-inflammatory cytokines. TRPA1 was also expressed in cells isolated from fetal disc tissue and in healthy juvenile samples, which pointed towards its involvement in disc development and/or maturation. FAST staining of tail motion segments of TRPA1 KO and TRPV1 KO mice suggested that TRPA1 might be involved in the homeostasis of GAG maintenance during the development of the IVD. Although these results corresponded to our findings in human IVD tissues/cells, they should be interpreted with caution, as differences between mice and human IVDs exist (e.g., the presence of notochordal cells in mice, degenerative status of mature mouse disc vs. human disc).

To evaluate the possible effects of TRPA1/TRPV1 activation in IVD cells, we used the TRPA1/TRPV1 agonist allyl isothiocyanate (AITC) [29]. AITC (or mustard oil) is commonly regarded as pro-inflammatory and nociceptive [30]. For example, TRPA1-deficient mice do not display acute pain-related behavior after the application of AITC to paws [31]. Our data indicated that the activation of TRPA1 may be the main mechanism for AITC-evoked increase in $[Ca^{2+}]_i$ in IVD cells. In the non-inflamed IVD cells (expressing TRPV1), 10 μM AITC stimulation was not associated with significant pro-inflammatory/catabolic effects, except for an increase in MMP1. In contrast, AITC-mediated regulation of gene expression of ADAMTS5, IL-8, and COL1A1 in cytokine-stimulated cells was likely to be TRPA1 dependent. The TRPA1 downstream effects may depend on agonist concentration, as previously shown by others [32,33] and in our study (reduced IL-8 in cells treated with 3 μM AITC vs. upregulated IL-8 in in cells treated with 10 μM AITC). The expression level of TRPA1 itself can be another reason for the observed differences in IL-8 release between IL-1β and TNF-α treated cells (lower relative TRPA1 expression in IL-1β-stimulated cells vs. TNF-α-stimulated cells). The downregulation of COL1A1 and upregulation of IL-8 in TNF-α, but not IL-1β-treated cells, may be related to lower expression of TRPV1. Altogether, our data suggested that TRPA1 might be involved in the regulation of ECM homeostasis. However, major limitation of this study is low sample number, which prevents definite conclusions.

DDD is considered to be similar to chronic arthritis, due to the fact that common mechanisms are involved in the progression of both diseases [12]. Similar to our findings, the expression of TRPA1 in primary human osteoarthritic (OA) chondrocytes increased upon stimulation with IL-1β, IL-17, LPS, and resistin [24]. Horvath et al. (2016) showed that the markers of chronic arthritis (chronic mechanical hypersensitivity, joint swelling, histopathological alterations, vascular leakage) were significantly reduced in TRPA1 KO mice (vs. wt), which indicated the involvement of TRPA1 in this disease [34]. A similar association of TRPV1 with chronic arthritis was previously demonstrated [35,36]. Interestingly, acute joint pain behaviors were not modified in TRPA1 KO mice [34]. The distinct roles of TRPA1 in chronic vs. acute arthritis could be attributed to a different distribution of TRPA1 (and possibly TRPV1) on sensory nerves and non-neuronal cells in these pathological conditions [34], e.g., due to the presence of pro-inflammatory cytokines. In this context, the modulation of inflammation itself can possibly regulate TRP channel activities (e.g., TRPV1 can be sensitized/desensitized by endogenous products of inflammation [37]). Importantly, pro-inflammatory cytokines (IL-1β and TNF-α) can cause $[Ca^{2+}]_i$ increase in OA chondrocytes [34], but likely not in IVD cells [34], which might be related to differences in the TRP channel expression/activation/function in OA and DDD.

Chronic inflammation in both OA and DDD is associated with neuronal plasticity, which is an important mechanism in the development and maintenance of chronic pain [38]. Our current study did not employ an IVD degeneration/pain model, and thus it did not test the involvement of sensory neurons. Future studies will focus on the interplay between TRPA1/TRPV1 in inflamed IVD cells and DRG neurons, as well as on more specific activation/inhibition experiments by gene editing, both in vitro and in experimental animals.

Although TRPA1/TRPV1 antagonists/agonists have reached clinical trials for the treatment of inflammatory and neuropathic pain [39–41], discrepancies as to whether and how these channels contribute to the underlying mechanisms of inflammatory and neuropathic hypersensitivity can still be found in the literature [42]. Some endogenous ligands of TRPA1 might not yet be discovered and it is still unclear how physiological loading, which is an important parameter in IVD health, regulates the activity of TRPA1. It is likely that TRPA1 activation may be protective under certain circumstances and/or in particular cell types, possibly including the IVD. The protective anti-inflammatory effects of TRPA1 were recently demonstrated in mouse model of colitis [43], with TRPA1 KO mice having a significantly higher 'Disease Activity Index' and levels of pro-inflammatory neuropeptides and cytokines in the distal colon [43]. Another study showed that both the colonic and systemic administration of AITC and capsazepine (another TRPA1 agonist) induced a profound, body-wide TRPA1-mediated desensitization of nociception in mice [44]. The authors suggested that systemic desensitization through TRPA1 might provide a novel strategy for the medicinal treatment of various chronic inflammatory and pain states [44], which possibly included DDD.

Concerning other TRP channels that were regulated by an inflammatory environment in this study, increased TRPV4 expression/signaling in the IVD has been associated with decreased tissue osmolarity and the production of pro-inflammatory cytokines [17]. Our study provided evidence that IL-1β itself can regulate gene expression of TRPV4 in IVD cells., the Gene expression of TRPC6 was shown to be reduced in IVD cells under microgravity [18], but elevated in IVDs with increasing degeneration grade [19]. In our study, TRPC6 was downregulated by IL-1β treatment. To explain these inconsistencies and their pathophysiological relevance, the activity, stability, and subcellular localization of TRPC6 will be investigated in the future. Possibly, the activity of TRPC6 may be regulated by exocytosis [45], while cytoplasmic calcium may influence its expression and degradation [46,47], the levels of which are dysregulated in degenerated IVD cells [48]. To our knowledge, this is the first study that reported the downregulation of TRPV1 and upregulation of TRPV2 by TNF-α in IVD cells. The expression of TRPV2 was shown upregulated in inflamed DRGs [49], where it possibly participated in calcitonin gene-related peptide (CGRP) release [50].

## 4. Materials and Methods

### 4.1. Subjects

#### 4.1.1. Non-Degenerated Human IVD Tissue

cDNA was synthesized from the non-degenerated human IVD cells, (gift provided by Prof. Lisbet Haglund from the Department of Surgery at McGill University, Canada), and prepared as previously described [51]. Informed consent for tissue collection was obtained from family members and the study was approved through the local ethics committee (A04-M53-08B).

#### 4.1.2. Human IVD Tissue

25 human degenerated lumbar IVD samples were used for direct tissue analysis. These biopsies were obtained from 20 donors [mean age = 46.2 years (16–76 years); nine male and eleven female] undergoing elective spinal surgery. IVD samples were intraoperatively separated into annulus fibrosus (AF, $n$ = 11) and nucleus pulposus (NP, $n$ = 14), followed by macroscopic tissue evaluation. The assessment of the disease state was performed using Pfirrmann and Modic grading. Demographic details are shown in Table 3—Tissue and [19]. An additional 30 lumbar degenerated IVDs, removed during surgeries for disc herniation (DH) or degenerative disc disease (DDD), were used for preparation of primary cell cultures. All of the biopsies were obtained with patient's informed consents. The Ethics Committee of Cantons Zurich and Lucerne approved the study (#1007). Demographic details are shown in Table 3—Cells.

**Table 3.** Donors used for tissue and cell culture analyses. DH = herniation, DDD = degenerative disc disease, AF = annulus fibrosus, NP = nucleus pulposus, uk = unknown.

| Tissue | | | | | | | |
|---|---|---|---|---|---|---|---|
| Donor | Age | Gender | Pathology | Tissue | Level | Grade | Experiments |
| T1 | 30 | m | DDD | AF, NP | L4/5 | II | qPCR |
| T2 | 46 | f | DH | AF, NP | L5/S1 | III | qPCR |
| T3 | 34 | m | DH | AF, NP | L5/S1 | III | qPCR |
| T4 | 46 | f | DH | AF, NP | L4-S1 | V | qPCR |
| T5 | 59 | f | DDD | AF, NP | L5/S1 | V | qPCR |
| T6 | 62 | f | DDD | AF | L5/S1 | V | qPCR |
| T7 | 66 | f | DH | AF | L4/5 | II | qPCR |
| T8 | 53 | m | DH | NP | L5/S1 | II | qPCR |
| T9 | 59 | m | DH | NP | L4/L5 | II | qPCR |

**Table 3.** *Cont.*

| Tissue | | | | | | | |
|---|---|---|---|---|---|---|---|
| Donor | Age | Gender | Pathology | Tissue | Level | Grade | Experiments |
| T10 | 52 | m | DDD | NP | L4/L5 | III | qPCR |
| T11 | 64 | f | DDD | NP | L4/L5 | IV | qPCR |
| T12 | 76 | f | DH | NP | L4/L5 | III | qPCR |
| T13 | 16 | f | DH | NP | L4/L5 | III | qPCR |
| T14 | 31 | m | DDD | AF | L4/5 L5/S1 | IV | qPCR |
| T15 | 54 | f | DH | NP | L5/S1 | II | qPCR |
| T16 | 33 | m | DH | AF | L5/S1 | II | qPCR |
| T17 | 70 | f | DDD | NP | L4/5 | IV | qPCR |
| T18 | 39 | m | DH | AF | L5/S1 | III | qPCR |
| T19 | 28 | m | DH | NP | L5/S1 | II | qPCR |
| T20 | 21 | f | DDD | AF | L4/5 | III | qPCR |
| **Cells** | | | | | | | |
| Donor | Age | Gender | Pathology | Tissue | Level | Grade | Experiments |
| C1 | 44 | F | uk | uk | L3/L4 | uk | qPCR |
| C2 | 82 | M | uk | uk | L5/S1 | uk | qPCR |
| C3 | 28 | M | uk | uk | L5/S1 | uk | qPCR |
| C4 | uk | uk | uk | uk | uk | uk | qPCR, ELISA |
| C5 | uk | uk | uk | uk | uk | uk | qPCR, ELISA |
| C6 | 39 | M | DDD, DH | Mix | L4/L5 | IV | ELISA |
| C7 | 58 | M | DDD, DH | Mix | L5/S1 | IV | qPCR, ELISA |
| C8 | 46 | F | DH | - | L5/S1 | IV | qPCR, ELISA |
| C9 | 52 | M | DDD, DH | Mix | L5/S1 | V | qPCR, ELISA |
| C10 | 46 | F | DDD, DH | Mix | L4/L5 | IV | qPCR, Ca imaging |
| C11 | 58 | M | uk | Mix | L4/L5 | IV | Ca imaging |
| C12 | 31 | M | DDD, DH | Mix | L5/S1 | IV | qPCR |
| C13 | 46 | M | DH | Mix | C5/C6 | III | qPCR |
| C14 | 40 | F | DH | Mix | L4/L5 | III | qPCR |
| C15 | 40 | M | DH | NP | L4/L5 | III | qPCR |
| C16 | 66 | F | DH | NP | L4/L5 | III | Ca imaging |
| C17 | uk | uk | uk | uk | uk | uk | qPCR, ELISA |
| C18 | 50 | F | DH | Mix | L5/S1 | IV | qPCR |
| C19 | 42 | M | DH | Mix | L5/S1 | V | qPCR |
| C20 | 68 | F | listhesis | Mix | L4/L5 | III | qPCR |
| C21 | 38 | M | DH | Mix | L5/S1 | III | qPCR, ELISA |
| C22 | 41 | F | DH | Mix | L4/5 | III | qPCR, ELISA |
| C23 | 42 | M | DH | NP | L5/S1 | IV | qPCR, ELISA |
| C24 | 41 | F | DH | Mix | L5/S1 | III | qPCR, ELISA |
| C25 | 45 | M | DH | NP | L4/L5 | IV | qPCR, ELISA |
| C26 | 71 | uk | DDD | Mix | L4/5 | III | qPCR, ELISA |
| C27 | 55 | F | DH | NP | L5/6 | I | qPCR, ELISA |
| C28 | 55 | F | DH | NP | L5/6 | II | Immuno |
| C29 | 55 | M | DH | Mix | L5/S1 | II | Immuno |
| C30 | 58 | M | DH | Mix | L4/5 | IV | Immuno |
| C31 | 34 | M | - | NP, AF | L1/2-L2/3-L3/4 | | qPCR |
| C32 | 55 | F | - | NP, AF | L1/2 | III | qPCR |
| C33 | 52 | M | - | NP | L1-L5 | I | qPCR |
| C34 | 17 | M | - | NP | T12-S1 | I | qPCR |
| LC1 | 16 | M | - | mix | uk | II | qPCR |
| LFC2 | Fetal IVD cells, Male, 16 weeks: p5 | | | | | | qPCR |
| LFC3 | Fetal IVD cells, Male, 14 weeks: p4 | | | | | | qPCR |
| LFC4 | Fetal IVD cells, Male, 14 weeks: p4 | | | | | | qPCR |

#### 4.1.3. Human Fetal IVD Cells

The cell cultures were derived from biopsies that were obtained in accordance with the Ethics Committee of the University Hospital of Lausanne (Ethics Protocol 51/01) and following the Federal Transplantation Program guidelines. The cell banks are managed in the Department Biobank following the regulations of the Biobank for clinical research for both the fetal and adult tissue. Specific biopsies consisting of a spinal unit of three vertebra and two discs representing tissue of 6.5 mm × 5 mm in size were obtained from fetal tissue following the voluntary interruption of pregnancy at 14–16 weeks of gestation. Biopsies were first rinsed in 1% penicillin/strepotomycin in phosphate buffer saline (PBS) and adjacent soft tissue that was delicately dissected from the main disc tissue. We thereafter prepared one IVD and two adjacent vertebrae from a fetal spine unit; primary cultures used Dulbecco's Modified Eagle's medium (DMEM) (41966-029, Gibco, Waltham, MA, USA) that was supplemented with 10% fetal bovine serum and 100 mM L-Glutamine (25030-024, Gibco). Culture conditions were at 37 °C under 5% $CO_2$. One juvenile patient was also used. Cell cultures from the juvenile patient following discectomy were established as above, except type II collagenase digestion, was implemented. Cells were expanded and stored frozen in liquid nitrogen at passage 1 or 2 and the cells were thawed, expanded in monolayer at passage 3, and used for analysis following passage 4. Table 3 shows donor details.

#### 4.1.4. Knock-Out Mice

Tails from C57BL/6 TRPA1 wild-type (WT) and knock out (KO) mice and 15 TRPV1 WT and KO mice were used (Table 4). The mice were divided into two groups: young (two, four months old) and mature (seven months old) and euthanized in the context of other research activities. C57BL/6 mice were use as control for TRPV1 KO as this strain was backcrossed 10× to the C57BL/6 background. Immediately after euthanasia with pentobarbital (100 mg/kg i.p.), the tails were dissected, skinned, rinsed in PBS, and fixed in 4% paraformaldehyde solution (Szkarabeusz Kft., Pecs, Hungary) in 0.1 M phosphate buffer. After two days, the fixed tails were washed with PBS and then placed into 10% EDTA (E6758, Sigma, St. Louis, MO, USA) exchanged every two days. After 10 days, the decalcified samples were washed in PBS and placed into 70% ethanol (51976, Sigma) at 4 °C until paraffin embedding. Paraffin blocks were then used to prepare 5 µm sections.

**Table 4.** Mouse spines used for FAST staining.

| Mice | 2 Months Old | 4 Months Old | 7 Months Old |
|---|---|---|---|
| TRPA1 WT | $n = 5$ | - | $n = 5$ |
| TRPA1 KO | $n = 5$ | - | $n = 5$ |
| TRPV1 KO | - | $n = 5$ | $n = 5$ |
| C57BL/6 | - | $n = 5$ | - |

### 4.2. Cell Culture

#### 4.2.1. 2D Cell Culture

IVD tissue was diced to around 1 $mm^3$ pieces and treated with a mixture of 0.3% dispase (04942078001, Roche Diagnostics, Mannheim, Germany) and 0.2% collagenase (17454, SERVA Electrophoresis, Heidelberg, Germany) in PBS at 37 °C for 4–8 h to isolate the cells. After the incubation, the suspension was filtered through a 70 µm cell strainer (542070, Greiner Bio-One, Kremsmünster, Austria), centrifuged at 700× $g$ for 5 min, and resuspended in DMEM/F12 (31330-038, Thermo Fisher Scientific, Waltham, MA, USA) supplemented with 10% fetal calf serum (FCS, F7524, Merck, Darmstadt, Germany), 100 units/mL penicillin, 100 µg/mL streptomycin, and 250 ng/mL amphotericin B (15240-062, Gibco, Carlsbad, CA, USA). The primary IVD cells were expanded in monolayer in a standard cell culture incubator (5% $CO_2$, 37 °C) up to passage 3 before being used in the experiments.

4.2.2. 3D Cell Culture

The IVD cells were detached using 1.5% trypsin (15090-046, Thermo Fisher Scientific) and seeded in 1.2% alginate (71238-50G, Sigma, St. Louis, MO, USA) at a density of 4 × $10^6$ IVD cells per 1 mL alginate, as described previously [52]. Briefly, the cells-alginate mixture was dropped into 102 mM calcium chloride solution (1.02382, Merck) while using a sterile syringe and needle and left for 8 min to polymerize under gentle stirring until beads were formed. After washing with 0.9% NaCl (1.06404, Merck) and PBS (3 × 1 min), the beads were transferred into six well plates and pre-cultured for seven days.

4.2.3. Cell Viability of 3D Cell Culture

Cell viability in the alginate beads was tested using calcein/ethidium homodimer staining. Staining solution was prepared by mixing culture media with 2 µM ethidium homodimer (46043, Sigma) and 2 µM calcein-AM (17783, Sigma). 200 µL/well of the staining solution was added into a 96-well plate containing beads (one bead per culture condition per well) and then incubated for 30 min. Subsequently, the beads were gently squeezed between cover slips and three photos were randomly captured with a fluorescence microscope (Olympus IX51, Tokyo, Japan) at the wavelength of 515 nm (calcein: living cells) and 620 nm (ethidium: dead cells). The cell numbers in each image were counted by ImageJ ver.1.51j8 and averaged. Cell viability is shown as the number of living cells per total cells.

*4.3. Treatments*

4.3.1. 2D Cell Cultures

Table 5 shows all treatments. Experiments were conducted in culture media without antibiotics and FCS (= experimental media). For gene expression analysis and ELISA, the IVD cells were seeded into T25 flasks (3.5 × $10^5$ cells/flask) or six-well plates (3 × $10^5$ cells/well). For immunofluorescence, 1 × $10^5$ cells were seeded into the wells of chambered slides (155380, Thermo Fisher Scientific). For Calcium imaging, the 4 × $10^4$ cells were seeded into 96-well plates in triplicates and incubated for 18 h. The next day, complete media was changed to the experimental media. After 2 h in experimental media, the cells were exposed to 5 or 10 ng/mL recombinant TNF-α (315-01A, PeproTech, Umkirch, Germany) or IL-1β (200-01B, PeproTech) for 18 h. Non-stimulated cells were used as the controls. To investigate the effects of TRPA1 and TRPV1 activation, 3 and 10 µM allyl isothiocyanate (AITC, TRPA1 agonist, 377430, Sigma) was used either in untreated cells or in IL1-β and TNF-α treated cells (focus on TRPA1, whose expression is increased in an inflammatory environment). The EC50 value for AITC that is reported in the literature is approximately 3 µM, while AITC concentrations higher than 10 µM may cause channel desensitization [32,33].

4.3.2. 3D Cell Cultures

The experiments were conducted in culture media without antibiotics and with FCS. On day 7, cells in alginate beads were stimulated with 5 ng/mL IL-1β and collected after 24 h (day 1), eight days (day 8), and 15 days (day 15) (Table 5). Culture media for the latter group was exchanged on day 8, with new 5 ng/mL IL-1β. At the end of the experiment, the cells were liberated from the beads in 1.9 mL of 55 mM sodium citrate solution (71406, Sigma) and centrifuged at 700× *g* for 5 min. Cell pellets were used for subsequent analyses.

**Table 5.** Details of cell culture treatments.

| Compound | Catalog Number | Function | Concentration | Exposure Time | Experiment |
| --- | --- | --- | --- | --- | --- |
| TNF-α | 315-01A PeproTech | Inflammatory cytokine | 5, 10 ng/mL | 18 h | qPCR Ca imaging |
| IL-1β | 200-01B PeproTech | Inflammatory cytokine | 5, 10 ng/mL | 18 h | qPCR Ca imaging |
| | | | 5 ng/mL | 1, 8, 15 days | qPCR (from 3D) |
| | | | 5 ng/mL | 18 h | Immuno |
| AITC | 377430 Sigma | TRPA1 agonist | 3 μM, 10 μM | 18 h | qPCR, ELISA |
| | | | 100 μM | during the measurement | Ca imaging |

### 4.4. Analyses

#### 4.4.1. Gene Expression Analysis of IVD Tissue

RNA extraction from IVD tissue and the following steps were performed according to [19]. For cDNA synthesis, two micrograms of RNA were used in a total volume of 60 μL, while using the reverse transcription kit (4374966, Applied Biosystems, Foster City, CA, USA). For samples with lower yields, the reverse transcription was conducted at reduced concentrations. cDNA (10 ng/well) was mixed with TaqMan Fast Universal PCR Master Mix and TaqMan primers (Table 6) to quantify gene expression. The obtained $C_t$ values were analyzed by a comparative method and displayed as $2^{-dCt}$ values, with GAPDH as housekeeping gene.

#### 4.4.2. FAST Staining

To study the glycosaminoglycan (GAG) contents in IVD, a multi-dye histological staining using Fast green, Alcian blue, Saffranin-O, and tartrazine was performed on IVD tissue sections accordingly [53]. In brief, parafilm embedded tissue sections were first dewaxed in xylene and then rehydrated in a stepwise gradient of ethanol. The IVD sections were first stained with 1% Alcian blue 8GX (A3157, Sigma) pH 1.0 for 1 min, followed by 0.1% Saffranin-O (S8884, Sigma) for 3 min. Saffranin-O reddish colour differentiation was performed in 25% ethanol for 15 s. The tissue sections were then stained in 0.08% Tartrazine (T0388, Sigma) with 0.25% acetic acid for 45 s and finally counterstained by 0.01% Fast green (F7258, Sigma) solution for 5 min. Sections were finally air-dried, mounted in DePeX (BDH Laboratory; Poole, UK), and examined under a Nikon Eclipse 80i microscope (Tokyo, Japan).

#### 4.4.3. Gene Expression Analysis of IVD Cells

RT-qPCR was performed to analyze the expression of target genes (Table 6). The cells were lysed with 1 mL Trizol (Genezol, GZR200, Geneaid biotech, New Taipei City, Taiwan) and RNA was isolated according to the manufacturer's recommendations. Briefly, after adding chloroform (132950, Sigma), the samples were mixed well for 15 s, left for 5 min at RT, and centrifuged (12,000× $g$ for 15 min). Supernatants were carefully transferred to new RNase free tubes, 500 μL of isopropanol (20842, VWR chemicals, Fontenay-sous-Bois, France) was added and mixed well. After 5 min, the samples were centrifuged again (12,000× $g$ for 15 min). The pellets were washed with 75% ethanol (1.00983, Merck) at 7500× $g$ for 5 min and then dissolved in RNase free water (10977, Invitrogen, Carlsbad, CA, USA). The purity and concentration of the resulting RNA were measured using the NanoDrop (ND-1000, Thermo Fisher Scientific). 1 μg of total RNA was reverse-transcribed to cDNA using a reverse transcription kit (4374966, Applied Biosystems, Foster City, CA, USA). qPCR of the mixture of primers/probes (Table 6) and master mix (4367846, Applied Biosystems) was performed on the CFX96 Real-Time System (Bio-Rad Laboratories, Hercules, CA, USA). The amplification program was as follows: heating at 95 °C for 10 min; 40 cycles of 95 °C for 1 s and 60 °C for 20 s. The relative expression level was calculated by the dd$C_t$ method. For normalization purposes, the samples with undetectable expression were assigned $C_t$ value 40. The results are shown as fold change, relative to control or relative to cytokine treatment.

**Table 6.** Target genes and assay identification (ID) numbers of corresponding TaqMan primers (TaqMan Gene Expression Assays; Thermo Fisher Scientific). TRP = transient receptor potential; MMP = matrix metalloproteinase; TIMP = tissue inhibitor of matrix metalloproteinase; ADAMTS = a disintegrin and metalloproteinase with thrombospondin motifs; COX-2 = cyclooxygenase-2; NGF = nerve growth factor. IL-6 = interleukin 6; IL-8 = interleukin 8; HKG = housekeeping gene.

| Target Gene | Assay ID | Putative Association with Inflammation |
|---|---|---|
| TBP | Hs00427620_m1 | HKG in the cell culture study and fetal cells |
| GAPDH | Hs02758991_g1 | HKG in the tissue study |
| TRPA1 | Hs00175798_m1 | inflammatory pain [12] |
| TRPC1 | Hs00608195_m1 | bladder inflammation (neuronal) [54] |
| TRPC3 | Hs00162985_m1 | inflammatory pain [55] |
| TRPC6 | Hs00988479_m1 | IVD inflammation (putative) [18,19] |
| TRPV1 | Hs00218912_m1 | neuroinflammation [20] |
| TRPV2 | Hs00901648_m1 | inflammation in DRG [49] |
| TRPV4 | Hs01099348_m1 | lung inflammation [56] |
| TRPV6 | Hs00367960_m1 | association with TNF-α [57] |
| TRPM2 | Hs01066091_m1 | inflammatory and neuropathic pain [58] |
| TRPM7 | Hs00559080_m1 | inflammation in colitis [59] |
| IL-6 | Hs00174131_m1 | inflammation mediator [60] |
| IL-8 | Hs00174103_m1 | inflammation mediator [60] |
| MMP1 | Hs00233958_m1 | cleaves mainly collagens (I, II, III) [61] |
| MMP3 | Hs00968305_m1 | cleaves proteoglycans and collagens (II, III) [61] |
| ADAMTS4 | Hs00192708_m1 | cleaves mainly aggrecan [61] |
| ADAMTS5 | Hs01095518_m1 | cleaves mainly aggrecan [61] |
| TIMP1 | Hs00234278_m1 | inhibits MMPs (1, 3) and ADAMTS (4) [61] |
| TIMP2 | Hs01092512_g1 | inhibits all MMPs [61] |
| COX-2 | Hs00153133_m1 | pain mediator [62] |
| NGF | Hs00171458_m1 | nerve ingrowth [38] |
| COL2A1 | Hs00264051_m1 | ECM constituent |
| COL1A1 | Hs00164004_m1 | ECM constituent |
| Aggrecan | Hs00153936_m1 | ECM constituent |

#### 4.4.4. Immunofluorescence

The cells were seeded into the wells of chambered slides, washed with PBS, fixed with ice cold methanol (10 min), and blocked with 5% normal goat serum in PBS (1 h at RT). A primary antibody recognizing the N-terminus of the human TRPA1 protein (NB110-40763, Novus Biologicals) was diluted in 1% normal goat serum in PBS (1:200) (G9023, Sigma) and applied 1h under agitation at RT. Cells without primary antibody were used as immunospecificity control. After washing with PBS (3 × 5 min), a secondary antibody that was diluted in 1% normal goat serum (1:200) (Cy2 anti-rabbit IgG, 111-225-144, Jackson Immuno Research) was applied for 1 h at RT under agitation. Next, cells were washed with PBS (3 × 5 min), coverslipped with 1–2 drops of antifade mounting medium with DAPI (VECTASHIELD; H-1200), and imaged with a fluorescence microscope (Olympus IX51).

#### 4.4.5. Enzyme-Linked Immunosorbent Assay (ELISA)

To quantify the release of inflammatory markers from IVD cells, the cell culture medium was collected and analyzed with IL-6 and IL-8 ELISA kit, according to the manufacturer's protocol (IL-6 555220, IL-8 555244, BD Biosciences, San Jose, CA, USA). 96-well plates were coated with capture antibody overnight. After washing, the wells were blocked in assay diluent, washed, loaded with samples or human recombinant IL-6 or IL-8 protein standards, and incubated for 2 h at RT. After washing, detection antibody and streptavidin-horseradish peroxidase (HRP) were applied for 1 h. Next, the plates were washed and substrate solution was added. After 30 min, the reaction was stopped by kit stop solution and the absorbance was measured at 450 nm, with 570 nm correction. The IL-6 and IL-8 concentrations were calculated based on the standard curve. IL-6 and IL-8 in culture media are shown relative to the cytokine treatment.

#### 4.4.6. $[Ca^{2+}]_i$ Imaging

Fura-2 QBT™ Calcium Kit was used to measure the increase in intracellular calcium (R8197, Molecular Devises, San Jose, CA, USA). Briefly, culture media was replaced with Fura-2-AM kit solution and the cells were incubated for 1 h. Basal Fura-2 fluorescence was recorded for 5 min using a plate reader (infinite M200 pro, Tecan, Männedorf, Switzerland) at an excitation wavelength of 340 and 380 nm and an emission wavelength of 510 nm. After five cycles, the cells were exposed in 100 μM of AITC (to activate disc cells within a short timeline) and the measurement was continued. Ionomycin (13909, Sigma) was used as a positive control for channel stimulation. Data is shown as the ratio of 340/380 wavelengths.

#### 4.4.7. Statistical Analysis

Statistical analysis was performed in GraphPad Prism 8.0.0. Groups were compared using the Kruskal–Wallis nonparametric test followed by Dunn's multiple comparison test. Data is shown as mean ± SD. $p < 0.05$ was considered to be statistically significant (* $p < 0.05$, ** $p < 0.01$, *** $p < 0.001$).

## 5. Conclusions

To our knowledge, this is the first study that demonstrated a cytokine-dependent increase in the gene expression of TRPA1, TRPV2, and TRPV4 and a decrease in the gene expression of TRPC6 and TRPV1 in human IVD cells. Although TRPA1 and TRPV1 are commonly associated with inflammatory pain, their activation in inflamed IVD cells did not have profound pro-inflammatory and catabolic effects. Instead, TRPA1 expression and activation was associated with ECM metabolism. Future studies will use targeted gene editing techniques to elucidate the exact role of TRPA1/TRPV1 in DDD.

**Supplementary Materials:** Supplementary materials can be found at http://www.mdpi.com/1422-0067/20/7/1767/s1.

**Author Contributions:** T.K. performed cell culture experiments, calcium flux analysis and drafted part of the manuscript. J.Z. performed cell culture experiments. M.V. performed cell culture experiments. A.S. performed tissue analyses. W.K.T. and V.Y.L. performed and evaluated FAST staining. K.B. and Z.H. secured the mouse experiments. L.A.A. provided fetal IVD cell lysates. O.N.H. and J.K. provided human IVD samples and clinical expertise. O.K. performed cell culture experiments, supervised the study and drafted parts of the manuscript. K.W.-K. supervised the study and conceived funding. The authors declare no conflict of interest.

**Acknowledgments:** This study was funded by the Swiss National Science Foundation (SNF PP00P2_163678/1), the Spine Society of Europe (Eurospine 2016_4), and Hungarian grants GINOP-2.3.2.-15-2016-00048 "Stay Alive" and EFOP 3.6.2. "Live longer". We thank H. Greutert for technical assistance in the study.

**Conflicts of Interest:** The authors declare no conflict of interest.

## References


1. Pai, S.; Sundaram, L.J. Low back pain: An economic assessment in the United States. *Orthop. Clin. N. Am.* **2004**, *35*, 1–5. [CrossRef]
2. Ito, K.; Creemers, L. Mechanisms of intervertebral disk degeneration/injury and pain: A review. *Glob. Spine J.* **2013**, *3*, 145–152. [CrossRef] [PubMed]
3. Wuertz, K.; Vo, N.; Kletsas, D.; Boos, N. Inflammatory and catabolic signalling in intervertebral discs: The roles of NF-kappaB and MAP kinases. *Eur. Cell. Mater.* **2012**, *23*, 103–119. [CrossRef] [PubMed]
4. Vo, N.V.; Hartman, R.A.; Patil, P.R.; Risbud, M.V.; Kletsas, D.; Iatridis, J.C.; Hoyland, J.A.; Le Maitre, C.L.; Sowa, G.A.; Kang, J.D. Molecular mechanisms of biological aging in intervertebral discs. *J. Orthop. Res.* **2016**, *34*, 1289–1306. [CrossRef]
5. Johnson, Z.I.; Schoepflin, Z.R.; Choi, H.; Shapiro, I.M.; Risbud, M.V. Disc in Flames: Roles of Tnf-alpha AND IL-1 beta in Intervertebral Disc Degeneration. *Eur. Cell. Mater.* **2015**, *30*, 104–117. [CrossRef]
6. Zhang, J.M.; An, J. Cytokines, inflammation, and pain. *Int. Anesthesiol. Clin.* **2007**, *45*, 27–37. [CrossRef]
7. Kepler, C.K.; Markova, D.Z.; Hilibrand, A.S.; Vaccaro, A.R.; Risbud, M.V.; Albert, T.J.; Anderson, D.G. Substance P stimulates production of inflammatory cytokines in human disc cells. *Spine* **2013**, *38*, E1291–E1299. [CrossRef] [PubMed]
8. Freemont, A.J.; Peacock, T.E.; Goupille, P.; Hoyland, J.A.; OBrien, J.; Jayson, M.I.V. Nerve ingrowth into diseased intervertebral disc in chronic back pain. *Lancet* **1997**, *350*, 178–181. [CrossRef]
9. Stemkowski, P.L.; Noh, M.C.; Chen, Y.S.; Smith, P.A. Increased excitability of medium-sized dorsal root ganglion neurons by prolonged interleukin-1 exposure is $K^+$ channel dependent and reversible. *J. Physiol.* **2015**, *593*, 3739–3755. [CrossRef]
10. Wilkinson, M.F.; Earle, M.L.; Triggle, C.R.; Barnes, S. Interleukin-1beta, tumor necrosis factor-alpha, and LPS enhance calcium channel current in isolated vascular smooth muscle cells of rat tail artery. *FASEB J.* **1996**, *10*, 785–791. [CrossRef]
11. Gavenis, K.; Schumacher, C.; Schneider, U.; Eisfeld, J.; Mollenhauer, J.; Schmidt-Rohlfing, B. Expression of ion channels of the TRP family in articular chondrocytes from osteoarthritic patients: Changes between native and in vitro propagated chondrocytes. *Mol. Cell. Biochem.* **2009**, *321*, 135–143. [CrossRef] [PubMed]
12. Krupkova, O.; Zvick, J.; Wuertz-Kozak, K. The role of transient receptor potential channels in joint diseases. *Eur. Cell. Mater.* **2017**, *34*, 180–201. [CrossRef] [PubMed]
13. Patapoutian, A.; Tate, S.; Woolf, C.J. Transient receptor potential channels: Targeting pain at the source. *Nat. Rev. Drug Discov.* **2009**, *8*, 55–68. [CrossRef]
14. Wu, L.J.; Sweet, T.B.; Clapham, D.E. International Union of Basic and Clinical Pharmacology. LXXVI. Current progress in the mammalian TRP ion channel family. *Pharmacol. Rev.* **2010**, *62*, 381–404. [CrossRef]
15. Gees, M.; Colsoul, B.; Nilius, B. The Role of Transient Receptor Potential Cation Channels in $Ca^{2+}$ Signaling. *Cold Spring Harb. Perspect. Biol.* **2010**, *2*, a003962. [CrossRef] [PubMed]
16. Montell, C.; Birnbaumer, L.; Flockerzi, V.; Bindels, R.J.; Brudorf, E.A.; Caterina, M.J.; Clapham, D.E.; Harteneck, C.; Heller, S.; Julius, D.; et al. A unified nomenclature for the superfamily of TRP cation channels. *Mol. Cell* **2002**, *9*, 229–231. [CrossRef]
17. Walter, B.A.; Purmessur, D.; Moon, A.; Occhiogrosso, J.; Laudier, D.M.; Hecht, A.C.; Iatridis, J.C. Reduced tissue osmolarity increases TRPV4 expression and pro-inflammatory cytokines in intervertebral disc cells. *Eur. Cell. Mater.* **2016**, *32*, 123–136. [CrossRef]
18. Franco-Obregon, A.; Cambria, E.; Greutert, H.; Wernas, T.; Hitzl, W.; Egli, M.; Sekiguchi, M.; Boos, N.; Hausmann, O.; Ferguson, S.J.; et al. TRPC6 in simulated microgravity of intervertebral disc cells. *Eur. Spine J.* **2018**, *27*, 2621–2630. [CrossRef]
19. Sadowska, A.; Touli, E.; Hitzl, W.; Greutert, H.; Ferguson, S.J.; Wuertz-Kozak, K.; Hausmann, O.N. Inflammaging in cervical and lumbar degenerated intervertebral discs: Analysis of proinflammatory cytokine and TRP channel expression. *Eur. Spine J.* **2018**, *27*, 564–577. [CrossRef]
20. Marrone, M.C.; Morabito, A.; Giustizieri, M.; Chiurchiu, V.; Leuti, A.; Mattioli, M.; Marinelli, S.; Riganti, L.; Lombardi, M.; Murana, E.; et al. TRPV1 channels are critical brain inflammation detectors and neuropathic pain biomarkers in mice. *Nat. Commun.* **2017**, *8*, 15292. [CrossRef]
21. Fernandes, E.S.; Fernandes, M.A.; Keeble, J.E. The functions of TRPA1 and TRPV1: Moving away from sensory nerves. *Br. J. Pharmacol.* **2012**, *166*, 510–521. [CrossRef]

22. Kelly, S.; Chapman, R.J.; Woodhams, S.; Sagar, D.R.; Turner, J.; Burston, J.J.; Bullock, C.; Paton, K.; Huang, J.; Wong, A.; et al. Increased function of pronociceptive TRPV1 at the level of the joint in a rat model of osteoarthritis pain. *Ann. Rheum. Dis.* **2015**, *74*, 252–259. [CrossRef]
23. Yu, L.; Yang, F.; Luo, H.; Liu, F.Y.; Han, J.S.; Xing, G.G.; Wan, Y. The role of TRPV1 in different subtypes of dorsal root ganglion neurons in rat chronic inflammatory nociception induced by complete Freund's adjuvant. *Mol. Pain* **2008**, *4*, 61. [CrossRef]
24. Nummenmaa, E.; Hamalainen, M.; Moilanen, L.J.; Paukkeri, E.L.; Nieminen, R.M.; Moilanen, T.; Vuolteenaho, K.; Moilanen, E. Transient receptor potential ankyrin 1 (TRPA1) is functionally expressed in primary human osteoarthritic chondrocytes. *Arthr. Res. Ther.* **2016**, *18*, 185. [CrossRef]
25. Gouin, O.; L'Herondelle, K.; Lebonvallet, N.; Le Gall-Ianotto, C.; Sakka, M.; Buhe, V.; Plee-Gautier, E.; Carre, J.L.; Lefeuvre, L.; Misery, L.; et al. TRPV1 and TRPA1 in cutaneous neurogenic and chronic inflammation: Pro-inflammatory response induced by their activation and their sensitization. *Protein Cell* **2017**, *8*, 644–661. [CrossRef]
26. Sadowska, A.; Hitzl, W.; Jaszczuk, P.; Cherif, H.; Haglund, L.; Hausmann, O.; Wuertz-Kozak, K. Differential regulation of TRP channel gene expression by intervertebral disc degeneration and back pain. *Sci. Rep.* **2018**. in revision.
27. Flegel, C.; Schobel, N.; Altmuller, J.; Becker, C.; Tannapfel, A.; Hatt, H.; Gisselmann, G. RNA-Seq Analysis of Human Trigeminal and Dorsal Root Ganglia with a Focus on Chemoreceptors. *PLoS ONE* **2015**, *10*, e0128951. [CrossRef]
28. Asmar, A.; Barrett-Jolley, R.; Werner, A.; Kelly, R., Jr.; Stacey, M. Membrane channel gene expression in human costal and articular chondrocytes. *Organogenesis* **2016**, *12*, 94–107. [CrossRef]
29. Gees, M.; Alpizar, Y.A.; Boonen, B.; Sanchez, A.; Everaerts, W.; Segal, A.; Xue, F.; Janssens, A.; Owsianik, G.; Nilius, B.; et al. Mechanisms of transient receptor potential vanilloid 1 activation and sensitization by allyl isothiocyanate. *Mol. Pharmacol.* **2013**, *84*, 325–334. [CrossRef]
30. Everaerts, W.; Gees, M.; Alpizar, Y.A.; Farre, R.; Leten, C.; Apetrei, A.; Dewachter, I.; van Leuven, F.; Vennekens, R.; de Ridder, D.; et al. The Capsaicin Receptor TRPV1 Is a Crucial Mediator of the Noxious Effects of Mustard Oil. *Curr. Biol.* **2011**, *21*, 316–321. [CrossRef]
31. Guimaraes, M.Z.P.; Jordt, S.E. TRPA1: A Sensory Channel of Many Talents. In *TRP Ion Channel Function in Sensory Transduction and Cellular Signaling Cascades*; Liedtke, W.B., Heller, S., Eds.; CRC Press: Boca Raton, FL, USA, 2007; pp. 151–160.
32. The Electrophysiology Team at Nanion Technologies GmbH, Munich. TRPA1 Activation by Allyl Isothiocyanate Recorded on the Port-a-Patch®. Available online: https://www.nanion.de/en/products/port-a-patch/137-articles/413-trpa1-trpa1-activation-by-allyl-isothiocyanate-recorded-on-the-port-a-patch.html (accessed on 8 April 2019).
33. McNamara, C.R.; Mandel-Brehm, J.; Bautista, D.M.; Siemens, J.; Deranian, K.L.; Zhao, M.; Hayward, N.J.; Chong, J.A.; Julius, D.; Moran, M.M.; et al. TRPA1 mediates formalin-induced pain. *Proc. Natl. Acad. Sci. USA* **2007**, *104*, 13525–13530. [CrossRef]
34. Horvath, A.; Tekus, V.; Boros, M.; Pozsgai, G.; Botz, B.; Borbely, E.; Szolcsanyi, J.; Pinter, E.; Helyes, Z. Transient receptor potential ankyrin 1 (TRPA1) receptor is involved in chronic arthritis: In vivo study using TRPA1-deficient mice. *Arthr. Res. Ther.* **2016**, *18*, 6. [CrossRef]
35. Barton, N.J.; McQueen, D.S.; Thomson, D.; Gauldie, S.D.; Wilson, A.W.; Salter, D.M.; Chessell, I.P. Attenuation of experimental arthritis in TRPV1R knockout mice. *Exp. Mol. Pathol.* **2006**, *81*, 166–170. [CrossRef]
36. Szabo, A.; Helyes, Z.; Sandor, K.; Bite, A.; Pinter, E.; Nemeth, J.; Banvolgyi, A.; Bolcskei, K.; Elekes, K.; Szolcsanyi, J. Role of transient receptor potential vanilloid 1 receptors in adjuvant-induced chronic arthritis: In vivo study using gene-deficient mice. *J. Pharmacol. Exp. Ther.* **2005**, *314*, 111–119. [CrossRef]
37. Schumacher, M.A. Transient Receptor Potential Channels in Pain and Inflammation: Therapeutic Opportunities. *Pain Pract.* **2010**, *10*, 185–200. [CrossRef]
38. Aoki, Y.; Nakajima, A.; Ohtori, S.; Takahashi, H.; Watanabe, F.; Sonobe, M.; Terajima, F.; Saito, M.; Takahashi, K.; Toyone, T.; et al. Increase of nerve growth factor levels in the human herniated intervertebral disc: Can annular rupture trigger discogenic back pain? *Arthr. Res. Ther.* **2014**, *16*, R159. [CrossRef]
39. Chen, J.; Hackos, D.H. TRPA1 as a drug target—Promise and challenges. *Naunyn Schmiedebergs Arch. Pharmacol.* **2015**, *388*, 451–463. [CrossRef]

40. De Petrocellis, L.; Moriello, A.S. Modulation of the TRPV1 channel: Current clinical trials and recent patents with focus on neurological conditions. *Recent Pat. CNS Drug Discov.* **2013**, *8*, 180–204. [CrossRef]
41. Botz, B.; Bolcskei, K.; Helyes, Z. Challenges to develop novel anti-inflammatory and analgesic drugs. *Wiley Interdiscip. Rev. Nanomed. Nanobiotechnol.* **2017**, *9*, e1427. [CrossRef]
42. Lehto, S.G.; Weyer, A.D.; Youngblood, B.D.; Zhang, M.; Yin, R.; Wang, W.; Teffera, Y.; Cooke, M.; Stucky, C.L.; Schenkel, L.; et al. Selective antagonism of TRPA1 produces limited efficacy in models of inflammatory- and neuropathic-induced mechanical hypersensitivity in rats. *Mol. Pain* **2016**, *12*. [CrossRef]
43. Kun, J.; Szitter, I.; Kemeny, A.; Perkecz, A.; Kereskai, L.; Pohoczky, K.; Vincze, A.; Godi, S.; Szabo, I.; Szolcsanyi, J.; et al. Upregulation of the transient receptor potential ankyrin 1 ion channel in the inflamed human and mouse colon and its protective roles. *PLoS ONE* **2014**, *9*, e108164. [CrossRef] [PubMed]
44. Kistner, K.; Siklosi, N.; Babes, A.; Khalil, M.; Selescu, T.; Zimmermann, K.; Wirtz, S.; Becker, C.; Neurath, M.F.; Reeh, P.W.; et al. Systemic desensitization through TRPA1 channels by capsazepine and mustard oil—A novel strategy against inflammation and pain. *Sci. Rep.* **2016**, *6*, 28621. [CrossRef] [PubMed]
45. Cayouette, S.; Lussier, M.P.; Mathieu, E.L.; Bousquet, S.M.; Boulay, G. Exocytotic insertion of TRPC6 channel into the plasma membrane upon G(q) protein-coupled receptor activation. *J. Biol. Chem.* **2004**, *279*, 7241–7246. [CrossRef]
46. Qu, Z.W.; Wang, Y.Q.; Li, X.; Wu, L.; Wang, Y.Z. TRPC6 expression in neurons is differentially regulated by NR2A-and NR2B-containing NMDA receptors. *J. Neurochem.* **2017**, *143*, 282–293. [CrossRef] [PubMed]
47. Xu, L.; Chen, Y.Q.; Yang, K.; Wang, Y.F.; Tian, L.C.; Zhang, J.; Wang, E.W.; Sun, D.J.; Lu, W.J.; Wang, J. Chronic Hypoxia Increases TRPC6 Expression and Basal Intracellular $Ca^{2+}$ Concentration in Rat Distal Pulmonary Venous Smooth Muscle. *PLoS ONE* **2014**, *9*, e112007. [CrossRef] [PubMed]
48. Pritchard, S.; Erickson, G.R.; Guilak, F. Hyperosmotically induced volume change and calcium signaling in intervertebral disk cells: The role of the actin cytoskeleton. *Biophys. J.* **2002**, *83*, 2502–2510. [CrossRef]
49. Shimosato, G.; Amaya, F.; Ueda, M.; Tanaka, Y.; Decosterd, I.; Tanaka, M. Peripheral inflammation induces up-regulation of TRPV2 expression in rat DRG. *Pain* **2005**, *119*, 225–232. [CrossRef] [PubMed]
50. Qin, N.; Neeper, M.P.; Liu, Y.; Hutchinson, T.L.; Lubin, M.L.; Flores, C.M. TRPV2 is activated by cannabidiol and mediates CGRP release in cultured rat dorsal root ganglion neurons. *J. Neurosci.* **2008**, *28*, 6231–6238. [CrossRef] [PubMed]
51. Krock, E.; Rosenzweig, D.H.; Currie, J.B.; Bisson, D.G.; Ouellet, J.A.; Haglund, L. Toll-like Receptor Activation Induces Degeneration of Human Intervertebral Discs. *Sci. Rep.* **2017**, *7*, 17184. [CrossRef]
52. Krupkova, O.; Sekiguchi, M.; Klasen, J.; Hausmann, O.; Konno, S.; Ferguson, S.J.; Wuertz-Kozak, K. Epigallocatechin 3-gallate suppresses interleukin-1beta-induced inflammatory responses in intervertebral disc cells in vitro and reduces radiculopathic pain in rats. *Eur. Cell. Mater.* **2014**, *28*, 372–386. [CrossRef] [PubMed]
53. Leung, V.Y.; Chan, W.C.; Hung, S.C.; Cheung, K.M.; Chan, D. Matrix remodeling during intervertebral disc growth and degeneration detected by multichromatic FAST staining. *J. Histochem. Cytochem.* **2009**, *57*, 249–256. [CrossRef]
54. Boudes, M.; Uvin, P.; Pinto, S.; Freichel, M.; Birnbaumer, L.; Voets, T.; de Ridder, D.; Vennekens, R. Crucial role of TRPC1 and TRPC4 in cystitis-induced neuronal sprouting and bladder overactivity. *PLoS ONE* **2013**, *8*, e69550. [CrossRef]
55. Alkhani, H.; Ase, A.R.; Grant, R.; O'Donnell, D.; Groschner, K.; Seguela, P. Contribution of TRPC3 to store-operated calcium entry and inflammatory transductions in primary nociceptors. *Mol. Pain* **2014**, *10*, 43. [CrossRef]
56. Pairet, N.; Mang, S.; Fois, G.; Keck, M.; Kuhnbach, M.; Gindele, J.; Frick, M.; Dietl, P.; Lamb, D.J. TRPV4 inhibition attenuates stretch-induced inflammatory cellular responses and lung barrier dysfunction during mechanical ventilation. *PLoS ONE* **2018**, *13*, e0196055. [CrossRef]
57. Hummel, D.M.; Fetahu, I.S.; Groschel, C.; Manhardt, T.; Kallay, E. Role of proinflammatory cytokines on expression of vitamin D metabolism and target genes in colon cancer cells. *J. Steroid Biochem. Mol. Biol.* **2014**, *144 Pt A*, 91–95. [CrossRef]
58. Haraguchi, K.; Kawamoto, A.; Isami, K.; Maeda, S.; Kusano, A.; Asakura, K.; Shirakawa, H.; Mori, Y.; Nakagawa, T.; Kaneko, S. TRPM2 Contributes to Inflammatory and Neuropathic Pain through the Aggravation of Pronociceptive Inflammatory Responses in Mice. *J. Neurosci.* **2012**, *32*, 3931–3941. [CrossRef]

59. Ramachandran, R.; Hyun, E.; Zhao, L.N.; Lapointe, T.K.; Chapman, K.; Hirota, C.L.; Ghosh, S.; McKemy, D.D.; Vergnolle, N.; Beck, P.L.; et al. TRPM8 activation attenuates inflammatory responses in mouse models of colitis. *Proc. Natl. Acad. Sci. USA* **2013**, *110*, 7476–7481. [CrossRef]
60. Wuertz, K.; Haglund, L. Inflammatory Mediators in Intervertebral Disk Degeneration and Discogenic Pain. *Glob. Spine J.* **2013**, *3*, 175–184. [CrossRef]
61. Brew, K.; Nagase, H. The tissue inhibitors of metalloproteinases (TIMPs): An ancient family with structural and functional diversity. *Biochim. Biophys. Acta Mol. Cell Res.* **2010**, *1803*, 55–71. [CrossRef]
62. Hilario, M.O.; Terreri, M.T.; Len, C.A. Nonsteroidal anti-inflammatory drugs: Cyclooxygenase 2 inhibitors. *J. Pediatr.* **2006**, *82* (Suppl. 5), S206–S212. [CrossRef]

International Journal of
*Molecular Sciences*

*Article*

# Antagonism of Transient Receptor Potential Ankyrin Type-1 Channels as a Potential Target for the Treatment of Trigeminal Neuropathic Pain: Study in an Animal Model

**Chiara Demartini [1,*], Rosaria Greco [1], Anna Maria Zanaboni [1,2], Oscar Francesconi [3], Cristina Nativi [3], Cristina Tassorelli [1,2] and Kristof Deseure [4]**

1 Laboratory of Neurophysiology of Integrative Autonomic Systems, Headache Science Center, IRCCS Mondino Foundation, via Mondino 2, 27100 Pavia, Italy; rosaria.greco@mondino.it (R.G.); annamaria.zanaboni@mondino.it (A.M.Z.); cristina.nativi@unifi.it (C.T.)
2 Department of Brain and Behavioral Sciences, University of Pavia, via Bassi 21, 27100 Pavia, Italy
3 Department of Chemistry 'Ugo Schiff', University of Florence, Via della Lastruccia 3-13, 50019 Sesto Fiorentino (FI), Italy; oscar.francesconi@unifi.it (O.F.); cristina.nativi@unifi.it (C.N.)
4 Department of Medicine, Laboratory for Pain Research, University of Antwerp, Universiteitsplein 1, 2610 Wilrijk, Belgium; kristof.deseure@uantwerpen.be
* Correspondence: chiara.demartini@mondino.it; Tel.: +39-(0382)-380255

Received: 17 September 2018; Accepted: 23 October 2018; Published: 25 October 2018

**Abstract:** Transient receptor potential ankyrin type-1 (TRPA1) channels are known to actively participate in different pain conditions, including trigeminal neuropathic pain, whose clinical treatment is still unsatisfactory. The aim of this study was to evaluate the involvement of TRPA1 channels by means of the antagonist ADM_12 in trigeminal neuropathic pain, in order to identify possible therapeutic targets. A single treatment of ADM_12 in rats 4 weeks after the chronic constriction injury of the infraorbital nerve (IoN-CCI) significantly reduced the mechanical allodynia induced in the IoN-CCI rats. Additionally, ADM_12 was able to abolish the increased levels of TRPA1, calcitonin gene-related peptide (CGRP), substance P (SP), and cytokines gene expression in trigeminal ganglia, cervical spinal cord, and medulla induced in the IoN-CCI rats. By contrast, no significant differences between groups were seen as regards CGRP and SP protein expression in the pars caudalis of the spinal nucleus of the trigeminal nerve. ADM_12 also reduced TRP vanilloid type-1 (TRPV1) gene expression in the same areas after IoN-CCI. Our findings show the involvement of both TRPA1 and TRPV1 channels in trigeminal neuropathic pain, and in particular, in trigeminal mechanical allodynia. Furthermore, they provide grounds for the use of ADM_12 in the treatment of trigeminal neuropathic pain.

**Keywords:** neuropathic pain; trigeminal system; allodynia; TRPA1; TRPV1

## 1. Introduction

Trigeminal neuralgia (TN) is a rare condition characterized by paroxysmal attacks of sharp pain, frequently described as an "electric shock". Up to 50% of patients with trigeminal neuralgia also have continuous pain in the same territory, which results in greater diagnostic difficulties, higher disability, and lower response to medical and surgical treatments [1]. Three diagnostic categories of TN are identified by the recent classification of headache disorders: Classical (without apparent cause other than neurovascular compression), secondary (caused by an underlying neurological disorder), and idiopathic (no cause is found) [2]. TN has a negative impact on activities of daily living, with up to 45% of patients being absent from usual daily activities for 15 days or more, and one third suffering from mild-to-severe depression [3]. Medications for TN exist, but they are poorly tolerated or ineffective. For this reason, multiple surgical approaches have been developed, but a portion of patients are refractory to both medical and surgical approaches [4,5]. Hence, there is need for further investigation into the mechanisms underlying pain in TN in order to identify new, possibly more effective, therapeutic targets.

In recent years, transient receptor potential (TRP) channels have attracted much attention in the pain field. These channels are non-selective cation channel proteins, widely distributed in many tissues and cell types, localized in the plasma membrane and membranes of intracellular organelles [6]. The TRP ankyrin type-1 (TRPA1) channels, mainly expressed with the vanilloid type-1 (TRPV1), are localized in a subpopulation of C- and Aδ-fibers of neurons located in the dorsal root ganglia (DRG) and trigeminal ganglia (TG) that produce and release neuropeptides, such as substance P (SP), neurokinin A, and calcitonin gene-related peptide (CGRP) [7–9]. Many experimental studies, from genetic knockouts to pharmacological manipulation models, reported a critical involvement of TRPA1 channels in different aspects of pain [10] and a role in several models of nerve injury, such as the lumbar spinal nerve ligation [11], and sciatic nerve injury by chronic constriction or transection [12–14]. In these models, it was demonstrated that an up-regulation of TRPA1 is associated with mechanical and thermal hyperalgesia, a condition reversed by TRPA1 antagonists [15,16]. In a recent study, Trevisan and colleagues [17] reported that pain-like behaviors are mediated by the TRPA1 channel in an animal model of TN based on the constriction of the infraorbital nerve (IoN) via the increased oxidative stress by-products released from monocytes and macrophages that gather at the site of nerve injury.

The aim of this study was to further investigate the role of TRP channels in trigeminal neuropathic pain induced by the model of a chronic constriction injury of the IoN (IoN-CCI) [18]. More specifically we evaluated: (i) The modulatory effect of TRPA1 antagonism, by means of ADM_12 treatment, on IoN-CCI-induced allodynia; (ii) the levels of TRPA1 and TRPV1 mRNA in specific cerebral and peripheral areas involved in trigeminal sensitization, with particular attention to changes in expression levels of genes coding for CGRP, SP, and cytokines after TRPA1 antagonism; and (iii) the expression of CGRP and SP proteins in the Spinal Nucleus of trigeminal nerve pars caudalis (Sp5C).

## 2. Results

### 2.1. ADM_12 Effect on Behavioral Response

In agreement with Deseure and Hans [18], 5 days after surgery, the two groups of rats that underwent IoN-CCI displayed a lack of responsiveness to ipsilateral mechanical stimulation testing (MST) of the IoN territory (Figure 1A). At day 12, the hyporesponsiveness was recovering to be replaced at day +26 by a significant increase in the MST response score as compared to the two Sham groups (Figure 1A). On day +27, the administration of the TRPA1 antagonist treatment in operated rats (IoN-CCI2 group) reduced the response score of the mechanical stimulation compared to the IoN-CCI1 group (injected with saline) (Figure 1B); whereas, ADM_12 treatment in sham-operated rats (Sham2 group) did not change the mechanical response. It is of note that the response to MST in the IoN-CCI2 group was significantly different between day +26 (before ADM_12 injection) and +27 (after drug treatment) (Figure 1C).

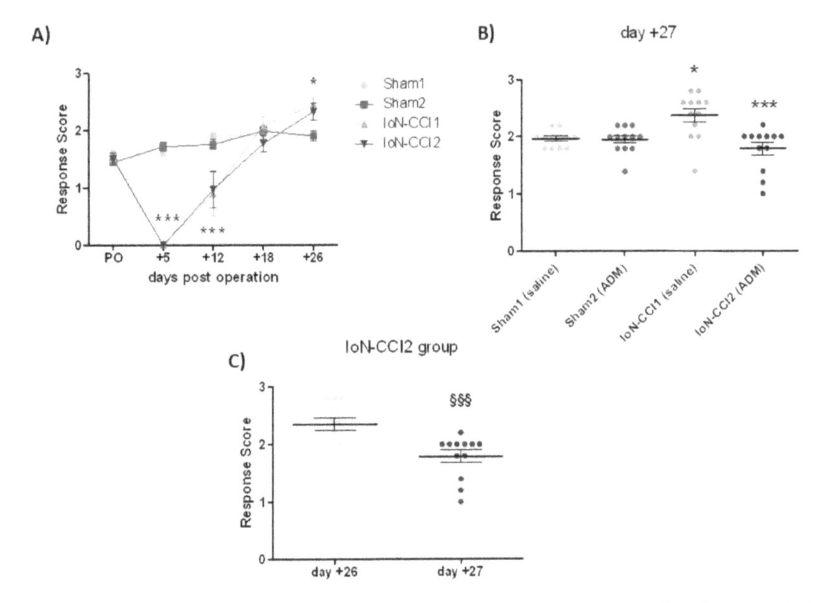


**Figure 1.** Mechanical stimulation testing (MST): (**A**) Mean response score to Von Frey hair stimulation of the ligated/sham infraorbital nerve (IoN) territory, on pre-operative day (PO) and on +5, +12, +18, and +26 days post operation. Data is expressed as mean ± SEM. Two-way ANOVA followed by Bonferroni post-hoc test, * $p < 0.05$ and *** $p < 0.001$ for chronic constriction injury of the infraorbital nerve (IoN-CCI) groups vs. Sham groups. Drug treatment effect on MST: (**B**) Mean response score to Von Frey hair stimulation on day +27, 1 h after ADM_12 (or saline) treatment. Data is expressed as mean ± SEM. One-way ANOVA followed by Tukey's Multiple Comparison Test, * $p < 0.05$ vs. Sham1 and Sham2, *** $p < 0.001$ vs. IoN-CCI1. (**C**) Comparison of the IoN-CCI2 group without treatment (day +26) and after ADM_12 treatment (on day +27). Data is expressed as mean ± SEM. Paired Student's $t$ test, §§§ $p < 0.001$ vs. day +26.

### 2.2. ADM_12 Effect on Gene Expression

The expression of *Trpa1*, calcitonin-related polypeptide alpha (*Calca*), and preprotachykinin-A, (*PPT-A*) was evaluated in the TG and cervical spinal cord (CSC) ipsilateral (ipsi) and contralateral (contra) to the IoN ligation, and in the medulla in toto. Because of the strong relationship between TRPA1 and TRPV1 channels [19–21], we also investigated the *Trpv1* mRNA expression levels in the same areas.

#### 2.2.1. *Trpa1* mRNA Expression

In the ipsilateral TG and CSC, and in medulla region, *Trpa1* mRNA expression levels were significantly increased in the IoN-CCI1 group compared with Sham1 and Sham2 groups (Figure 2). The increased mRNA levels were significantly reduced after treatment with ADM_12 in IoN-CCI rats (IoN-CCI2 group) in the same regions (Figure 2). ADM_12 administration did not provoke any changes in sham-operated rats.

A significant difference in mRNA levels, both in TG and CSC, was detected between sides in the IoN-CCI1 group; whereas, there was no difference between groups when comparing *Trpa1* mRNA levels on the contralateral side of TG and CSC (Figure 2A,B).

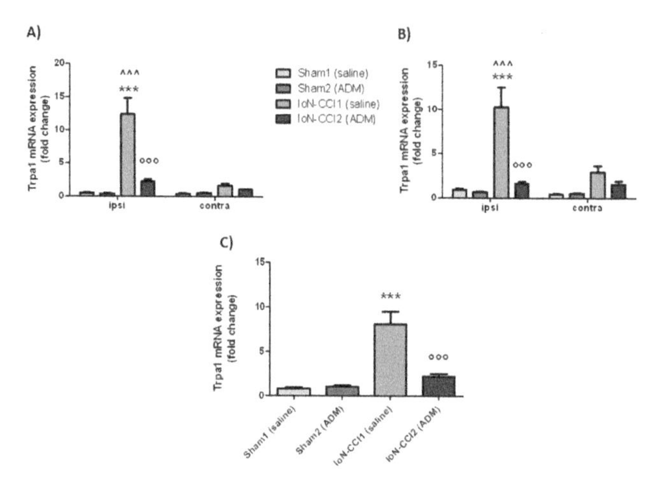


**Figure 2.** *Trpa1* mRNA expression in trigeminal ganglia (TGs) (**A**), cervical spinal cord (CSC) (**B**), and medulla (**C**). Data is expressed as mean + SEM. One way analysis of variance (ANOVA) followed by Tukey's Multiple Comparison Test or Two-way ANOVA followed by Bonferroni post-hoc test, *** $p < 0.001$ vs. Sham1 and Sham2 (ipsi), °°° $p < 0.001$ vs. IoN-CCI1 (ipsi), ^^^ $p < 0.001$ vs. IoN-CCI1 (contra).

### 2.2.2. *Trpv1* mRNA Expression

In the ipsilateral TG and CSC, and in the medulla region, *Trpv1* mRNA expression levels were significantly increased in the IoN-CCI1 group compared with Sham1 and Sham2 groups (Figure 3). The mRNA levels of *Trpv1* were also significantly higher in the IoN-CCI1 group in the contralateral CSC when compared to Sham groups (Figure 3B), though this increase was less marked than the increase observed on the ipsilateral side. The increased mRNA levels in these areas were significantly reduced by ADM_12 treatment in CCI rats (IoN-CCI2 group) in ipsilateral TG and CSC, and in medulla in toto (Figure 3). ADM_12 administration did not provoke any changes in sham-operated rats.

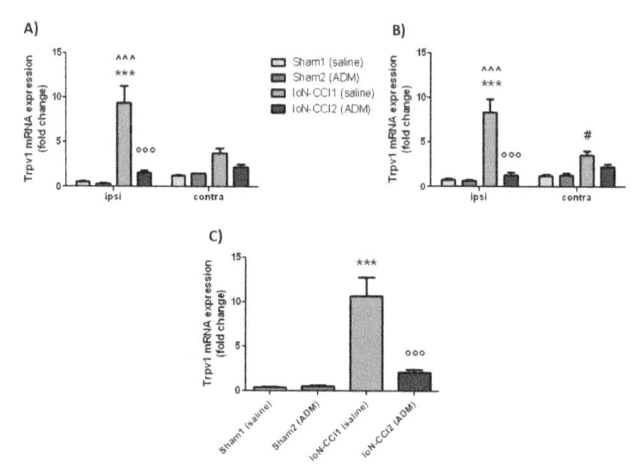


**Figure 3.** *Trpv1* mRNA expression in TGs (**A**), CSC (**B**), and medulla (**C**). Data is expressed as mean + SEM. One way analysis of variance (ANOVA) followed by Tukey's Multiple Comparison Test or Two-way ANOVA followed by Bonferroni post-hoc test, *** $p < 0.001$ vs. Sham1 and Sham 2 (ipsi), °°° $p < 0.001$ vs. IoN-CCI1 (ipsi), ^^^ $p < 0.001$ vs. IoN-CCI1 (contra), # $p < 0.05$ vs. Sham1 and Sham2 (contra).

A significant difference was seen between the ipsi- and contralateral side (both in TG and CSC) in the IoN-CCI1 (Figure 3A,B).

### 2.2.3. *Calca* mRNA Expression

In the ipsilateral TG and CSC, and in the medulla region, *Calca* mRNA expression levels were significantly increased in the IoN-CCI1 group compared with Sham1 and Sham2 groups (Figure 4). Moreover, *Calca* mRNA levels in IoN-CCI1 and IoN-CCI2 groups were also significantly increased in the contralateral TG as compared to Sham groups (Figure 4A). The increased mRNA levels were significantly reduced after treatment with ADM_12 in IoN-CCI2 rats in ipsilateral TG and CSC, and in medulla in toto (Figure 4). ADM_12 administration did not provoke any changes in sham-operated rats (Figure 4).

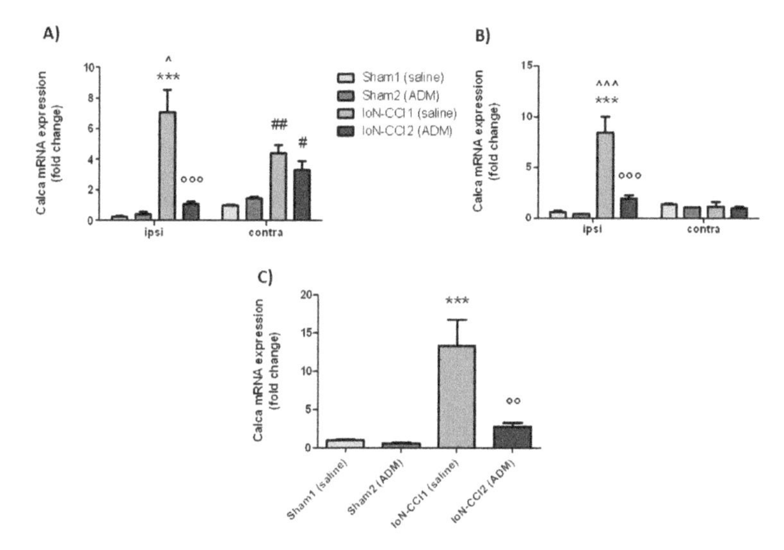


**Figure 4.** *Calca* mRNA expression in TGs (**A**), CSC (**B**), and medulla (**C**). Data is expressed as mean + SEM. One way analysis of variance (ANOVA) followed by Tukey's Multiple Comparison Test or Two-way ANOVA followed by Bonferroni post-hoc test, *** $p < 0.001$ vs. Sham1 and Sham2 (ipsi), °° $p < 0.01$ and °°° $p < 0.001$ vs. IoN-CCI1 (ipsi), ^ $p < 0.05$ and ^^^ $p < 0.001$ vs. IoN-CCI1 (contra), # $p < 0.05$ and ## $p < 0.01$ vs. Sham1 and Sham2 (contra).

A significant difference was seen between the ipsi- and contralateral side (both in TG and CSC) in the IoN-CCI1 group (Figure 4A,B).

### 2.2.4. *PPT-A* mRNA Expression

In the ipsilateral TG and CSC, and in the medulla region, *PPT-A* mRNA expression levels were significantly increased in the IoN-CCI1 group compared with Sham1 and Sham2 groups (Figure 5). The increased mRNA levels were significantly reduced after treatment with ADM_12 in IoN-CCI rats (IoN-CCI2 group) in the same regions (Figure 5). ADM_12 administration did not cause any changes in sham-operated rats.

A significant difference was seen between the ipsi- and contralateral side (both in TG and CSC) in the IoN-CCI1 group, as well as in the IoN-CCI2 group at the TG level; whereas, there was no difference between groups on the contralateral side of TG and CSC (Figure 5A,B).

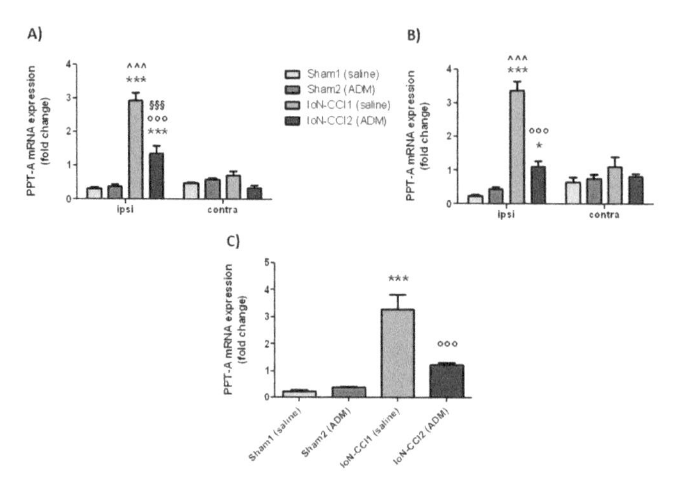


**Figure 5.** *PPT-A* mRNA expression in TGs (**A**), CSC (**B**), and medulla (**C**). Data is expressed as mean + SEM. One way analysis of variance (ANOVA) followed by Tukey's Multiple Comparison Test or Two-way ANOVA followed by Bonferroni post-hoc test, * $p < 0.05$ and *** $p < 0.001$ vs. Sham1 and Sham2 groups (ipsi), °°° $p < 0.001$ vs. IoN-CCI1 group (ipsi), ^^^ $p < 0.001$ vs. IoN-CCI1 group (contra), §§§ $p < 0.001$ vs. IoN-CCI2 (contra).

### 2.2.5. *IL-1beta*, *IL-6*, and *TNF-alpha* mRNA Expression

Since the effects of the surgery, and consequently of the TRPA1 antagonist, on the transcript levels were seen mainly at the ipsilateral side, the cytokines mRNA expression was not evaluated contralaterally.

Interleukin *(IL)-1beta*, *IL-6*, and tumor necrosis factor *(TNF)-alpha* mRNA expression levels were significantly increased in all the areas under evaluation in the IoN-CCI1 group compared with Sham1 and Sham2 groups (Figure 6). Such increases were significantly reduced after treatment with ADM_12 in IoN-CCI rats (IoN-CCI2 group) in the same regions (Figure 6).

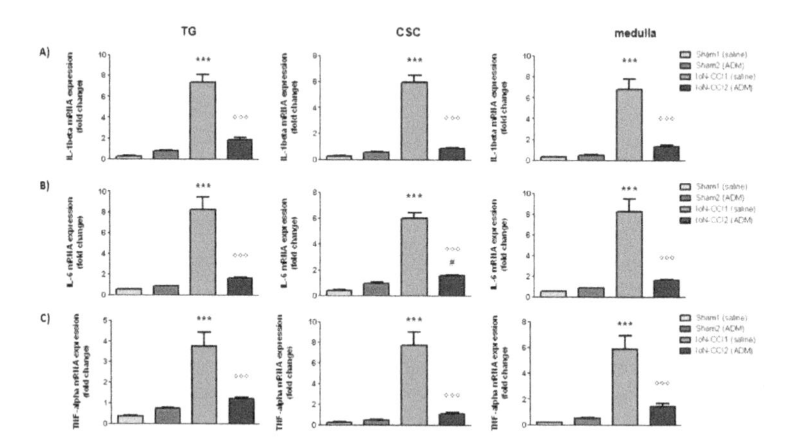


**Figure 6.** mRNA expression of *IL-1beta* (**A**), *IL-6* (**B**), and *TNF-alpha* (**C**) in ipsilateral TG and CSC, and in medulla in toto. Data are expressed as mean + SEM. One way analysis of variance (ANOVA) followed by Tukey's Multiple Comparison Test, *** $p < 0.001$ vs. Sham1 and Sham2, °°° $p < 0.001$ vs. IoN-CCI1, # $p < 0.05$ vs. Sham1.

### 2.3. ADM_12 Effect on Neuropeptide Protein Expression

CGRP and SP protein expression was evaluated in Sp5C on both sides. A slight, but not significant difference in the density of immunoreactive fibers for CGRP and SP protein was observed between the ipsilateral and contralateral side in both the IoN-CCI1 and IoN-CCI2 groups (Figure 7). No significant change was seen between sham and operated rats (Figure 7). ADM_12 administration did not induce any change in CGRP and SP expression either in sham or in CCI operated rats (Figure 7).

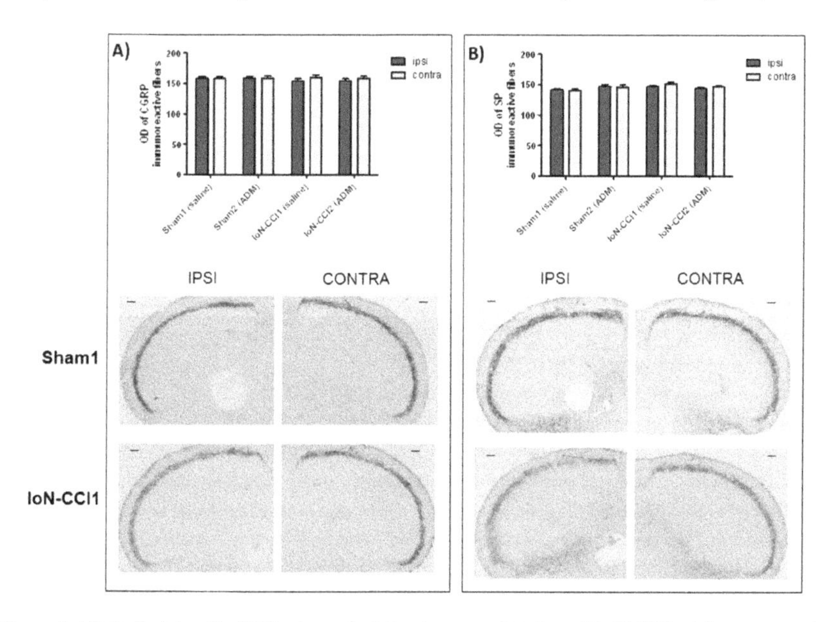


**Figure 7.** (**A**) Optical density (OD) values of calcitonin gene-related peptide (CGRP) with representative photomicrographs of CGRP immunoreactive fibers in the spinal nucleus of trigeminal nerve pars caudalis (Sp5C) ipsilateral (ipsi) and contralateral (contra) of Sham1 and IoN-CCI1 groups. (**B**) OD values of substance P (SP) with representative photomicrographs of SP immunoreactive fibers in the Sp5C ipsilateral (ipsi) and contralateral (contra) of Sham1 and IoN-CCI1 groups. Data is expressed as mean + SEM. Two-way analysis of variance (ANOVA) followed by Bonferroni post-hoc test. Scale bar: 100 μm.

## 3. Discussion

The pathways of trigeminal neuropathic pain are poorly understood. Experimental evidences suggest a strong involvement of TRPA1 in different patterns of neuropathic pain, and recently its role was also demonstrated in a trigeminal neuropathic pain model [17].

Here we evaluated the role of TRPA1 channels in an animal model of trigeminal neuropathic pain (IoN-CCI model), investigating the effects of the TRPA1 antagonist ADM_12 on mechanical allodynia, and neurochemical and transcriptional changes.

ADM_12 was previously shown to revert *in vivo* the Oxaliplatin-induced neuropathy [22]. At the trigeminal level, ADM_12 was able to reduce orofacial pain in a model of temporomandibular joint inflammation [23], and to counteract trigeminal hyperalgesia in a model of migraine pain, together with decreased *Trpa1* and neuropeptide mRNA expression levels in specific areas implicated in trigeminal pain [24].

### 3.1. Behavioral Response

Infraorbital nerve injury in rats leads to the development, in the ipsilateral side, of a hyporesponsiveness to mechanical stimulation within the first week post operation, followed by a hyperresponsiveness, that according to several studies [18,25], reflects a condition of mechanical allodynia. This biphasic response is probably related to the demyelination process, occurring in the early post-operative period, and remyelination process, that occurs in the late post-operative period [26]. Compared to the above cited papers [18,25,26], the time needed in this study to develop allodynia was somewhat longer. This may have been the result of small differences in the degree of nerve constriction; indeed, different degrees of IoN constriction have been shown to produce different time courses in isolated face grooming behavior [27], and this can also be true for mechanical allodynia.

The allodynic response of operated rats was abolished after treatment with the TRPA1 antagonist ADM_12, suggesting that the blockade of TRPA1 channels located on the trigeminal afferents prevented the release of neuropeptides (CGRP and SP) [28,29], thus resulting in a reduced neurogenic inflammation, and ultimately nociceptor sensitization [30]. Accordingly, Wu and colleagues reported an increase in TRPA1 protein, as well as TRPV1 channels, in the Sp5C region of rats that underwent IoN-CCI surgery [31], confirming their involvement in this process. An additional mechanism is represented by the reduction in the release of pro-inflammatory factors via the inhibition of TRPA1 located on glial cells in the nervous system, as suggested by our results, in which we observed a reduction of the *IL-1beta*, *IL-6*, and *TNF-alpha* transcripts that possibly parallel protein expression [32–34], which could account for reduced glial cells activation [32]; or via the inhibition of TRPA1 located on non-neuronal cells, such as keratinocytes and macrophages, in the tissues surrounding the damaged nerve [35]. Pro-inflammatory mediators released in the tissues that surround the damaged nerve, and glial cell activation, are indeed known to play a crucial role in the pathophysiology of neuropathic pain [36,37]. Glial activation and pro-inflammatory cytokines are associated with the onset of neuropathic pain symptoms such as allodynia or hyperalgesia [38–42].

The involvement of TRPA1 in mechanosensation has been extensively studied; both genetic deletion of TRPA1 and its pharmacological blockade abrogate mechanical pain-like behaviors [17,43,44]. Recently, Trevisan and colleagues [17] confirmed the critical role played by TRPA1 channels in mechanical allodynia induced by trigeminal neuropathic pain; conversely, in a model of sciatic nerve injury, Lehto and co-workers [45] reported a non-significant involvement of these channels in the mechanical sensitivity. On the other hand, other authors showed that TRPA1 blockade attenuated mechanical hypersensitivity following spinal injury [46,47] or neuropathic pain induced by chemotherapeutic agents [48,49]. Altogether these observations suggest that mechanical allodynia might be differently mediated by TRPA1 channels depending on the type of pain, site of damage, or distribution profile in TG and DRGs [50]. Moreover, the different responses observed in the experimental models could also be related to the different TRPA1 antagonists used, that may inhibit the channel through binding at different sites, with specific regulatory mechanisms [51].

### 3.2. *Trpa1* and *Trpv1* mRNA Expression

Chronic constriction injury of the IoN produced a marked increase in the *Trpa1* and *Trpv1* mRNA expression in central and peripheral areas ipsilaterally, and a slight increase even at the contralateral side, compared to the sham group. This contralateral increase is probably due to activation of inflammatory processes occurring after nerve injury, which can also affect the contralateral side [52]. The elevated TRP transcripts are accompanied by increased *IL-1beta*, *IL-6*, and *TNF-alpha* mRNA levels in the medulla region, and ipsilateral TG and CSC. It is known that TRPA1 and TRPV1 channels can be sensitized by inflammatory agents, causing up-regulation of these channels [53–55]. For example, *Trpa1* expression has been shown to be up-regulated by TNF-alpha and IL-1 alpha via transcriptional factor hypoxia-inducible factor-1$\alpha$ [56]. Similarly, TNF-alpha can up-regulate TRPV1 protein and mRNA in DRG and TG neurons [57,58]; one of the suggested pathways for *Trpv1* regulation is the p38 mitogen-activated protein kinase pathway [59], which may also be partly involved in *Trpa1* expression [60]. As regards TRPA1, its activation seems to depend on the activation of the nuclear factor-κB signaling pathway [61].

Furthermore, an important role in neuropathic pain seems to be played by oxidative stress [62–64], whose components can directly activate TRPA1 channels [65], thereby contributing to inflammation in a TRPA1-dependent manner. Indeed, it was recently found that trigeminal neuropathic pain behaviors were mediated by TRPA1 targeted by oxidative stress by-products released from monocytes and macrophages surrounding the site of the nerve injury [17].

In agreement with our study, an up-regulation of *Trpa1* and *Trpv1* mRNA levels, as well as protein levels in TG, DRGs, and dorsal horns, has been seen in different models of neuropathic pain [11,47,60,66–71]. Increased mRNA levels may reflect an increase in functional TRPA1 and TRPV1 channels [72,73].

The increased mRNA levels detected in our experiments in CSC and medulla may have different origins: *Trpv1* mRNA undergoes bidirectional axon transport along primary afferents [74], and the same could be true for *Trpa1*, since both TRPV1 and TRPA1 are (co-)expressed, not only on peripheral, but also on central terminals of primary afferent neurons where their activation can lead to the release of transmitters that promote the sensitization of postsynaptic pain transmission pathways [75–78]. In addition, *Trpv1* mRNA could originate from GABAergic interneurons and glial cells in the rat dorsal horn, which are known to express TRPV1 [69,79].

Systemic administration of ADM_12 markedly reduced the mRNA expression levels of both TRPs induced by IoN ligation. The effect of drug treatment on mRNA transcripts is likely to be due to an indirect effect rather than a direct one. It can be reasonably hypothesized that the effect of ADM_12 on TRPA1 mRNA expression is indirectly due to the blockade of the channel, located either on neuronal and non-neuronal cells, which is followed by two events. On one side, the reduction of calcium ($Ca^{2+}$) entry provokes a reduced activation of second messenger ($Ca^{2+}$ dependent) molecules (e.g., via the phospholipase C/$Ca^{2+}$ signaling pathway and Ca($^{2+}$)/calmodulin-dependent protein kinase II [CaMKII]) and interfering with the $Ca^{2+}$-interacting proteins [80,81], with the consequent reduction in transcriptional rate; for example, through the CaMK—cAMP response element-binding protein (CaMK—CREB) cascade. The other event that follows TRPA1 antagonism is the reduction in neuropeptide (CGRP and SP) release [28,29], and pro-inflammatory agents from neuronal fibers and non-neuronal cells. In this frame, we hypothesize that ADM_12 may break off a self-feeding loop in which TRPA1 channels are directly activated or sensitized by $Ca^{2+}$ [51,81], endogenous substances produced by intracellular $Ca^{2+}$ elevation [82], and pro-inflammatory molecules [83–85], and indirectly by the activation of nociceptive fibers caused by neuropeptide-induced neuroinflammation.

Moreover, we can also speculate that since TRPA1 and TRPV1 functions may be influenced by each other [20,86,87], a re-organization in the expression and nature of these channels after nerve injury [88–90] enabled ADM_12 to modulate TRPV1 channels as well. Although a physical interaction between these two channels may be questionable, even if some studies described it *in vitro* [19,21], many studies reported a functional interaction between them [20,86,91–93]. For instance, Masuoka et al. [87] showed in DRG neurons that TRPA1 channels suppress TRPV1 channel activity, possibly through the regulation of basal intracellular calcium concentration, and that the TRPA1 sensitization, induced by inflammatory agents, enhance TRPV1-mediated currents [87].

These observations, including our data, show a relationship between these two TRP channels, although more information and studies are needed to understand the precise mechanisms of this putative interaction.

### 3.3. Neuropeptide Expression

After nerve injury, an inflammatory process leads to the release of many pro-inflammatory mediators, which participate in peripheral sensitization, promoting an excessive release of neurotransmitters [94]. Together with the inflammatory process, neuropeptides and degenerative changes affecting the nervous fibers are also crucial peripheral mechanisms [95].

In our experimental setting, the mRNA expression levels of genes coding for CGRP (*Calca*) and SP (*PTT-A*) markedly increased in the central areas containing the Sp5C, as well as in the TG ipsilateral to the IoN ligation. Interestingly, *Calca* mRNA expression in IoN ligated rats was also elevated on the contralateral TG. It has been shown that projections from the TG reach the medullary and cervical dorsal horns on both sides [96,97], and that unilateral TG stimulation activates neurons in both ipsi- and contralateral Sp5C [98,99].

One of the mechanisms that could contribute to neuropeptide expression is the CaMK—CREB cascade, which is probably triggered following TRP channel activation [100], and that may represent the target mechanism for the observed inhibitory effect of ADM_12 on the mRNA expression of CGRP and SP. The blockade of TRP channels, which co-localize with CGRP and SP in the trigeminal neurons [7,101], can inhibit *Calca* and *PPT-A* mRNA expression, thus reducing the neuropeptide release and the trigeminal sensitization process. The data supports the pivotal involvement of CGRP and SP in the delivery and transmission of pain sensation to the central nervous system, and their role in trigeminal pain syndrome. In fact, an increased concentration of neuropeptides was found in the cerebrospinal fluid and venous blood of patients with trigeminal neuralgia compared to healthy controls [102,103].

In this frame, it was quite surprising that we did not detect any significant difference in neuropeptide protein expression at the Sp5C level, neither among groups, nor between sides. Lynds and co-workers [104] reported no differences in neuropeptide (CGRP and SP) levels between ipsi- and contralateral TG two weeks after IoN transection injury, while Xu and colleagues [105] described a reduction of CGRP and SP protein levels in the ipsilateral caudal medulla eight days after partial IoN ligation. Taken together, these findings suggest that in our model the neuropeptide release at central sites might have taken place at early time points after surgery, and therefore went undetected since we only measured it on day +27, or alternatively, that CGRP and SP are mostly involved at the peripheral terminals [26]. These apparently contrasting findings prompt the need for specifically targeted studies in order to investigate in more depth the role of neuropeptide release in central and peripheral sites in this model of trigeminal neuropathic pain.

### 3.4. Limitations of the Study and Future Perspectives

We evaluated changes of behavioral responses and mRNA expression after a short period (1 h) of drug exposure. This approach may be questionable, however there are many studies that support our observations. For instance, the mRNA expression of metabotropic glutamate receptors was found to be upregulated 1 h after treatment in mice DRG neurons [106]. Ambalavanar et al. [107] were able to detect changes in CGRP mRNA levels in rat's TG even 30 minutes after complete Freund's adjuvant injection. Furthermore, Nesic and co-workers [108] reported change in the mRNA signal of cytokines 1 h after treatment with MK-801, a NMDA receptor antagonist, in the spinal cord of rats subjected to spinal cord injury.

Nevertheless, to elucidate and confirm the present findings, additional experiments with different techniques are necessary. It will be interesting to evaluate in this model the effects of ADM_12 at later time points, as well as after chronic treatment. Another limitation of the present study is the absence of a time course of the expression of CRGP and SP. This was motivated by the ethical and organizational need to keep the number of rats as low as possible. However, based on the present findings, it seems important to address in future studies the parallel evaluation of mRNA and protein expression of CGRP and SP in order to elucidate more clearly the role of these neuropeptides in peripheral and central sites.

## 4. Materials and Methods

### 4.1. Animals

Male Sprague-Dawley rats (Charles River, weighing 225–250 g at arrival) were used following the International Association for the Study of Pain (IASP)'s guidelines for pain research in animals [109]. Animals were housed in groups of 2 with water and food available ad libitum, and kept in a colony room (humidity: 45 ± 5%; room temperature: 21 ± 1 °C). Rats were kept under a reversed 12:12 h dark/light cycle (lights on at 20 h). All procedures were in accordance with the European Convention for Care and Use of Laboratory Animals, and were approved by the Ethical Committee for Animal Testing (*Ethische Commissie Dierproeven*, ECD) of the University of Antwerp (number 2017-16, approval 20/02/2017).

Rats were allowed to acclimate for 8 days to the housing conditions before the surgery; they were habituated to the behavioral test procedure daily for three days before pre-operative testing. Habituation and testing were conducted in a darkened room (light provided by a 60 W red light bulb suspended 1 m above the observation area) with a 45 dB background noise.

### 4.2. Surgery

The IoN-CCI was performed as previously described [18,25,27]. Rats were anaesthetized with pentobarbital (60 mg/kg, intraperitoneally (i.p.)) and treated with atropine (0.1 mg/kg, i.p.). The surgery was performed under direct visual control using a Zeiss operation microscope (×10–25). The rat's head was fixed in a stereotaxic frame and a mid-line scalp incision was made, exposing the skull and nasal bone. The edge of the orbit was dissected free, and the orbital contents were deflected with a cotton-tipped wooden rod to give access to the left IoN, which was loosely ligated with two chromic catgut ligatures (5-0) (2 mm apart). The scalp incision was closed using polyester sutures (4-0; Ethicon, Johnson & Johnson, Belgium). In sham operated rats, the IoN was exposed using the same procedure, but the nerve was not ligated.

### 4.3. Mechanical Stimulation Testing (MST)

Baseline data were obtained 1 day before surgery. Following surgery, rats were tested on post-operative days +5, +12, +18, +26, and +27 (Figure 1). A graded series of five Von Frey hairs (0.015 g, 0.127 g, 0.217 g, 0.745 g, and 2.150 g) (Pressure Aesthesiometer®, Stoelting Co, Chicago, IL, USA) were applied by an experimenter who was blind to animal and treatment groups, within the IoN territory, near the center of the vibrissal pad [25,110–113]. Von Frey hairs were applied in an ascending order of intensity either ipsi- or contralaterally. The scoring system described by Vos [25] was used to evaluate the rats' response to the stimulation (Table 1). For each rat, and at every designated time, a mean score for the five von Frey filaments was determined.

**Table 1.** Response categories with the corresponding score values.

| SCORE | TYPE OF RESPONSE |
|---|---|
| 0 | no response |
| 1 | detection: the rat turns the head toward the stimulating object and the stimulus object is then explored |
| 2 | withdrawal reaction: the rat turns the head slowly away or pulls it briskly backward when the stimulation is applied; sometimes a single face wipe ipsilateral to the stimulated area occurs |
| 3 | escape/attack: the rat avoids further contact with the stimulus object, either passively by moving its body away from the stimulating object to assume a crouching position against the cage wall, or actively by attacking the stimulus object, making biting and grabbing movements |
| 4 | asymmetric face grooming: the rat displays an uninterrupted series of at least three face-wash strokes directed toward the stimulated facial area |

### 4.4. Drug and Experimental Plan

The TRPA1 antagonist ADM_12, synthesized in the Laboratory of Prof. Cristina Nativi (University of Florence, Italy) and characterized by a high binding constant versus TRPA1 [23], was dissolved in saline and administered intraperitoneally (i.p.) at the dose of 30 mg/kg in a volume of 1 ml/kg [22–24].

The animals were randomly allocated in four groups of 12 animals each and assigned to different experimental sets, as shown in Table 2.

**Table 2.** Experimental groups and number (N) of animals per group that underwent the mechanical stimulation test (MST). The samples of the subsets were processed for the real time polymerase chain reaction (RT-PCR) or immunohistochemistry (IHC).

| EXPERIMENTAL GROUPS | Surgery | Treatment on Day +27 | MST | RT-PCR | IHC |
|---|---|---|---|---|---|
| Sham1 | Sham | saline | N = 12 | N = 6 | N = 6 |
| Sham2 | Sham | ADM_12 | N = 12 | N = 6 | N = 6 |
| IoN-CCI1 | IoN-CCI | saline | N = 12 | N = 6 | N = 6 |
| IoN-CCI2 | IoN-CCI | ADM_12 | N = 12 | N = 6 | N = 6 |

On day +27, sham and operated rats were treated with ADM_12 or saline 1 h prior to the MST (Figure 8). The timing was chosen on the basis of previous studies reporting a significant effect of acute ADM_12 treatment on behavioral responses [22–24]. At the end of the behavioral test, each rat was sacrificed with an i.p. overdose of pentobarbital (150 mg/kg). A subset of 6 rats per experimental group served for the detection of gene expression levels by means of real time polymerase chain reaction (RT-PCR); another subset of 6 animals per experimental group underwent the immunohistochemical evaluation of protein expression (Table 2).

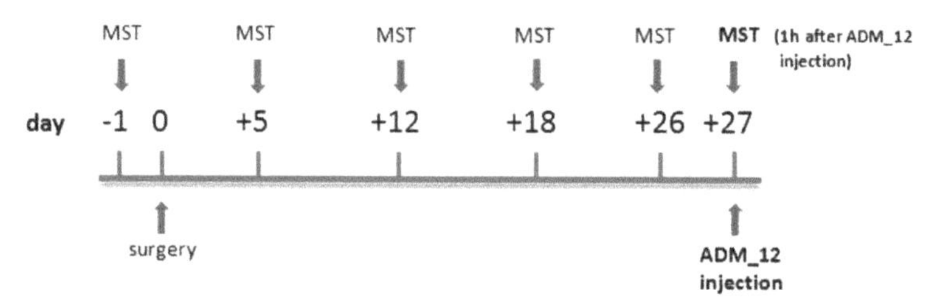


**Figure 8.** Schematic representation of the experimental design.

### 4.5. Real Time-PCR

The trigeminal ganglia (TG), cervical spinal cord (CSC, C1-C2 level), and medulla (bregma, −13.30 to −14.60 mm; Paxinos and Watson 4th edition), containing the Sp5C, of each animal were quickly removed after completing the MST on day +27 and frozen at –80 °C. Samples were then processed to evaluate the expression levels of the genes encoding for TRPA1 (*Trpa1*), TRPV1 (*Trpv1*), CGRP (*Calca*), SP (*PPT-A*), IL-1beta (*IL-1beta*), IL-6 (*IL-6*), and TNF-alpha (*TNF-alpha*). mRNA levels were analyzed by RT-PCR, as previously described [24,114,115]. After tissue homogenization by means of ceramic beads (PRECELLYS, Berthin Pharma, Montigny-le-Bretonneux, France), total RNA was extracted with TRIzol®reagent (Invitrogen, Carlsbad, California, USA) and quantified by measuring the absorbance at 260/280 nm using a nanodrop spectrophotometer (Euroclone, Pero (MI), Italy). Following cDNA generation with the iScript cDNA Synthesis kit (BIO-RAD, Hercules, California, USA), gene expression was analyzed using the Fast Eva Green supermix (BIO-RAD). Primer sequences were obtained from the AutoPrime software (http://www.autoprime.de/AutoPrimeWeb) (Table 3). The amplification was performed through two-step cycling (95–60°C) for 45 cycles with a Light Cycler 480 Instrument RT-PCR Detection System (Roche, Milan, Italy). The expression of the housekeeping gene, glyceraldehyde 3-phosphate dehydrogenase (*GAPDH*), remained constant in all the experimental groups considered. All samples were assayed in triplicate.

**Table 3.** Primer sequences.

| Gene | Forward Primer | Reverse Primer |
|---|---|---|
| *GAPDH* | AACCTGCCAAGTATGATGAC | GGAGTTGCTGTTGAAGTCA |
| *Trpa1* | CTCCCCGAGTGCATGAAAGT | TGCATATACGCGGGGATGTC |
| *Trpv1* | CTTGCTCCATTTGGGGTGTG | CTGGAGGTGGCTTGCAGTTA |
| *Calca* | CAGTCTCAGCTCCAAGTCATC | TTCCAAGGTTGACCTCAAAG |
| *PPT-A* | GCTCTTTATGGGCATGGTC | GGGTTTATTTACGCCTTCTTTC |
| *IL-1beta* | CTTCCTTGTGCAAGTGTCTG | CAGGTCATTCTCCTCACTGTC |
| *IL-6* | TTCTCTCCGCAAGAGACTTC | GGTCTGTTGTGGGTGGTATC |
| *TNF-alpha* | CCTCACACTCAGATCATCTTCTC | CGCTTGGTGGTTTGCTAC |

### 4.6. Immunohistochemistry

According to Terayama et al. [116] and Panneton et al. [117], the central afferent innervations of the IoN are mostly distributed in (but not restricted to) the dorsal and lateral part of the Sp5C, projecting to all the laminae. The pattern of CGRP and SP protein related to the painful component of the IoN was investigated in the superficial laminae of the Sp5C.

Immediately after the MST test, animals were anaesthetized and perfused transcardially with phosphate buffered saline (PBS) and 4% paraformaldehyde. The medullary segment containing the Sp5C, between +1 and −5 mm from the obex, was removed and post-fixed for 24 h in the same fixative; subsequently, samples were transferred in solutions of sucrose at increasing concentrations (up to 30%) during the following 72 h. All samples were cut transversely at 30 μm on a freezing sliding microtome. CGRP and SP protein expression was evaluated using the free-floating immunohistochemical technique, as previously reported [24]. For CGRP we used an anti-rabbit antibody (Santa Cruz Biotechnology, Santa Cruz, CA, USA) at a dilution of 1:3200, and an anti-rabbit antibody (Chemicon, Temecula, CA, USA) at a dilution of 1:5000 for SP; both primary antibodies were incubated for 24 h at room temperature. After incubation at room temperature with the secondary biotinylated antibody (Vector Laboratories, Burlingame, CA, USA) and the avidin-biotin complex (Vectastain, Vector Laboratories), sections were stained with the peroxidase substrate kit DAB (3′3′-diaminobenzidine tetrahydrochloride) (Vector Laboratories, Burlingame, CA, USA).

The area covered by CGRP and SP immunoreactive fibers in the Sp5C ipsilateral and contralateral to the surgery (12 sections per animal), was expressed as optical density (OD) values [24,114,118], acquired using an AxioSkop 2 microscope (Zeiss) and a computerized image analysis system (AxioCam, Zeiss, Göttingen, Germany) equipped with dedicated software (AxioVision Rel 4.2, Zeiss, Göttingen, Germany). All sections were averaged and reported as the mean + SEM of OD values.

### 4.7. Statistical Evaluation

Data from recent studies [18,27] was used to calculate the required number of animals per experimental group to obtain a statistical power of 0.80 at an alpha level of 0.05, and a difference of at least 20% in behavioral responses after IoN-CCI surgery. The calculations were done using software (Lenth RV. Java Applets for Power and Sample Size) retrieved on 8 April 2013, from http://www.stat.uiowa.edu/~rlenth/Power, which estimated a sample size of 12 rats per experimental group.

Statistical analysis was performed with the GraphPad Prism program (GraphPad Software, San Diego, California, USA). In the MST, for each rat and at every designated time, a mean score for the five Von Frey hairs was determined. The IoN-CCI rats were compared to the sham-operated rats. For mRNA levels, results were analyzed using the ΔΔCt method to compare expression of genes of interest with that of *GAPDH*, used as control gene. All data was tested for normality using the Kolmogorov–Smirnov normality test and considered normal. Differences between groups, or between ipsilateral and contralateral sides, were analyzed by one-way analysis of variance (ANOVA) followed by Tukey's Multiple Comparison Test, or by means of two-way ANOVA followed by Bonferroni post-hoc test, respectively. Differences between two groups were analyzed by the Paired student's t test. A probability level of less than 5% was regarded as significant.

## 5. Conclusions

Antagonism of the TRPA1 channel by means of ADM_12 attenuates experimentally-induced mechanical allodynia [17,119] in a reliable animal model of trigeminal neuropathic pain. Allodynia is one of the major clinical features of trigeminal neuropathic pain [120,121], thus the modulation of the TRPA1 channel may represent a suitable therapeutic target [122,123], and ADM_12 a possible tool, in trigeminal neuropathic pain management. As a corollary, our data also suggests a possible role for TRPV1 channels in the behavioral and biomolecular responses related to trigeminal neuropathic pain. Further exploration of the mechanisms underlying the antinociceptive effects of TRPA1, and studies directed to better understand the relationship between TRPA1 and TRPV1, would improve our understanding of the complex nociceptive processing in trigeminal neuropathic pain.

**Author Contributions:** C.D.: Conceptualization, Investigation, Formal analysis, Writing—original draft; R.G.: Conceptualization, Writing—review & editing; A.M.Z.: Investigation; O.F.: Methodology; C.N.: Writing—review & editing; C.T.: Funding acquisition, Writing—review & editing; K.D.: Conceptualization, Supervision, Investigation, Writing—review & editing. All authors read and approved the final manuscript.

**Funding:** This research was funded by a grant of the Italian Ministry of Health (Ricerca Corrente, 2016) to the IRCCS Mondino Foundation.

**Acknowledgments:** The authors would like to thank Stefania Ceruti for her precious suggestions during the writing of the manuscript.

**Conflicts of Interest:** The authors declare no conflict of interest.

## Abbreviations

| | |
|---|---|
| CaMKII | Ca($^{2+}$)/calmodulin-dependent protein kinase II |
| Calca | calcitonin-related polypeptide alpha |
| CGRP | calcitonin gene-related peptide |
| CREB | cAMP response element-binding protein |
| CSC | cervical spinal cord |
| DRG | dorsal root ganglia |
| GAPDH | glyceraldehyde 3-phosphate dehydrogenase |
| IL | interleukin |
| IoN | infraorbital nerve |
| IoN-CCI | chronic constriction injury of the infraorbital nerve |
| MST | mechanical stimulation testing |
| OD | optical density |
| PPT-A | preprotachykinin-A |
| RT-PCR | real time polymerase chain reaction |
| SP | substance P |
| Sp5C | spinal nucleus of trigeminal nerve pars caudalis |
| TG | trigeminal ganglia |
| TN | trigeminal neuralgia |
| TNF-alpha | tumor necrosis factor alpha |
| TRP | transient receptor potential |

## References

1. Cruccu, G. Trigeminal Neuralgia. *Continuum* **2017**, *23*, 396–420. [CrossRef] [PubMed]
2. Headache Classification Committee of the International Headache Society (IHS). The International Classification of Headache Disorders, 3rd edition. *Cephalalgia* **2018**, *38*, 1–211. [CrossRef] [PubMed]

3. Zakrzewska, J.M.; Wu, J.; Mon-Williams, M.; Phillips, N.; Pavitt, S.H. Evaluating the impact of trigeminal neuralgia. *Pain* **2017**, *158*, 1166–1174. [CrossRef] [PubMed]
4. Obermann, M.; Katsarava, Z. Update on trigeminal neuralgia. *Expert Rev. Neurother.* **2009**, *9*, 323–329. [CrossRef] [PubMed]
5. Zakrzewska, J.M.; Akram, H. Neurosurgical interventions for the treatment of classical trigeminal neuralgia. *Cochrane Database Syst. Rev.* **2011**, *9*, CD007312. [CrossRef] [PubMed]
6. Dong, X.P.; Wang, X.; Xu, H. TRP channels of intracellular membranes. *J. Neurochem.* **2010**, *113*, 313–328. [CrossRef] [PubMed]
7. Story, G.M.; Peier, A.M.; Reeve, A.J.; Eid, S.R.; Mosbacher, J.; Hricik, T.R.; Earley, T.J.; Hergarden, A.C.; Andersson, D.A.; Hwang, S.W.; et al. ANKTM1, a TRP-like channel expressed in nociceptive neurons, is activated by cold temperatures. *Cell* **2003**, *112*, 819–829. [CrossRef]
8. Bhattacharya, M.R.; Bautista, D.M.; Wu, K.; Haeberle, H.; Lumpkin, E.A.; Julius, D. Radial stretch reveals distinct populations of mechanosensitive mammalian somatosensory neurons. *Proc. Natl. Acad. Sci. USA* **2008**, *105*, 20015–20020. [CrossRef] [PubMed]
9. Quartu, M.; Serra, M.P.; Boi, M.; Poddighe, L.; Picci, C.; Demontis, R.; Del Fiacco, M. TRPV1 receptor in the human trigeminal ganglion and spinal nucleus: Immunohistochemical localization and comparison with the neuropeptides CGRP and SP. *J. Anat.* **2016**, *229*, 755–767. [CrossRef] [PubMed]
10. Jardín, I.; López, J.J.; Diez, R.; Sánchez-Collado, J.; Cantonero, C.; Albarrán, L.; Woodard, G.E.; Redondo, P.C.; Salido, G.M.; Smani, T.; et al. TRPs in Pain Sensation. *Front Physiol.* **2017**, *8*, 392. [CrossRef] [PubMed]
11. Obata, K.; Katsura, H.; Mizushima, T.; Yamanaka, H.; Kobayashi, K.; Dai, Y.; Fukuoka, T.; Tokunaga, A.; Tominaga, M.; Noguchi, K. TRPA1 induced in sensory neurons contributes to cold hyperalgesia after inflammation and nerve injury. *J. Clin. Investig.* **2005**, *115*, 2393–2401. [CrossRef] [PubMed]
12. Katsura, H.; Obata, K.; Mizushima, T.; Yamanaka, H.; Kobayashi, K.; Dai, Y.; Fukuoka, T.; Tokunaga, A.; Sakagami, M.; Noguchi, K. Antisense knock down of TRPA1, but not TRPM8, alleviates cold hyperalgesia after spinal nerve ligation in rats. *Exp. Neurol.* **2006**, *200*, 112–123. [CrossRef] [PubMed]
13. Caspani, O.; Zurborg, S.; Labuz, D.; Heppenstall, P.A. The contribution of TRPM8 and TRPA1 channels to cold allodynia and neuropathic pain. *PLoS ONE* **2009**, *4*, e7383. [CrossRef] [PubMed]
14. Staaf, S.; Oerther, S.; Lucas, G.; Mattsson, J.P.; Ernfors, P. Differential regulation of TRP channels in a rat model of neuropathic pain. *Pain* **2009**, *144*, 187–199. [CrossRef] [PubMed]
15. Eid, S.R.; Crown, E.D.; Moore, E.L.; Liang, H.A.; Choong, K.C.; Dima, S.; Henze, D.A.; Kane, S.A.; Urban, M.O. HC-030031, a TRPA1 selective antagonist, attenuates inflammatory- and neuropathyinduced mechanical hypersensitivity. *Mol. Pain* **2008**, *4*, 48. [CrossRef] [PubMed]
16. Chen, J.; Joshi, S.K.; DiDomenico, S.; Perner, R.J.; Mikusa, J.P.; Gauvin, D.M.; Segreti, J.A.; Han, P.; Zhang, X.F.; Niforatos, W.; et al. Selective blockade of TRPA1 channel attenuates pathological pain without altering noxious cold sensation or body temperature regulation. *Pain* **2011**, *152*, 1165–1172. [CrossRef] [PubMed]
17. Trevisan, G.; Benemei, S.; Materazzi, S.; De Logu, F.; De Siena, G.; Fusi, C.; Fortes Rossato, M.; Coppi, E.; Marone, I.M.; Ferreira, J.; et al. TRPA1 mediates trigeminal neuropathic pain in mice downstream of monocytes/macrophages and oxidative stress. *Brain* **2016**, *139*, 1361–1377. [CrossRef] [PubMed]
18. Deseure, K.; Hans, G.H. Chronic Constriction Injury of the Rat's Infraorbital Nerve (IoN-CCI) to Study Trigeminal Neuropathic Pain. *J. Vis. Exp.* **2015**, 103. [CrossRef] [PubMed]
19. Staruschenko, A.; Jeske, N.A.; Akopian, A.N. Contribution of TRPV1-TRPA1 interaction to the single channel properties of the TRPA1 channel. *J. Biol. Chem.* **2010**, *285*, 15167–15177. [CrossRef] [PubMed]
20. Akopian, A.N. Regulation of nociceptive transmission at the periphery via TRPA1-TRPV1 interactions. *Curr. Pharm. Biotechnol.* **2011**, *12*, 89–94. [CrossRef] [PubMed]
21. Fischer, M.J.; Balasuriya, D.; Jeggle, P.; Goetze, T.A.; McNaughton, P.A.; Reeh, P.W.; Edwardson, J.M. Direct evidence for functional TRPV1/TRPA1 heteromers. *Pflugers Arch.* **2014**, *466*, 2229–2241. [CrossRef] [PubMed]
22. Fragai, M.; Comito, G.; Di Cesare Mannelli, L.; Gualdani, R.; Calderone, V.; Louka, A.; Richichi, B.; Francesconi, O.; Angeli, A.; Nocentini, A.; et al. Lipoyl-Homotaurine Derivative (ADM_12) Reverts Oxaliplatin-Induced Neuropathy and Reduces Cancer Cells Malignancy by Inhibiting Carbonic Anhydrase IX (CAIX). *J. Med. Chem.* **2017**, *60*, 9003–9011. [CrossRef] [PubMed]
23. Gualdani, R.; Ceruti, S.; Magni, G.; Merli, D.; Di Cesare Mannelli, L.; Francesconi, O.; Richichi, B.; la Marca, G.; Ghelardini, C.; Moncelli, M.R.; et al. Lipoic-based TRPA1/TRPV1 antagonist to treat orofacial pain. *ACS Chem. Neurosci.* **2015**, *6*, 380–385. [CrossRef] [PubMed]

24. Demartini, C.; Tassorelli, C.; Zanaboni, A.M.; Tonsi, G.; Francesconi, O.; Nativi, C.; Greco, R. The role of the transient receptor potential ankyrin type-1 (TRPA1) channel in migraine pain: Evaluation in an animal model. *J. Headache Pain* **2017**, *18*, 9. [CrossRef] [PubMed]
25. Vos, B.P.; Strassman, A.M.; Maciewicz, R.J. Behavioral evidence of trigeminal neuropathic pain following chronic constriction injury to the rat's infraorbital nerve. *J. Neurosci.* **1994**, *14*, 2708–2723. [CrossRef] [PubMed]
26. Costa, G.M.F.; de Oliveira, A.P.; Martinelli, P.M.; da Silva Camargos, E.R.; Arantes, R.M.E.; de Almeida-Leite, C.M. Demyelination/remyelination and expression of interleukin-1β, substance P, nerve growth factor, and glial-derived neurotrophic factor during trigeminal neuropathic pain in rats. *Neurosci. Lett.* **2016**, *612*, 210–218. [CrossRef] [PubMed]
27. Deseure, K.; Hans, G. Behavioral study of non-evoked orofacial pain following different types of infraorbital nerve injury in rats. *Physiol. Behav.* **2015**, *138*, 292–296. [CrossRef] [PubMed]
28. Nakamura, Y.; Une, Y.; Miyano, K.; Abe, H.; Hisaoka, K.; Morioka, N.; Nakata, Y. Activation of transient receptor potential ankyrin 1 evokes nociception through substance P release from primary sensory neurons. *J. Neurochem.* **2012**, *120*, 1036–1047. [CrossRef] [PubMed]
29. Benemei, S.; Fusi, C.; Trevisan, G.; Geppetti, P. The TRPA1 channel in migraine mechanism and treatment. *Br. J. Pharmacol.* **2014**, *171*, 2552–2567. [CrossRef] [PubMed]
30. Gold, M.S.; Gebhart, G.F. Nociceptor sensitization in pain pathogenesis. *Nat. Med.* **2010**, *16*, 1248–1257. [CrossRef] [PubMed]
31. Wu, C.; Xie, N.; Lian, Y.; Xu, H.; Chen, C.; Zheng, Y.; Chen, Y.; Zhang, H. Central antinociceptive activity of peripherally applied botulinum toxin type A in lab rat model of trigeminal neuralgia. *SpringerPlus* **2016**, *5*, 431. [CrossRef] [PubMed]
32. Latrémolière, A.; Mauborgne, A.; Masson, J.; Bourgoin, S.; Kayser, V.; Hamon, M.; Pohl, M. Differential implication of proinflammatory cytokine interleukin-6 in the development of cephalic versus extracephalic neuropathic pain in rats. *J. Neurosci.* **2008**, *28*, 8489–8501. [CrossRef] [PubMed]
33. Choi, B.M.; Lee, S.H.; An, S.M.; Park, D.Y.; Lee, G.W.; Noh, G.J. The time-course and RNA interference of TNF-α, IL-6, and IL-1β expression on neuropathic pain induced by L5 spinal nerve transection in rats. *Korean J. Anesthesiol.* **2015**, *68*, 159–169. [CrossRef] [PubMed]
34. Zhang, B.; Yu, Y.; Aori, G.; Wang, Q.; Kong, D.; Yang, W.; Guo, Z.; Zhang, L. Tanshinone IIA Attenuates Diabetic Peripheral Neuropathic Pain in Experimental Rats via Inhibiting Inflammation. *Evid.-Based Complement. Altern. Med.* **2018**, *2018*, 2789847. [CrossRef] [PubMed]
35. Atoyan, R.; Shander, D.; Botchkareva, N.V. Non-neuronal expression of transient receptor potential type A1 (TRPA1) in human skin. *J. Invest. Dermatol.* **2009**, *129*, 2312–2315. [CrossRef] [PubMed]
36. Thacker, M.A.; Clark, A.K.; Marchand, F.; McMahon, S.B. Pathophysiology of peripheral neuropathic pain: Immune cells and molecules. *Anesth. Analg.* **2007**, *105*, 838–847. [CrossRef] [PubMed]
37. Mika, J.; Zychowska, M.; Popiolek-Barczyk, K.; Rojewska, E.; Przewlocka, B. Importance of glial activation in neuropathic pain. *Eur. J. Pharmacol.* **2013**, *716*, 106–119. [CrossRef] [PubMed]
38. Ledeboer, A.; Sloane, E.M.; Milligan, E.D.; Frank, M.G.; Mahony, J.H.; Maier, S.F.; Watkins, L.R. Minocycline attenuates mechanical allodynia and proinflammatory cytokine expression in rat models of pain facilitation. *Pain* **2005**, *115*, 71–83. [CrossRef] [PubMed]
39. Zhuang, Z.Y.; Gerner, P.; Woolf, C.J.; Ji, R.R. ERK is sequentially activated in neurons, microglia, and astrocytes by spinal nerve ligation and contributes to mechanical allodynia in this neuropathic pain model. *Pain* **2005**, *114*, 149–159. [CrossRef] [PubMed]
40. Lees, J.G.; Duffy, S.S.; Moalem-Taylor, G. Immunotherapy targeting cytokines in neuropathic pain. *Front. Pharmacol.* **2013**, *4*, 142. [CrossRef] [PubMed]
41. Zanjani, T.M.; Sabetkasaei, M.; Karimian, B.; Labibi, F.; Farokhi, B.; Mossafa, N. The attenuation of pain behaviour and serum interleukin-6 concentration by nimesulide in a rat model of neuropathic pain. *Scand. J. Pain* **2010**, *1*, 229–234. [CrossRef] [PubMed]
42. Piao, Z.G.; Cho, I.H.; Park, C.K.; Hong, J.P.; Choi, S.Y.; Lee, S.J.; Lee, S.; Park, K.; Kim, J.S.; Oh, S.B. Activation of glia and microglial p38 MAPK in medullary dorsal horn contributes to tactile hypersensitivity following trigeminal sensory nerve injury. *Pain* **2006**, *121*, 219–231. [CrossRef] [PubMed]
43. Kerstein, P.C.; del Camino, D.; Moran, M.M.; Stucky, C.L. Pharmacological blockade of TRPA1 inhibits mechanical firing in nociceptors. *Mol. Pain* **2009**, *5*, 19. [CrossRef] [PubMed]

44. Kwan, K.Y.; Glazer, J.M.; Corey, D.P.; Rice, F.L.; Stucky, C.L. TRPA1 modulates mechanotransduction in cutaneous sensory neurons. *J. Neurosci.* **2009**, *29*, 4808–4819. [CrossRef] [PubMed]
45. Lehto, S.G.; Weyer, A.D.; Youngblood, B.D.; Zhang, M.; Yin, R.; Wang, W.; Teffera, Y.; Cooke, M.; Stucky, C.L.; Schenkel, L.; et al. Selective antagonism of TRPA1 produces limited efficacy in models of inflammatory- and neuropathic-induced mechanical hypersensitivity in rats. *Mol. Pain* **2016**, *12*. [CrossRef] [PubMed]
46. Wei, H.; Koivisto, A.; Saarnilehto, M.; Chapman, H.; Kuokkanen, K.; Hao, B.; Huang, J.L.; Wang, Y.X.; Pertovaara, A. Spinal transient receptor potential ankyrin 1 channel contributes to central pain hypersensitivity in various pathophysiological conditions in the rat. *Pain* **2011**, *152*, 582–591. [CrossRef] [PubMed]
47. Park, J.; Zheng, L.; Acosta, G.; Vega-Alvarez, S.; Chen, Z.; Muratori, B.; Cao, P.; Shi, R. Acrolein contributes to TRPA1 up-regulation in peripheral and central sensory hypersensitivity following spinal cord injury. *J. Neurochem.* **2015**, *135*, 987–997. [CrossRef] [PubMed]
48. Materazzi, S.; Fusi, C.; Benemei, S.; Pedretti, P.; Patacchini, R.; Nilius, B.; Prenen, J.; Creminon, C.; Geppetti, P.; Nassini, R. TRPA1 and TRPV4 mediate paclitaxel-induced peripheral neuropathy in mice via a glutathione-sensitive mechanism. *Pflugers Arch.* **2012**, *463*, 561–569. [CrossRef] [PubMed]
49. Trevisan, G.; Materazzi, S.; Fusi, C.; Altomare, A.; Aldini, G.; Lodovici, M.; Patacchini, R.; Geppetti, P.; Nassini, R. Novel Therapeutic Strategy to Prevent Chemotherapy-Induced Persistent Sensory Neuropathy By TRPA1 Blockade. *Cancer Res.* **2013**, *73*, 3120–3131. [CrossRef] [PubMed]
50. Vandewauw, I.; Owsianik, G.; Voets, T. Systematic and quantitative mRNA expression analysis of TRP channel genes at the single trigeminal and dorsal root ganglion level in mouse. *BMC Neurosci.* **2013**, *14*, 21. [CrossRef] [PubMed]
51. Paulsen, C.E.; Armache, J.P.; Gao, Y.; Cheng, Y.; Julius, D. Structure of the TRPA1 ion channel suggests regulatory mechanisms. *Nature* **2015**, *525*, 552. [CrossRef] [PubMed]
52. Jancalek, R. Signaling mechanisms in mirror image pain pathogenesis. *Ann. Neurosci.* **2011**, *18*, 123–127. [CrossRef] [PubMed]
53. Amaya, F.; Oh-hashi, K.; Naruse, Y.; Iijima, N.; Ueda, M.; Shimosato, G.; Tominaga, M.; Tanaka, Y.; Tanaka, M. Local inflammation increases vanilloid receptor 1 expression within distinct subgroups of DRG neurons. *Brain Res.* **2003**, *963*, 190–196. [CrossRef]
54. Devesa, I.; Planells-Cases, R.; Fernández-Ballester, G.; González-Ros, J.M.; Ferrer-Montiel, A.; Fernández-Carvajal, A. Role of the transient receptor potential vanilloid 1 in inflammation and sepsis. *J. Inflamm. Res.* **2011**, *4*, 67–81. [CrossRef] [PubMed]
55. Diogenes, A.; Akopian, A.N.; Hargreaves, K.M. NGF up-regulates TRPA1: Implications for orofacial pain. *J. Dent. Res.* **2007**, *86*, 550–555. [CrossRef] [PubMed]
56. Hatano, N.; Itoh, Y.; Suzuki, H.; Muraki, Y.; Hayashi, H.; Onozaki, K.; Wood, I.C.; Beech, D.J.; Muraki, K. Hypoxia-inducible factor-1$\alpha$ (HIF1$\alpha$) switches on transient receptor potential ankyrin repeat 1 (TRPA1) gene expression via a hypoxia response element-like motif to modulate cytokine release. *J. Biol. Chem.* **2012**, *287*, 31962–31972. [CrossRef] [PubMed]
57. Hensellek, S.; Brell, P.; Schaible, H.G.; Bräuer, R.; Segond von Banchet, G. The cytokine TNFalpha increases the proportion of DRG neurones expressing the TRPV1 receptor via the TNFR1 receptor and ERK activation. *Mol. Cell Neurosci.* **2007**, *36*, 381–391. [CrossRef] [PubMed]
58. Khan, A.A.; Diogenes, A.; Jeske, N.A.; Henry, M.A.; Akopian, A.; Hargreaves, K.M. Tumor necrosis factor alpha enhances the sensitivity of rat trigeminal neurons to capsaicin. *Neuroscience* **2008**, *155*, 503–509. [CrossRef] [PubMed]
59. Ikeda-Miyagawa, Y.; Kobayashi, K.; Yamanaka, H.; Okubo, M.; Wang, S.; Dai, Y.; Yagi, H.; Hirose, M.; Noguchi, K. Peripherally increased artemin is a key regulator of TRPA1/V1 expression in primary afferent neurons. *Mol. Pain* **2015**, *11*, 8. [CrossRef] [PubMed]
60. Yamamoto, K.; Chiba, N.; Chiba, T.; Kambe, T.; Abe, K.; Kawakami, K.; Utsunomiya, I.; Taguchi, K. Transient receptor potential ankyrin 1 that is induced in dorsal root ganglion neurons contributes to acute cold hypersensitivity after oxaliplatin administration. *Mol. Pain* **2015**, *11*, 69. [CrossRef] [PubMed]
61. Kang, J.; Ding, Y.; Li, B.; Liu, H.; Yang, X.; Chen, M. TRPA1 mediated aggravation of allergic contact dermatitis induced by DINP and regulated by NF-$\kappa$B activation. *Sci. Rep.* **2017**, *7*, 43586. [CrossRef] [PubMed]

62. Kim, D.; You, B.; Jo, E.K.; Han, S.K.; Simon, M.I.; Lee, S.J. NADPH oxidase 2-derived reactive oxygen species in spinal cord microglia contribute to peripheral nerve injury-induced neuropathic pain. *Proc. Natl. Acad. Sci. USA* **2010**, *107*, 14851–14856. [CrossRef] [PubMed]
63. Kim, H.K.; Park, S.K.; Zhou, J.L.; Taglialatela, G.; Chung, K.; Coggeshall, R.E.; Chung, J.M. Reactive oxygen species (ROS) play an important role in a rat model of neuropathic pain. *Pain* **2004**, *111*, 116–124. [CrossRef] [PubMed]
64. Naik, A.K.; Tandan, S.K.; Dudhgaonkar, S.P.; Jadhav, S.H.; Kataria, M.; Prakash, V.R.; Kumar, D. Role of oxidative stress in pathophysiology of peripheral neuropathy and modulation by N-acetyl-L-cysteine in rats. *Eur. J. Pain* **2006**, *10*, 573–579. [CrossRef] [PubMed]
65. Andersson, D.A.; Gentry, C.; Moss, S.; Bevan, S. Transient receptor potential A1 is a sensory receptor for multiple products of oxidative stress. *J. Neurosci.* **2008**, *28*, 2485–2494. [CrossRef] [PubMed]
66. Hudson, L.J.; Bevan, S.; Wotherspoon, G.; Gentry, C.; Fox, A.; Winter, J. VR1 protein expression increases in undamaged DRG neurons after partial nerve injury. *Eur. J. Neurosci.* **2001**, *13*, 2105–2114. [CrossRef] [PubMed]
67. Fukuoka, T.; Tokunaga, A.; Tachibana, T.; Dai, Y.; Yamanaka, H.; Noguchi, K. VR1, but not P2X(3), increases in the spared L4 DRG in rats with L5 spinal nerve ligation. *Pain* **2002**, *99*, 111–120. [CrossRef]
68. Frederick, J.; Buck, M.E.; Matson, D.J.; Cortright, D.N. Increased TRPA1, TRPM8, and TRPV2 expression in dorsal root ganglia by nerve injury. *Biochem. Biophys. Res. Commun.* **2007**, *358*, 1058–1064. [CrossRef] [PubMed]
69. Kim, Y.H.; Back, S.K.; Davies, A.J.; Jeong, H.; Jo, H.J.; Chung, G.; Na, H.S.; Bae, Y.C.; Kim, S.J.; Kim, J.S.; et al. TRPV1 in GABAergic interneurons mediates neuropathic mechanical allodynia and disinhibition of the nociceptive circuitry in the spinal cord. *Neuron* **2012**, *74*, 640–647. [CrossRef] [PubMed]
70. Urano, H.; Ara, T.; Fujinami, Y.; Hiraoka, B.Y. Aberrant TRPV1 expression in heat hyperalgesia associated with trigeminal neuropathic pain. *Int. J. Med. Sci.* **2012**, *9*, 690–697. [CrossRef] [PubMed]
71. Quartu, M.; Carozzi, V.A.; Dorsey, S.G.; Serra, M.P.; Poddighe, L.; Picci, C.; Boi, M.; Melis, T.; Del Fiacco, M.; Meregalli, C.; et al. Bortezomib treatment produces nocifensive behavior and changes in the expression of TRPV1, CGRP, and substance P in the rat DRG, spinal cord, and sciatic nerve. *Biomed. Res. Int.* **2014**, *2014*, 180428. [CrossRef] [PubMed]
72. Schwartz, E.S.; Christianson, J.A.; Chen, X.; La, J.H.; Davis, B.M.; Albers, K.M.; Gebhart, G.F. Synergistic role of TRPV1 and TRPA1 in pancreatic pain and inflammation. *Gastroenterology* **2011**, *140*, 1283–1291. [CrossRef] [PubMed]
73. Nummenmaa, E.; Hämäläinen, M.; Moilanen, L.J.; Paukkeri, E.L.; Nieminen, R.M.; Moilanen, T.; Vuolteenaho, K.; Moilanen, E. Transient receptor potential ankyrin 1 (TRPA1) is functionally expressed in primary human osteoarthritic chondrocytes. *Arthritis Res. Ther.* **2016**, *18*, 185. [CrossRef] [PubMed]
74. Tohda, C.; Sasaki, M.; Konemura, T.; Sasamura, T.; Itoh, M.; Kuraishi, Y. Axonal transport of VR1 capsaicin receptor mRNA in primary afferents and its participation in inflammation-induced increase in capsaicin sensitivity. *J. Neurochem.* **2001**, *76*, 1628–1635. [CrossRef] [PubMed]
75. Kosugi, M.; Nakatsuka, T.; Fujita, T.; Kuroda, Y.; Kumamoto, E. Activation of TRPA1 channel facilitates excitatory synaptic transmission in substantia gelatinosa neurons of the adult rat spinal cord. *J. Neurosci.* **2007**, *27*, 4443–4451. [CrossRef] [PubMed]
76. Koivisto, A.; Chapman, H.; Jalava, N.; Korjamo, T.; Saarnilehto, M.; Lindstedt, K.; Pertovaara, A. TRPA1: A transducer and amplifier of pain and inflammation. *Basic Clin. Pharmacol. Toxicol.* **2014**, *114*, 50–55. [CrossRef] [PubMed]
77. Kim, Y.S.; Chu, Y.; Han, L.; Li, M.; Li, Z.; LaVinka, P.C.; Sun, S.; Tang, Z.; Park, K.; Caterina, M.J.; et al. Central terminal sensitization of TRPV1 by descending serotonergic facilitation modulates chronic pain. *Neuron* **2014**, *81*, 873–887. [CrossRef] [PubMed]
78. Yeo, E.J.; Cho, Y.S.; Paik, S.K.; Yoshida, A.; Park, M.J.; Ahn, D.K.; Moon, C.; Kim, Y.S.; Bae, Y.C. Ultrastructural analysis of the synaptic connectivity of TRPV1-expressing primary afferent terminals in the rat trigeminal caudal nucleus. *J. Comp. Neurol.* **2010**, *518*, 4134–4146. [CrossRef] [PubMed]
79. Doly, S.; Fischer, J.; Salio, C.; Conrath, M. The vanilloid receptor-1 is expressed in rat spinal dorsal horn astrocytes. *Neurosci. Lett.* **2004**, *357*, 123–126. [CrossRef] [PubMed]
80. Mandadi, S.; Armati, P.J.; Roufogalis, B.D. Protein kinase C modulation of thermo-sensitive transient receptor potential channels: Implications for pain signaling. *J. Nat. Sci. Biol. Med.* **2011**, *2*, 13–25. [CrossRef] [PubMed]

81. Zurborg, S.; Yurgionas, B.; Jira, J.A.; Caspani, O.; Heppenstall, P.A. Direct activation of the ion channel TRPA1 by $Ca^{2+}$. *Nat. Neurosci.* **2007**, *10*, 277–279. [CrossRef] [PubMed]
82. Takahashi, N.; Mizuno, Y.; Kozai, D.; Yamamoto, S.; Kiyonaka, S.; Shibata, T.; Uchida, K.; Mori, Y. Molecular characterization of TRPA1 channel activation by cysteine-reactive inflammatory mediators. *Channels* **2008**, *2*, 287–298. [CrossRef] [PubMed]
83. Bandell, M.; Story, G.M.; Hwang, S.W.; Viswanath, V.; Eid, S.R.; Petrus, M.J.; Earley, T.J.; Patapoutian, A. Noxious cold ion channel TRPA1 is activated by pungent compounds and bradykinin. *Neuron* **2004**, *41*, 849–857. [CrossRef]
84. Bautista, D.M.; Jordt, S.E.; Nikai, T.; Tsuruda, P.R.; Read, A.J.; Poblete, J.; Yamoah, E.N.; Basbaum, A.I.; Julius, D. TRPA1 mediates the inflammatory actions of environmental irritants and proalgesic agents. *Cell* **2006**, *124*, 1269–1282. [CrossRef] [PubMed]
85. Wang, S.; Dai, Y.; Fukuoka, T.; Yamanaka, H.; Kobayashi, K.; Obata, K.; Cui, X.; Tominaga, M.; Noguchi, K. Phospholipase C and protein kinase A mediate bradykinin sensitization of TRPA1: A molecular mechanism of inflammatory pain. *Brain* **2008**, *131*, 1241–1251. [CrossRef] [PubMed]
86. Lee, L.Y.; Hsu, C.C.; Lin, Y.J.; Lin, R.L.; Khosravi, M. Interaction between TRPA1 and TRPV1: Synergy on pulmonary sensory nerves. *Pulm. Pharmacol. Ther.* **2015**, *35*, 87–93. [CrossRef] [PubMed]
87. Masuoka, T.; Kudo, M.; Yamashita, Y.; Yoshida, J.; Imaizumi, N.; Muramatsu, I.; Nishio, M.; Ishibashi, T. TRPA1 Channels Modify TRPV1-Mediated Current Responses in Dorsal Root Ganglion Neurons. *Front. Physiol.* **2017**, *8*, 272. [CrossRef] [PubMed]
88. Matzner, O.; Devor, M. Hyperexcitability at sites of nerve injury depends on voltage-sensitive $Na^+$ channels. *J. Neurophysiol.* **1994**, *72*, 349–359. [CrossRef] [PubMed]
89. McCallum, J.B.; Wu, H.E.; Tang, Q.; Kwok, W.M.; Hogan, Q.H. Subtype-specific reduction of voltage-gated calcium current in medium-sized dorsal root ganglion neurons after painful peripheral nerve injury. *Neuroscience* **2011**, *179*, 244–255. [CrossRef] [PubMed]
90. Waxman, S.G.; Kocsis, J.D.; Black, J.A. Type III $Na^+$ channel mRNA is expressed in embryonic but not adult spinal sensory neurons, and is re-expressed following axotomy. *J. Neurophysiol.* **1994**, *72*, 466–470. [CrossRef] [PubMed]
91. Ruparel, N.B.; Patwardhan, A.M.; Akopian, A.N.; Hargreaves, K.M. Desensitization of transient receptor potential ankyrin 1 (TRPA1) by the TRP vanilloid 1-selective cannabinoid arachidonoyl-2 chloroethanolamine. *Mol. Pharmacol.* **2011**, *80*, 117–123. [CrossRef] [PubMed]
92. Andrade, E.L.; Meotti, F.C.; Calixto, J.B. TRPA1 antagonists as potential analgesic drugs. *Pharmacol. Ther.* **2012**, *133*, 189–204. [CrossRef] [PubMed]
93. Honda, K.; Shinoda, M.; Furukawa, A.; Kita, K.; Noma, N.; Iwata, K. TRPA1 contributes to capsaicin-induced facial cold hyperalgesia in rats. *Eur. J. Oral Sci.* **2014**, *122*, 391–396. [CrossRef] [PubMed]
94. Kim, K.H.; Kim, J.I.; Han, J.A.; Choe, M.A.; Ahn, J.H. Upregulation of neuronal nitric oxide synthase in the periphery promotes pain hypersensitivity after peripheral nerve injury. *Neuroscience* **2011**, *8*, 367–378. [CrossRef] [PubMed]
95. Costa, G.M.F.; Leite, C.M.A. Trigeminal neuralgia: Peripheral and central mechanisms. *Rev. Dor São Paulo.* **2015**, *16*, 297–301. [CrossRef]
96. Pfaller, K.; Arvidsson, J. Central distribution of trigeminal and upper cervical primary afferents in the rat studied by anterograde transport of horseradish peroxidase conjugated to wheat germ agglutinin. *J. Comp. Neurol.* **1988**, *268*, 91–108. [CrossRef] [PubMed]
97. Jacquin, M.F.; Chiaia, N.L.; Rhoades, R.W. Trigeminal projections to contralateral dorsal horn: Central extent, peripheral origins, and plasticity. *Somatosens. Mot. Res.* **1990**, *7*, 153–183. [CrossRef] [PubMed]
98. Ingvardsen, B.K.; Laursen, H.; Olsen, U.B.; Hansen, A.J. Possible mechanism of c-fos expression in trigeminal nucleus caudalis following cortical spreading depression. *Pain* **1997**, *72*, 407–415. [CrossRef]
99. Samsam, M.; Coveñas, R.; Csillik, B.; Ahangari, R.; Yajeya, J.; Riquelme, R.; Narváez, J.A.; Tramu, G. Depletion of substance P, neurokinin A and calcitonin gene-related peptide from the contralateral and ipsilateral caudal trigeminal nucleus following unilateral electrical stimulation of the trigeminal ganglion; a possible neurophysiological and neuroanatomical link to generalized head pain. *J. Chem. Neuroanat.* **2001**, *21*, 161–169. [CrossRef] [PubMed]
100. Nakanishi, M.; Hata, K.; Nagayama, T.; Sakurai, T.; Nishisho, T.; Wakabayashi, H.; Hiraga, T.; Ebisu, S.; Yoneda, T. Acid activation of Trpv1 leads to an up-regulation of calcitonin gene-related peptide expression

in dorsal root ganglion neurons via the CaMK-CREB cascade: A potential mechanism of inflammatory pain. *Mol. Biol. Cell* **2010**, *21*, 2568–2577. [CrossRef] [PubMed]
101. Huang, D.; Li, S.; Dhaka, A.; Story, G.M.; Cao, Y.Q. Expression of the transient receptor potential channels TRPV1, TRPA1 and TRPM8 in mouse trigeminal primary afferent neurons innervating the dura. *Mol. Pain* **2012**, *8*, 66. [CrossRef] [PubMed]
102. Qin, Z.L.; Yang, L.Q.; Li, N.; Yue, J.N.; Wu, B.S.; Tang, Y.Z.; Guo, Y.N.; Lai, G.H.; Ni, J.X. Clinical study of cerebrospinal fluid neuropeptides in patients with primary trigeminal neuralgia. *Clin. Neurol. Neurosurg.* **2016**, *143*, 111–115. [CrossRef] [PubMed]
103. Strittmatter, M.; Grauer, M.; Isenberg, E.; Hamann, G.; Fischer, C.; Hoffmann, K.H.; Blaes, F.; Schimrigk, K. Cerebrospinal fluid neuropeptides and monoaminergic transmitters in patients with trigeminal neuralgia. *Headache* **1997**, *37*, 211–216. [CrossRef] [PubMed]
104. Lynds, R.; Lyu, C.; Lyu, G.W.; Shi, X.Q.; Rosén, A.; Mustafa, K.; Shi, T.J.S. Neuronal plasticity of trigeminal ganglia in mice following nerve injury. *J. Pain Res.* **2017**, *10*, 349–357. [CrossRef] [PubMed]
105. Xu, M.; Aita, M.; Chavkin, C. Partial infraorbital nerve ligation as a model of trigeminal nerve injury in the mouse: Behavioral, neural, and glial reactions. *J. Pain* **2008**, *9*, 1036–1048. [CrossRef] [PubMed]
106. Chiechio, S.; Copani, A.; De Petris, L.; Morales, M.E.; Nicoletti, F.; Gereau, R.W., 4th. Transcriptional regulation of metabotropic glutamate receptor 2/3 expression by the NF-kappaB pathway in primary dorsal root ganglia neurons: A possible mechanism for the analgesic effect of L-acetylcarnitine. *Mol. Pain* **2006**, *2*, 20. [CrossRef] [PubMed]
107. Ambalavanar, R.; Dessem, D.; Moutanni, A.; Yallampalli, C.; Yallampalli, U.; Gangula, P.; Bai, G. Muscle inflammation induces a rapid increase in calcitonin gene-related peptide (CGRP) mRNA that temporally relates to CGRP immunoreactivity and nociceptive behavior. *Neuroscience* **2006**, *143*, 875–884. [CrossRef] [PubMed]
108. Nesic, O.; Svrakic, N.M.; Xu, G.Y.; McAdoo, D.; Westlund, K.N.; Hulsebosch, C.E.; Ye, Z.; Galante, A.; Soteropoulos, P.; Tolias, P.; et al. DNA microarray analysis of the contused spinal cord: Effect of NMDA receptor inhibition. *J. Neurosci. Res.* **2002**, *68*, 406–423. [CrossRef] [PubMed]
109. Zimmerman, M. Ethical guidelines for investigations of experimental pain in conscious animals. *Pain* **1983**, *16*, 109–110. [CrossRef]
110. Deseure, K.; Koek, W.; Colpaert, F.C.; Adriaensen, H. The 5-HT(1A) receptor agonist F 13640 attenuates mechanical allodynia in a rat model of trigeminal neuropathic pain. *Eur. J. Pharmacol.* **2002**, *456*, 51–57. [CrossRef]
111. Deseure, K.; Koek, W.; Adriaensen, H.; Colpaert, F.C. Continuous administration of the 5-hydroxytryptamine1A agonist (3-Chloro-4-fluoro-phenyl)-[4-fluoro-4-[[(5-methyl-pyridin-2-ylmethyl) -amino]-methyl]piperidin-1-yl]-methadone (F 13640) attenuates allodynia-like behavior in a rat model of trigeminal neuropathic pain. *J. Pharmacol. Exp. Ther.* **2003**, *306*, 505–514. [PubMed]
112. Deseure, K.R.; Adriaensen, H.F.; Colpaert, F.C. Effects of the combined continuous administration of morphine and the high-efficacy 5-HT1A agonist, F 13640 in a rat model of trigeminal neuropathic pain. *Eur. J. Pain* **2004**, *8*, 547–554. [CrossRef] [PubMed]
113. Deseure, K.; Bréand, S.; Colpaert, F.C. Curative-like analgesia in a neuropathic pain model: Parametric analysis of the dose and the duration of treatment with a high-efficacy 5-HT(1A) receptor agonist. *Eur. J. Pharmacol.* **2007**, *568*, 134–141. [CrossRef] [PubMed]
114. Greco, R.; Demartini, C.; Zanaboni, A.M.; Redavide, E.; Pampalone, S.; Toldi, J.; Fülöp, F.; Blandini, F.; Nappi, G.; Sandrini, G.; et al. Effects of kynurenic acid analogue 1 (KYNA-A1) in nitroglycerin-induced hyperalgesia: Targets and anti-migraine mechanisms. *Cephalalgia* **2017**, *37*, 1272–1284. [CrossRef] [PubMed]
115. Greco, R.; Ferrigno, A.; Demartini, C.; Zanaboni, A.; Mangione, A.S.; Blandini, F.; Nappi, G.; Vairetti, M.; Tassorelli, C. Evaluation of ADMA-DDAH-NOS axis in specific brain areas following nitroglycerin administration: Study in an animal model of migraine. *J. Headache Pain* **2015**, *16*, 560. [CrossRef] [PubMed]
116. Terayama, R.; Nagamatsu, N.; Ikeda, T.; Nakamura, T.; Rahman, O.I.; Sakoda, S.; Shiba, R.; Nishimori, T. Differential expression of Fos protein after transection of the rat infraorbital nerve in the trigeminal nucleus caudalis. *Brain Res.* **1997**, *768*, 135–146. [CrossRef]
117. Panneton, W.M.; Pan, B.; Gan, Q. Somatotopy in the Medullary Dorsal Horn As a Basis for Orofacial Reflex Behavior. *Front. Neurol.* **2017**, *8*, 522. [CrossRef] [PubMed]

118. Greco, R.; Tassorelli, C.; Sandrini, G.; Di Bella, P.; Buscone, S.; Nappi, G. Role of calcitonin gene-related peptide and substance P in different models of pain. *Cephalalgia* **2008**, *28*, 114–126. [CrossRef] [PubMed]
119. Green, D.; Ruparel, S.; Gao, X.; Ruparel, N.; Patil, M.; Akopian, A.; Hargreaves, K. Central activation of TRPV1 and TRPA1 by novel endogenous agonists contributes to mechanical allodynia and thermal hyperalgesia after burn injury. *Mol. Pain* **2016**, *12*. [CrossRef] [PubMed]
120. Ossipov, M.H.; Lai, J.; Malan, T.P., Jr.; Porreca, F. Spinal and supraspinal mechanisms of neuropathic pain. *Ann. N. Y. Acad. Sci.* **2000**, *909*, 12–24. [CrossRef] [PubMed]
121. Zakrzewska, J.M. Differential diagnosis of facial pain and guidelines for management. *Br. J. Anaesth.* **2013**, *111*, 95–104. [CrossRef] [PubMed]
122. Garrison, S.R.; Stucky, C.L. The dynamic TRPA1 channel: A suitable pharmacological pain target? *Curr. Pharm. Biotechnol.* **2011**, *12*, 1689–1697. [CrossRef] [PubMed]
123. Chen, J.; Hackos, D.H. TRPA1 as a drug target—promise and challenges. *N.-S. Arch. Pharmacol.* **2015**, *388*, 451–463. [CrossRef] [PubMed]

International Journal of
*Molecular Sciences*

*Review*

# The Emerging Role of Mechanosensitive Piezo Channels in Migraine Pain

**Adriana Della Pietra [1], Nikita Mikhailov [1] and Rashid Giniatullin [1,2,*]**

1 A. I. Virtanen Institute for Molecular Sciences, University of Eastern Finland, 70211 Kuopio, Finland; adriande@student.uef.fi (A.D.P.); nikita.mikhailov@uef.fi (N.M.)
2 Laboratory of Neurobiology, Kazan Federal University, 420008 Kazan, Russia
* Correspondence: rashid.giniatullin@uef.fi

Received: 12 December 2019; Accepted: 19 January 2020; Published: 21 January 2020

**Abstract:** Recently discovered mechanosensitive Piezo channels emerged as the main molecular detectors of mechanical forces. The functions of Piezo channels range from detection of touch and pain, to control of the plastic changes in different organs. Recent studies suggested the role of Piezo channels in migraine pain, which is supposed to originate from the trigeminovascular nociceptive system in meninges. Interestingly, migraine pain is associated with such phenomenon as mechanical hypersensitivity, suggesting enhanced mechanotransduction. In the current review, we present the data that propose the implication of Piezo channels in migraine pain, which has a distinctive pulsatile character. These data include: (i) distribution of Piezo channels in the key elements of the trigeminovascular nociceptive system; (ii) the prolonged functional activity of Piezo channels in meningeal afferents providing a mechanistical basis for mechanotransduction in nociceptive nerve terminals; (iii) potential activation of Piezo channels by shear stress and pulsating blood flow; and (iv) modulation of these channels by emerging chemical agonists and modulators, including pro-nociceptive compounds. Achievements in this quickly expanding field should open a new road for efficient control of Piezo-related diseases including migraine and chronic pain.

**Keywords:** Piezo channels; mechanotransduction; pain; migraine; CGRP

---

## 1. Introduction

Touch, proprioception, and nociception are the fundamental senses mediated by activation of mechanosensitive ion channels, which are expressed in various types of sensory neurons. The detection of mechanical forces is mediated by different ion channels such as mechanosensitive ion channels (MSCs), potassium K2P channels, TMEM63/OSCA, and TMC1/2 [1]. In the current review, we will focus on the potential role of the recently discovered mechanosensitive Piezo1/2 channels in the nociceptive signaling in migraine. Migraine pain remains poorly understood mainly because of lack of mechanistic explanations for the initial steps in generation of pain signals in the nociceptive system.

Migraine, a very common neurological disorder, is characterized by severe and long-lasting headache associated also with mechanical hypersensitivity and allodynia (pain induced by normally non-painful touch, which, during long-lasting migraine attack, is not limited to the head) [2]. Mechanosensitive ion channels most likely mediate mechanical hypersensitivity in the peripheral or central parts of the nociceptive system. However, nature of these ion channels remains unknown. Therefore, mechanosensitive Piezo channels, recently detected in human trigeminal ganglia [3,4], are the most probable candidates to mediate typical symptoms of migraine, such as mechanical hypersensitivity and pulsating type of migraine pain [5].

## 2. Complex Structure of Gigantic Piezo Channels

Piezo channels are the family of mechanotransducers composed by two nonselective cationic channels known as Piezo1 and Piezo2 with a relatively homologous structure (Figure 1) [6–9]. Piezo channels are 2500 amino acids long proteins with up to 38 transmembrane segments per monomer. These monomers, combined together, build up the functional homo-trimers in the cell membrane [10]. Notably, these channels do not share sequence or structural homology with other above-mentioned mechanosensitive channels. This makes Piezo channels to be a completely new molecular target for therapeutic interventions, with their own profile of preferential physical triggers, chemical agonists, and specific modulators [11].

The molecule of Piezo1 channel is organized as a gigantic 'three peripheral blade-like structure' (Figure 1A), three 90-Å-long intracellular beam–resembling components bridging the blades together [10]. The central pore, formed by the C-terminals, has a main role in determining the channel conductance and ion selectivity [10]. The intracellular beam, part of the central cap, seems to be the perfect intermediate structure for mechanical transduction from the periphery to the central ion-conducting pore [7].

The uncommon structure of Piezo channels suggests their unique role in mechanical transduction [11], including various important functions in sensory neurons [12].

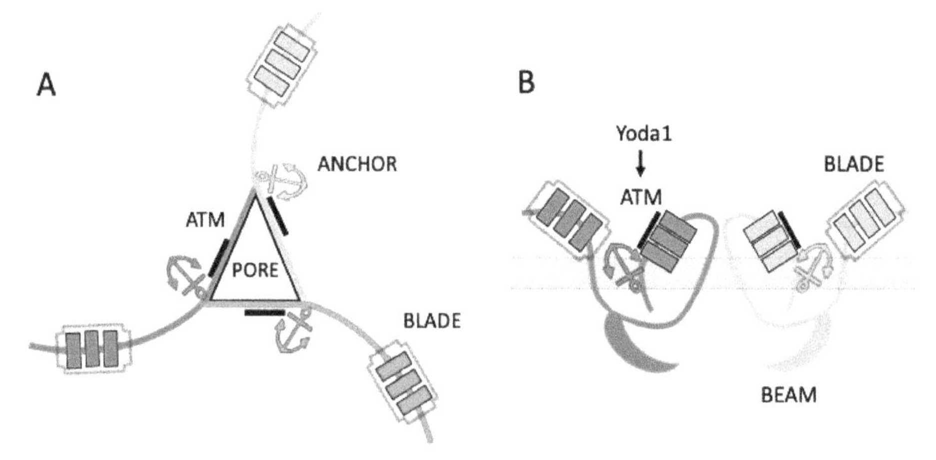


**Figure 1.** Schematic presentation of the Piezo channel. (**A**) Bird view of the Piezo channel with peripheral blade-like structures located in three subunits forming the trimeric functional unit with the central pore, blades and anchor regions. (**B**) Side view of the Piezo channel located in the lipid cell membrane. The following elements of the single subunit are presented: the intracellular beams, the C-terminals with the anchor regions and the ATM region containing the Yoda1 binding site and the extracellular blades.

## 3. Functional Properties and the Role of Piezo1 vs. Piezo2 in Nociception

After discovery, isolation and cloning of Piezo1 and Piezo2 channels [8], the next challenging step would be to describe the functional properties of these channels in different cell types [9]. The natural activation of Piezo1 and Piezo2 channels can be achieved by different types of mechanical stimulation: stretching, pulling, pushing, exposure to hypo- or hyper-osmotic solutions, and flow-induced shear stress [8,13,14]. The kinetic properties of Piezo channels, studied by recording of the whole cell-inward currents in HEK cells, showed the fast activation and inactivation of these channels with recovery in the range of hundred milliseconds [6].

The comparison of two subtypes of Piezo channels showed that Piezo2 has a faster kinetics, and typically mediates a rapid membrane response, whereas Piezo1 channels are characterized by slower kinetics [8]. Thus, Piezo2 are more specified for detection of transient mechanical forces, whereas Piezo1 can react to more persistent activation [15]. Notably, these properties are sensitive to the channel's environment in the membrane. Thus, the reduction of cholesterol in the cell membrane largely slowed down the inactivation kinetics of Piezo1 channels [16], which can support the persistence of signaling via this receptor type.

Consistent with fast kinetics, Piezo2 channels mediate the short-lasting mechanosensitive processes such as touch [6]. Piezo2 channels, implicated in this sensory modality, have been found in dorsal root ganglion (DRG) neurons [6]. Mechanical pain represents a different sensory modality, which is also expected to be mechanistically linked either to Piezo1 or Piezo2 channels if they are expressed in neurons mediating somatic and visceral pain. However, there are some controversies regarding expression of Piezo channel subtypes in nociceptive neurons. Indeed, one study showed the presence of Piezo2 but not Piezo1 channel in DRG neurons [17]. Another group [18] presented the evidence for the expression not only of Piezo2 but also of Piezo1 in DRG neurons. They found Piezo2 transcripts in sensory neurons of different sizes, including the largest diameter neurons mediating touch and proprioception. In contrast, Piezo1 was preferentially expressed in small size nociceptive neuron suggesting their role in pain [18]. Co-expression of Piezo1 and Piezo2 in the same neurons raises an interesting issue of functional interactions between these channels. The functional interplay between Piezo1 and Piezo2 channels was analyzed in the study where they found that the deletion of Piezo2 from the low-threshold mechanoreceptors in mice impaired touch but surprisingly sensitized mechanical pain, suggesting a negative interaction between Piezo1 and Piezo2 [19].

Our studies indicated that both Piezo1 and Piezo2 channels are expressed in trigeminal sensory neurons [5], which innervate head and face tissues and implicated in generation of migraine pain. Most of electrophysiological studies of Piezo channels in sensory neurons were performed by recording signaling from somas of these cells, representing a surrogate model of nerve terminals. However, more physiologically relevant approach to study the role of Piezo channels in migraine pain would be the recording of electrical spiking activity directly from the trigeminal nerve terminals in brain meninges [20]. Cranial meninges, comprising abundant blood vessels that are densely innervated by somatic and autonomous nerves, represent the so-called 'trigeminovascular system', which is considered as the origin site of primary headaches including migraine [21]. Application of the pro-inflammatory compounds to meninges induces mechanical sensitization of meningeal nociceptors [22], suggesting involvement of the professional mechanotransducers such as Piezo1 channels in this phenomenon. The repetitive nociceptive traffic in trigeminal neurons is a likely reason for mechanical hypersensitivity and allodynia, typical for migraine pathology [23]. Mechanical hypersensitivity can be directly linked to activation of Piezo channels in peripheral neurons whereas allodynia, mostly a central phenomenon, nevertheless, can also start from excessive and repetitive activation of peripheral Piezo channels in primary afferents. Taken together, these studies can serve as a background for the hypothesis on the role of Piezo channels in migraine.

## 4. Unusual Chemical Activation of Piezo Channels

Since the discovery of Piezo mechanotransducers, the main tools to activate Piezo channels were the different types of mechanical stimulation. Unexpectedly, a very efficient alternative approach to activate Piezo channels has been recently found: that is the compound called Yoda1 [24]. The small lipid soluble molecule Yoda1 is able to activate specifically Piezo1 but not Piezo2 channels [24]. Yoda1 interacts with the C-terminal of the Piezo1 protein in the region of 1961–2063 amino acids, also known as the Agonist Transduction Motif (ATM) [25] (Figure 1B). The recent molecular dynamic simulations identified the Yoda1 binding pocket located in the domain approximately 40 Å away from the central pore [26]. Although the Piezo1 channel has three interacting monomers, interestingly, the binding of Yoda1 to only one subunit is already enough to open the ion channel [25], which provides a rationale for the high sensitivity of Piezo1 to this chemical agonist.

The other Piezo1 agonist, Jedi1/2, acts on the blade-beam structure inducing activation of the channel from the peripheral extracellular side [10]. However, it remains to be discovered if there are any endogenous molecules, which, similarly to the synthetic Yoda1 or Jedi1/2 compounds, can activate and/or sensitize Piezo1 channels in the healthy or disease states.

The discovery of chemical agonists of Piezo channels opened a new toolbox to investigate the function of mechanotransduction in different tissues, especially, when the traditional mechanical stimulation is not applicable or when the aim is to provide a widespread activation of Piezo channels in multiple targets. Thus, we found that the application of Yoda1 to the extended receptive field of meningeal afferents induced a massive and prolonged activation of trigeminal nerve fibers [5]. The nociceptive effect of Yoda1 in this study was reproduced by the similar prolonged activation of trigeminal mechanosensitive receptors by the hypo-osmotic solution [5]. Moreover, Yoda1 stimulation triggered the release of CGRP (calcitonin gene-related peptide) from these trigeminal nerve fibers. CGRP, the main migraine mediator, is known as a powerful promoter of meningeal inflammation and sensitization of trigeminal neurons [27–29].

These findings largely supported the proposed role of Piezo channels in peripheral mechanisms of migraine pain. Consistent with the pro-nociceptive role of Piezo channels activated by Yoda1 found in our study, Wang et al. [18] showed that Yoda1 induced a mechanical hyperalgesia with the prolonged time-course which is also typical for migraine pain.

## 5. Puzzling Phenomenon of Pulsatile Pain: Role of Piezo?

The headache phase of a migraine attack is characterized by pulsating (throbbing) type of headache as a specific symptom of migraine pain [30]. This puzzling migraine phenomenon has been attracting attention of many migraine researchers and already served as a basis for the famous Wollf's vascular theory of migraine headache [31]. This popular theory was supported by the clear correlation between the level of pulsations of the temporal artery and the intensity of headache during migraine attack in the patient treated with ergots [31]. Vascular mechanisms of migraine pain were also supported by the ability of the migraine mediator neuropeptide CGRP to produce the dilation of cranial vessels [32,33]. However, the role of vessels as the contributors to headache during migraine attack is still debated. As noted by Waeber and Moskowitz [34], the idea of abnormal dilatation of intracranial blood vessels leading to mechanical excitation of sensory fibers that innervate these vessels has never been validated. It is worth noting that there are also observations that the simple vasodilation of cranial vessels induced by the neuropeptide VIP is not enough to trigger migraine attack per se [35]. These findings added more intriguing aspects to the puzzling mechanisms of pulsating migraine pain. Notably, the 'vascular theory' suggested only the general explanation of pulsating migraine pain due to regular dilations of cranial vessels but it did not propose a molecular mechanism supporting this view. The discovery of Piezo mechanotransducers in trigeminal neurons suggested an attractive possibility to test these specific transducers as the sensors of the mechanical forces generated by pulsating vessels. Indeed, our recent investigation of Piezo channels in meningeal sensory nerve fibers allowed us to suggest a new model of mechanosensation in meninges during migraine attack [5].

## 6. New Model of Mechanosensation in Meninges during Migraine Attack

Based on the concept of the meningeal trigeminovascular system (TGVS) as the initial site for the generation of migraine headache [21], we propose the following model potentially explaining the mechanism of pulsating migraine pain. Figure 2 shows that in interictal state (or in healthy subjects), meningeal nerves express a plethora of pain transducing channels such as mechanosensitive Piezo1 and Piezo2 receptors [5], along with capsaicin-sensitive TRPV1 [36] and ATP-activated P2X3 receptors [37], which all are in low-active non-sensitized state. The main key components of the TGVS, such as meningeal vessels and neighboring nerves, in this state, have a low chance to interact to each other either physically or chemically and a low probability to generate pain signals.

The key event in migraine attack, is the release of the main migraine mediator CGRP [33], which has multiple actions in the TGVS (Figure 2, right). Thus, CGRP induces dilation of meningeal vessels, it promotes local neurogenic inflammation and degranulates dural mast cells, which, in turn, release several pro-inflammatory and pro-nociceptive compounds such as serotonin, histamine, ATP, prostaglandins, and nitric oxide [38–40]. Notably, most of these compounds are able to increase the sensitivity of meningeal afferents to mechanical stimuli [22,41]. Among other compounds released from mast cells, serotonin appeared to be the most strong and fast trigger of nociceptive spiking in nerve terminals [42,43]. Apart from the immediate firing of nociceptors, serotonin also promotes neuronal sensitization and local inflammation directly or through the additional release of CGRP [42]. The inflammation induced by CGRP, substance P and mediators of mast cells can sensitize not only peripheral but also central neurons, expanding the enhanced mechanical sensitivity to extracranial body regions, presented as the phenomenon of allodynia [22,44].

Notably, Piezo1 channels are expressed not only in neurons but also in vessels. Vasodilatation, shear stress and enhanced pulse waves in dilated vessels can activate mechanoreceptors in endothelial cells, triggering, via pannexins, ATP release (Figure 2) [45]. ATP is a strong promoter of nociceptive firing in meninges by itself [37,43] but it is also a trigger of mast cell degranulation promoting further release of the pro-inflammatory and pro-nociceptive compounds to meninges.

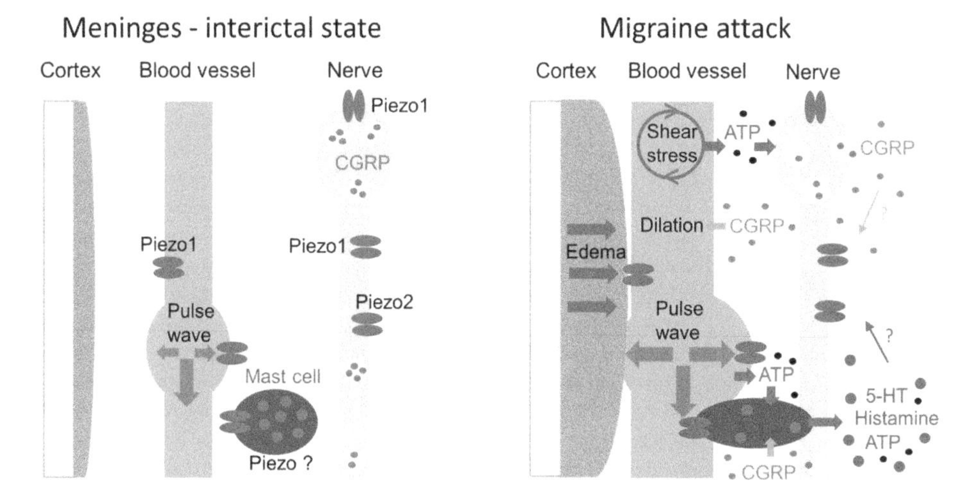


**Figure 2.** Schematic presentation of the key elements of the trigeminovascular system comprising meningeal blood vessels, local mast cells and trigeminal nerve fibers before and during migraine attack. **Left:** In the interictal state, before attack, there are only slight pulsations of meningeal vessels with the minimal activation of vascular Piezo1 channels or Piezo1 and Piezo2 channels in nerve fibers. **Right:** During migraine attack, which is often associated with brain oedema and CGRP-induced dilation of vessels, the extracellular space is reduced, promoting more close contact between pulsating vessels, nerves and nearby mast cells. The shear stress in dilated vessels and enhanced vascular pulsations promote mechanosensitive ATP release from the endothelial cells. Mechanical stimulation of calcium permeable Piezo channels in nerve fibers by pulsating vessels promotes neuronal CGRP release. CGRP and ATP can degranulate mast cells directly. In addition, the fraction of mast cells contacting vessels, is directly mechanically activated by blood pulsations. Activation of mast cells induces release of a plethora of pro-nociceptive compounds such as serotonin, histamine, leukotrienes, prostaglandins, ATP, and nitric oxide, further exciting the nociceptive fibers and promoting more CGRP release. All these pro-inflammatory compounds, in long run, together with CGRP, promote neuroinflammation, neuronal sensitization leading to long-lasting pulsating pain.

Approximately one third of migraine cases is represented by migraine with aura. The pathophysiological mechanism underlying migraine aura is a phenomenon called 'cortical spreading depression' (CSD) [46,47]. CSD is a wave of strong depolarization of cortical neurons and glial cells, leading to meningeal neurogenic inflammation, involving the neuropeptide CGRP and substance P, ATP, and the mast cells activation [41,48]. Brain oedema, associated with CSD [46,49] compresses the extracellular space, which assists in the formation of more close contact between blood vessels and meningeal nerve fibers. Furthermore, CSD can increase the brain volume and raise the intracranial pressure [46]. Taken together, these factors, along with strongly pulsating vessels, should facilitate activation of mechanosensitive channels, such as Piezo ones, in nerve fibers [5].

Since meningeal tissues are protected by the skull from the external mechanical forces, the only source for activation of mechanosensitive Piezo channels in meningeal nerves are internal triggers, such as pulsating dural vessels. These pulsating dilated vessels can provide the regular Piezo-mediated excitation of nerve fibers processed towards the high-pain centers and perceived as a pulsating migraine pain. Notably, in migraine condition, there is an increased expansibility of arteries [50], which is also consistent with our hypothesis as the factor supporting the increased amplitude of pulsating waves. The exaggerated mechanical sensitivity of the trigeminal nerves during migraine attack can explain also the painful sensitivity to slight movements of the head or to cough [41].

## 7. How to Alleviate Migraine Pain Through Piezo Channels?

Accumulating evidence suggest that Piezo channels represent an attractive new molecular target to block pathological pain conditions via inhibition of these membrane's mechanotransducers. Likewise, the idea of targeting Piezo channel for the novel type of analgesia might be extended to migraine pain. Given the pro-nociceptive role of Piezo channels in activation of meningeal afferents [5], the attractive approach would be to alleviate migraine pain through the direct or indirect inhibition of Piezo channels in trigeminal neurons. The antagonists of Piezo channels are of special interest as they potentially can block excessive activation of nociceptors in migraine conditions. However, currently available ligands of Piezo channels (Table 1) cannot directly serve as potential medicines for this neurological disorder. This is because the list of blockers of Piezo channels is limited and most of them are not specific (Table 1). Thus, there are Piezo channel inhibitors, such as the antagonist Dooku1, the blockers neurotoxin GsMTx4 and gadolinium [51]. Unfortunately, Dooku1 has a partial blocking activity in some cell types, whereas GsMTx4 or gadolinium are not specific for Piezo channels [51]. The main issue is that even if the specific blocker of Piezo1 is found, Piezo1 channels are not limited to trigeminal neurons but expressed in other cell types such as vessels (as mentioned above). Therefore, for the translational purposes, in migraine pain the alternative promising approach would be to find out the functional partners of these mechanosensitive channels or accessory proteins, which control the function of Piezo proteins. Although a purified Piezo1 protein retains the channel activity when reconstructed in artificial lipid bilayers [52]; in the natural environment, Piezo channels are likely under control of membrane lipids and certain intracellular messengers. The beforementioned functional interactions between Piezo1 and Piezo2, in trigeminal neurons, are of special interest but the underlying mechanism of such interactions requires further investigations. Likewise, little is known so far about the intracellular modulatory pathways for Piezo channels. However, it has been shown that, in response to phospholipase C (PLC) activation, both Piezo1 and Piezo2 channels could be inhibited by depletion of plasma membrane phosphoinositides [53]. The activation of PLC was supported by calcium influx via TRPV1 receptors, which are highly expressed in meningeal trigeminal afferents [36]. This finding suggests a negative functional crosstalk between Piezo and TRPV1 channels, which both are enriched in meningeal nociceptive fibers generating migraine pain [5,36]. Thus, our data showed that a fraction of trigeminal nerve fibers in meninges responded by nociceptive firing both to the TRPV1 agonist capsaicin and to the Piezo1 agonist Yoda1 [5]. The latter fact suggests that in trigeminal afferents in meninges there is a specific profile of nociceptive receptors, which can functionally interact.

In contrast, in DRG sensory neurons, which mediate somatic pain, Piezo1 has a limited co-expression with TRPV1 channels [18].

An interesting family of potential Piezo modulators has been identified in the study showing that margaric acid, a saturated fatty acid that makes membrane stiffer, inhibits Piezo1 channels, whereas the poly-unsaturated fatty acid docosahexaenoic acid delays the inactivation of these channels [16] (Table 1). A dietary strategy to diminish the increased activity of Piezo channels in hemolytic anemia suggested by these authors could be extended to the excessive nociception via Piezo channels in migraine pain states associated with mechanical hyperalgesia. In particular, food enriched by margaric acid or other similarly actin fatty acids looks like a promising approach for these aims.

Another open question, which has a translational perspective in migraine, is whether Piezo channels are modulated by the migraine mediators such as CGRP, serotonin, or NO. One study proposed the role of Epac1–Piezo2 axis in sensory neurons for the development of mechanical allodynia during neuropathic pain [54]. Epac, known as the sensor of cAMP, contains the conserved cAMP-binding domain and the level of cAMP can be raised in trigeminal neurons by the main migraine mediator CGRP [27,33]. However, all these pathways, activated by migraine mediators, which potentially can control Piezo channels, were not studied yet in migraine pathology.

Taken together, the pro-nociceptive Piezo channels in the trigeminovascular system represent a new promising target for the therapeutic interventions in migraine. Thus, the analgesic effect in migraine pain, can be approached either by the targeted delivery to the TGVS of the novel potent Piezo blockers, or by the inactivation of these mechanotransducers by modulators acting via changes in the lipid membrane environment.

**Table 1.** Main agonists, antagonists and modulators of Piezo1 channels.

| **Agonists** | | **Antagonists** | | | **Modulators** | |
|---|---|---|---|---|---|---|
| Yoda1 | Jedi1/2 | Dooku1 | Gadolinium | GsMTx4 | Margaric acid (saturated) | Docosahexaenoic acid (unsaturated) |
| Selective | | Selective | Nonselective | | Nonselective | |
| Acting sites | | | | | | |
| ATM region in C-terminus [24,25] | L15-16/L19-20 regions [10] | Yoda1 binding site [51] | Ion channel pore [51] | | Changes in membrane lipid environment [16] | |
| | | | | | accelerated inactivation | reduced inactivation |

## 8. Summary and Outlook

Migraine is a common neurological disorder with many intractable cases. Despite the many decades of active investigations of this disorder, the molecular mechanisms implicated in the initial steps of migraine pain remain unclear. The recent studies suggesting Piezo channels as the vascular and neuronal sensors of intracranial mechanical forces present a new view on molecular processes implicated in meningeal nociception leading to migraine headache. These mechanosensitive mechanisms can underlie the worst migraine's symptoms such as mechanical hyperalgesia and pulsating pain. Identification of novel molecular targets in meningeal trigeminovascular nociceptive system such as Piezo channels suggests the new approach to control this devastating neurological condition.

**Funding:** This research was funded by the Finnish Academy, grant number 325392.

**Conflicts of Interest:** The review was conducted in the absence of any commercial or financial relationships that could be construed as a potential conflict of interest.

## Abbreviations

| | |
|---|---|
| MSCs | Mechanosensitive ion channels |
| K2P | Two-pore-domain potassium channels |
| TMEM63 | Transmembrane protein 63 |
| OSCA | Hyperosmolality-gated calcium-permeable channels |
| TMC1/2 | Transmembrane channel-like protein 1/2 |
| DRG | Dorsal Root Ganglion |
| CSD | Cortical spreading depression |
| CGRP | Calcitonin gene-related peptide |
| NO | Nitric Oxides |
| Epac | Exchange protein activated by cAMP |
| cAMP | Cyclic Adenosine Monophosphate |
| TGVS | Trigeminovascular system |

## References


1. Douguet, D.; Honore, E. Mammalian mechanoelectrical transduction: Structure and function of force-gated ion channels. *Cell* **2019**, *179*, 340–354. [CrossRef]
2. Benatto, M.T.; Florencio, L.L.; Carvalho, G.F.; Dach, F.; Bigal, M.E.; Chaves, T.C.; Bevilaqua-grossi, D. Cutaneous allodynia is more frequent in chronic migraine, and its presence and severity seems to be more associated with the duration of the disease. *Arq. Neuropsiquiatr.* **2017**, *75*, 153–159. [CrossRef]
3. LaPaglia, D.M.; Sapio, M.R.; Burbelo, P.D.; Thierry-Mieg, J.; Thierry-Mieg, D.; Raithel, S.J.; Ramsden, C.E.; Iadarola, M.J.; Mannes, A.J. RNA-Seq investigations of human post-mortem trigeminal ganglia. *Cephalalgia* **2018**, *38*, 912–932. [CrossRef]
4. Nguyen, M.Q.; Wu, Y.; Bonilla, L.S.; Von Buchholtz, L.J.; Nicholas, J.; Ryba, P. Diversity amongst trigeminal neurons revealed by high throughput single cell sequencing. *PLoS ONE* **2017**, *12*, e0185543. [CrossRef]
5. Mikhailov, N.; Leskinen, J.; Fagerlund, I.; Poguzhelskaya, E.; Giniatullina, R.; Gafurov, O.; Malm, T.; Karjalainen, T.; Gröhn, O.; Giniatullin, R. Mechanosensitive meningeal nociception via Piezo channels: Implications for pulsatile pain in migraine? *Neuropharmacology* **2019**, *149*, 113–123. [CrossRef]
6. Bagriantsev, S.N.; Gracheva, E.O.; Gallagher, P.G. Piezo proteins: Regulators of mechanosensation and other cellular processes. *J. Biol. Chem.* **2014**, *289*, 31673–31681. [CrossRef] [PubMed]
7. Zhao, Q.; Zhou, H.; Chi, S.; Wang, Y.; Wang, J.; Geng, J.; Wu, K.; Liu, W.; Zhang, T.; Dong, M.Q.; et al. Structure and mechanogating mechanism of the Piezo1 channel. *Nature* **2018**, *554*, 487–492. [CrossRef] [PubMed]
8. Coste, B.; Mathur, J.; Schmidt, M.; Earley, T.J.; Ranade, S.; Petrus, M.J.; Dubin, A.E.; Patapoutian, A. Piezo1 and Piezo2 are essential components of distinct mechanically activated cation channels. *Science* **2010**, *330*, 55–60. [CrossRef] [PubMed]
9. Gottlieb, P.A. A tour de force: The discovery, properties, and function of Piezo channels. *Curr. Top. Membr.* **2017**, *79*, 1–36. [PubMed]
10. Wang, Y.; Chi, S.; Guo, H.; Li, G.; Wang, L.; Zhao, Q.; Rao, Y.; Zu, L.; He, W.; Xiao, B. A lever-like transduction pathway for long-distance chemical- and mechano-gating of the mechanosensitive Piezo1 channel. *Nat. Commun.* **2018**, *9*, 1300. [CrossRef] [PubMed]
11. Gottlieb, P.A.; Sachs, F. Properties of a cation selective mechanical channel. *Landes Bioscence* **2012**, *6*, 1–6.
12. Anderson, E.O.; Schneider, E.R.; Bagriantsev, S.N. Piezo2 in cutaneous and proprioceptive mechanotransduction in vertebrates. *Curr. Top. Membr.* **2017**, *79*, 197–217.
13. Ge, J.; Li, W.; Zhao, Q.; Li, N.; Chen, M.; Zhi, P.; Li, R.; Gao, N.; Xiao, B.; Yang, M. Architecture of the mammalian mechanosensitive Piezo1 channel. *Nature* **2015**, *527*, 64–69. [CrossRef] [PubMed]
14. Li, J.; Hou, B.; Tumova, S.; Muraki, K.; Bruns, A.; Ludlow, M.J.; Sedo, A.; Hyman, A.J.; McKeown, L.; Young, R.S.; et al. Piezo1 integration of vascular architecture with physiological force. *Nature* **2014**, *515*, 279–282. [CrossRef] [PubMed]
15. Lewis, A.H.; Cui, A.F.; McDonald, M.F.; Grandl, J. Transduction of repetitive mechanical stimuli by Piezo1 and Piezo2 ion channels. *Cell Rep.* **2017**, *19*, 2572–2585. [CrossRef]

16. Romero, L.O.; Massey, A.E.; Mata-Daboin, A.D.; Sierra-Valdez, F.J.; Chauhan, S.C.; Cordero-Morales, J.F.; Vásquez, V. Dietary fatty acids fine-tune Piezo1 mechanical response. *Nat. Commun.* **2019**, *10*, 1–14. [CrossRef]
17. Ranade, S.S.; Woo, S.; Dubin, A.E.; Moshourab, R.A.; Wetzel, C.; Petrus, M.; Mathur, J.; Bégay, V.; Coste, B.; Mainquist, J.; et al. Piezo2 is the major transducer of mechanical forces for touch sensation in mice. *Nature* **2014**, *516*, 121–125. [CrossRef]
18. Wang, J.; La, J.; Hamill, O.P. Piezo1 is selectively expressed in small diameter mouse DRG neurons distinct from neurons strongly expressing TRPV1. *Front. Mol. Neurosci.* **2019**, *12*, 1–15. [CrossRef]
19. Zhang, M.; Wang, Y.; Geng, J.; Zhou, S.; Xiao, B. Mechanically activated Piezo channels mediate touch and suppress acute mechanical pain response in mice. *Cell Rep.* **2019**, *26*, 1419–1431. [CrossRef]
20. Bolay, H.; Reuter, U.; Dunn, A.K.; Huang, Z.; Boas, D.A.; Moskowitz, M.A. Intrinsic brain activity triggers trigeminal meningeal afferents in a migraine model. *Nat. Med.* **2002**, *8*, 136–142. [CrossRef]
21. Messlinger, K. Migraine: Where and how does the pain originate? *Exp. Brain Res.* **2009**, *196*, 179–193. [CrossRef] [PubMed]
22. Yan, J.; Wei, X.; Bischoff, C.; Edelmayer, R.M.; Dussor, G. Research submission pH-evoked dural afferent signaling is mediated by ASIC3 and is sensitized by mast cell mediators. *Headache* **2013**, *53*, 1250–1261. [CrossRef]
23. Noseda, R.; Burstein, R. Migraine pathophysiology: Anatomy of the trigeminovascular pathway and associated neurological symptoms, CSD, sensitization and modulation of pain. *Pain* **2013**, *154*, 44–53. [CrossRef] [PubMed]
24. Syeda, R.; Xu, J.; Dubin, A.E.; Coste, B.; Mathur, J.; Huynh, T.; Matzen, J.; Lao, J.; Tully, D.C.; Engels, I.H.; et al. Chemical activation of the mechanotransduction channel Piezo1. *Elife* **2015**, *4*, 1–11. [CrossRef]
25. Lacroix, J.J.; Botello-Smith, W.M.; Luo, Y. Probing the gating mechanism of the mechanosensitive channel Piezo1 with the small molecule Yoda. *Nat. Commun.* **2018**, *9*, 2029. [CrossRef]
26. Botello-smith, W.M.; Jiang, W.; Lacroix, J.J.; Luo, Y. A mechanism for the activation of the mechanosensitive Piezo1 channel by the small molecule Yoda1. *Nat. Commun.* **2019**, *10*, 4503. [CrossRef] [PubMed]
27. Fabbretti, E.; Arco, M.D.; Fabbro, A.; Simonetti, M.; Nistri, A.; Giniatullin, R. Delayed upregulation of ATP P2X 3 receptors of trigeminal sensory neurons by calcitonin gene-related peptide. *J. Neurosci.* **2006**, *26*, 6163–6171. [CrossRef]
28. D'Arco, M.; Giniatullin, R.; Simonetti, M.; Fabbro, A.; Nair, A.; Nistri, A. Neutralization of nerve growth factor induces plasticity of ATP-sensitive P2X 3 receptors of nociceptive trigeminal ganglion neurons. *J. Neurosci.* **2007**, *27*, 8190–8201. [CrossRef] [PubMed]
29. Giniatullin, R.; Nistri, A.; Fabbretti, E. Molecular mechanisms of sensitization of pain-transducing P2X 3 receptors by the migraine mediators CGRP and NGF. *Mol. Neurobiol.* **2008**, *37*, 83–90. [CrossRef]
30. Vincent, M.; Wang, S. Headache classification committee of the International Headache Society (IHS) The International Classification of Headache Disorders, 3rd edition. *Cephalalgia* **2018**, *38*, 1–211.
31. Wolff, H.G. *Wolff's Headache and Other Head Pain*, 1st ed.; Oxford University Press: Oxford, UK, 1948.
32. Edvinsson, L. CGRP as the target of new migraine therapies—Successful translation from bench to clinic. *Nat. Rev. Neurol.* **2018**, *14*, 338–350. [CrossRef] [PubMed]
33. Hargreaves, R.; Olesen, J. Calcitonin gene-related peptide modulators–The history and renaissance of a new migraine drug class. *Headache Curr.* **2019**, *59*, 951–970. [CrossRef] [PubMed]
34. Waeber, C.; Moskowitz, M.A. Migraine as an inflammatory disorder. *Neurology* **2005**, *64*, S9–S15. [CrossRef] [PubMed]
35. Rahmann, A.; Wienecke, T.; Hansen, J.M.; Fahrenkrug, J.; Olesen, J.; Ashina, M. Vasoactive intestinal peptide causes marked cephalic vasodilation, but does not induce migraine. *Cephalalgia* **2008**, *28*, 226–236. [CrossRef]
36. Zakharov, A.; Vitale, C.; Kilinc, E.; Koroleva, K.; Fayuk, D.; Shelukhina, I. Hunting for origins of migraine pain: Cluster analysis of spontaneous and capsaicin-induced firing in meningeal trigeminal nerve fibers. *Front. Cell. Neurosci.* **2015**, *9*, 1–14. [CrossRef]
37. Yegutkin, G.G.; Guerrero-toro, C.; Kilinc, E.; Koroleva, K. Nucleotide homeostasis and purinergic nociceptive signaling in rat meninges in migraine-like conditions. *Purinergic Signal.* **2016**, *12*, 561–574. [CrossRef]
38. Levy, D. Migraine pain, meningeal inflammation, and mast cells. *Curr. Pain Headache Rep.* **2009**, *13*, 237–240. [CrossRef]

39. Mekori, Y.A.; Metcalfe, D.D. Mast cells in innate immunity. *Immunol. Rev.* **2000**, *173*, 131–140. [CrossRef] [PubMed]
40. Demartini, C.; Greco, R.; Maria, A.; Sances, G.; De Icco, R.; Borsook, D.; Tassorelli, C.; Foundation, I.C.M. Progress in neurobiology nitroglycerin as a comparative experimental model of migraine pain: From animal to human and back. *Prog. Neurobiol.* **2019**, *177*, 15–32. [CrossRef] [PubMed]
41. Strassman, A.M.; Raymond, S.A.; Burstein, R. Sensitization of meningeal sensory neurons and the origin of headaches. *Nature* **1996**, *384*, 560–564. [CrossRef] [PubMed]
42. Kilinc, E.; Guerrero-toro, C.; Zakharov, A.; Vitale, C.; Gubert-olive, M.; Koroleva, K.; Timonina, A.; Luz, L.L.; Shelukhina, I.; Giniatullina, R.; et al. Neuropharmacology serotonergic mechanisms of trigeminal meningeal nociception: Implications for migraine pain. *Neuropharmacology* **2017**, *116*, 160–173. [CrossRef] [PubMed]
43. Koroleva, K.; Gafurov, O.; Guselnikova, V.; Nurkhametova, D. Meningeal mast cells contribute to ATP-induced nociceptive firing in trigeminal nerve terminals: Direct and indirect purinergic mechanisms triggering migraine pain. *Front. Cell. Neurosci.* **2019**, *13*, 1–7. [CrossRef] [PubMed]
44. Burstein, R.; Yamamura, H.; Malick, A.; Strassman, A.M. Chemical stimulation of the intracranial dura induces enhanced responses to facial stimulation in brain stem trigeminal neurons. *J. Neurophysiol.* **2017**, *79*, 964–982. [CrossRef] [PubMed]
45. Wang, S.; Chennupati, R.; Kaur, H.; Iring, A.; Wettschureck, N.; Offermanns, S. Endothelial cation channel PIEZO1 controls blood pressure by mediating flow-induced ATP release. *J. Clin. Investig.* **2016**, *126*, 4527–4536. [CrossRef] [PubMed]
46. Somjen, G.G. Mechanisms of spreading depression and hypoxic spreading depression-like depolarization. *Physiol. Rev.* **2001**, *81*, 1065–1096. [CrossRef] [PubMed]
47. Aristides, A.P.L. Spreading depression of activity in the cerebral cortex. *J. Neurophysiol.* **1944**, *7*, 359–390.
48. Karatas, H.; Erdener, S.E.; Gursoy-Ozdemir, Y.; Lule, S.; Eren-Koçak, E.; Sen, Z.D.; Dalkara, T. Spreading depression triggers headache by activating neuronal Panx1 channels. *Science* **2013**, *350*, 1092–1096. [CrossRef]
49. Anderson, T.R.; Andrew, R.D. Spreading depression: Imaging and blockade in the rat neocortical brain slice. *J. Neurophysiol.* **2002**, *88*, 2713–2725. [CrossRef]
50. Viola, S.; Viola, P.; Buongarzone, M.P.; Fiorelli, L.; Litterio, P. The increased distensibility of the wall of cerebral arterial network may play a role in the pathogenic mechanism of migraine headache. *Neurol. Sci.* **2014**, *35*, 163–166. [CrossRef]
51. Ridone, P.; Vassalli, M.; Martinac, B. Piezo1 mechanosensitive channels: What are they and why are they important. *Biophys. Rev.* **2019**, *11*, 795–805. [CrossRef]
52. Ranade, S.S.; Qiu, Z.; Woo, S.-H.; Hur, S.S.; Murthy, S.E.; Cahalan, S.M.; Xu, J.; Mathur, J.; Bandell, M.; Coste, B.; et al. Piezo1, a mechanically activated ion channel, is required for vascular development in mice. *Proc. Natl. Acad. Sci. USA* **2014**, *111*, 10347–10352. [CrossRef] [PubMed]
53. Borbiro, I.; Rohacs, T. Regulation of Piezo channels by cellular signaling pathways. *Curr. Top. Membr.* **2017**, *79*, 245–261. [PubMed]
54. Eijkelkamp, N.; Linley, J.E.; Torres, J.M.; Bee, L.; Dickenson, A.H.; Gringhuis, M.; Minett, M.S. A role for Piezo2 in EPAC1-dependent mechanical allodynia. *Nat. Commun.* **2013**, *4*, 1682. [CrossRef] [PubMed]

International Journal of
*Molecular Sciences*

*Review*

# Migraine: Experimental Models and Novel Therapeutic Approaches

**Giuseppe Tardiolo, Placido Bramanti and Emanuela Mazzon ***

IRCCS Centro Neurolesi "Bonino Pulejo", 98124 Messina, Italy; giuseppe.tardiolo@irccsme.it (G.T.); placido.bramanti@irccsme.it (P.B.)
* Correspondence: emanuela.mazzon@irccsme.it; Tel.: +39-090-6012-8172

Received: 19 April 2019; Accepted: 13 June 2019; Published: 15 June 2019

**Abstract:** Migraine is a disorder affecting an increasing number of subjects. Currently, this disorder is not entirely understood, and limited therapeutic solutions are available. Migraine manifests as a debilitating headache associated with an altered sensory perception that may compromise the quality of life. Animal models have been developed using chemical, physical or genetic modifications, to evoke migraine-like hallmarks for the identification of novel molecules for the treatment of migraine. In this context, experimental models based on the use of chemicals as nitroglycerin or inflammatory soup were extensively used to mimic the acute state and the chronicity of the disorder. This manuscript is aimed to provide an overview of murine models used to investigate migraine pathophysiology. Pharmacological targets as 5-HT and calcitonin gene-related peptide (CGRP) receptors were evaluated for their relevance in the development of migraine therapeutics. Drug delivery systems using nanoparticles may be helpful for the enhancement of the brain targeting and bioavailability of anti-migraine drugs as triptans. In conclusion, the progresses in migraine management have been reached with the development of emerging agonists of 5-HT receptors and novel antagonists of CGRP receptors. The nanoformulations may represent a future perspective in which already known anti-migraine drugs showed to better exert their therapeutic effects.

**Keywords:** migraine; animal models; experimental approaches

## 1. Introduction

Migraine is a complex neurological disease considered the primary headache disorder leading to disabling conditions [1]. The last International Classification of Headache Disorders (3rd edition) describes migraine as a recurrent headache disorder manifesting as a unilateral and throbbing headache with pain intensity from moderate to severe [2]. Common symptoms observed are photophobia and phonophobia, nausea and/or vomiting [2]. Migraine can be distinguished in episodic migraine (EM) and chronic migraine (CM). EM is defined whether the headache days per month are less than 15, and CM, whether headache days are equal or more than 15 for a period of more than three months. Moreover, migraine appears to occur more in women than in men [3]. In addition, it is estimated that about 30% of patients experience an aura that consists in a short period of visual, sensory, or motor disturbances [4]. Mechanisms involved in migraine are not yet entirely clarified. To date, it is thought that the genesis of pain occurs by activation of the trigeminovascular system (TGVS). This system is composed of the cranial vasculature, the trigeminal nerve and the trigeminal nucleus caudalis (TNC). TGVS plays an important role as a major control center in regulating the cerebral blood flow and it is believed as a key conduit for pain transmission [5]. The activation of trigeminal sensory nerve endings induces the release of vasoactive agents, such as calcitonin gene-related peptide (CGRP), substance P and neurokinin A, resulting in vasodilation and dural plasma extravasation, leading to neurogenic inflammation [5]. Current migraine treatments regard the use of drugs aimed to decrease the frequency, severity, and last of migraine attacks [6]. In case of mild attacks, medications such as acetaminophen and aspirin are used, whereas triptans or dihydroergotamine (DHE) are used for the treatment of a moderate to severe migraine [7,8]. The current pharmacological treatment used in migraine is summarized in Table 1. Our current knowledge about the pathophysiology of this complex disorder is based mostly on animal models developed to study the nociceptive pathways of the TGVS and their ascending projections to the brainstem and the diencephalic nuclei. These models are mostly based on modifications operated in order to try mimicking headache symptoms, requiring manipulations to activate the trigeminal nerve or dural nociceptors [5,9,10]. Although animal models show shortcomings per se, due to difficulties in reflecting all hallmarks of migraine, they have been used as a screening tool for the development of novel anti-migraine drugs, in which serotonin (5-HT) and CGRP receptors have importantly contributed as targets, showing to be involved in pathways that result in headache attacks [11]. Furthermore, drug delivery systems based on the use of formulations composed of nanoparticles could be considered a new attempt to improve the effects of drugs used in migraine treatments. The present manuscript has been generated using PubMed as source of information, with the aim to offer an overview of murine models developed to study migraine pathophysiology. For the promising results obtained in migraine treatment, in this manuscript the pharmacological targets 5-HT and CGRP receptors were evaluated. In addition, the database ClinicalTrials.gov [12] has been used as source of information for the new emerging treatments using agonists of 5-HT receptors and new antagonists of CGRP receptors. Validated animal models of migraine are summarized to provide an overview. At last, new therapeutic strategies and nanoparticles tested in experimental models performed on rats or mice in the last years were considered.

**Table 1.** Current pharmacological treatment in migraine.

| Medication | Mechanism of Action | Potential Common Side Effects | Ref. |
| --- | --- | --- | --- |
| NSAIDs (nonsteroidal anti-inflammatory drugs) | Inhibition of the synthesis of prostanoids; cyclooxygenase enzymes (COX-1 and/or COX-2) inhibitors. | Nausea or vomiting, dyspepsia and diarrhea. | [13,14] |
| Ergotamine and dihydroergotamine (DHE) | Agonist to: 5-HT -1B, -1D and -1F receptors; D -1, -2, dopamine receptors. Partial agonists to alpha -1A, -1B, 1D adrenergic receptors. | Nausea and vomiting. | [13,14] |
| beta-blockers (metoprolol, propranolol, timolol) | beta -1, -2, -3, adrenergic receptors blockers. | Bradycardia, hypotension, nausea, diarrhea, bronchospasm, dyspnea, fatigue, insomnia, and dizziness. | [13,14] |
| Triptans | Agonists to 5-HT -1B,1D receptors | Dizziness, fatigue, dry mouth, flushing, feeling hot or cold, chest pain. | [13,14] |
| Opioids | Agonist μ opioid receptors | Constipation, itchiness, and nausea. | [13,14] |
| Antiepileptic drugs (divalproex sodium, valproate sodium, topiramate) | Sodium channel blockade Calcium channel blockade GABA agonism/potentiation | Nausea, vomiting, dizziness, diarrhea, drowsiness, constipation and dry mouth. | [13,14] |
| Antiemetics (metoclopramide, prochlorperazine) | Dopamine receptor antagonists (at D1, D2, D3 and D4 receptors) | Fatigue, constipation, ringing in the ears, dry mouth, restlessness, and muscle spasms. | [13,14] |

## 2. Pharmacological Targets in Migraine Treatment: 5-HT and CGRP Receptors

Research advances are increasingly focusing on the development of anti-migraine drugs. In this context, 5-HT and CGRP receptors have been mainly investigated. 5-HT receptor agonists and CGRP receptor antagonists are providing an important contribution in emerging treatments aimed to counteract migraine attacks [15]. The new emerging treatments, approved or still looking for approval by FDA, are summarized in Table 2.

Serotonin or 5-hydroxytryptamine (5-HT) is a monoamine neurotransmitter, widely distributed both centrally and peripherally, in the human body. It is primarily found in the enteric nervous system located in the gastrointestinal tract. It is also produced in the central nervous system (CNS), specifically in the raphe nuclei where neurons containing 5-HT have been observed [16]. 5-HT biological functions are multiple and complex and its eventual response mainly depends on the nature of the 5-HT receptors implicated [17]. Seven types of 5-HT receptors have been identified, respectively named: 5-$HT_1$, 5-$HT_2$, 5-$HT_3$, 5-$HT_4$, 5-$HT_5$, 5-$HT_6$ and 5-$HT_7$. 5-$HT_1$ and 5-$HT_2$ receptors are respectively divided into the following subtypes: 5-$HT_{1A}$, 5-$HT_{1B}$, 5-$HT_{1D}$, 5-$HT_{1E}$, 5-$HT_{1F}$ and 5-$HT_{2A}$, 5-$HT_{2B}$, and 5-$HT_{2C}$ [18,19]. Receptors functional characteristics depend on their molecular structure, being G-protein-coupled receptors such as: 5-$HT_1$, 5-$HT_2$, 5-$HT_4$, 5-$HT_{5A}$, 5-$HT_6$ and 5-$HT_7$, or integral to an ion channel such as 5-$HT_3$ [20]. All 5-$HT_1$ receptor subtypes are coupled to Gi/o, a protein that predominantly inhibits adenylyl cyclase activity, with consequent inhibition of the release of neurotransmitters and reduction in neuronal firing [17,20]. Triptans are agonists of 5-$HT_{1B/1D}$ receptors and showed to be efficacious in the acute treatment of migraine, thus providing an indirect proof suggesting the involvement of serotonin in the pathogenesis of migraine [21,22]. An increased production of 5-HT seems to occur in migraineurs brain compared to control subjects, that might lead to cortical hyperexcitability [23]. A reduction in blood concentrations of 5-HT was observed in subjects with migraine in the absence of aura during the headache phase [24]. Another study indicates that low 5-$HT_4$ receptor binding, that suggests a high 5-HT concentration in the brain, can be considered a trait marker of migraineurs rather than a risk factor that can lead to a conversion from EM to CM. Higher concentrations of 5-HT could lead to an enhancement of the susceptibility to migraine attacks. Therefore, a reduction of 5-HT concentrations could reveal effective to treat migraine, indicating that further studies involving other 5-HT receptor subtypes and modulation of cerebral 5-HT concentrations in migraine subjects are required [25]. The increased understanding of migraine pathophysiology recently resulted in developing novel molecules that are under investigation as emerging options to be used in therapeutics. The novel ditans are serotonin 5-$HT_{1F}$ receptor agonists, a class of drugs that differentiate from the known triptans in showing high selectivity at receptors. The ditan Lasmiditan is currently under investigation in migraine acute treatment [15].

CGRP is a multifunctional neuropeptide that appears to play an important part in migraine mechanism and its origin was predicted observing the alternative splicing of the calcitonin gene [26]. This neuropeptide is known as one of the most potent vasodilators [27]. Two isoforms have been characterized, α-CGRP and β-CGRP. The isoform α is principally expressed in primary sensory neurons, whereas the isoform β is mainly found in intrinsic enteric neurons [27]. The mature form of this neuropeptide is composed of 37 amino acids, and its expression has been particularly noticed in sensory neurons of the dorsal root ganglia (DRG) and trigeminal ganglion (TG) [26,27]. The mature form is stored in vesicles localized in the terminal region of central and peripheral nerve endings. There, their content may be secreted in the dorsal spinal cord or in various peripheral tissues, especially surrounding blood vessels which may modulate vascular tone [28]. In addition, the presence of nociceptors network positive to CGRP in rodent and human meningeal vessels has been observed [29,30], and about 40–50% of TG neurons are positive to CGRP [31,32]. Moreover, in areas of the CNS, such as hypothalamus, thalamus, periaqueductal gray, superior and inferior colliculi, amygdala, trigeminocervical complex (TCC) and the cerebellum, the expression of CGRP has been observed [33,34]. These mentioned brain areas may be associated with migraine pathophysiology, considering the capability of CGRP to change synaptic and neuronal activity at the TCC, and transmission of nociceptive signals to the thalamus and cortical areas [35,36]. The structure of CGRP receptor is a complex of proteins composed as follows: a G-protein-coupled receptor named the calcitonin receptor-like receptor (CLR) [37]; a single transmembrane accessory protein named receptor activity-modifying protein 1 (RAMP1) [38] (needed to establish the binding of CGRP to CLR), and the receptor component protein (RCP) [39] that characterizes the G-protein associated with the receptor. The expression of the subunits that compose the CGRP receptor complex has been observed in peripheral and central sites [32], e.g., in cell bodies in TG, in the periaqueductal grey, and in the TNC [40,41]. Nevertheless, it is not yet completely established whether the assembly of all these subunits composing the fully functional receptor form is performed in these anatomical structures. Despite this, the expression of the functional form of CGRP-receptor complex has been observed in vascular smooth muscle cells in arteries and arterioles (also in those of the cranial circulation), as suggested, both in vitro and in vivo [27], by the powerful vasodilatory effect of CGRP in these blood vessels. To date, CGRP is considered a new important pharmacological target for migraine treatment [15]. CGRP antagonists act by inhibiting vasodilation and neurogenic inflammation through release blockage of CGRP in the migraine pathway [42].

Emerging CGRP receptor antagonists, such as ubrogepant and rimegepant, are currently under assessment in therapeutics [15]. In addition, the development of new CGRP receptor or ligand antagonists is ongoing, in which monoclonal antibodies (mAbs) such as fremanezumab, galcanezumab and erenumab (all approved by FDA in 2018 for EM or CM), are opening a new approach in therapeutic strategy, representing a valuable support to the solutions already available [15].

**Table 2.** New emerging treatments in migraine.

| New Drugs | FDA Evaluation | Administration | Mechanism of Action | Dosage | Response to Treatment | Common AEs | Ref. |
|---|---|---|---|---|---|---|---|
| Erenumab | Approved | SC | IgG2 CGRP receptor blocker | 70 or 140 mg monthly | The percentage of patients with at least a 50% reduction in migraine days per month, was about 40% and 50% for both dosages in EM and CM. | Injection site pain, URI, fatigue, Nasopharyngitis, constipation, nausea. | [43,44] |
| Fremanezumab | | | IgG2 CGRP ligand antagonist | 225 mg monthly or 675 mg every 3 months | For both dosages the percentage of patients with at least a 50% reduction in migraine days per month, was about 40% of CM and 45 % for EM. | Injection site erythema, injection site induration, diarrhea, anxiety, depression. | [45] |
| Galcanezumab | | | IgG4 CGRP ligand antagonist | 240 mg loading dose (2 subsequent injections of 120 mg) | The percentage of patients with at least a 50% reduction in migraine days per month was about 30% of CM and 60 % for EM for both dosages. | Nasopharyngitis, URI, diarrhea, injection site pruritus, injection site erythema. | [46] |
| Eptinezumab | Not yet approved | IV | IgG1 CGRP ligand antagonist | Dosage ranges 30 to 300 mg | For EM, decreased the average number of migraine-days (30 mg = −4.0; 100 mg = −3.9; and 300 mg = −4.3). For CM, the mean change in migraine days was −8.2 in the 300-mg group. | Nausea, influenza, dizziness, fatigue, URI, UTI. | [47] |
| Rimegepant | | Oral | CGRP receptor agonist | Dosage ranges 10 to 600 mg | At 2 h post dose, freedom from the most bothersome associated symptoms (MBS) and freedom from pain were reached. Improvements in functional disability were observed, with many patients reporting normal function. | Nausea, UTI. | [48,49] |
| Ubrogepant | | | CGRP receptor agonist | Dosage ranges 50 to 100 mg | For both doses, the percentage of patients free from pain at 2 h post administration was about 20% and the percentage of patients free from MBS at 2 h post administration was about 40%. In addition, phonophobia and photophobia resolution at 2 h post administration was reached. | Nausea, somnolence, dizziness. | [50] |
| Lasmiditan | | | 5-$HT_{1F}$ receptor agonist | Dosage ranges 50 to 200 mg | The percentage of patients free from headache 2 h post dose was about 30% for 50 and 100 mg, and about 40% for 200 mg. The percentage of patients free from MBS 2 h post dose was about 40% and 45% for 50 and 100 mg, and about 50% for 200 mg. In addition, photophobia and phonophobia resolution was reached. | Dizziness, somnolence, paresthesia, fatigue, and nausea. | [51] |

## 3. Animal Models of Migraine

Animal models have been developed with the aim to comprehend migraine disorder, even though all of them show shortcomings. To be considered reliable, animal models should display a similar etiology and phenotype to human migraine. Although this disorder is considered complex with a variable phenotype, there is currently no animal model able to replicate all its features [52]. In this paragraph, we consider murine models developed to study the migraine disorder. In Table 3, we summarized the current animal models of migraine in which rats or mice were used.

Currently, animal models focus on activation of nerves in TCC. Nociceptors are located in the terminal structures of meningeal and trigeminovascular afferents deriving from the ophthalmic division of the trigeminal nerve that innervate intracranial structures sensitive to pain, such as the dura mater and meningeal vasculature, large cerebral arteries and the paranasal sinuses. Headaches similar to migraine can be caused by stimulation of nerves that innervate these structures [53,54]. Animal models based on chemical provocations that use different vasodilating agents are probably the most investigated in preclinical research. The administration of a mix of inflammatory mediators, named "inflammatory soup", e.g., using a mixture composed by prostaglandin (PGE2), histamine, 5-HT and bradykinin, has been used to stimulate meningeal and trigeminovascular nociceptors [55]. In this model, the inflammatory soup can be administered by injection using a micro-catheter placed into the cisterna magna through the atlanto occipital membrane of animals that received anesthetics. This injection of chemicals induces the activation of the primary sensory fibers supplying the meninges. In addition, topical application on the dura mater of rats is also used causing a reversible cephalic mechanical sensitivity [56–58]. Potential limitations of this technique are linked to the use of chemicals that can compromise the functionality of the blood-brain-barrier (BBB), resulting in activation of central sites directly rather than synaptically by the activation of meningeal afferent fibers [52].

Another model widely used, based on nitroglycerin (NTG), nitric oxide donor glycerol trinitrate, intravenously or intraperitoneally administered, has been studied as method to provoke migraine-like pain [52]. In rodents, hyperalgesia provoked by NTG has been used to develop a model to study sensory hypersensitivity associated with migraine [59,60]. The infusion of NTG in mice caused thermal and mechanical allodynia, symptom reversed by sumatriptan [59]. Moreover, in a transgenic mouse model of familial migraine that was studied, animals that expressed a human migraine gene, casein kinase 1δ, in which a major sensitivity to hyperalgesia induced by NTG, compared to controls, was observed [61]. In addition, NTG-administered in mice was able to induce aversion to light and increased meningeal blood flow [60,62]. This mouse model was used to develop a test able to model the progression of the disorder from an acute to a chronic condition by means intermittent injection of NTG. This modality of treatment evoked acute hyperalgesia and developed a progressive basal hypersensitivity to mechanical stimulation [63]. Both acute hyperalgesia and hypersensitivity were blocked by topiramate and propranolol, whereas hyperalgesia was inhibited by sumatriptan, suggesting that this model may be considered in the screening of novel therapies in migraine treatment [64]. The use of NTG in rodents may effectively model migraine-like symptoms [65], even though doses, type of administrations and the determination of the time of observation need to be carefully monitored when animal models based on NTG are adopted for the study of the TGVS [52].

Medication overuse headache (MOH) is a condition that in predisposed subjects affected by migraine or tension-type headache typically occurs. In these conditions, an overusing of drugs such as triptans and opioids, e.g., for more than 10 days per month over 3 months, may transform the headache from an episodic to a chronic state. In rodents, MOH may be used as a model in which repeated administration of these drugs induced a long-lasting state of latent sensitization [66]. It has been reported that TEV-48125, a humanized CGRP antibody, inhibited cutaneous allodynia both induced by bright light stress and NO donor in a MOH model, in which rodents were prior treated with sumatriptan or morphine, thus highlighting the importance of this model in discovering migraine medications [67].

In recent decades, several gene mutations have been correlated with some forms of severe and rare migraine by the use of an approach based on classic linkage analysis. This analysis contributed to sustain the hypothesis that an inherited trait is implicated in this disorder. A translational step offered by transgenic mouse technologies, in supporting the understanding about the pathophysiology of migraine and the evaluation of new therapeutic targets, is represented by behavioral characterization of genetic models of migraine [68]. The familial hemiplegic migraine (FHM) is a rare monogenic migraine, in which affected subjects show a severe hemiplegic aura accompanied by weakness perceived on one side of the body. Specifically, three gene mutations were identified to date, CACNA1A (FHM1), ATP1A2 (FHM2), and SCNA1A (FHM3), encoding for subunits of ion channels and transporters that show to have a role in neurotransmission [69]. These mutations, respectively encode for subunits of voltage-gated calcium channels, sodium-potassium ATPases, and voltage-gated sodium channels [70]. In this frame, genetic models of migraine have been created by inserting human mutated genes in the mouse genome obtaining knock-in (KI) mouse models, respectively, two of FHM1 [71,72], and one of FHM2 [73]. The humanized FHM1 KI mouse models contain gain of function missense mutations (R192Q or S218L) in the CACNA1A gene, one of the most investigated genes whose product is the pore-forming α1A subunit of Cav2.1 channels (P/Q type) (55). S218L mice showed a phenotype similar to a severe clinical phenotype of patients showing the same mutation (FHM1 S218L) [72]. In R192Q mice, decreased CGRP-immunoreactivity was observed when compared to controls, in cells of TG and in the superficial laminae of the TCC, suggesting that alteration in the expression of CGRP is induced by FHM-1 CACNA1A mutation [74]. These models showed decreased neuronal response to nociceptive activation of the TGVS in comparison with controls [75], suggesting that they show different reactions to common nociceptive signals observed in migraine. In addition, enhanced susceptibility in showing spontaneous pain behaviors correlated to nociceptive headache and photophobia was highlighted [76,77]. Only one KI mouse model for FHM2 expressed the loss of function W887R missense mutation in ATP1A2 [73], whereas no FHM3 KI mouse model has been developed yet.

In a study based on a rodent model, during the process of evaluating baseline periorbital von Frey thresholds, a male rat affected by spontaneous episodic trigeminal allodynia was discovered [78]. This characteristic was noticed by episodic alteration of periorbital pain threshold. The mating demonstrated that this trait is inheritable in both sexes and with a diversity of phenotypes. Some animals showed a similarity to chronic migraineurs, with thresholds lower than normal for more than two weeks per month. Other animals showed a similarity to episodic migraine, manifesting periods of normal thresholds and periods of lower thresholds. Chemicals such as sumatriptan, ketorolac and DHE were tested in order to validate this model, temporarily reversing the pain thresholds. Furthermore, the treatment with valproic acid for a period of one month blocked spontaneous changes in trigeminal allodynia. After the discontinuation of the treatments, the animals returned to the initial baseline. This study might be considered a unique model of spontaneous allodynia with phenotypes similar to migraine, providing a possible predictive model for drug development and for the investigation of the pathophysiology of spontaneous episodic trigeminal pain disorders [78]. Research based on these genetic models offers an advantage due to the fact that animals are manipulated in order to have predefined phenotypes similar as possible to migraine features. Nonetheless, the disadvantages might regard the common polygenic forms of this disorder in the general population, in which these specific mutations might not be relevant.

Migraine aura and cortical spreading depression (CSD) are transient neurological deficits highlighted in around 30% of migraineurs [79]. CSD is a phenomenon described as a slow wave of depolarization of neuronal and glial cells in the cortex that can be induced experimentally. In rodent models, injections of KCl showed to initiate this phenomenon in the cortex. CSD visualization is operated measuring the electrical activity of cortical neurons by means implanted microelectrodes or changes in cerebral blood flow through laser doppler flowmetry [52]. Furthermore, CSD can be also induced by electrical or mechanical stimulation of the cortex [80]. Progresses in the study of CSD mechanisms have focused to elucidate whether it plays a part in provoking the trigeminovascular activation, resulting in migraine triggering. In rodents, studies based on imaging showed that CSD induces vasodilation of meningeal blood vessels [81], and enhanced neuronal activation at the level of trigemino nuclear complex and in higher cerebral areas of the trigeminal pain pathway [82]. Considering that migraine frequently occurs without aura, the mechanisms of CSD require further studies to be better clarified and for drug discovering in migraine aura treatment.

**Table 3.** Current animal models of migraine.

| Animal Models | Route of Administration | Description | Response to Treatment | Ref. |
|---|---|---|---|---|
| Inflammatory soup | Dural cannulation | Mechanical hyperalgesia; reduced locomotor activities; nociceptive behavior; unilateral hind paw; facial grooming; anxiety- and depression-like behaviors; altered CGRP-related genes in the TG and TNC. | Zolmitriptan reduced nociceptive behaviors. Ketorolac reduced the nociceptive behavior, ipsilateral hind paw and facial grooming. Amitriptyline reversed the allodynia and decreased depression- and anxiety-like behaviors. | [56,57,83,84] |
| NTG | Intravenously or intraperitoneally | Mechanical hyperalgesia; thermal and mechanical allodynia; photophobia; meningeal blood flow; reduced locomotor activities; facial expressions of pain. | Propranolol, topiramate, and amiloride inhibited mechanical hyperalgesia. Valproic acid were ineffective. Sumatriptan inhibited hypoactivity and grimace scale scores in rats, but resulted in hyperalgesia. | [64,85–87] |
| MOH | Intraperitoneally | Mechanical hyperalgesia; CSD-related bioelectrical alterations; activation of inflammatory markers in the TG. | Topiramate blocked the enhanced Fos expression in the TNC and inhibited cutaneous allodynia. | [66,88,89] |
| Genetic model | - | Spontaneous episodic trigeminal allodynia. | Valproic acid prevented the spontaneous changes in trigeminal allodynia. | [78] |
| Transgenic models (FHM-1, FHM-2) | - | Photophobia; unilateral head grooming; lateralized winking/blinking. | Rizatriptan reduced head grooming. | [73,77] |
| CSD model | Injections of KCl on the cerebral cortex, or electrical or mechanical stimulation of the cortex | Induced dural activation, vasodilation of meningeal blood vessels. Enhanced neuronal activation at the level of trigemino nuclear complex and in higher cerebral areas of the trigeminal pain pathway. | Valproate, topiramate, propranolol, amitriptyline and methysergide were shown to suppress SD susceptibility. Lamotrigine was also shown to block KCl-induced CSD. | [90,91] |

## 4. New Therapeutic Strategies in Experimental Models

The first therapeutic approach to migraine is generally symptomatic and aimed to alleviate acute pain, and medications are more effective whether quickly administered. Notwithstanding, there are advances in the development of anti-migraine drugs, and there is a growing need in researching novel therapeutic approaches aimed to treat more effectively migraine compared to actual treatments. In this context, we therefore consider studies on animal models developed with this perspective in this study. The most relevant findings of new compounds and nanoparticles in experimental models are summarized in Table 4.

In a recent study, Moye et al. [92] studied the efficacy of SNC80, a δ opioid receptor (DOR) agonist, in mouse models that replicated different headache disorders. In these models, mice were managed in order to induce CM, post-traumatic headache (PTH), MOH, and opioid-induced hyperalgesia (OIH) [92]. In CM model, mice received NTG by the intraperitoneally intermittent administration. In PTH, mice received isoflurane to be mildly anesthetized and then underwent the closed head weight-drop method in order to induce mild traumatic brain injury, and two weeks after PTH was modelled by low NTG dose intraperitoneally. To model MOH and OIH, animals received intraperitoneally treatment using respectively sumatriptan or morphine. In CM model, animals treated with NTG showed basal peripheral and cephalic hypersensitivity. To evaluate the effect of the activation of DOR, an acute treatment of SNC80 was performed 24 h after the last injection of NTG. This treatment showed a relevant attenuation of peripheral and cephalic allodynia compared to controls, indicating that pain associated with CM was blocked by DOR activation. In PTH model, basal peripheral and cephalic hypersensitivity were developed in mice treated with NTG compared to controls. Twenty-four hours after the last NTG injection, cephalic allodynia was inhibited by performing an acute SNC80 treatment, indicating that also in this case, the pain associated with PTH was attenuated by DOR activation. In MOH model, basal hind paw and cephalic hypersensitivity were developed in mice treated with chronic administration of sumatriptan. Twenty-four hours after the final injection of medication, mice received an acute treatment with SNC80 that resulted in allodynia attenuation, suggesting that MOH induced by overuse of sumatriptan can be inhibited by DOR activation. In OIH model, mice received chronic treatment with morphine, showing basal hind paw and cephalic hypersensitivity, an effect that was also observed 18–24 h after the last drug injection. After, SNC80 was administered resulting in allodynia effect attenuation induced by morphine treatment. Furthermore, it has been observed that chronic daily administration of SNC80 causes a limited form of MOH, less severe in comparison with mice treated with sumatriptan. These results suggest that DOR agonists might represent a novel therapeutic approach in the treatment in diverse headache disorders showing a different etiology [92].

Pradhan et al., have investigated the therapeutic potential of δ-opioid receptor agonists in mouse migraine models induced with acute and chronic doses of NTG [93]. Animals were treated with three different δ-opioid receptor agonists, SNC80, ARM390 or JNJ20788560, about 1 h 30 min following NTG injection. These receptor agonists were able to significantly reduce NTG-evoked hyperalgesia. In addition, a model of migraine aura was induced by continuous application of KCl in order to evaluate the effects of SNC80 on evoked CSD. SNC80 in the 1 h time interval following administration, was able to reduce the number of CSD events. These data showed the therapeutic ability of the δ-opioid receptor as a promising therapeutic target for migraine [93].

Hoelig et al., studied the effectiveness of NOX-L41, a CGRP-neutralizing mirror-image (L-) aptamer (termed Spiegelmer), in a rat model of electrically evoked meningeal plasma protein extravasation (PPE) [94]. In this study, the authors have tested a Spiegelmer, a molecule synthesized chemically consisting of mirror-image oligonucleotide, which is able to bind to a pharmacologically relevant target molecule. Animals received NOX-L41 as a single dose intravenously or subcutaneously, showing a plasma half-life of 8 h. Furthermore, by means pharmacodynamic studies, after a single administration an extravasation of NOX-L41 from blood vessels in the dura mater and inhibition of neurogenic meningeal PPE for at least 18 h was observed. The Spiegelmer action consisted in binding CGRP, a neuropeptide that promotes meningeal vasodilation, inhibiting CGRP in signaling at its receptor. NOX-L41 showed increased affinity and selectivity for both isoforms, α and β, of human CGRP than a previous studied NOX-C89. The capacity of NOX-L41 to extravasate in surrounding tissues of dural circulation, in order to interact with perivascular vasodilating agents, results in the need to develop compounds able to antagonize CGRP, preventing neurogenic inflammation. This study suggests further the involvement of CGRP in neurogenic PPE, indicating NOX-L41 as a future potential compound for the treatment or prevention of migraine [94].

In recent times, the development of drug delivery systems using nanoparticles has received an increased attention from the scientific community. In this context, Girotra et al. [95–99] conducted studies on nanoparticulates formulations with the aim to increase the brain targeting of pharmacological anti-migraine drugs. Here, we consider some of their investigations.

A dual therapeutic approach was performed on rats and mice in the development of brain-targeted rizatriptan benzoate-loaded solid lipid nanoparticles (RB-SLNs), with the purpose to improve the drug potentiality in counteracting migraine [95]. The formulations have been elaborated by fabricating optimized solutions to obtain particle sizes showing sufficiently high entrapment efficiency and drug release in about 8 h. To evaluate the typical symptoms related to migraine, acetic acid-induced writhing test and light/dark box model were respectively used to induce hyperalgesia and light aversive behavior. Animals that received optimized RB-SLNs by oral administration showed a decrease in migraine-related hallmarks. After 2 h of oral drug treatment, pharmacodynamic evaluations showed that in rats, the brain uptake potential of optimized RB-SLNs was about 18.43-folds greater compared to the free form of the pure drug, whereas in mice showed to cross the BBB resulting in an improvement of its anti-migraine effectiveness. These findings indicate that RB-SLNs showed an improvement in brain target ability, thus providing a potential approach for migraine management [95].

In another study, an innovative approach has been conceived by developing chitosan solid lipid nanoparticles (SLN) that contained sumatriptan succinate (SS) [96]. The SLN formulations optimized in brain targeting. The optimization of the formulations was accomplished by multi-level design factorial in order to obtain a minimize size particles with a high entrapment efficiency and drug concentrations. Rats received the formulations, previously dispersed in deionized water, by oral administration. Behavioral studies indicating a reduction in hyperalgesia in the acetic acid induced writhing test and reduced aversion to light in light/dark box model. The treatment with formulations showed a major availability of SS in the brain in comparison with controls. These results suggest that formulations orally administered consisting in hydrophilic drug SS, loaded in chitosan SLN, were able to cross the BBB, allowing the drug in exerting its pharmacological activity in the brain. Considering their data, nanoparticulate drug delivery systems might represent a future approach to cross the BBB and to improve brain targeting of medications in migraine therapeutics [96].

A further study has been conducted on another formulation for brain targeting of SS, with the aim to evaluate the optimal therapeutic effect of the drug in migraine. For this purpose, nanoparticulate drug delivery system using poly (butyl cyanoacrylate) (PBCA) and bovine serum albumin linked with apolipoprotein E3 (BSA-ApoE) was used [97]. SS was incorporated in the BSA-ApoE NPs and compared with the same drug loaded polysorbate 80 coated optimized PBCA NPs to determine the brain uptake potential of these formulations. The central composite design was used for the formulation of PBCA NPs optimized with minimum particle size, maximum entrapment efficiency along with the sustained drug release. Also, in this study, animals were treated and assessed as described in the previous investigations, and behavioral studies showed similar improvements. The treatments with the nanoformulations prepared in this study showed a high brain/plasma drug ratio 2 h after the oral drug administration. The data obtained by the authors suggest that BSA-ApoE NPs showed a better activity than polysorbate 80 coated PBCA NPs for brain targeting of SS. This technique might offer a perspective, as improved therapeutic approach for the treatment of migraine [97].

In another study concerning the use of nanoparticles, Poly (D,L Lactide-co-Glycolide) (PLGA)/ poloxamer nanoparticles (NPs) of the hydrophilic medication zolmitriptan were developed [98]. Randomized factorial design to obtain the critical quality characteristics of minimized particle size and maximized encapsulation efficiency was applied. To determine the brain uptake potential, rats received optimized zolmitriptan encapsulated PLGA NPs as oral administration, and at the different time points, plasma and brain samples were collected. Acetic acid induced writhing test and light/dark box model were respectively used to induce hyperalgesia and aversion to light in mice. After the treatment, the in vivo studies for determining the brain uptake potential showed a 14.13-fold increase in the drug delivered to the brain from the NPs as compared to the free drug. Behavioral tests showed a decrease in the number of writhings, and a significant reduction of light aversion compared to controls. These data suggest that PLGA NPs containing zolmitriptan may represent a future tool in providing systems to cross the BBB for drug delivery that can exert an enhanced anti-migraine effect [98].

Another investigation has been performed using pharmacophore modeling [99]. This technique led to the identification of nystatin as compound active against the receptors iGluR5 kainate receptor (1VSO), CGRP (3N7R), $\beta_2$ adrenoceptor (3NYA) and Dopamine $D_3$ (3PBL). Following this result, brain targeted chitosan nanoparticles containing nystatin were prepared, later intraperitoneally administered in rats in order to evaluate brain targeting efficacy. Confocal laser scanning microscopy showed a higher nanoparticles accumulation in the brain than liver and spleen. After treatment with nystatin nanoformulations, behavioral tests performed in mice showed a reduction in hyperalgesia, photophobia and phonophobia compared to controls. This study represents the first approach of the therapeutic potential of nystatin nanoformulations, suggesting their future application in the treatment of migraine [99].

Wang et al., developed a novel ester derivative of gastrodin (Gas), termed Gas-D, and studied its effectiveness in a model of NTG induced migraine in rats [100]. Gas is a compound obtained by *Gastrodiae Rhizoma* (known in China, also as Tianma), and it has already been used for the treatment of migraine. Rats received Gas-D intragastrically, and subsequently treated by NTG subcutaneously 1 h after the final treatment. The pretreatment with Gas-D showed a reduction in head-scratching behavior, previously induced by NTG treatment. Results obtained from this study require further pharmacokinetic and pharmacodynamic investigations, suggesting that Gas-D, thanks to its anti-migraine effect, might be considered as a future candidate for migraine treatment [100].

Zhao et al., performed a comparative study evaluating two traditional Chinese drugs, gastrodin and ligustrazine, in a rat model of nociceptive durovascular trigeminal activation [101]. In this study, Chinese medications were compared to two Western approaches with propranolol and levetiracetam. Animals underwent surgical procedures for drug administration by femoral vein cannulation. An electrode was applied onto the dura mater above the middle meningeal artery and used to record the electrical stimuli in the TCC. When a reliable baseline to dural electrical stimulation was established, either gastrodin, levetiracetam, ligustrazine, or propranolol was administered. The treatment showed that gastrodin was able to inhibit nociceptive dural-evoked neuronal firing in the TCC, whereas ligustrazine showed no relevant effect on spontaneous activity in the TCC. To perform a comparison with the Chinese drugs, the established migraine preventive propranolol and the ineffective compound levetiracetam were used. As a result, propranolol showed a significant inhibition of dural-evoked responses, whereas the use of levetiracetam showed no effect. Their data suggest that gastrodin showed potential as an anti-migraine treatment, and on the contrary, ligustrazine appeared less promising. Therefore, these findings suggest further investigations about the use of gastrodin in migraine treatment. In addition, the results indicate the usefulness in exploring the traditional Chinese medicine approaches as signposts in developing new drugs for migraine [101].

In the last years, further experimental models have been developed using different approaches. Electrical stimulations have been used in experimental models with the purpose to better understand the mechanisms of migraine. Recently, Zhang et al. developed a repetitive electrical stimulation rat model [102]. In this study, the authors showed a dynamic model that upon stimuli of the dura mater, the TG begins to increase the production of the vasoactive neuropeptide pituitary adenylate cyclase-activating peptide (PACAP). PACAP is released from periphery terminals of the TG to innervating areas such as the dura mater, leading to vasodilation. PACAP is transported through central terminals to the TNC, where PACAP binds to PACAP-preferring type 1 (PAC1) receptor and triggers the excitation of nociceptive neurons with a consequent further increase in PACAP. The new repetitive electrical stimulation model established by stimulating the dura mater in conscious rats can simulate the chronification of frequent onset of acute migraine, from the perspective of cutaneous allodynia and nociceptive behaviors. PACAP appears to have a part in the pathogenesis of migraine potentially via PAC1 receptor. PACAP is co-expressed with CGRP, and therefore shows the potential to be considered a new therapeutic target for migraine [102]. The use of electrical stimulations led to the development of some clinical trials in which their use for the treatment of migraine has been assessed [103,104].

Furthermore, the effect of electroacupuncture (EA) pretreatment has been investigated. Pei et al. conducted a study regarding the use of EA in rats [105]. In this study, the authors used a conscious rat model of migraine induced by repeated electrical stimulation of the dura mater. Animals treated with EA showed an increase in exploratory, locomotor and eating/drinking behavior and a reduction in freezing-like resting and grooming behavior. In animals that received dural stimulation, an increase of c-Fos neurons in the periaqueductal grey, raphe magnus nucleus, and TNC was observed. This study showed that EA pretreatment may reduce behavioral responses to electrical stimulation of the dura mater in a rat model of recurrent migraine. These findings suggest that EA pretreatment may improve migraine-like symptoms by altering the descending pain modulatory system. Notwithstanding, further molecular and electrophysiological research is needed to better comprehend the central mechanisms of EA treatment of migraine [105].

Further emerging therapeutic targets are the acid-sensing ion channels (ASICs), which are considered neuronal proton sensors. Amiloride (a non-specific ASIC blocker) is a compound that showed to have benefic effects in animal models of migraine. Verkest et al., investigated the involvement of the ASIC1-subtype in cutaneous allodynia [106]. The authors conducted an investigation on effects of systemic administrations of amiloride and mambalgin-1 (a specific inhibitor of ASIC1a- and ASIC1b-containing channels) on cephalic and extra-cephalic mechanical sensitivity. The treatment was performed on a rat model of acute and CM induced by intraperitoneally administration of isosorbide dinitrate. The systemic administration of these compounds reversed cephalic and extra-cephalic acute cutaneous mechanical allodynia, whereas a single administration caused a delay in the subsequent establishment of chronic allodynia. Established chronic allodynia was also reversed by both mambalgin-1 and amiloride. A single daily administration of mambalgin-1 also showed to have a preventive effect on allodynia chronification. Pharmacological results obtained in this study suggest the involvement of peripheral ASIC1-containing channels in cutaneous allodynia and in its chronification. Furthermore, the results indicate the therapeutic potential of ASIC1 inhibitors in acute and prophylactic migraine treatment [106].

The involvement of TWIK-related spinal cord $K^+$ (TRESK) channels has also been investigated. These channels are expressed in TG and DRG neurons and are the major background $K^+$ channels in primary afferent neurons. Mutations in TRESK channels have been associated with familial and sporadic migraine. Nevertheless, whether enhanced TRESK channel activity would reduce the excitability of primary afferent neurons has not been evaluated. Guo et al. [107] observed that the over-expression of TRESK subunits lead to an increase in background $K^+$ currents, a reduction of input resistance, and a reduction in the excitability of small-diameter TG neurons. The overexpression of TRESK subunits inhibits capsaicin-evoked spikes in TG neurons, suggesting that a TRESK-specific channel opener may exhibit analgesic effect via reducing the excitability of primary afferent neurons [107].

Lengyel et al., investigated in in vitro the effects of chemically modified analogs of cloxyquin, tested on TRESK and other $K_{2P}$ channels. Cloxyquin is known as a specific activator of TRESK ($K_{2P}$18.1, TWIK-related spinal cord $K^+$ channel) background potassium channel. In this recent study, among the modified analogues of cloxyquin used, the authors identified A2764, a selective inhibitor of TRESK that can inhibit TRESK in native cells, leading to cell depolarization and increased excitability. This compound may be of use to probe the role of TRESK channel in migraine and nociception [108].

**Table 4.** New compounds and nanoparticles in experimental models.

| Experimental Models | Compounds | Type of Administration | Response to Treatment | Ref. |
|---|---|---|---|---|
| CM (NTG-induced) | SNC80 | IP | The administration of the compound showed in all four models to induce a reduction of peripheral and cephalic allodynia | [92] |
| PTH (closed head weight drop method combined to NTG) | | | | |
| MOH (sumatriptan-induced) | | | | |
| OIH (morphine-induced) | | | | |
| NTG model | SNC80 or ARM390 or JNJ20788560 | IP | Reduction of NTG-evoked hyperalgesia | [93] |
| | Gas-D | IG | Reduced head-scratching behavior | [100] |
| CSD (KCl-evoked) | SNC80 | IP | Reduced number of CSD events | [93] |
| Electrically evoked meningeal PPE | NOX-L41 | IV or SC | Inhibition of neurogenic meningeal PPE | [94] |
| Behavioral studies | RB-SLNs | Orally | The brain uptake potential was 18.43-folds higher with respect to the pure drug in its free form, 2 h post the drug administration. Higher effectiveness in minimizing the number of writhings than the standard pure drug group. Enhanced time spent by animals in the light compartment of the light/dark box model ($p < 0.001$). | [95] |
| | SS-chitosan SLNs | | The brain uptake potential was 4.54-folds increase in drug targeted to brain, compared to plasma, after 2 h of drug administration. A reduction of the number of writhings ($p < 0.001$) and enhanced time spent in lit box of light/dark box model ($p < 0.001$) compared to control groups was observed. | [96] |
| | SS-BSA-ApoE NPs | | The brain uptake potential of SS was 12.67-folds higher compared to controls, 2 h post drug administration. Reduced writhings events compared to control groups. Enhanced tolerance to light in the light compartment of the light/dark box model compared to controls. | [97] |
| | ZNPs | | An increase of 14.13-folds of drug that reached the brain compared to the pure drug was observed. The treatment reduced significantly the number of writhings compared to control ($p < 0.001$). Significant reduction ($p < 0.001$) of photophobia was achieved by enhancing the time spent in lit compartment of the light/dark box model. | [98] |
| | Nystatin-NPs | IP | Major accumulation of NPs in the brain than the other organs considered i.e., liver and spleen, indicating that nanoformulation was successful in reaching the brain through i.p. administration. The nanoformulation induced a decrease in the number of writhings in the acetic acid induced writhings test compared to controls ($p < 0.001$). The time spent in lit compartment by animals treated with Nystatin-NPs was higher than controls ($p < 0.001$), indicating the successful brain targeting through its nanoformulation. | [99] |
| Model of nociceptive durovascular trigeminal activation | Gastrodin, ligustrazine | IV | Gastrodin showed to inhibit nociceptive dural-evoked neuronal firing in the TCC. Ligustrazine showed no relevant effect on spontaneous activity in the TCC. | [101] |

## 5. Conclusions

Migraine is a disorder with a multifactorial etiopathogenesis in which the complicated pharmacological management requires further efforts to develop more efficacious therapies. Research advances strongly contributed in expanding our understanding of the pathways involved in its complex pathophysiology. Animal models have been developed on the base of clinical observations and some of them represent valuable predictive tools for the identification of anti-migraine drugs. Among the animal models developed to date, chemical provocations models based on the use of NTG or inflammatory soup have been the most widely used models to induce hyperalgesia and inflammation. The opioid receptor agonists such as SNC80, ARM390 or JNJ20788560, are revealing to be effective in counteracting hyperalgesia and CSD, respectively induced by chemical provocation using NTG and KCl. Nanoparticulate drug delivery systems might represent novel avenues to improve drug efficacy in brain targeting. In these novel formulations, triptans are encapsulated in nanoparticulates in order to better exert their pharmacological activity in brain by crossing the BBB. Furthermore, new emerging classes of medications, including 5-HT receptor agonists (ditans), CGRP receptor antagonists (gepants) and receptor or ligand antagonists (mAbs) are opening further options in therapeutics for EM and CM. Despite the medications already used in migraine therapeutics, further efforts are required to improve research in the translational pharmacological approach, from animal models to humans, in order to develop new therapeutic strategies.

**Author Contributions:** G.T. wrote the manuscript; P.B. supervised the paper; E.M. conceived the manuscript.

**Funding:** This work was supported by current research funds 2019 of IRCCS "Centro Neurolesi Bonino-Pulejo", Messina, Italy.

**Conflicts of Interest:** The authors declare no conflict of interest.

## Abbreviations

| | |
|---|---|
| AEs | Adverse Effects |
| SC | Subcutaneous |
| IV | Intravenous |
| IP | Intraperitoneal |
| IG | Intragastrically |
| URI | Upper Respiratory Infection |
| UTI | Urinary Tract Infection |
| EM | Episodic Migraine |
| CM | Chronic Migraine |
| TGVS | Trigeminovascular System |
| TNC | Trigeminal nucleus caudalis |
| CGRP | Calcitonin gene-related peptide |
| DHE | Dihydroergotamine |
| 5-HT | Serotonin |
| CNS | Central nervous system |
| DRG | Dorsal root ganglia |
| TG | Trigeminal ganglion |
| TCC | Trigeminocervical complex |
| CLR | Calcitonin receptor-like receptor |
| RAMP1 | Receptor activity-modifying protein 1 |
| RCP | Receptor component protein |
| mAbs | Monoclonal antibodies |
| PGE | Prostaglandin |
| BBB | Blood-brain-barrier |

| | |
|---|---|
| NTG | Nitroglycerin |
| MOH | Medication overuse headache |
| FHM | Familial hemiplegic migraine |
| KI | Knock-in |
| CSD | Cortical spreading depression |
| DOR | Delta opioid receptor |
| PTH | Post-traumatic headache |
| OIH | Opioid-induced hyperalgesia |
| PPE | Plasma protein extravasation |
| RB-SLNs | Rizatriptan benzoate-loaded solid lipid nanoparticles |
| ZNPs | Zolmitriptan nanoparticles |
| SS | Sumatriptan succinate |
| SLN | Solid lipid nanoparticles |
| NPs | Nanoparticles |
| PBCA | Poly (butyl cyanoacrylate) |
| BSA-ApoE | Bovine serum albumin linked with apolipoprotein E3 |
| Gas | Gastrodin |

## References


1. Goadsby, P.J.; Lipton, R.B.; Ferrari, M.D. Migraine–current understanding and treatment. *N. Engl. J. Med.* **2002**, *346*, 257–270. [CrossRef] [PubMed]
2. Headache Classification Committee of the International Headache Society. The international classification of headache disorders, 3rd edition (beta version). *Cephalalgia Int. J. Headache* **2013**, *33*, 629–808. [CrossRef] [PubMed]
3. Buse, D.C.; Greisman, J.D.; Baigi, K.; Lipton, R.B. Migraine progression: A systematic review. *Headache* **2018**, *59*, 306–338. [CrossRef] [PubMed]
4. Charles, A. The evolution of a migraine attack—A review of recent evidence. *Headache* **2013**, *53*, 413–419. [CrossRef] [PubMed]
5. Noseda, R.; Burstein, R. Migraine pathophysiology: Anatomy of the trigeminovascular pathway and associated neurological symptoms, csd, sensitization and modulation of pain. *Pain* **2013**, *154* (Suppl. 1), S44–S53. [CrossRef] [PubMed]
6. Silberstein, S.D.; Holland, S.; Freitag, F.; Dodick, D.W.; Argoff, C.; Ashman, E. Evidence-based guideline update: Pharmacologic treatment for episodic migraine prevention in adults: Report of the quality standards subcommittee of the american academy of neurology and the american headache society. *Neurology* **2012**, *78*, 1337–1345. [CrossRef]
7. Pringsheim, T.; Davenport, W.J.; Marmura, M.J.; Schwedt, T.J.; Silberstein, S. How to apply the ahs evidence assessment of the acute treatment of migraine in adults to your patient with migraine. *Headache* **2016**, *56*, 1194–1200. [CrossRef] [PubMed]
8. Bigal, M.E.; Serrano, D.; Buse, D.; Scher, A.; Stewart, W.F.; Lipton, R.B. Acute migraine medications and evolution from episodic to chronic migraine: A longitudinal population-based study. *Headache* **2008**, *48*, 1157–1168. [CrossRef]
9. Akerman, S.; Holland, P.R.; Goadsby, P.J. Diencephalic and brainstem mechanisms in migraine. *Nat. Rev. Neurosci.* **2011**, *12*, 570–584. [CrossRef]
10. Pietrobon, D.; Moskowitz, M.A. Pathophysiology of migraine. *Annu. Rev. Physiol.* **2013**, *75*, 365–391. [CrossRef]
11. Gupta, S.; Villalon, C.M. The relevance of preclinical research models for the development of antimigraine drugs: Focus on 5-HT(1B/1D) and cgrp receptors. *Pharmacol. Ther.* **2010**, *128*, 170–190. [CrossRef] [PubMed]
12. ClinicalTrials.gov. Available online: https://clinicaltrials.gov/ (accessed on 20 May 2019).
13. Kalra, A.A.; Elliott, D. Acute migraine: Current treatment and emerging therapies. *Ther. Clin. Risk Manag.* **2007**, *3*, 449–459. [PubMed]
14. Toldo, I.; De Carlo, D.; Bolzonella, B.; Sartori, S.; Battistella, P.A. The pharmacological treatment of migraine in children and adolescents: An overview. *Expert Rev. Neurother.* **2012**, *12*, 1133–1142. [CrossRef] [PubMed]

15. Peters, G.L. Migraine overview and summary of current and emerging treatment options. *Am. J. Manag. Care* **2019**, *25*, S23–S34. [PubMed]
16. Dahlstroem, A.; Fuxe, K. Evidence for the existence of monoamine-containing neurons in the central nervous system. I. Demonstration of monoamines in the cell bodies of brain stem neurons. *Acta Physiol. Scand. Suppl.* **1964**, *232*, 231–255.
17. Hannon, J.; Hoyer, D. Molecular biology of 5-HT receptors. *Behav. Brain Res.* **2008**, *195*, 198–213. [CrossRef]
18. Hoyer, D.; Humphrey, P.P. Nomenclature and classification of transmitter receptors: An integrated approach. *J. Recept. Signal Transduct. Res.* **1997**, *17*, 551–568. [CrossRef] [PubMed]
19. Saxena, P.R.; De Vries, P.; Heiligers, J.P.; Bax, W.A.; Maassen VanDenBrink, A.; Yocca, F.D. Bms-181885, a 5-HT1B/1D receptor ligand, in experimental models predictive of antimigraine activity and coronary side-effect potential. *Eur. J. Pharmacol.* **1998**, *351*, 329–339. [CrossRef]
20. Hoyer, D.; Clarke, D.E.; Fozard, J.R.; Hartig, P.R.; Martin, G.R.; Mylecharane, E.J.; Saxena, P.R.; Humphrey, P.P. International union of pharmacology classification of receptors for 5-hydroxytryptamine (serotonin). *Pharmacol. Rev.* **1994**, *46*, 157–203. [PubMed]
21. Villalon, C.M.; Centurion, D.; Valdivia, L.F.; De Vries, P.; Saxena, P.R. An introduction to migraine: From ancient treatment to functional pharmacology and antimigraine therapy. *Proc. West. Pharmacol. Soc.* **2002**, *45*, 199–210.
22. Tfelt-Hansen, P.; De Vries, P.; Saxena, P.R. Triptans in migraine: A comparative review of pharmacology, pharmacokinetics and efficacy. *Drugs* **2000**, *60*, 1259–1287. [CrossRef] [PubMed]
23. Chugani, D.C.; Niimura, K.; Chaturvedi, S.; Muzik, O.; Fakhouri, M.; Lee, M.L.; Chugani, H.T. Increased brain serotonin synthesis in migraine. *Neurology* **1999**, *53*, 1473–1479. [CrossRef] [PubMed]
24. Nagata, E.; Shibata, M.; Hamada, J.; Shimizu, T.; Katoh, Y.; Gotoh, K.; Suzuki, N. Plasma 5-hydroxytryptamine (5-HT) in migraine during an attack-free period. *Headache* **2006**, *46*, 592–596. [CrossRef] [PubMed]
25. Deen, M.; Hougaard, A.; Hansen, H.D.; Svarer, C.; Eiberg, H.; Lehel, S.; Knudsen, G.M.; Ashina, M. Migraine is associated with high brain 5-ht levels as indexed by 5-HT4 receptor binding. *Cephalalgia Int. J. Headache* **2019**, *39*, 526–532. [CrossRef] [PubMed]
26. Amara, S.G.; Arriza, J.L.; Leff, S.E.; Swanson, L.W.; Evans, R.M.; Rosenfeld, M.G. Expression in brain of a messenger rna encoding a novel neuropeptide homologous to calcitonin gene-related peptide. *Science* **1985**, *229*, 1094–1097. [CrossRef]
27. Brain, S.D.; Grant, A.D. Vascular actions of calcitonin gene-related peptide and adrenomedullin. *Physiol. Rev.* **2004**, *84*, 903–934. [CrossRef] [PubMed]
28. Rosenfeld, M.G.; Mermod, J.J.; Amara, S.G.; Swanson, L.W.; Sawchenko, P.E.; Rivier, J.; Vale, W.W.; Evans, R.M. Production of a novel neuropeptide encoded by the calcitonin gene via tissue-specific rna processing. *Nature* **1983**, *304*, 129–135. [CrossRef]
29. Tsai, S.H.; Tew, J.M.; McLean, J.H.; Shipley, M.T. Cerebral arterial innervation by nerve fibers containing calcitonin gene-related peptide (CGRP): I. Distribution and origin of cgrp perivascular innervation in the rat. *J. Comp. Neurol.* **1988**, *271*, 435–444. [CrossRef]
30. Edvinsson, L.; Gulbenkian, S.; Barroso, C.P.; Cunha e Sa, M.; Polak, J.M.; Mortensen, A.; Jorgensen, L.; Jansen-Olesen, I. Innervation of the human middle meningeal artery: Immunohistochemistry, ultrastructure, and role of endothelium for vasomotility. *Peptides* **1998**, *19*, 1213–1225. [CrossRef]
31. Tajti, J.; Uddman, R.; Moller, S.; Sundler, F.; Edvinsson, L. Messenger molecules and receptor mrna in the human trigeminal ganglion. *J. Auton. Nerv. Syst.* **1999**, *76*, 176–183. [CrossRef]
32. Eftekhari, S.; Salvatore, C.A.; Calamari, A.; Kane, S.A.; Tajti, J.; Edvinsson, L. Differential distribution of calcitonin gene-related peptide and its receptor components in the human trigeminal ganglion. *Neuroscience* **2010**, *169*, 683–696. [CrossRef] [PubMed]
33. Hokfelt, T.; Arvidsson, U.; Ceccatelli, S.; Cortes, R.; Cullheim, S.; Dagerlind, A.; Johnson, H.; Orazzo, C.; Piehl, F.; Pieribone, V.; et al. Calcitonin gene-related peptide in the brain, spinal cord, and some peripheral systems. *Ann. N. Y. Acad. Sci.* **1992**, *657*, 119–134. [CrossRef] [PubMed]
34. Van Rossum, D.; Hanisch, U.K.; Quirion, R. Neuroanatomical localization, pharmacological characterization and functions of cgrp, related peptides and their receptors. *Neurosci. Biobehav. Rev.* **1997**, *21*, 649–678. [CrossRef]
35. Storer, R.J.; Akerman, S.; Goadsby, P.J. Calcitonin gene-related peptide (CGRP) modulates nociceptive trigeminovascular transmission in the cat. *Br. J. Pharmacol.* **2004**, *142*, 1171–1181. [CrossRef] [PubMed]

36. Goadsby, P.J. Recent advances in understanding migraine mechanisms, molecules and therapeutics. *Trends Mol. Med.* **2007**, *13*, 39–44. [CrossRef]
37. Aiyar, N.; Rand, K.; Elshourbagy, N.A.; Zeng, Z.; Adamou, J.E.; Bergsma, D.J.; Li, Y. A cdna encoding the calcitonin gene-related peptide type 1 receptor. *J. Boil. Chem.* **1996**, *271*, 11325–11329. [CrossRef] [PubMed]
38. McLatchie, L.M.; Fraser, N.J.; Main, M.J.; Wise, A.; Brown, J.; Thompson, N.; Solari, R.; Lee, M.G.; Foord, S.M. Ramps regulate the transport and ligand specificity of the calcitonin-receptor-like receptor. *Nature* **1998**, *393*, 333–339. [CrossRef] [PubMed]
39. Ma, W.; Chabot, J.G.; Powell, K.J.; Jhamandas, K.; Dickerson, I.M.; Quirion, R. Localization and modulation of calcitonin gene-related peptide-receptor component protein-immunoreactive cells in the rat central and peripheral nervous systems. *Neuroscience* **2003**, *120*, 677–694. [CrossRef]
40. Oliver, K.R.; Wainwright, A.; Edvinsson, L.; Pickard, J.D.; Hill, R.G. Immunohistochemical localization of calcitonin receptor-like receptor and receptor activity-modifying proteins in the human cerebral vasculature. *J. Cereb. Blood Flow Metab.* **2002**, *22*, 620–629. [CrossRef]
41. Lennerz, J.K.; Ruhle, V.; Ceppa, E.P.; Neuhuber, W.L.; Bunnett, N.W.; Grady, E.F.; Messlinger, K. Calcitonin receptor-like receptor (CLR), receptor activity-modifying protein 1 (RAMP1), and calcitonin gene-related peptide (CGRP) immunoreactivity in the rat trigeminovascular system: Differences between peripheral and central CGRP receptor distribution. *J. Comp. Neurol.* **2008**, *507*, 1277–1299. [CrossRef]
42. Tepper, S.J. History and review of anti-calcitonin gene-related peptide (CGRP) therapies: From translational research to treatment. *Headache* **2018**, *58* (Suppl. 3), 238–275. [CrossRef] [PubMed]
43. Lattanzi, S.; Brigo, F.; Trinka, E.; Vernieri, F.; Corradetti, T.; Dobran, M.; Silvestrini, M. Erenumab for preventive treatment of migraine: A systematic review and meta-analysis of efficacy and safety. *Drugs* **2019**, *79*, 417–431. [CrossRef] [PubMed]
44. Dodick, D.W.; Ashina, M.; Brandes, J.L.; Kudrow, D.; Lanteri-Minet, M.; Osipova, V.; Palmer, K.; Picard, H.; Mikol, D.D.; Lenz, R.A. Arise: A phase 3 randomized trial of erenumab for episodic migraine. *Cephalalgia Int. J. Headache* **2018**, *38*, 1026–1037. [CrossRef] [PubMed]
45. Dodick, D.W.; Silberstein, S.D.; Bigal, M.E.; Yeung, P.P.; Goadsby, P.J.; Blankenbiller, T.; Grozinski-Wolff, M.; Yang, R.; Ma, Y.; Aycardi, E. Effect of fremanezumab compared with placebo for prevention of episodic migraine: A randomized clinical trial. *Jama* **2018**, *319*, 1999–2008. [CrossRef]
46. Skljarevski, V.; Matharu, M.; Millen, B.A.; Ossipov, M.H.; Kim, B.K.; Yang, J.Y. Efficacy and safety of galcanezumab for the prevention of episodic migraine: Results of the evolve-2 phase 3 randomized controlled clinical trial. *Cephalalgia Int. J. Headache* **2018**, *38*, 1442–1454. [CrossRef] [PubMed]
47. Xu, D.; Chen, D.; Zhu, L.N.; Tan, G.; Wang, H.J.; Zhang, Y.; Liu, L. Safety and tolerability of calcitonin-gene-related peptide binding monoclonal antibodies for the prevention of episodic migraine—A meta-analysis of randomized controlled trials. *Cephalalgia Int. J. Headache* **2019**. [CrossRef] [PubMed]
48. Biohaven Pharmaceuticals. Rimegepant. Available online: biohavenpharma.com/rimegepant-for-acute-treatment-ofmigraine (accessed on 25 October 2018).
49. *Biohaven Announces Robust Clinical Data with Single Dose Rimegepant That Defines Acute and Durable Benefits to Patients: The First Oral CGRP Receptor Antagonist to Deliver Positive Data on Pain Freedom and Most Bothersome Symptom in Two Pivotal Phase 3 Trials in Acute Treatment of Migraine [News Release]*; Biohaven Pharmaceuticals: Los Angeles, CA, USA, 2018. Available online: www.biohavenpharma.com/investors/newsevents/press-releases/04-23-2018 (accessed on 25 October 2018).
50. Martelletti, P.; Giamberardino, M.A. Advances in orally administered pharmacotherapy for the treatment of migraine. *Expert Opin. Pharmacother.* **2019**, *20*, 209–218. [CrossRef] [PubMed]
51. Kuca, B.; Silberstein, S.D.; Wietecha, L.; Berg, P.H.; Dozier, G.; Lipton, R.B.; Group, C.M.-S. Lasmiditan is an effective acute treatment for migraine: A phase 3 randomized study. *Neurology* **2018**, *91*, e2222–e2232. [CrossRef] [PubMed]
52. Andreou, A.P.; Summ, O.; Charbit, A.R.; Romero-Reyes, M.; Goadsby, P.J. Animal models of headache: From bedside to bench and back to bedside. *Expert Rev. Neurother.* **2010**, *10*, 389–411. [CrossRef]
53. Olesen, J.; Burstein, R.; Ashina, M.; Tfelt-Hansen, P. Origin of pain in migraine: Evidence for peripheral sensitisation. *Lancet Neurol.* **2009**, *8*, 679–690. [CrossRef]
54. Bogduk, N. Anatomy and physiology of headache. *Biomed. Pharmacother.* **1995**, *49*, 435–445. [CrossRef]
55. Munro, G.; Jansen-Olesen, I.; Olesen, J. Animal models of pain and migraine in drug discovery. *Drug Discov. Today* **2017**, *22*, 1103–1111. [CrossRef] [PubMed]

56. Oshinsky, M.L.; Gomonchareonsiri, S. Episodic dural stimulation in awake rats: A model for recurrent headache. *Headache* **2007**, *47*, 1026–1036. [CrossRef] [PubMed]
57. Melo-Carrillo, A.; Lopez-Avila, A. A chronic animal model of migraine, induced by repeated meningeal nociception, characterized by a behavioral and pharmacological approach. *Cephalalgia Int. J. Headache* **2013**, *33*, 1096–1105. [CrossRef] [PubMed]
58. Boyer, N.; Dallel, R.; Artola, A.; Monconduit, L. General trigeminospinal central sensitization and impaired descending pain inhibitory controls contribute to migraine progression. *Pain* **2014**, *155*, 1196–1205. [CrossRef] [PubMed]
59. Bates, E.A.; Nikai, T.; Brennan, K.C.; Fu, Y.H.; Charles, A.C.; Basbaum, A.I.; Ptacek, L.J.; Ahn, A.H. Sumatriptan alleviates nitroglycerin-induced mechanical and thermal allodynia in mice. *Cephalalgia Int. J. Headache* **2010**, *30*, 170–178. [CrossRef] [PubMed]
60. Markovics, A.; Kormos, V.; Gaszner, B.; Lashgarara, A.; Szoke, E.; Sandor, K.; Szabadfi, K.; Tuka, B.; Tajti, J.; Szolcsanyi, J.; et al. Pituitary adenylate cyclase-activating polypeptide plays a key role in nitroglycerol-induced trigeminovascular activation in mice. *Neurobiol. Dis.* **2012**, *45*, 633–644. [CrossRef] [PubMed]
61. Brennan, K.C.; Bates, E.A.; Shapiro, R.E.; Zyuzin, J.; Hallows, W.C.; Huang, Y.; Lee, H.Y.; Jones, C.R.; Fu, Y.H.; Charles, A.C.; et al. Casein kinase idelta mutations in familial migraine and advanced sleep phase. *Sci. Transl. Med.* **2013**, *5*, 183ra156. [CrossRef]
62. Greco, R.; Meazza, C.; Mangione, A.S.; Allena, M.; Bolla, M.; Amantea, D.; Mizoguchi, H.; Sandrini, G.; Nappi, G.; Tassorelli, C. Temporal profile of vascular changes induced by systemic nitroglycerin in the meningeal and cortical districts. *Cephalalgia Int. J. Headache* **2011**, *31*, 190–198. [CrossRef]
63. Moye, L.S.; Pradhan, A.A.A. Animal model of chronic migraine-associated pain. *Curr. Protoc. Neurosci.* **2017**, *80*, 9–60.
64. Tipton, A.F.; Tarash, I.; McGuire, B.; Charles, A.; Pradhan, A.A. The effects of acute and preventive migraine therapies in a mouse model of chronic migraine. *Cephalalgia Int. J. Headache* **2016**, *36*, 1048–1056. [CrossRef] [PubMed]
65. Erdener, S.E.; Dalkara, T. Modelling headache and migraine and its pharmacological manipulation. *Br. J. Pharmacol.* **2014**, *171*, 4575–4594. [CrossRef] [PubMed]
66. De Felice, M.; Ossipov, M.H.; Wang, R.; Lai, J.; Chichorro, J.; Meng, I.; Dodick, D.W.; Vanderah, T.W.; Dussor, G.; Porreca, F. Triptan-induced latent sensitization: A possible basis for medication overuse headache. *Ann. Neurol.* **2010**, *67*, 325–337. [CrossRef] [PubMed]
67. Kopruszinski, C.M.; Xie, J.Y.; Eyde, N.M.; Remeniuk, B.; Walter, S.; Stratton, J.; Bigal, M.; Chichorro, J.G.; Dodick, D.; Porreca, F. Prevention of stress- or nitric oxide donor-induced medication overuse headache by a calcitonin gene-related peptide antibody in rodents. *Cephalalgia Int. J. Headache* **2017**, *37*, 560–570. [CrossRef] [PubMed]
68. Romero-Reyes, M.; Akerman, S. Update on animal models of migraine. *Curr. Pain Headache Rep.* **2014**, *18*, 462. [CrossRef]
69. Chen, S.P.; Tolner, E.A.; Eikermann-Haerter, K. Animal models of monogenic migraine. *Cephalalgia Int. J. Headache* **2016**, *36*, 704–721. [CrossRef]
70. Tolner, E.A.; Houben, T.; Terwindt, G.M.; de Vries, B.; Ferrari, M.D.; van den Maagdenberg, A.M. From migraine genes to mechanisms. *Pain* **2015**, *156* (Suppl. 1), S64–S74. [CrossRef]
71. Van den Maagdenberg, A.M.; Pietrobon, D.; Pizzorusso, T.; Kaja, S.; Broos, L.A.; Cesetti, T.; van de Ven, R.C.; Tottene, A.; van der Kaa, J.; Plomp, J.J.; et al. A cacna1a knockin migraine mouse model with increased susceptibility to cortical spreading depression. *Neuron* **2004**, *41*, 701–710. [CrossRef]
72. Van den Maagdenberg, A.M.; Pizzorusso, T.; Kaja, S.; Terpolilli, N.; Shapovalova, M.; Hoebeek, F.E.; Barrett, C.F.; Gherardini, L.; van de Ven, R.C.; Todorov, B.; et al. High cortical spreading depression susceptibility and migraine-associated symptoms in ca(v)2.1 s218l mice. *Ann. Neurol.* **2010**, *67*, 85–98. [CrossRef]
73. Leo, L.; Gherardini, L.; Barone, V.; De Fusco, M.; Pietrobon, D.; Pizzorusso, T.; Casari, G. Increased susceptibility to cortical spreading depression in the mouse model of familial hemiplegic migraine type 2. *PLoS Genet.* **2011**, *7*, e1002129. [CrossRef]

74. Mathew, R.; Andreou, A.P.; Chami, L.; Bergerot, A.; van den Maagdenberg, A.M.; Ferrari, M.D.; Goadsby, P.J. Immunohistochemical characterization of calcitonin gene-related peptide in the trigeminal system of the familial hemiplegic migraine 1 knock-in mouse. *Cephalalgia Int. J. Headache* **2011**, *31*, 1368–1380. [CrossRef] [PubMed]
75. Park, J.; Moon, H.; Akerman, S.; Holland, P.R.; Lasalandra, M.P.; Andreou, A.P.; Ferrari, M.D.; van den Maagdenberg, A.M.; Goadsby, P.J. Differential trigeminovascular nociceptive responses in the thalamus in the familial hemiplegic migraine 1 knock-in mouse: A fos protein study. *Neurobiol. Dis.* **2014**, *64*, 1–7. [CrossRef] [PubMed]
76. Langford, D.J.; Bailey, A.L.; Chanda, M.L.; Clarke, S.E.; Drummond, T.E.; Echols, S.; Glick, S.; Ingrao, J.; Klassen-Ross, T.; Lacroix-Fralish, M.L.; et al. Coding of facial expressions of pain in the laboratory mouse. *Nat. Methods* **2010**, *7*, 447–449. [CrossRef] [PubMed]
77. Chanda, M.L.; Tuttle, A.H.; Baran, I.; Atlin, C.; Guindi, D.; Hathaway, G.; Israelian, N.; Levenstadt, J.; Low, D.; Macrae, L.; et al. Behavioral evidence for photophobia and stress-related ipsilateral head pain in transgenic cacna1a mutant mice. *Pain* **2013**, *154*, 1254–1262. [CrossRef] [PubMed]
78. Oshinsky, M.L.; Sanghvi, M.M.; Maxwell, C.R.; Gonzalez, D.; Spangenberg, R.J.; Cooper, M.; Silberstein, S.D. Spontaneous trigeminal allodynia in rats: A model of primary headache. *Headache* **2012**, *52*, 1336–1349. [CrossRef]
79. Rasmussen, B.K.; Olesen, J. Migraine with aura and migraine without aura: An epidemiological study. *Cephalalgia Int. J. Headache* **1992**, *12*, 221–228, discussion 186. [CrossRef] [PubMed]
80. Eikermann-Haerter, K.; Moskowitz, M.A. Animal models of migraine headache and aura. *Curr. Opin. Neurol.* **2008**, *21*, 294–300. [CrossRef]
81. Bolay, H.; Reuter, U.; Dunn, A.K.; Huang, Z.; Boas, D.A.; Moskowitz, M.A. Intrinsic brain activity triggers trigeminal meningeal afferents in a migraine model. *Nat. Med.* **2002**, *8*, 136–142. [CrossRef]
82. Cui, Y.; Toyoda, H.; Sako, T.; Onoe, K.; Hayashinaka, E.; Wada, Y.; Yokoyama, C.; Onoe, H.; Kataoka, Y.; Watanabe, Y. A voxel-based analysis of brain activity in high-order trigeminal pathway in the rat induced by cortical spreading depression. *NeuroImage* **2015**, *108*, 17–22. [CrossRef]
83. Zhang, M.; Liu, Y.; Zhao, M.; Tang, W.; Wang, X.; Dong, Z.; Yu, S. Depression and anxiety behaviour in a rat model of chronic migraine. *J. Headache Pain* **2017**, *18*, 27. [CrossRef]
84. Stucky, N.L.; Gregory, E.; Winter, M.K.; He, Y.Y.; Hamilton, E.S.; McCarson, K.E.; Berman, N.E. Sex differences in behavior and expression of cgrp-related genes in a rodent model of chronic migraine. *Headache* **2011**, *51*, 674–692. [CrossRef] [PubMed]
85. Sufka, K.J.; Staszko, S.M.; Johnson, A.P.; Davis, M.E.; Davis, R.E.; Smitherman, T.A. Clinically relevant behavioral endpoints in a recurrent nitroglycerin migraine model in rats. *J. Headache Pain* **2016**, *17*, 40. [CrossRef]
86. Pradhan, A.A.; Smith, M.L.; McGuire, B.; Tarash, I.; Evans, C.J.; Charles, A. Characterization of a novel model of chronic migraine. *Pain* **2014**, *155*, 269–274. [CrossRef] [PubMed]
87. Kim, S.J.; Yeo, J.H.; Yoon, S.Y.; Kwon, S.G.; Lee, J.H.; Beitz, A.J.; Roh, D.H. Differential development of facial and hind paw allodynia in a nitroglycerin-induced mouse model of chronic migraine: Role of capsaicin sensitive primary afferents. *Boil. Pharm. Bull.* **2018**, *41*, 172–181. [CrossRef]
88. De Felice, M.; Ossipov, M.H.; Wang, R.; Dussor, G.; Lai, J.; Meng, I.D.; Chichorro, J.; Andrews, J.S.; Rakhit, S.; Maddaford, S.; et al. Triptan-induced enhancement of neuronal nitric oxide synthase in trigeminal ganglion dural afferents underlies increased responsiveness to potential migraine triggers. *Brain J. Neurol.* **2010**, *133*, 2475–2488. [CrossRef] [PubMed]
89. Green, A.L.; Gu, P.; De Felice, M.; Dodick, D.; Ossipov, M.H.; Porreca, F. Increased susceptibility to cortical spreading depression in an animal model of medication-overuse headache. *Cephalalgia Int. J. Headache* **2014**, *34*, 594–604. [CrossRef]
90. Ayata, C.; Jin, H.; Kudo, C.; Dalkara, T.; Moskowitz, M.A. Suppression of cortical spreading depression in migraine prophylaxis. *Ann. Neurol.* **2006**, *59*, 652–661. [CrossRef] [PubMed]
91. Bogdanov, V.B.; Chauvel, V.; Multon, S.; Makarchuk, M.Y.; Schoenen, J. Preventive antimigraine drugs differentially affect kcl-induced cortical spreading depression in rat. *Cephalalgia Int. J. Headache* **2009**, *29*, 131.
92. Moye, L.S.; Tipton, A.F.; Dripps, I.; Sheets, Z.; Crombie, A.; Violin, J.D.; Pradhan, A.A. Delta opioid receptor agonists are effective for multiple types of headache disorders. *Neuropharmacology* **2019**, *148*, 77–86. [CrossRef] [PubMed]

93. Pradhan, A.A.; Smith, M.L.; Zyuzin, J.; Charles, A. Delta-opioid receptor agonists inhibit migraine-related hyperalgesia, aversive state and cortical spreading depression in mice. *Br. J. Pharmacol.* **2014**, *171*, 2375–2384. [CrossRef] [PubMed]
94. Hoehlig, K.; Johnson, K.W.; Pryazhnikov, E.; Maasch, C.; Clemens-Smith, A.; Purschke, W.G.; Vauleon, S.; Buchner, K.; Jarosch, F.; Khiroug, L.; et al. A novel CGRP-neutralizing spiegelmer attenuates neurogenic plasma protein extravasation. *Br. J. Pharmacol.* **2015**, *172*, 3086–3098. [CrossRef] [PubMed]
95. Girotra, P.; Singh, S.K. Multivariate optimization of rizatriptan benzoate-loaded solid lipid nanoparticles for brain targeting and migraine management. *AAPS PharmSciTech* **2017**, *18*, 517–528. [CrossRef] [PubMed]
96. Hansraj, G.P.; Singh, S.K.; Kumar, P. Sumatriptan succinate loaded chitosan solid lipid nanoparticles for enhanced anti-migraine potential. *Int. J. Boil. Macromol.* **2015**, *81*, 467–476. [CrossRef] [PubMed]
97. Girotra, P.; Singh, S.K. A comparative study of orally delivered pbca and apoe coupled bsa nanoparticles for brain targeting of sumatriptan succinate in therapeutic management of migraine. *Pharm. Res.* **2016**, *33*, 1682–1695. [CrossRef] [PubMed]
98. Girotra, P.; Singh, S.K.; Kumar, G. Development of zolmitriptan loaded plga/poloxamer nanoparticles for migraine using quality by design approach. *Int. J. Boil. Macromol.* **2016**, *85*, 92–101. [CrossRef]
99. Girotra, P.; Thakur, A.; Kumar, A.; Singh, S.K. Identification of multi-targeted anti-migraine potential of nystatin and development of its brain targeted chitosan nanoformulation. *Int. J. Boil. Macromol.* **2017**, *96*, 687–696. [CrossRef]
100. Wang, P.H.; Zhao, L.X.; Wan, J.Y.; Zhang, L.; Mao, X.N.; Long, F.Y.; Zhang, S.; Chen, C.; Du, J.R. Pharmacological characterization of a novel gastrodin derivative as a potential anti-migraine agent. *Fitoterapia* **2016**, *109*, 52–57. [CrossRef]
101. Zhao, Y.; Martins-Oliveira, M.; Akerman, S.; Goadsby, P.J. Comparative effects of traditional chinese and western migraine medicines in an animal model of nociceptive trigeminovascular activation. *Cephalalgia Int. J. Headache* **2018**, *38*, 1215–1224. [CrossRef]
102. Zhang, Q.; Han, X.; Wu, H.; Zhang, M.; Hu, G.; Dong, Z.; Yu, S. Dynamic changes in cgrp, pacap, and pacap receptors in the trigeminovascular system of a novel repetitive electrical stimulation rat model: Relevant to migraine. *Mol. Pain* **2019**, *15*, 1744806918820452. [CrossRef]
103. Tepper, S.J.; Rezai, A.; Narouze, S.; Steiner, C.; Mohajer, P.; Ansarinia, M. Acute treatment of intractable migraine with sphenopalatine ganglion electrical stimulation. *Headache* **2009**, *49*, 983–989. [CrossRef]
104. Barloese, M.C.; Jurgens, T.P.; May, A.; Lainez, J.M.; Schoenen, J.; Gaul, C.; Goodman, A.M.; Caparso, A.; Jensen, R.H. Cluster headache attack remission with sphenopalatine ganglion stimulation: Experiences in chronic cluster headache patients through 24 months. *J. Headache Pain* **2016**, *17*, 67. [CrossRef] [PubMed]
105. Pei, P.; Liu, L.; Zhao, L.; Cui, Y.; Qu, Z.; Wang, L. Effect of electroacupuncture pretreatment at gb20 on behaviour and the descending pain modulatory system in a rat model of migraine. *Acupunct. Med. J. Br. Med. Acupunct. Soc.* **2016**, *34*, 127–135. [CrossRef] [PubMed]
106. Verkest, C.; Piquet, E.; Diochot, S.; Dauvois, M.; Lanteri-Minet, M.; Lingueglia, E.; Baron, A. Effects of systemic inhibitors of acid-sensing ion channels 1 (ASIC1) against acute and chronic mechanical allodynia in a rodent model of migraine. *Br. J. Pharmacol.* **2018**, *175*, 4154–4166. [CrossRef] [PubMed]
107. Guo, Z.; Cao, Y.Q. Over-expression of TRESK $K^+$ channels reduces the excitability of trigeminal ganglion nociceptors. *PLoS ONE* **2014**, *9*, e87029. [CrossRef] [PubMed]
108. Lengyel, M.; Erdelyi, F.; Pergel, E.; Balint-Polonka, A.; Dobolyi, A.; Bozsaki, P.; Dux, M.; Kiraly, K.; Hegedus, T.; Czirjak, G.; et al. Chemically modified derivatives of the activator compound cloxyquin exert inhibitory effect on tresk (k2p18.1) background potassium channel. *Mol. Pharmacol.* **2019**, *95*, 652–660. [CrossRef] [PubMed]

International Journal of
*Molecular Sciences*

*Review*

# Ion Channels Involved in Substance P-Mediated Nociception and Antinociception

**Chu-Ting Chang [1], Bo-Yang Jiang [1] and Chih-Cheng Chen [1,2,*]**

1 Institute of Biomedical Sciences, Academia Sinica, Taipei 115, Taiwan; chuting82@gmail.com (C.-T.C.); qwert985432@gmail.com (B.-Y.J.)
2 Taiwan Mouse Clinic, National Comprehensive Mouse Phenotyping and Drug Testing Center, Taipei 115, Taiwan
* Correspondence: chih@ibms.sinica.edu.tw; Tel.: +886-2-2652-3917

Received: 5 March 2019; Accepted: 27 March 2019; Published: 30 March 2019

**Abstract:** Substance P (SP), an 11-amino-acid neuropeptide, has long been considered an effector of pain. However, accumulating studies have proposed a paradoxical role of SP in anti-nociception. Here, we review studies of SP-mediated nociception and anti-nociception in terms of peptide features, SP-modulated ion channels, and differential effector systems underlying neurokinin 1 receptors (NK1Rs) in differential cell types to elucidate the effect of SP and further our understanding of SP in anti-nociception. Most importantly, understanding the anti-nociceptive SP-NK1R pathway would provide new insights for analgesic drug development.

**Keywords:** substance P; NK1R; nociception; anti-nociception; pain

## 1. Background

Substance P (SP) was first described by von Euler and Gaddum, in 1931 [1]. The authors observed an unknown substance that stimulated contraction of the intestine ex vivo. This substance was identified and Euler and Gaddum named it substance P, from the bottle containing it, labeled P1, P2 etc., meaning "powder". In the 1970s, SP was homogeneously purified by Chang and Leeman [2] and was later determined to be an 11 amino-acid peptide, H-Arg Pro Lys Pro Gln Gln Phe Phe Gly Leu Met-$NH_2$ (RPKPQQFFGLM), with an amidation at the C-terminus [3]. SP belongs to the tachykinin family and serves as a neurotransmitter and a neuromodulator. It is encoded by preprotachykinin-1 (or tachykinin 1 (*TAC1*)) that produces SP and neurokinin A via alternative slicing and post-translational modifications [4].

SP is widely distributed in the human body, especially in nervous systems and inflammatory cells. A general assumption of the SP action describes SP as a neuropeptide released from pain-sensing fibers (nociceptors) to increase pain sensitivity through its actions in the dorsal horn of the spinal cord [5]. SP also triggers proinflammatory cytokine release resulting in inflammation, vasodilation and plasma extravasation [6]. Although considerable evidence indicates that SP transmits pain signaling and serves as a mediator of pain [7–9], accumulating studies reveal that SP also has an anti-nociceptive effect [10,11]. Interestingly, evidence showing the antinociceptive role of SP has been built over several years and can be dated back to 1976 [10]. The discrepant function of SP is believed to depend on the cell type regarding the expression of neurokinin receptors with unique underlying effector systems modulating differential ion channels.

## 2. SP-Mediated Signaling

There are three tachykinin receptors: Neurokinin 1 (NK1), 2 (NK2), and 3 (NK3). SP preferentially binds to NK1 receptor (NK1R). NKRs are G-protein-coupled receptors located in both the central nervous system (CNS) and peripheral nervous system (PNS). G proteins have G$\alpha$, G$\beta$, and G$\gamma$ subunits. The G$\alpha$ subunit is classified into 5 families ($G_s$, $G_i$, $G_o$, $G_{q/11}$, and $G_{12/13}$). In most cases, NK1R is coupled to the pertussis toxin-insensitive $G_{q/11}$ cascade [12,13]; nevertheless, crosstalk with other G proteins such as $G_{12/13}$ [14], $G_o$, and $G_s$ [15] has been reported. In $G_{q/11}$-mediated signaling, phospholipase C is activated by $G_{\beta\gamma}$ binding, which in turn results in hydrolysis of membrane phospholipid PtdIns(4,5)P2 to form diacylglycerol (DAG) and inositol triphosphate (IP3). IP3 further triggers calcium release in cytoplasm from sarcoplasmic reticulum, whereas DAG activates protein kinase C (PKC) and results in calcium influx from the L-type calcium channel in the plasma membrane [16]. The increase in calcium ion in the cytoplasm further leads to an array of cell responses.

There are two natural forms of NK1R: A full-length receptor with 407 amino acids and a truncated receptor with 311 amino acids, lacking 96 amino acids in the carboxyl terminal [17,18]. The human *NK1R* gene contains five exons. The truncated version is generated when the intron between exons 4 and 5 is not removed that encounters a premature stop codon before the start of exon 5. These two types of NK1Rs have differential features. The binding affinity of SP is 10 times less to the truncated NK1R than the full-length form. Furthermore, the underlying signaling of the two different NK1Rs differs. The carboxyl terminus of full-length NK1R is a crucial element for G-protein coupling. The truncated form of NK1R impairs the ability of G-protein binding and results in G-protein-independent mechanisms. Indeed, the truncated form of NK1R fails to interact with β-arrestin, an important protein mediating the desensitization and internalization of activated G-protein-coupled receptors [19]. Furthermore, activation of the truncated NK1R has different effects on calcium mobilization, phosphorylation of PKCδ, extracellular signal-regulated kinase 1/2, and regulation of interleukin 8 mRNA expression as compared with the full-length NK1R [20].

The two NK1Rs are differentially distributed in peripheral tissues and in the nervous system. In the brain, full-length form is abundantly expressed in striatum, caudate nucleus, putamen, globus pallidus, nucleus accumbent, and hypothalamus; whereas the truncated NK1R expression is relatively low in the brain, and most represented in the PNS and peripheral tissues including heart, lung, prostate, and bone [21]. To sum up, the different NK1R subtypes could trigger differential effector systems and further result in distinct cellular responses in different tissues and organs.

## 3. SP and Pain

Pain is an unpleasant sensation and can be divided into two major categories: Acute and chronic pain. Acute pain lasts only short time and can be ameliorated over time. Acute pain helps us avoid the physical damage to our body and serves as a warning signal. In contrast, chronic pain has no biological function and is a disease itself rather than symptom of a disease. Chronic pain usually lasts for more than six months. People with chronic pain represent about 10.1% to 55.2% of the population in various countries according to epidemiological study [22]. People with chronic pain often show other symptoms, such as fatigue, sleep disorder, memory problems, anxiety, and depression, which greatly affect their quality of life.

Chronic pain can be further divided into two subtypes: Inflammatory and neuropathic pain. Inflammatory pain is related to tissue injuries, which lead to inflammatory reactions. In contrast, neuropathic pain results from nerve injury and neuronal sensitization in the CNS and PNS. However, inflammatory pain and neuropathic pain are also closely correlated. Nerve injuries result in tissue inflammatory responses, which lead to inflammatory pain. Likewise, tissue inflammation triggers an inflammatory cascade releasing a variety of inflammatory mediators such as SP, calcitonin gene-related peptide, and neurokinin A to damage nerves [23,24]. Clinical studies have demonstrated that many forms of chronic pain have mixed components of inflammation and neuropathy [25].

Substantial evidence suggests that SP is the key element in neurogenic inflammation and has an important role in eliciting pain sensation in both the CNS and PNS. In the CNS, SP results in central sensitization by activating excitatory post-synaptic potential [26]. In the PNS, releasing SP from peripheral nociceptive nerve fibers can cause neurogenic inflammation in the skin [27,28]. In addition, people with fibromyalgia, a chronic pain disorder, showed elevated SP level in cerebrospinal fluid [29,30]. Chemical ablation of neurons expressing SP receptors in lamina I [31] and genetic disruption of the encoding gene of SP [32] or its receptor [33] reduced pain responses. Furthermore, a recent study showed that selectively ablating *TAC1*-expressing neurons in the spinal cord abolished the sustained pain but not the reflexive defensive response [34]. Together, these studies support that SP is an important signal molecule in pain transmission.

Thus, many studies have focused on developing selective NK1R antagonists as a potential analgesic drug. Although several preclinical studies showed an anti-nociceptive effect of NK1R antagonists, most clinical trials failed to show analgesia effects. The reasons for failure of most NK1R antagonists in clinical trials are still elusive. Several reasons for discrepancy between preclinical and clinical results have been proposed such as species differences in NKRs distribution [35], species differences in affinities to antagonists for NK1R [36–38], and the ability of animal models in predicting clinical pain [39]. The discovery of an antinociceptive effect of SP may also explain in part the failure of those clinical trials [10]. One hypothesis is that neurons innervating distinct locations may respond differently to SP—some neurons excited by SP and some inhibited by SP. Thus, elucidation of SP-mediated responses in different tissues and organs becomes a crucial step in developing specific NK1 antagonists as analgesic drugs.

## 4. SP-Mediated Anti-Nociception

The antinociceptive effect of SP was first reported in 1976 [10]. Although SP had been identified for more than four decades, in the early time, the impure natural SP contaminated by bradykinin or some other kinin-like compounds impeded SP research. In the 1970s, SP was homogenously synthesized. With purely synthetic SP, the effect of SP could be clearly deciphered. Stewart et al. first reported that SP treatment by intracerebral and intraperitoneal injection could produce naloxone-reversible analgesia, and the site of action was in the CNS [9]. Subsequently, several studies confirmed the SP-mediated anti-nociception via opioid receptors in the CNS, and SP seemed to be a regulatory peptide to normalize the responses to pain stimuli. In 1978, Frederickson et al. claimed that a small amount of SP (1.25 to 5 ng per mouse) by intracerebroventricular injection produced a naloxone-reversed anti-nociception effect in mice [32]. However, higher doses (>50 ng per mouse) caused hyperalgesia. The authors also reported that although the C-terminus of SP (SP6–11) is very similar to that of endogenous opioid peptides, neither SP nor the SP6–11 acted on opioid receptors. At low doses, SP triggered the release of endorphins but at higher doses, directly excited neurons in the brain [40]. In 1980, Oehme et al. suggested that SP produced naloxone-reversed analgesia in mice with high sensitivity to thermal stimulation but induced hyperalgesia in mice with low sensitivity to thermal stimulation [41]. In addition, SP has been found to effectively reduce neuropathic pain [42] and inflammatory pain [43]. To sum up, these studies demonstrated that SP can regulates opioid-dependent analgesic effects in distinct cell types, probably via different receptors.

Because it has also been reported that the C-terminus of SP is sufficient for biological activities in nociception [44], a certain active fragment of SP may be essential for anti-nociception. Indeed, evidence has shown that the N-terminal and C-terminal domains of SP have opposite functions. Hall and Stewart demonstrated that the N-terminus of SP (SP1–7) was related to naloxone-reversible anti-nociceptive and anti-aggressive actions, whereas the C-terminus (SP7-11) was thought to mediate pain transmission [45,46]. Furthermore, Skilling et al. showed that N-terminus of SP (SP1–7) but not the C-terminus (SP5–11) inhibited the release of excitatory neurotransmitters into spinal-cord extracellular fluid, which was reversed by naloxone [47]. These studies agree with the findings of the naloxone-reversible analgesic effect by SP in the CNS and PNS. The 11 amino-acid SP may be cleaved by enzymatic degradation into differential fragments. Those fragments could interact with differential receptors to induce an anti-nociceptive effect possibly via release of met-enkephalin or other endogenous opioid peptides [48–50].

Other evidence has shown that SP could act on NK1R to modulate opioid receptors [51,52]. Bowman and colleagues showed that SP increased the recycling of mu-opioid receptors in sensory neurons and led to elevated sensitivity of opioids [53]. Together, these studies suggest that the anti-nociceptive role of SP could act via opioid signaling.

## 5. SP-Mediated Anti-Nociception in Muscle

Accumulating evidence has shown a role for SP in anti-nociception in the PNS, especially in muscle. Despite much evidence indicating that SP can cause cutaneous pain, applying SP to muscle induced neither neurogenic inflammation nor painful perception in humans and rats [54–56]. In contrast, Lin et al. showed that SP had an anti-nociceptive role in muscle rather than causing pain [57]. With whole-cell patch clamp recordings on dissociated muscle-afferent dorsal root ganglion (DRG) neurons, the authors revealed that SP attenuated the acid sensing ion channel 3 (ASIC3)-induced inward current by enhancing M-channel-like potassium current. ASIC3 is a voltage-independent sodium channel activated by the extracellular protons. It has been found as a molecular determinant involved in pain-associated tissue acidosis [58,59]. As well, a recent study showed that ASIC3 can detect extracellular acidification and also respond to mechanical stimuli [60,61].

The in vivo antinociceptive role of SP was demonstrated in a rodent model of chronic widespread muscle pain induced by dual intramuscular acid injections, one of the fibromyalgia pain models developed by Sluka et al., in 2001 [62]. Two injections, separated by one to five days, of pH-4 acidic saline in the unilateral gastrocnemius muscle in rodents produced chronic and bilateral mechanical hyperalgesia of hind paws and muscle that required activation of ASIC3 [63]. Blocking ASIC3 activation, at the first or the second or both acid injections, abolished the induction and development of chronic muscle hyperalgesia. Furthermore, in the dual acid-injection model, the first acid injection could depolarize ASIC3-expressing muscle nociceptors and also simultaneously trigger SP release, which further enhanced the M-channel-like potassium current to attenuate ASIC3-induced depolarization in gastrocnemius muscle-afferent DRG neurons. In mice lacking *TAC1* (no SP and neurokinin A production), chronic pain could be induced by a single acid injection, which suggests that the anti-nociceptive effect was produced by the first acid injection but was diminished with the second injection [57]. The reason for ineffective SP in a second acid injection is still unclear and requires further investigation.

Regarding the acid-induced anti-nociception via SP release, the other important question is what types of acid sensors contribute to the release of SP as an anti-nociceptive acid sensor. The anti-nociceptive acid sensors are still unknown. Although previous study indicated that an acid sensor other than ASIC3 and transient receptor potential cation channel subfamily V member 1 (TRPV1) could trigger SP release [64], the possibility of the co-contribution of ASIC3, TRPV1, and other acid sensors such as other ASIC subtypes and/or proton-sensing G protein-coupled receptors is still not excluded. A recent study demonstrated that low-level laser therapy (LLLT) was effective in reducing mechanical hyperalgesia in the dual acid-injection model. The analgesic mechanism is associated with activation of TRPV1 to release SP in muscle [65]. This study provides new insights regarding the involvement of TRPV1 in acid-mediated anti-nociception. Furthermore, it reveals the involvement of SP in LLLT analgesia, which is widely used in pain control for musculoskeletal pain in the field of physical medicine and rehabilitation. In light of the antinociceptive role of SP in muscle, NK1R agonists might be promising candidates for pain relief in intractable musculoskeletal pain, such as fibromyalgia.

## 6. Ion Channels Involved in SP Signaling

SP can modulate a variety of ion channels (Table 1) resulting in an increase or decrease of neuronal excitability [66]. In most studies, SP excites neurons by increasing the function of excitatory ion channels and decreasing that of inhibitory ion channels. For example, SP has been shown to excite neurons by elevating the conductance of sodium channels and decreasing that of potassium channels in locus coeruleus neurons [67]. SP also inhibits inwardly rectifying $K^+$ channels in nucleus basalis neurons via $G_{q/11}$ [68,69], and inhibits $Ca^{2+}$-activated potassium channels [$I_{K(Ca)}$] in stellate ganglion neurons via pertussis toxin-insensitive G proteins [70]. Other studies showed that SP can inhibit the N-type calcium channel in sympathetic neurons via pertussis toxin-insensitive G proteins [71,72]. The above studies suggest that SP mainly modulates ion channel activity via the G-protein-dependent pathway. However, non-G-protein effector systems are also reported in SP-mediated signaling. Lu and colleagues revealed that SP-induced increase of sodium conductance was mediated by activating the sodium ion-permeable cation channel complex of NALCN (sodium leak channel, non-selective) and UNC-80 in mouse hippocampal and ventral tegmental area neurons independent of G-protein but mediated by Src family tyrosine kinase [73]. Accordingly, SP can modulate diverse channels and activate the neurons by G-protein-dependent or -independent signaling.

A few studies showed that SP hyperpolarizes neurons in the PNS. SP hyperpolarized vagal sensory neurons of ferrets by inducing a $Ca^{2+}$-dependent outward potassium current [74]. SP decreased non-selective cation channel conductance in outer hair cells of guinea pig cochlea [75]. SP enhanced the M-type potassium current independent of G-protein but dependent on tyrosine kinase in half of muscle-afferent DRG neurons [57]. Similarly, an SP-mediated $G_{i/o}$-dependent pathway could augment the M-type potassium current in DRG neurons and trigeminal ganglion (TG) neurons [76]. Finally, SP could inhibit T-type calcium channels in DRG and TG neurons [77]. Together, SP can modulate a variety of ion channels via different signaling pathways, which are cell type-specific.

**Table 1.** Ion channels modulated by substance P.

| Channel Types | Effector System | Cell Types | Species | Effects on Current | Outcomes | References |
|---|---|---|---|---|---|---|
| NMDAR | PKC | Spinal dorsal horn neurons, spinal thalamic neurons | Rat, monkey | ↑ | Pro-nociception | [78–81] |
| $K_{ir}$ | $G_{q/11}$ and PLC-β1 | Locus coeruleus neurons, nucleus basalis neurons | Rat | ↓ | | [67–69] |
| $K_{Ca}$ | PTX-insensitive G protein | Stellate ganglion neurons | Guinea pig | ↓ | | [70] |
| N-type $Ca^{2+}$ channel | PTX-insensitive G protein | Stellate ganglion neurons, superior cervical ganglion neurons, sympathetic neurons | Guinea pig, rat, frog | ↓ | | [70–72] |
| NALCN | Src family kinases | Hippocampal and ventral tegmental area neurons | Mice | ↑ | | [73] |
| TRP3C | Via NK2R | HEK293 | | ↑ | | [82] |
| Nav1.8 | PKCε | DRG | Rat | ↑ | Pro-nociception | [83] |
| TRPV1 | PKCε | DRG | Rat | ↑ | Pro-nociception | [84–86] |
| L, N type calcium channel | PKC | DRG | Rat | ↑ | | [87] |
| Low threshold potassium channel (Kv4) | | DRG | Rat | ↓ | Pro-nociception | [88] |
| P2X3 | | TG | Rat | ↑ | Pro-nociception | [89] |
| $GABA_AR$ | $G_{i/o}$ | Spinal dorsal horn neurons | Rat | ↑ | Anti-nociception | [90] |
| Glycine receptor | $G_{i/o}$ | Spinal dorsal horn neurons | Rat | ↑ | Anti-nociception | [90] |
| M-type potassium channel | Tyrosine kinase | DRG | Mice | ↑ | Anti-nociception | [57] |
| M-type potassium channel | $G_{i/o}$ | DRG, TG | Rat | ↑ | Anti-nociception | [76] |
| T-type calcium channel | $G_{i/o}$ | DRG | Rat | ↓ | Anti-nociception | [77] |
| $K_{Ca}$ | | Vagal sensory neurons | Ferret | ↑ | | [74] |
| $I_h$ | | Vagal sensory neurons | Ferret | ↓ | | [91] |
| Non-selective cation channel | PTX-insensitive G protein | Outer hair cells of cochlea | Guinea pig | ↓ | | [75] |

Abbreviations: NMDAR, *N*-methyl-*D*-aspartate receptor; $K_{ir}$, inward rectifier potassium channel; $K_{Ca}$, calcium-activated potassium channel; $I_h$, hyperpolarization-activated channel; PLC, phospholipase C; PTX, pertussis toxin; PKC, $Ca^{2+}$/phospholipid-dependent protein kinase; DRG, dorsal root ganglion; TG, trigeminal ganglion; Symbols: ↑, increase; ↓, decrease.

## 7. Ion Channels Involved in SP-Mediated Nociception

SP has a well-known role in the transmission of nociceptive information in the spinal cord. In the supraspinal level, microinjection of SP in the rostral ventromedial medulla (RVM) has also been found to induce hyperalgesia via descending facilitation mechanisms in a glutamate- and $GABA_A$-dependent manner [92,93]. Furthermore, SP release is increased in the spinal dorsal horn after peripheral nociceptive stimulation [94]. Overall, the amount of nociceptive neurotransmitter released by primary afferent nerve terminals determines the level of pain. The release of SP in the spinal dorsal horn would interact with glutamate to enhance peripheral inputs. Previously, SP and glutamate were found to co-exist in small-diameter DRG neurons and their nerve terminals in the spinal dorsal horn [95]. Glutamate acts as the molecule transmitting the fast excitatory signal, whereas SP modulates relatively slow excitatory synapse responses [96]. Besides, SP signaling can enhance the *N*-methyl-*D*-aspartate (NMDA) channel function leading to greater pain sensitivity [78–81]. Thus, the two transmitters, glutamate and SP, are considered to interact and convey the nociceptive information in the spinal cord.

In the peripheral system, SP is an important element in neurogenic inflammation causing extravasation and sensory neuron sensitization. During inflammatory processes, inflammatory cells and peripheral nerve terminals release SP, which, in turn, modulates a variety of ion channels rendering sensitization of sensory neurons in an autocrine or paracrine manner. In the PNS, SP mainly exists in the small sensory nociceptors. Release of SP can act on NK1R via differential intracellular mechanisms to potentiate the channel activities of TRPV1 [84–86], Nav1.8 [83], and L- and N-type calcium channels [87] in a subset of small-diameter DRG neurons, thereby resulting in hyperalgesia. SP could also decease the activity of low-threshold potassium channel (kv4) in capsaicin-sensitive DRG neurons and thus sensitize the nociceptors [88]. In the orofacial region, SP can potentiate the P2X3 receptor in TG neurons, leading to elevated pain sensitivity [89]. In summary, SP predominantly acts on peripheral sensory small neurons (presumably nociceptors) to excite the neurons, thereby increasing nociceptive responses.

## 8. Ion Channels Involved in SP-Mediated Anti-Nociception

The role of SP in anti-nociception has been confirmed in both the CNS and the PNS. In the CNS, the analgesic effect of SP is mainly associated with opioid-dependent pathways, although other studies also demonstrated the involvement of $GABA_AR$ and glycine receptors in the lamina V of the spinal cord [90,97].

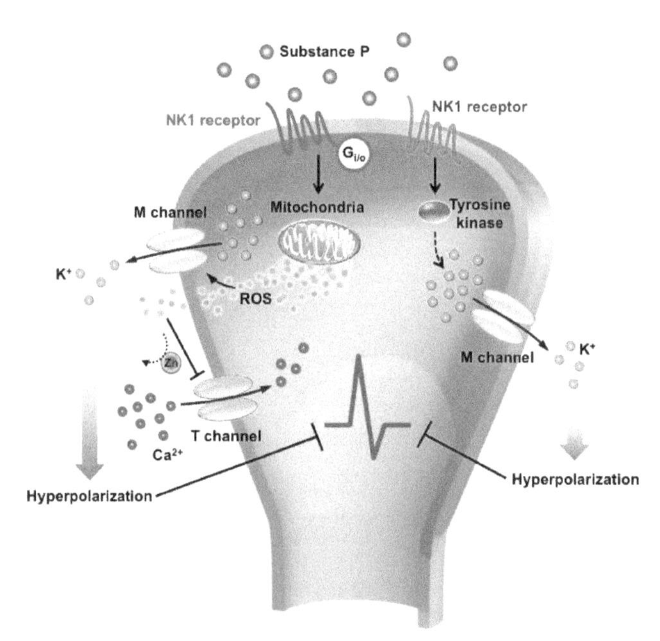


**Figure 1.** Schematic diagram of substance P-mediated signaling and ion channels in the peripheral sensory neurons. Release of substance P (SP) in the nerve terminal acts on neurokinin 1 receptor (NK1R) via two different effector systems modulating M-type $K^+$ and T-type $Ca^{2+}$ channels. First, activated NK1R coupled to tyrosine kinase augments the M-type potassium channels, resulting in neuronal hyperpolarization. Second, activated NK1R coupled to $G_{i/o}$ triggers reactive oxygen species (ROS) release from mitochondria simultaneously to augment M-type potassium channels and inhibit T-type calcium channels, which inhibits neural firing in peripheral sensory neurons.

In the PNS, SP can act on NK1R via either tyrosine kinase or $G_{i/o}$ effector system to potentiate the inhibitory M-type potassium channels and inhibit excitatory T-type calcium channels in a specific subset of sensory neurons [76,77]. In muscle afferent DRG neurons, the SP-NK1R signaling is coupled with tyrosine kinase to enhance M-type potassium channel activity [57]. Enhancing the M-type potassium current and inhibiting the T-type calcium current results in dampening neuronal excitability, in turn further leading to an anti-nociceptive effect (Figure 1).

## 9. M-type Potassium Channels

M-type potassium channels are voltage-gated potassium channels (M for muscarine) encoded by *KCNQ* genes. *KCNQ* genes encode five Kv7 subunits (Kv7.1–7.5) [98]. Kv7.2, 7.3, 7.4, and 7.5 are expressed in the nervous system. The M-type potassium channel mainly features Kv7.2 and Kv7.3, although other subunits can also contribute to the formation of M channels in some locations.

M channels were first discovered in sympathetic neurons [99]. When muscarinic agonists activate the muscarinic acetylcholine receptors, sympathetic neurons become more responsive to the synaptic inputs and can become burst firing, rather than fire a single spike. This situation is due to the suppression of a unique channel, which led to the name "M" channel.

M-type channels are able to conduct a non-inactivating outward current with a threshold of about −60 to −80 mV and regulated by many neurotransmitters such as SP and bradykinin. The biophysical features of this current include slow activation and deactivation potassium current. Because M channels can be opened near the resting membrane potential, they also play a role in clamping the resting membrane potential, so they are ideally suited to control neuronal excitability.

M channels are functionally expressed in various central and peripheral neurons including hippocampal [100] and DRG neurons [101]. *KCNQ2, 3,* and *5* are variably expressed in sensory neurons, including small-diameter nociceptors and large-diameter proprioceptors. Because pathological pain such as inflammatory and neuropathic pain features for neuronal hypersensitivity responds to nociceptive inputs, M-type potassium channels are considered a potential analgesic target in controlling nociceptive excitability. Opening of M channels results in hyperpolarization of the neurons, which decreases cell membrane excitability. Accordingly, the M-channel openers retigabine and flupirtine can effectively attenuate muscle pain [102] and inflammatory pain [101].

In 2012, Lin and colleagues showed that SP-enhanced M channel activity in muscle-afferent DRG neurons via an unconventional signal pathway by activating NK1R coupled with phosphotyrosine kinase to attenuate the mechanical hyperalgesia [57]. Then, Gamper's group demonstrated that SP could augment the M channel in a subset DRG neurons mainly by acting on NK1R via Gi/o and a redox-dependent pathway [76]. The similar result but discrepant cellular mechanisms could be due to the different subset of DRG neurons examined. The SP-enhanced M current via the phosphotyrosine kinase pathway is prominent on medium to large DRG neurons expressing ASIC3 innervating muscle, whereas the SP-enhanced M current via the Gi/o pathway acts mainly on small neurons expressing TRPV1 innervating skin.

These studies provide insight into SP-mediated anti-nociception in local tissues. Activating the SP-NK1R pathway specifically in muscle afferents becomes an attractive therapeutic target to treat chronic pain. Although the pro-nociceptive effect of SP in the spinal cord has been well documented, local peripheral application of SP can be diluted and has little effect on the spinal cord. Thus, targeting the SP-NK1R pathway in the PNS creates a strategy for pain relief without producing severe side effects in the CNS.

## 10. T-type Calcium Channels

T-type calcium channels (Cav3) are low-voltage activated calcium channels opened with small membrane depolarization and can generate $Ca^{2+}$-dependent burst firing, pacemaker activity, and low-amplitude neuronal oscillation [103]. In contrast to high-threshold L-type calcium channels (long-lasting calcium channels), T-type calcium channels only mediate a "transient" calcium influx, with the voltage threshold about −60 with slower activation. T channels also form a window current in the resting membrane potential controlling the sub-threshold neuronal excitability.

The α1 subunit consists of the major channel pore of T-type calcium channels, allowing for calcium influx through it. The channel pore-forming α subunits include Cav3.1, Cav3.2, and Cav3.3 encoded by calcium voltage-gated channel subunit alphal G (CACNA1G), H, and I, respectively. T-type calcium channels were first found in small sensory neurons by Carbone and Lux in 1984 [104]. Subsequently, functional expression of T-type channels was confirmed in nociceptive DRG neurons, with Cav3.2 as the predominant isoform. Nelson et al. further characterized the expression profile of T channels and indicated that a subtype of DRG neurons highly expressed T channels, named T-rich cells, which were also highly sensitive to capsaicin and ATP stimuli [105]. Those T-rich cells are mainly small-diameter neurons with narrower APs and after depolarizing potentials during the action potential falling phase. Altering gating parameters or directly increasing the amplitude of the T current can result in burst firing and neuronal excitability [105]. Inhibiting the T-type channels can effectively reduce the neuropathic and inflammatory pain [106–108].

Recently, Huang et al. demonstrated that T channels were modulated by SP [77]. The authors found that SP release in damaging tissue would act on NK1R via Gi/o to release mitochondria reactive oxygen species, which further inhibited the T channels. This study revealed SP-mediated anti-nociception in a subtype of small-diameter sensory neurons. Further studies are required to determine whether medium- to large-diameter sensory neurons also show the same modulation of SP in T-type calcium channels.

## 11. Conclusions

Accumulating evidence has shown SP-mediated anti-nociception in both the CNS and PNS. However, SP has a mixed effect of pro-nociception and anti-nociception in the nervous system. Especially, blocking NK1R signaling in the CNS can result in adverse outcomes. In contrast, the anti-nociceptive effect of SP in the PNS has been well characterized, especially in muscle. Applying SP to muscle-afferent DRG neurons enhances the M-channel outward current, which further hyperpolarizes the neurons. Accordingly, locally targeting the SP-NK1R pathway on muscle afferent neurons could be a promising analgesic strategy.

**Author Contributions:** C.-C.C. composed the idea and wrote the manuscript. C.-T.C. drafted the manuscript and figure. B.-Y.J. prepared the information of table.

**Funding:** This work was supported by a grant from the Institute of Biomedical Sciences, Academia Sinica (IBMS-CRC107-P01), and grants from the Ministry of Science and Technology, Taiwan (MOST105-2320-B-001-018-MY3, MOST107-2319-B-001-002, and MOST108-2321-B-001-005).

**Acknowledgments:** We thank the staff of the Medical Art Room at Institute of Biomedical Sciences, Academia Sinica, for the scientific illustration in Figure 1.

**Conflicts of Interest:** The authors declare no conflict of interest.

## References

1. Euler, U.S.; Gaddum, J.H. An unidentified depressor substance in certain tissue extracts. *J. Physiol.* **1931**, *72*, 74–87. [CrossRef]
2. Chang, M.M.; Leeman, S.E. Isolation of a sialogogic peptide from bovine hypothalamic tissue and its characterization as substance P. *J. Biol. Chem.* **1970**, *245*, 4784–4790. [PubMed]
3. Chang, M.M.; Leeman, S.E.; Niall, H.D. Amino-acid sequence of substance P. *Nat. New Biol.* **1971**, *232*, 86–87. [CrossRef] [PubMed]
4. Krause, J.E.; Chirgwin, J.M.; Carter, M.S.; Xu, Z.S.; Hershey, A.D. Three rat preprotachykinin mRNAs encode the neuropeptides substance P and neurokinin A. *Proc. Natl. Acad. Sci. USA* **1987**, *84*, 881–885. [CrossRef]
5. Jessell, T.M. Substance P in Nociceptive Sensory Neurons. In *Substance P in the Nervous System*; Porter, R., O'Connor, M., Eds.; John Wiley & Sons: Hoboken, NJ, USA, 1982.
6. Rameshwar, P.; Gascon, P.; Ganea, D. Immunoregulatory effects of neuropeptides. Stimulation of interleukin-2 production by substance p. *J. Neuroimmunol.* **1992**, *37*, 65–74. [CrossRef]

7. Potter, G.D.; Guzman, F.; Lim, R.K. Visceral pain evoked by intra-arterial injection of substance P. *Nature* **1962**, *193*, 983–984. [CrossRef] [PubMed]
8. Jessell, T.M. Neurotransmitters and CNS disease. Pain. *Lancet* **1982**, *2*, 1084–1088. [CrossRef]
9. Iversen, L. Substance P equals pain substance? *Nature* **1998**, *392*, 334–335. [CrossRef] [PubMed]
10. Stewart, J.M.; Getto, C.J.; Neldner, K.; Reeve, E.B.; Krivoy, W.A.; Zimmermann, E. Substance P and analgesia. *Nature* **1976**, *262*, 784–785. [CrossRef]
11. Hall, M.E.; Stewart, J.M. Substance P and antinociception. *Peptides* **1983**, *4*, 31–35. [CrossRef]
12. Macdonald, S.G.; Dumas, J.J.; Boyd, N.D. Chemical cross-linking of the substance P (NK-1) receptor to the alpha subunits of the G proteins Gq and G11. *Biochemistry* **1996**, *35*, 2909–2916. [CrossRef]
13. Meza, U.; Thapliyal, A.; Bannister, R.A.; Adams, B.A. Neurokinin 1 receptors trigger overlapping stimulation and inhibition of CaV2.3 (R-type) calcium channels. *Mol. Pharmacol.* **2007**, *71*, 284–293. [CrossRef]
14. Meshki, J.; Douglas, S.D.; Lai, J.P.; Schwartz, L.; Kilpatrick, L.E.; Tuluc, F. Neurokinin 1 receptor mediates membrane blebbing in HEK293 cells through a Rho/Rho-associated coiled-coil kinase-dependent mechanism. *J. Biol. Chem.* **2009**, *284*, 9280–9289. [CrossRef]
15. Roush, E.D.; Kwatra, M.M. Human substance P receptor expressed in Chinese hamster ovary cells directly activates G(alpha q/11), G(alpha s), G(alpha o). *FEBS Lett.* **1998**, *428*, 291–294. [CrossRef]
16. Khawaja, A.M.; Rogers, D.F. Tachykinins: Receptor to effector. *Int. J. Biochem. Cell Biol.* **1996**, *28*, 721–738. [CrossRef]
17. Fong, T.M.; Anderson, S.A.; Yu, H.; Huang, R.R.; Strader, C.D. Differential activation of intracellular effector by two isoforms of human neurokinin-1 receptor. *Mol. Pharmacol.* **1992**, *41*, 24–30. [PubMed]
18. Baker, S.J.; Morris, J.L.; Gibbins, I.L. Cloning of a C-terminally truncated NK-1 receptor from guinea-pig nervous system. *Brain Res. Mol. Brain Res.* **2003**, *111*, 136–147. [CrossRef]
19. DeFea, K.A.; Vaughn, Z.D.; O'Bryan, E.M.; Nishijima, D.; Dery, O.; Bunnett, N.W. The proliferative and antiapoptotic effects of substance P are facilitated by formation of a beta -arrestin-dependent scaffolding complex. *Proc. Natl. Acad. Sci. USA* **2000**, *97*, 11086–11091. [CrossRef]
20. Lai, J.P.; Lai, S.; Tuluc, F.; Tansky, M.F.; Kilpatrick, L.E.; Leeman, S.E.; Douglas, S.D. Differences in the length of the carboxyl terminus mediate functional properties of neurokinin-1 receptor. *Proc. Natl. Acad. Sci. USA* **2008**, *105*, 12605–12610. [CrossRef] [PubMed]
21. Caberlotto, L.; Hurd, Y.L.; Murdock, P.; Wahlin, J.P.; Melotto, S.; Corsi, M.; Carletti, R. Neurokinin 1 receptor and relative abundance of the short and long isoforms in the human brain. *Eur. J. Neurosci.* **2003**, *17*, 1736–1746. [CrossRef] [PubMed]
22. Harstall, C.; Ospina, M. How prevalent is chronic pain. *Pain Clin. Updates* **2003**, *11*, 1–4.
23. Tracey, D.J.; Walker, J.S. Pain Due to Nerve Damage—Are Inflammatory Mediators Involved. *Inflamm. Res.* **1995**, *44*, 407–411. [CrossRef]
24. Moalem, G.; Tracey, D.J. Immune and inflammatory mechanisms in neuropathic pain. *Brain Res. Rev.* **2006**, *51*, 240–264. [CrossRef] [PubMed]
25. Bennett, G.J. Can we distinguish between inflammatory and neuropathic pain? *Pain Res. Manag.* **2006**, *11*, 11A–15A. [CrossRef]
26. De Koninck, Y.; Henry, J.L. Substance P-mediated slow excitatory postsynaptic potential elicited in dorsal horn neurons in vivo by noxious stimulation. *Proc. Natl. Acad. Sci. USA* **1991**, *88*, 11344–11348. [CrossRef] [PubMed]
27. Holzer, P. Neurogenic vasodilatation and plasma leakage in the skin. *Gen. Pharmacol.* **1998**, *30*, 5–11. [CrossRef]
28. Otsuka, M.; Yoshioka, K. Neurotransmitter functions of mammalian tachykinins. *Physiol. Rev.* **1993**, *73*, 229–308. [CrossRef] [PubMed]
29. Russell, I.J.; Orr, M.D.; Littman, B.; Vipraio, G.A.; Alboukrek, D.; Michalek, J.E.; Lopez, Y.; MacKillip, F. Elevated cerebrospinal fluid levels of substance P in patients with the fibromyalgia syndrome. *Arthritis Rheumatol.* **1994**, *37*, 1593–1601. [CrossRef]
30. Vaeroy, H.; Helle, R.; Forre, O.; Kass, E.; Terenius, L. Elevated CSF levels of substance P and high incidence of Raynaud phenomenon in patients with fibromyalgia: New features for diagnosis. *Pain* **1988**, *32*, 21–26. [CrossRef]
31. Mantyh, P.W.; Rogers, S.D.; Honore, P.; Allen, B.J.; Ghilardi, J.R.; Li, J.; Daughters, R.S.; Lappi, D.A.; Wiley, R.G.; Simone, D.A. Inhibition of hyperalgesia by ablation of lamina I spinal neurons expressing the substance P receptor. *Science* **1997**, *278*, 275–279. [CrossRef] [PubMed]

32. Cao, Y.Q.; Mantyh, P.W.; Carlson, E.J.; Gillespie, A.M.; Epstein, C.J.; Basbaum, A.I. Primary afferent tachykinins are required to experience moderate to intense pain. *Nature* **1998**, *392*, 390–394. [CrossRef] [PubMed]
33. De Felipe, C.; Herrero, J.F.; O'Brien, J.A.; Palmer, J.A.; Doyle, C.A.; Smith, A.J.; Laird, J.M.; Belmonte, C.; Cervero, F.; Hunt, S.P. Altered nociception, analgesia and aggression in mice lacking the receptor for substance P. *Nature* **1998**, *392*, 394–397. [CrossRef] [PubMed]
34. Huang, T.; Lin, S.H.; Malewicz, N.M.; Zhang, Y.; Zhang, Y.; Goulding, M.; LaMotte, R.H.; Ma, Q. Identifying the pathways required for coping behaviours associated with sustained pain. *Nature* **2019**, *565*, 86–90. [CrossRef] [PubMed]
35. Rigby, M.; O'Donnell, R.; Rupniak, N.M. Species differences in tachykinin receptor distribution: Further evidence that the substance P (NK1) receptor predominates in human brain. *J. Comp. Neurol.* **2005**, *490*, 335–353. [CrossRef] [PubMed]
36. Gitter, B.D.; Waters, D.C.; Bruns, R.F.; Mason, N.R.; Nixon, J.A.; Howbert, J.J. Species differences in affinities of non-peptide antagonists for substance P receptors. *Eur. J. Pharmacol.* **1991**, *197*, 237–238. [CrossRef]
37. Beresford, I.; Birch, P.; Hagan, R.; Ireland, S. Investigation into species variants in tachykinin NK1 receptors by use of the non-peptide antagonist, CP-96,345. *Br. J. Pharmacol.* **1991**, *104*, 292–293. [CrossRef] [PubMed]
38. Appell, K.C.; Fragale, B.J.; Loscig, J.; Singh, S.; Tomczuk, B.E. Antagonists That Demonstrate Species-Differences in Neurokinin-1 Receptors. *Mol. Pharmacol.* **1992**, *41*, 772–778. [PubMed]
39. Mogil, J.S. Animal models of pain: Progress and challenges. *Nat. Rev. Neurosci.* **2009**, *10*, 283–294. [CrossRef]
40. Frederickson, R.C.; Burgis, V.; Harrell, C.E.; Edwards, J.D. Dual actions of substance P on nociception: Possible role of endogenous opioids. *Science* **1978**, *199*, 1359–1362. [CrossRef] [PubMed]
41. Oehme, P.; Hilse, H.; Morgenstern, E.; Gores, E. Substance-P—Does It Produce Analgesia or Hyperalgesia. *Science* **1980**, *208*, 305–307. [CrossRef] [PubMed]
42. Chung, E.; Yoon, T.G.; Kim, S.; Kang, M.; Kim, H.J.; Son, Y. Intravenous Administration of Substance P Attenuates Mechanical Allodynia Following Nerve Injury by Regulating Neuropathic Pain-Related Factors. *Biomol. Ther. (Seoul)* **2017**, *25*, 259–265. [CrossRef] [PubMed]
43. Parenti, C.; Arico, G.; Ronsisvalle, G.; Scoto, G.M. Supraspinal injection of Substance P attenuates allodynia and hyperalgesia in a rat model of inflammatory pain. *Peptides* **2012**, *34*, 412–418. [CrossRef] [PubMed]
44. Bury, R.W.; Mashford, M.L. Biological activity of C-terminal partial sequences of substance P. *J. Med. Chem.* **1976**, *19*, 854–856. [CrossRef] [PubMed]
45. Hall, M.E.; Stewart, J.M. Substance P and behavior: Opposite effects of N-terminal and C-terminal fragments. *Peptides* **1983**, *4*, 763–768. [CrossRef]
46. Stewart, J.M.; Hall, M.E.; Harkins, J.; Frederickson, R.C.; Terenius, L.; Hokfelt, T.; Krivoy, W.A. A fragment of substance P with specific central activity: SP(1-7). *Peptides* **1982**, *3*, 851–857. [CrossRef]
47. Skilling, S.R.; Smullin, D.H.; Larson, A.A. Differential effects of C- and N-terminal substance P metabolites on the release of amino acid neurotransmitters from the spinal cord: Potential role in nociception. *J. Neurosci.* **1990**, *10*, 1309–1318. [CrossRef]
48. Tang, J.; Chou, J.; Yang, H.Y.; Costa, E. Substance P stimulates the release of Met5-enkephalin-Arg6-Phe7 and Met5-enkephalin from rat spinal cord. *Neuropharmacology* **1983**, *22*, 1147–1150. [CrossRef]
49. Naranjo, J.R.; Arnedo, A.; De Felipe, M.C.; Del Rio, J. Antinociceptive and Met-enkephalin releasing effects of tachykinins and substance P fragments. *Peptides* **1986**, *7*, 419–423. [CrossRef]
50. Naranjo, J.R.; Sanchez-Franco, F.; Garzon, J.; del Rio, J. Analgesic activity of substance P in rats: Apparent mediation by met-enkephalin release. *Life Sci.* **1982**, *30*, 441–446. [CrossRef]
51. Komatsu, T.; Sasaki, M.; Sanai, K.; Kuwahata, H.; Sakurada, C.; Tsuzuki, M.; Iwata, Y.; Sakurada, S.; Sakurada, T. Intrathecal substance P augments morphine-induced antinociception: Possible relevance in the production of substance P N-terminal fragments. *Peptides* **2009**, *30*, 1689–1696. [CrossRef] [PubMed]
52. Dong, X.G.; Yu, L.C. Alterations in the substance P-induced anti-nociception in the central nervous system of rats after morphine tolerance. *Neurosci. Lett.* **2005**, *381*, 47–50. [CrossRef] [PubMed]
53. Bowman, S.L.; Soohoo, A.L.; Shiwarski, D.J.; Schulz, S.; Pradhan, A.A.; Puthenveedu, M.A. Cell-autonomous regulation of Mu-opioid receptor recycling by substance P. *Cell Rep.* **2015**, *10*, 1925–1936. [CrossRef]
54. Pedersen-Bjergaard, U.; Nielsen, L.B.; Jensen, K.; Edvinsson, L.; Jansen, I.; Olesen, J. Algesia and local responses induced by neurokinin A and substance P in human skin and temporal muscle. *Peptides* **1989**, *10*, 1147–1152. [CrossRef]

55. Babenko, V.V.; Graven-Nielsen, T.; Svensson, P.; Drewes, A.M.; Jensen, T.S.; Arendt-Nielsen, L. Experimental human muscle pain induced by intramuscular injections of bradykinin, serotonin, and substance P. *Eur. J. Pain* **1999**, *3*, 93–102. [CrossRef]
56. Jensen, K.; Tuxen, C.; Pedersen-Bjergaard, U.; Jansen, I. Pain, tenderness, wheal and flare induced by substance-P, bradykinin and 5-hydroxytryptamine in humans. *Cephalalgia* **1991**, *11*, 175–182. [CrossRef]
57. Lin, C.C.; Chen, W.N.; Chen, C.J.; Lin, Y.W.; Zimmer, A.; Chen, C.C. An antinociceptive role for substance P in acid-induced chronic muscle pain. *Proc. Natl. Acad. Sci. USA* **2012**, *109*, E76–E83. [CrossRef] [PubMed]
58. Wu, W.L.; Cheng, C.F.; Sun, W.H.; Wong, C.W.; Chen, C.C. Targeting ASIC3 for pain, anxiety, and insulin resistance. *Pharmacol. Ther.* **2012**, *134*, 127–138. [CrossRef]
59. Lin, J.H.; Hung, C.H.; Han, D.S.; Chen, S.T.; Lee, C.H.; Sun, W.Z.; Chen, C.C. Sensing acidosis: Nociception or sngception? *J. Biomed. Sci.* **2018**, *25*, 85. [CrossRef]
60. Lin, S.H.; Cheng, Y.R.; Banks, R.W.; Min, M.Y.; Bewick, G.S.; Chen, C.C. Evidence for the involvement of ASIC3 in sensory mechanotransduction in proprioceptors. *Nat. Commun.* **2016**, *7*, 11460. [CrossRef] [PubMed]
61. Cheng, Y.R.; Jiang, B.Y.; Chen, C.C. Acid-sensing ion channels: Dual function proteins for chemo-sensing and mechano-sensing. *J. Biomed. Sci.* **2018**, *25*, 46. [CrossRef] [PubMed]
62. Sluka, K.A.; Kalra, A.; Moore, S.A. Unilateral intramuscular injections of acidic saline produce a bilateral, long-lasting hyperalgesia. *Muscle Nerve* **2001**, *24*, 37–46. [CrossRef]
63. Sluka, K.A.; Price, M.P.; Breese, N.M.; Stucky, C.L.; Wemmie, J.A.; Welsh, M.J. Chronic hyperalgesia induced by repeated acid injections in muscle is abolished by the loss of ASIC3, but not ASIC1. *Pain* **2003**, *106*, 229–239. [CrossRef]
64. Chen, W.N.; Chen, C.C. Acid mediates a prolonged antinociception via substance P signaling in acid-induced chronic widespread pain. *Mol. Pain* **2014**, *10*, 30. [CrossRef]
65. Han, D.S.; Lee, C.H.; Shieh, Y.D.; Chen, C.C. Involvement of Substance P in the Analgesic Effect of Low-levle Laser Therapy in a Mouse Model of Chronic Widespread Muscle Pain. *Pain Med.* **2019**. [CrossRef]
66. Moraes, E.R.; Kushmerick, C.; Naves, L.A. Characteristics of dorsal root ganglia neurons sensitive to Substance P. *Mol. Pain* **2014**, *10*, 73. [CrossRef] [PubMed]
67. Shen, K.Z.; North, R.A. Substance P opens cation channels and closes potassium channels in rat locus coeruleus neurons. *Neuroscience* **1992**, *50*, 345–353. [CrossRef]
68. Nakajima, Y.; Nakajima, S.; Inoue, M. Pertussis toxin-insensitive G protein mediates substance P-induced inhibition of potassium channels in brain neurons. *Proc. Natl. Acad. Sci. USA* **1988**, *85*, 3643–3647. [CrossRef]
69. Takano, K.; Yasufuku-Takano, J.; Kozasa, T.; Singer, W.D.; Nakajima, S.; Nakajima, Y. Gq/11 and PLC-beta 1 mediate the substance P-induced inhibition of an inward rectifier K+ channel in brain neurons. *J. Neurophysiol.* **1996**, *76*, 2131–2136. [CrossRef]
70. Gilbert, R.; Ryan, J.S.; Horackova, M.; Smith, F.M.; Kelly, M.E. Actions of substance P on membrane potential and ionic currents in guinea pig stellate ganglion neurons. *Am. J. Physiol.* **1998**, *274*, C892–C903. [CrossRef] [PubMed]
71. Shapiro, M.S.; Hille, B. Substance P and somatostatin inhibit calcium channels in rat sympathetic neurons via different G protein pathways. *Neuron* **1993**, *10*, 11–20. [CrossRef]
72. Bley, K.R.; Tsien, R.W. Inhibition of $Ca^{2+}$ and $K^{+}$ channels in sympathetic neurons by neuropeptides and other ganglionic transmitters. *Neuron* **1990**, *4*, 379–391. [CrossRef]
73. Lu, B.; Su, Y.; Das, S.; Wang, H.; Wang, Y.; Liu, J.; Ren, D. Peptide neurotransmitters activate a cation channel complex of NALCN and UNC-80. *Nature* **2009**, *457*, 741–744. [CrossRef] [PubMed]
74. Jafri, M.S.; Weinreich, D. Substance P hyperpolarizes vagal sensory neurones of the ferret. *J. Physiol.* **1996**, *493 Pt 1*, 157–166. [CrossRef]
75. Kakehata, S.; Akaike, N.; Takasaka, T. Substance P decreases the non-selective cation channel conductance in dissociated outer hair cells of guinea pig cochlea. *Ann. N. Y. Acad. Sci.* **1993**, *707*, 476–479. [CrossRef]
76. Linley, J.E.; Ooi, L.; Pettinger, L.; Kirton, H.; Boyle, J.P.; Peers, C.; Gamper, N. Reactive oxygen species are second messengers of neurokinin signaling in peripheral sensory neurons. *Proc. Natl. Acad. Sci. USA* **2012**, *109*, E1578–E1586. [CrossRef] [PubMed]
77. Huang, D.; Huang, S.; Gao, H.; Liu, Y.; Qi, J.; Chen, P.; Wang, C.; Scragg, J.L.; Vakurov, A.; Peers, C.; et al. Redox-Dependent Modulation of T-Type Ca(2+) Channels in Sensory Neurons Contributes to Acute Anti-Nociceptive Effect of Substance P. *Antioxid. Redox Signal.* **2016**, *25*, 233–251. [CrossRef]

78. Rusin, K.I.; Bleakman, D.; Chard, P.S.; Randic, M.; Miller, R.J. Tachykinins Potentiate N-Methyl-D-Aspartate Responses in Acutely Isolated Neurons from the Dorsal Horn. *J. Neurochem.* **1993**, *60*, 952–960. [CrossRef]
79. Dougherty, P.M.; Willis, W.D. Enhancement of Spinothalamic Neuron Responses to Chemical and Mechanical Stimuli Following Combined Micro-Iontophoretic Application of *N*-Methyl-D-Aspartic Acid and Substance-P. *Pain* **1991**, *47*, 85–93. [CrossRef]
80. Randic, M.; Hecimovic, H.; Ryu, P.D. Substance P modulates glutamate-induced currents in acutely isolated rat spinal dorsal horn neurones. *Neurosci. Lett.* **1990**, *117*, 74–80. [CrossRef]
81. Rusin, K.I.; Ryu, P.D.; Randic, M. Modulation of Excitatory Amino-Acid Responses in Rat Dorsal Horn Neurons by Tachykinins. *J. Neurophysiol.* **1992**, *68*, 265–286. [CrossRef] [PubMed]
82. Oh, E.J.; Gover, T.D.; Cordoba-Rodriguez, R.; Weinreich, D. Substance P evokes cation currents through TRP channels in HEK293 cells. *J. Neurophysiol.* **2003**, *90*, 2069–2073. [CrossRef] [PubMed]
83. Cang, C.L.; Zhang, H.; Zhang, Y.Q.; Zhao, Z.Q. PKCepsilon-dependent potentiation of TTX-resistant Nav1.8 current by neurokinin-1 receptor activation in rat dorsal root ganglion neurons. *Mol. Pain* **2009**, *5*, 33. [CrossRef]
84. Zhang, H.; Cang, C.L.; Kawasaki, Y.; Liang, L.L.; Zhang, Y.Q.; Ji, R.R.; Zhao, Z.Q. Neurokinin-1 receptor enhances TRPV1 activity in primary sensory neurons via PKCepsilon: A novel pathway for heat hyperalgesia. *J. Neurosci.* **2007**, *27*, 12067–12077. [CrossRef] [PubMed]
85. Sculptoreanu, A.; Aura Kullmann, F.; de Groat, W.C. Neurokinin 2 receptor-mediated activation of protein kinase C modulates capsaicin responses in DRG neurons from adult rats. *Eur. J. Neurosci.* **2008**, *27*, 3171–3181. [CrossRef] [PubMed]
86. Lapointe, T.K.; Basso, L.; Iftinca, M.C.; Flynn, R.; Chapman, K.; Dietrich, G.; Vergnolle, N.; Altier, C. TRPV1 sensitization mediates postinflammatory visceral pain following acute colitis. *Am. J. Physiol.-Gastr. Liver Physiol.* **2015**, *309*, G87–G99. [CrossRef] [PubMed]
87. Sculptoreanu, A.; de Groat, W.C. Protein kinase C is involved in neurokinin receptor modulation of N- and L-type Ca2+ channels in DRG neurons of the adult rat. *J. Neurophysiol.* **2003**, *90*, 21–31. [CrossRef]
88. Sculptoreanu, A.; Artim, D.E.; de Groat, W.C. Neurokinins inhibit low threshold inactivating K+ currents in capsaicin responsive DRG neurons. *Exp. Neurol.* **2009**, *219*, 562–573. [CrossRef] [PubMed]
89. Park, C.K.; Bae, J.H.; Kim, H.Y.; Jo, H.J.; Kim, Y.H.; Jung, S.J.; Kim, J.S.; Oh, S.B. Substance P sensitizes P2X3 in nociceptive trigeminal neurons. *J. Dent. Res.* **2010**, *89*, 1154–1159. [CrossRef] [PubMed]
90. Nakatsuka, T.; Chen, M.; Takeda, D.; King, C.; Ling, J.; Xing, H.; Ataka, T.; Vierck, C.; Yezierski, R.; Gu, J.G. Substance P-driven feed-forward inhibitory activity in the mammalian spinal cord. *Mol. Pain* **2005**, *1*, 20. [CrossRef] [PubMed]
91. Jafri, M.S.; Weinreich, D. Substance P regulates Ih via a NK-1 receptor in vagal sensory neurons of the ferret. *J. Neurophysiol.* **1998**, *79*, 769–777. [CrossRef] [PubMed]
92. Lagraize, S.C.; Guo, W.; Yang, K.; Wei, F.; Ren, K.; Dubner, R. Spinal cord mechanisms mediating behavioral hyperalgesia induced by neurokinin-1 tachykinin receptor activation in the rostral ventromedial medulla. *Neuroscience* **2010**, *171*, 1341–1356. [CrossRef] [PubMed]
93. Brink, T.S.; Pacharinsak, C.; Khasabov, S.G.; Beitz, A.J.; Simone, D.A. Differential modulation of neurons in the rostral ventromedial medulla by neurokinin-1 receptors. *J. Neurophysiol.* **2012**, *107*, 1210–1221. [CrossRef] [PubMed]
94. Duggan, A.W.; Hendry, I.A.; Morton, C.R.; Hutchison, W.D.; Zhao, Z.Q. Cutaneous stimuli releasing immunoreactive substance P in the dorsal horn of the cat. *Brain Res.* **1988**, *451*, 261–273. [CrossRef]
95. Debiasi, S.; Rustioni, A. Glutamate and Substance-P Coexist in Primary Afferent Terminals in the Superficial Laminae of Spinal-Cord. *Proc. Natl. Acad. Sci. USA* **1988**, *85*, 7820–7824. [CrossRef]
96. Urban, L.; Randic, M. Slow excitatory transmission in rat dorsal horn: Possible mediation by peptides. *Brain Res.* **1984**, *290*, 336–341. [CrossRef]
97. Wu, L.J.; Xu, H.; Ko, S.W.; Yoshimura, M.; Zhuo, M. Feed-forward inhibition: A novel cellular mechanism for the analgesic effect of substance P. *Mol. Pain* **2005**, *1*, 34. [CrossRef] [PubMed]
98. Jentsch, T.J. Neuronal KCNQ potassium channels: Physiology and role in disease. *Nat. Rev. Neurosci.* **2000**, *1*, 21–30. [CrossRef] [PubMed]
99. Brown, D.A.; Adams, P.R. Muscarinic suppression of a novel voltage-sensitive K+ current in a vertebrate neurone. *Nature* **1980**, *283*, 673–676. [CrossRef] [PubMed]
100. Shah, M.; Mistry, M.; Marsh, S.J.; Brown, D.A.; Delmas, P. Molecular correlates of the M-current in cultured rat hippocampal neurons. *J. Physiol.* **2002**, *544*, 29–37. [CrossRef] [PubMed]

101. Passmore, G.M.; Selyanko, A.A.; Mistry, M.; Al-Qatari, M.; Marsh, S.J.; Matthews, E.A.; Dickenson, A.H.; Brown, T.A.; Burbidge, S.A.; Main, M.; et al. KCNQ/M currents in sensory neurons: Significance for pain therapy. *J. Neurosci.* **2003**, *23*, 7227–7236. [CrossRef]
102. Nielsen, A.N.; Mathiesen, C.; Blackburn-Munro, G. Pharmacological characterisation of acid-induced muscle allodynia in rats. *Eur. J. Pharmacol.* **2004**, *487*, 93–103. [CrossRef] [PubMed]
103. Huguenard, J.R. Low-threshold calcium currents in central nervous system neurons. *Annu. Rev. Physiol.* **1996**, *58*, 329–348. [CrossRef]
104. Carbone, E.; Lux, H.D. A Low Voltage-Activated, Fully Inactivating Ca-Channel in Vertebrate Sensory Neurons. *Nature* **1984**, *310*, 501–502. [CrossRef] [PubMed]
105. Nelson, M.T.; Joksovic, P.M.; Perez-Reyes, E.; Todorovic, S.M. The endogenous redox agent L-cysteine induces T-type Ca2+ channel-dependent sensitization of a novel subpopulation of rat peripheral nociceptors. *J. Neurosci.* **2005**, *25*, 8766–8775. [CrossRef] [PubMed]
106. Todorovic, S.M.; Meyenburg, A.; Jevtovic-Todorovic, V. Mechanical and thermal antinociception in rats following systemic administration of mibefradil, a T-type calcium channel blocker. *Brain Res* **2002**, *951*, 336–340. [CrossRef]
107. Berger, N.D.; Gadotti, V.M.; Petrov, R.R.; Chapman, K.; Diaz, P.; Zamponi, G.W. NMP-7 inhibits chronic inflammatory and neuropathic pain via block of Cav3.2 T-type calcium channels and activation of CB2 receptors. *Mol. Pain* **2014**, *10*, 77. [CrossRef] [PubMed]
108. Jarvis, M.F.; Scott, V.E.; McGaraughty, S.; Chu, K.L.; Xu, J.; Niforatos, W.; Milicic, I.; Joshi, S.; Zhang, Q.W.; Xia, Z.R. A peripherally acting, selective T-type calcium channel blocker, ABT-639, effectively reduces nociceptive and neuropathic pain in rats. *Biochem. Pharmacol.* **2014**, *89*, 536–544. [CrossRef] [PubMed]

International Journal of
*Molecular Sciences*

*Article*

# Electroacupuncture Stimulation Alleviates CFA-Induced Inflammatory Pain Via Suppressing P2X3 Expression

**Xuaner Xiang [1,†], Sisi Wang [1,†], Fangbing Shao [1], Junfan Fang [1], Yingling Xu [1], Wen Wang [1], Haiju Sun [1], Xiaodong Liu [2], Junying Du [1,*] and Jianqiao Fang [1,*]**

1 Key Laboratory of Acupuncture and Neurology of Zhejiang Province, Department of Neurobiology and Acupuncture Research, The Third Clinical Medical College, Zhejiang Chinese Medical University, Hangzhou 310053, China
2 Department of Anaesthesia and Intensive Care, The Chinese University of Hong Kong, Hong Kong SAR, China
* Correspondence: dujunying0706@zcmu.edu.cn (J.D.); fjq@zcmu.edu.cn (J.F.); Tel.: +86-1358-8842-157 (J.D.); +86-5718-6673-000 (J.F.)
† These authors contributed equally to this work.

Received: 6 June 2019; Accepted: 28 June 2019; Published: 2 July 2019

**Abstract:** Chronic inflammatory pain is one of the most common complaints that seriously affects patients' quality of life. Previous studies have demonstrated that the analgesic effect of electroacupuncture (EA) stimulation on inflammatory pain is related to its frequency. In this study, we focused on whether the analgesic effects of EA are related to the period of stimulation. Purinergic receptor P2X3 (P2X3) is involved in the pathological process underlying chronic inflammatory pain and neuropathic pain. We hypothesized that 100 Hz EA stimulation alleviated Freund's complete adjuvant (CFA) induced inflammatory pain via regulating P2X3 expression in the dorsal root ganglion (DRG) and/or spinal cord dorsal horn (SCDH). We also assumed that the analgesic effect of EA might be related to the period of stimulation. We found that both short-term (three day) and long-term (14 day) 100 Hz EA stimulation effectively increased the paw withdrawal threshold (PWT) and reversed the elevation of P2X3 in the DRG and SCDH of CFA rats. However, the analgesic effects of 100 Hz EA were not dependent on the period of stimulation. Moreover, P2X3 inhibition or activation may contribute to or attenuate the analgesic effects of 100 Hz EA on CFA-induced inflammatory pain. This result indicated that EA reduced pain hypersensitivity through P2X3 modulation.

**Keywords:** electroacupuncture; alleviates; inflammatory pain; DRG; SCDH; P2X3

## 1. Introduction

Inflammatory pain is a common clinical symptom that widely exists in various acute and chronic diseases [1]. However, its underlying mechanism is still unclear. At present, anti-inflammatory and analgesic drugs are commonly used in the clinic, but drug treatment is accompanied by gastrointestinal discomfort and other side effects [2].

Electroacupuncture (EA), a modified technique based on the theory of traditional manual acupuncture, has been widely used in clinical and scientific research [3,4]. Acupoints, frequency and the period of stimulation are three main elements that affect the effect of EA. Different frequencies of EA stimulation at the same acupoint may lead to different therapeutic effects, while the same frequency may also exert "pain-type" specific analgesic effects on different pathological conditions [5,6]. Our previous studies compared the analgesic effects of 2, 100 and 2/100 Hz EA on chronic pain models of inflammatory pain and neuropathic pain, e.g., Freund's complete adjuvant (CFA) injection and spared nerve injury (SNI), respectively. We found that 100 Hz exhibited the best analgesic effect on inflammatory pain [7]. However, whether the analgesic effects of EA are related to the period of stimulation is largely unknown.

P2X3 is purinoceptor with ion channel activity in response to extracellular adenosine triphosphate (ATP) [8,9]. P2X3 is highly expressed in small- and medium-sized nociceptors of dorsal root ganglions (DRGs) and has been associated with pain response in multiple animal pain models, including the bone cancer pain model, chronic constriction nerve injury (CCI) model, spared nerve injury (SNI) model, and the CFA model [10–13]. P2X3 can be found in the peripheral terminals where it senses the ATP leaked from damaged tissues or released from inflammatory cells [14]. In addition, P2X3 is distributed in the central terminals of nociceptors, indicating that P2X3 might play presynaptic roles in the spinal cord dorsal horn [15]. Indeed, blockage of P2X3 at the spinal cord level significantly reduced postsynaptic neuronal hyperexcitability in a model of bone cancer pain [16]. Previous studies demonstrated that the expression of P2X3 was upregulated by CFA induced inflammation and CCI in the DRG and spinal cord dorsal horn (SCDH), respectively [7,17]. Notably, EA stimulation was reported to attenuate CFA (100 Hz) and CCI (15 Hz) induced pain hypersensitivity [18,19]. We speculated that EA stimulation modulated P2X3 levels in pain pathways and then exerted analgesic actions on chronic inflammatory pain.

In this study, the analgesic effects of short-term and long-term 100 Hz EA stimulation on CFA-induced chronic inflammatory pain were detected and compared. The potential involvement of P2X3 underlying EA mediated analgesia was also explored.

## 2. Results

### 2.1. Both Short-Term and Long-Term EA Stimulation Attenuated CFA Induced Mechanical Allodynia

Rats were injected with CFA into the surface of the right hindpaw to induce persistent inflammatory pain. The paw withdrawal threshold (PWT) of the rats was significantly decreased after CFA injection, indicative of mechanical allodynia establishment. The pain hypersensitivity was sustained throughout the experiment ($P < 0.05$, Figure 1). With daily EA stimulation for 3 days, 100 Hz EA significantly increased the PWT on days 1 and 3 when compared with the CFA group and sham EA group at the same time points ($P < 0.05$, Figure 1C). With daily EA stimulation for 14 days, 100 Hz EA significantly increased the PWT on days 1, 3, 7, and 14 when compared with the CFA group and sham EA group at the same time point ($P < 0.05$, Figure 1D). Compared with the no acupuncture group (i.e., the CFA only group), sham EA had no significant effect on the PWT.

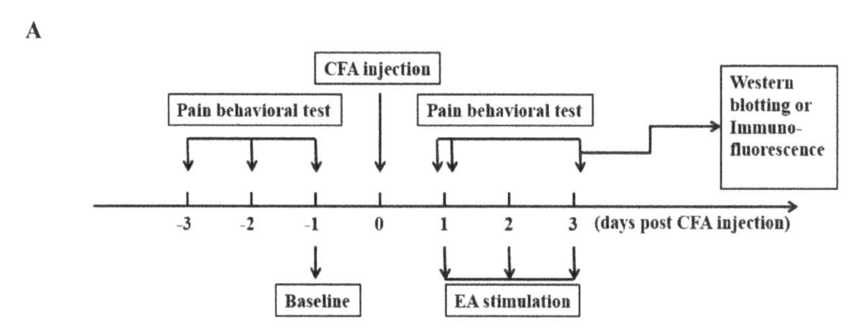


**Figure 1.** *Cont.*

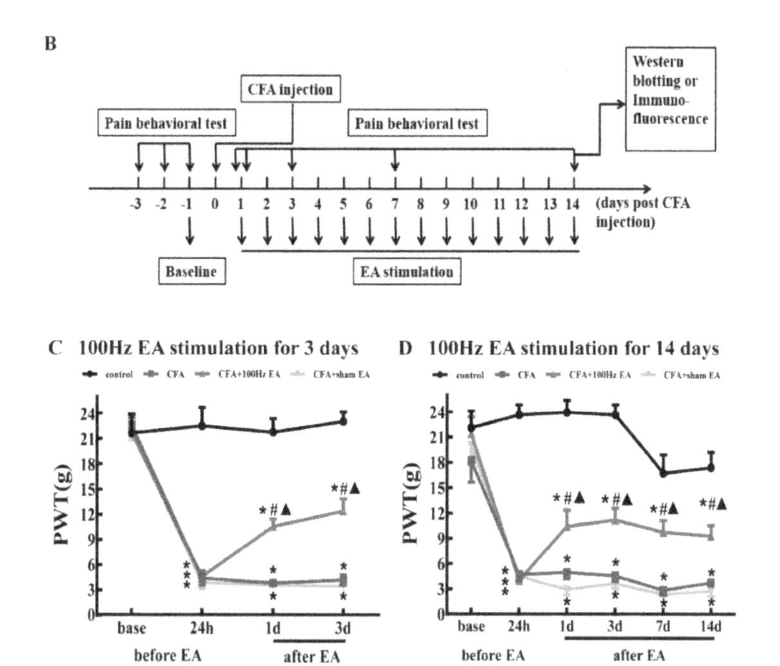


**Figure 1.** The analgesic effect of 3 days and 14 days 100 Hz electroacupuncture (EA) stimulation on the paw withdrawal threshold (PWT). (**A**) The procedure of short-term (3 days) high frequency EA stimulation experiment. (**B**) The procedure of long-term (14 days) high frequency EA stimulation experiment. (**C**) The analgesic effect of 3 days 100 Hz EA stimulation on the PWT. (**D**) The analgesic effect of 14 days 100 Hz EA stimulation on the PWT. Data are presented as the mean ± SEM, $n = 9$. * $P < 0.05$, compared with the control group; # $P < 0.05$, compared with the Freund's complete adjuvant (CFA) group; ▲ $P < 0.05$, compared with the CFA + sham EA group.

### 2.2. Both Short-Term and Long-Term 100 Hz EA Stimulation Reversed P2X3 Elevation in L4-6 DRG and SCDH after CFA Injection

To investigate the effects of the short-term and long-term 100 Hz EA stimulation on the expression of P2X3 in the DRG and SCDH following CFA injection, we used immunofluorescence and western blotting to measure the P2X3 protein levels in the DRG and SCDH in the control group, CFA group, 100 Hz EA group and sham EA group, 3 or 14 days after 100 Hz EA stimulation. As expected, CFA injection significantly increased the mean intensity of P2X3-ir in L4-6 DRG (Figure 2A,B, $P < 0.05$; Figure 4A,B, $P < 0.05$) and SCDH (Figure 3A,B, $P < 0.05$; Figure 5A,B, $P < 0.05$). Short-term and long-term 100 Hz EA stimulation significantly reduced the mean intensity of P2X3-ir in L4-6 DRG (Figure 2A,B, $P < 0.05$; Figure 4A,B, $P < 0.05$) and SCDH (Figure 3A,B, $P < 0.05$; Figure 5A,B, $P < 0.05$) when compared with the control group. In contract, sham EA had no observable effects. As shown in Figures 2C and 4C, P2X3 was mainly expressed in the small- and medium- diameter DRG neurons (diameter less than 35 μm). Neither short-term nor long-term 100 Hz EA changed the distribution of P2X3.

We also used western blotting to measure the P2X3 protein expression. CFA injection increased P2X3 protein expression in L4-6 DRG (Figure 2D,E, $P < 0.05$; Figure 4D,E, $P < 0.05$) and SCDH (Figure 3C,D, $P < 0.05$; Figure 5C,D, $P < 0.05$). Consistent with the immunofluorescence staining results, short-term and long-term 100 Hz EA stimulation significantly reversed paw inflammation induced P2X3 up-regulation in L4-6 DRG (Figure 2D,E, $P < 0.05$; Figure 4D,E, $P < 0.05$) and SCDH (Figure 3C,D, $P < 0.05$; Figure 5C,D, $P < 0.05$). Sham EA had no effect on P2X3 expression either in L4-6 DRG or SCDH.

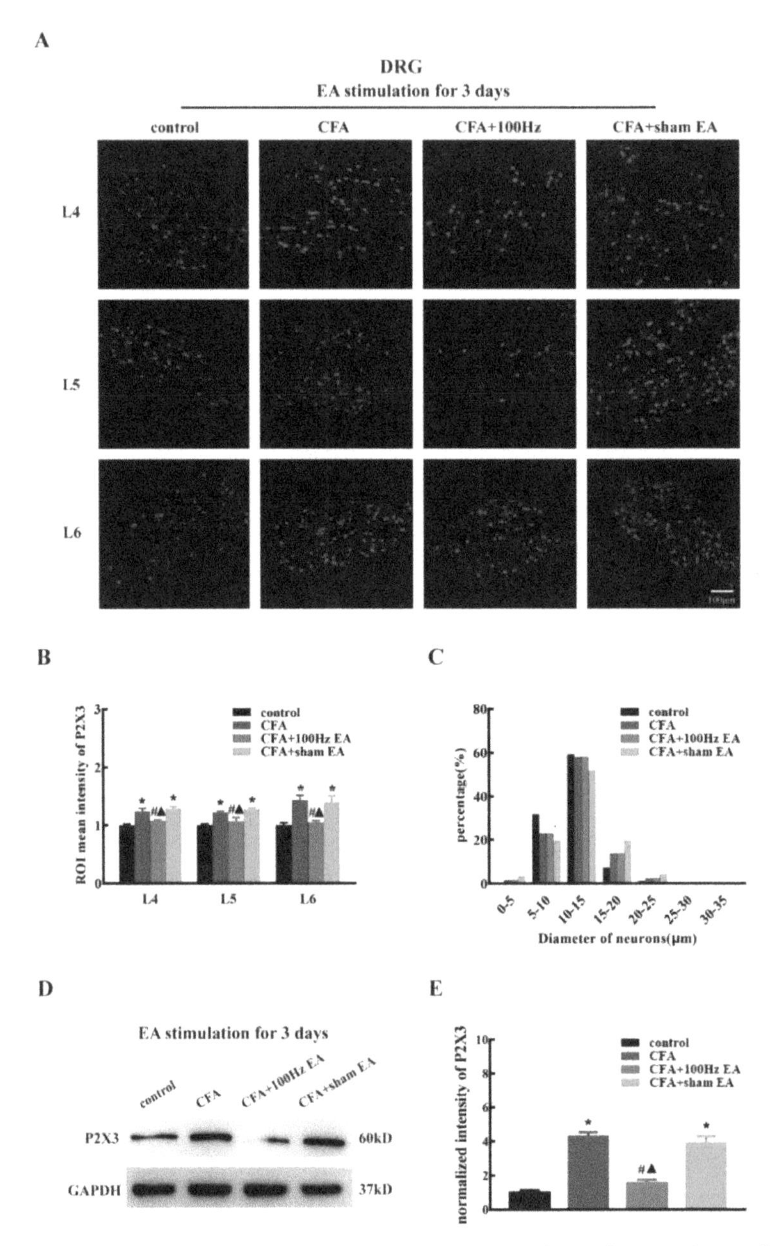


**Figure 2.** The 100 Hz EA stimulation for 3 days repressed the up-regulation of P2X3 in the L4-6 dorsal root ganglions (DRG) of inflamed rats. (**A**) Representative images of L4-6 DRG immunofluorescence staining from the control, CFA, CFA+100 Hz EA and CFA + sham EA groups. Scale bars = 100 μm. (**B**) Mean intensity analysis of P2X3-ir in L4-6 DRG in each group. Data are presented as the mean ± SEM, $n = 3$. (**C**) Size distribution of P2X3 in L4-6 DRG in different groups. (**D**) Representative western blotting images of L4-6 DRG in each group. E. Relative protein level of P2X3 in rat L4-6 DRG from different groups. Data are presented as the mean ± SEM, $n = 6$. * $P < 0.05$, compared with the control group; # $P < 0.05$, compared with the CFA group; ▲ $P < 0.05$, compared with the CFA + sham EA group.

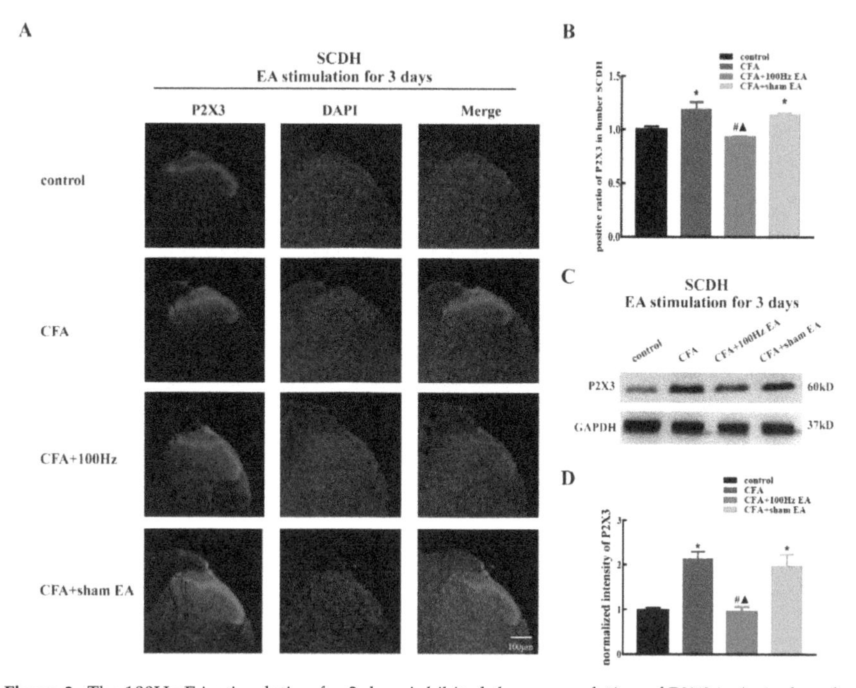


**Figure 3.** The 100Hz EA stimulation for 3 days inhibited the up-regulation of P2X3 in (spinal cord dorsal horn) SCDH of inflamed rats. (**A**). Representative images of SCDH immunofluorescence staining from the control, CFA, CFA + 100 Hz EA and CFA + sham EA groups. Scale bars = 100 μm. (**B**) Mean intensity analysis of P2X3-ir in SCDH in each group. Data are presented as the mean ± SEM, $n = 3$. (**C**) Representative western blotting images of SCDH in each group. (**D**) Relative protein level of P2X3 in SCDH from different groups. Data are presented as the mean ± SEM, $n = 6$. * $P < 0.05$, compared with the control group; # $P < 0.05$, compared with the CFA group; ▲ $P < 0.05$, compared with the CFA + sham EA group.

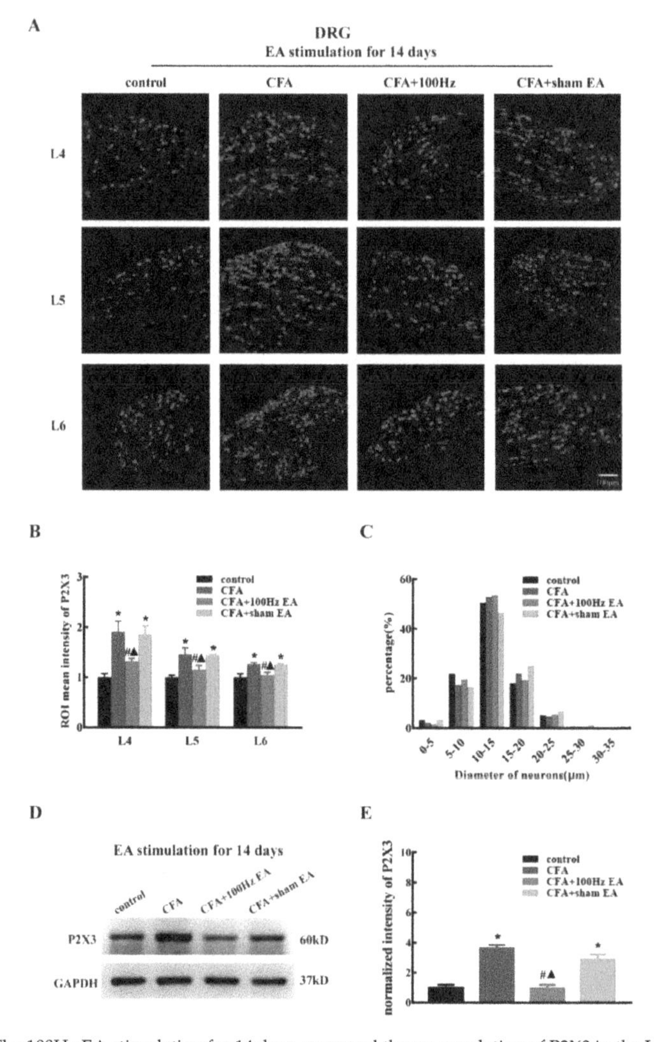


**Figure 4.** The 100Hz EA stimulation for 14 days repressed the up-regulation of P2X3 in the L4-6 DRG of inflamed rats. (**A**) Representative images of L4-6 DRG immunofluorescence staining from the control, CFA, CFA + 100 Hz EA and CFA + sham EA groups. Scale bars = 100 μm. (**B**) Mean intensity analysis of P2X3-ir in L4-6 DRG in each group. Data are presented as the mean ± SEM, $n = 3$. (**C**) Size distribution of P2X3 in L4-6 DRG in different groups. (**D**) Representative western blotting images of L4-6 DRG in each group. (**E**) Relative protein level of P2X3 in rat L4-6 DRG from different groups. Data are presented as the mean ± SEM, $n = 6$. * $P < 0.05$, compared with the control group; # $P < 0.05$, compared with the CFA group; ▲ $P < 0.05$, compared with the CFA + sham EA group.

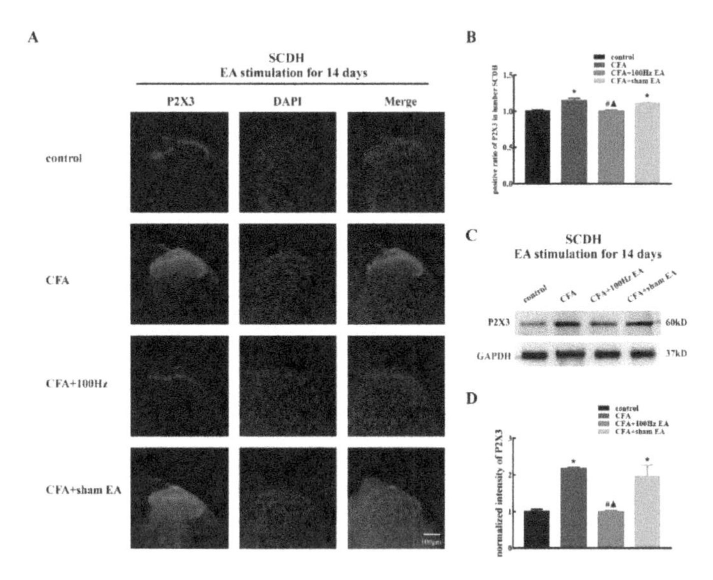


**Figure 5.** The 100 Hz EA stimulation for 14 days inhibited the up-regulation of P2X3 in the SCDH of inflamed rats. (**A**) Representative images of SCDH immunofluorescence staining from the control, CFA, CFA + 100 Hz EA and CFA + sham EA groups. Scale bars = 100 µm. (**B**) Mean intensity analysis of P2X3-ir in the SCDH in each group. Data are presented as the mean ± SEM, $n = 3$. (**C**) Representative western blotting images of SCDH in each group. (**D**) Relative protein level of P2X3 in the SCDH from different groups. Data are presented as the mean ± SEM, $n = 6$. * $P < 0.05$, compared with the control group; # $P < 0.05$, compared with the CFA group; ▲ $P < 0.05$, compared with the CFA + sham EA group.

### 2.3. *Short-Term and Long-Term EA Stimulation Exerted Comparable Effects on Pain Hypersensitivity and P2X3 Expression*

To investigate the effects of short-term and long-term EA stimulation, we compared the ratio of change in PWT and P2X3 expression in L4-6 DRG and SCDH. As shown in Figure 6A, there was no significant difference in the ratio of changes in PWT between EA stimulation for 3 and 14 days ($P > 0.05$). Additionally, there was no significant difference in the ratio of changes in P2X3 expression either in L4-6 DRG or SCDH between EA stimulation for 3 and 14 days (Figure 6B, $P > 0.05$).

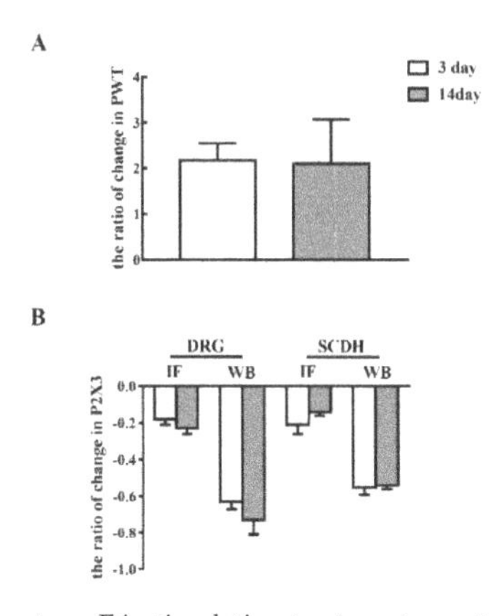


**Figure 6.** Short-term and long-term EA stimulation treatments exerted comparable effects on pain hypersensitivity and P2X3 expression. (**A**) The ratio of changes in PWT after 100 Hz EA stimulation for 3 and 14 days. (**B**) The ratio of changes in P2X3 after 100 Hz EA stimulation for 3 and 14 days. Data are presented as the mean ± SEM.

## *2.4. P2X3 Levels in L4-6 DRG and SCDH Are Involved in Chronic Inflammatory Pain*

Previous studies demonstrated that the P2X3 agonist αβ-me ATP might induce pain hypersensitivity in rats. In this study, intraplantar injection (i.pl.) or intrathecal injection (i.t.) of αβ-me ATP can induce mechanical hyperalgesia in normal rats. Compared with the control + vehicle, the PWTs of control + i.pl. α β-me ATP, and control + i.t. α β-me ATP both decreased rapidly after αβ-me ATP administration (Figure 7B,D, $P < 0.05$). In contrast, treatment with A317491 (P2X3 antagonist) via i.pl. or i.t., significantly alleviated CFA induced mechanical hypersensitivity (Figure 7B,D). These outcomes suggested that P2X3 activation is sufficient to cause pain hypersensitivity, and that P2X3 might play essential roles in the modulation of pain signal transmission from the DRG to the spinal cord.

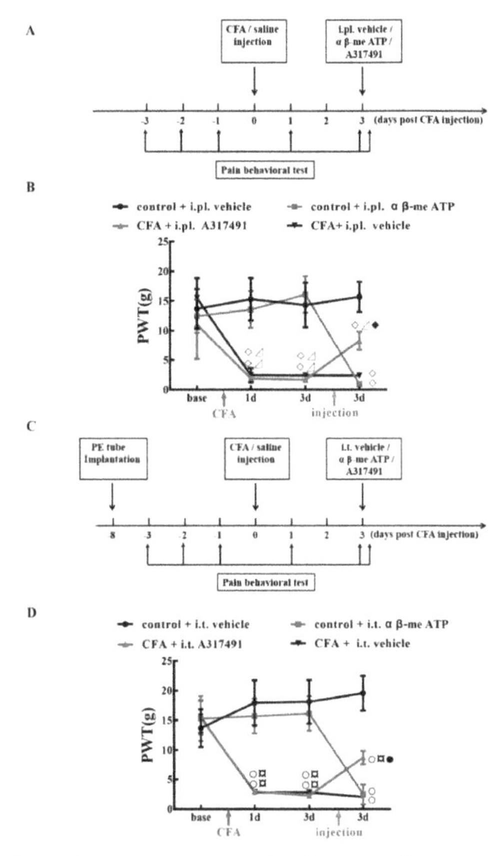


**Figure 7.** The effect of intraplantar injection or intrathecal injection of α β-me ATP (P2X3 agonist) and A317491 (P2X3 antagonist) on the PWT. (**A**) Schematic flow diagram of the intraplantar injection of α β-me ATP (P2X3 agonist) and A317491 (P2X3 antagonist). (**B**) The effect of intraplantar injection of α β-me ATP (P2X3 agonist) and A317491 (P2X3 antagonist) on the PWT. Data are presented as the mean ± SEM, $n = 5$. ◇ $P < 0.05$, compared with the control + intraplantar injection (i.pl.) vehicle; ⊿ $P < 0.05$, compared with the control + i.pl. α β-me ATP; ◆ $P < 0.05$, compared with the CFA + i.pl. vehicle group. (**C**) Schematic flow diagram of the intrathecal injection of α β-me ATP (P2X3 agonist) and A317491 (P2X3 antagonist). (**D**)The effect of intrathecal injection of α β-me ATP (P2X3 agonist) and A317491 (P2X3 antagonist) on the PWT. Data are presented as the mean ± SEM, $n = 5$. ○ $P < 0.05$, compared with the control + intrathecal injection (i.t.) vehicle; ◻ $P < 0.05$, compared with the control + i.t. α β-me ATP group; ● $P < 0.05$, compared with the CFA + i.t. vehicle.

### *2.5. P2X3 Inhibition Contributed to the Analgesic Effects of 100 Hz EA on CFA-Induced Inflammatory Pain*

To investigate whether P2X3 is involved in the analgesic effect of 100 Hz EA on chronic inflammatory pain, A317491 was co-administered via i.pl. and i.t. to observe the changes in PWT in rats of each group. Consistent with the findings in Figure 7, i.pl. and i.t. of A317491 exerted significant analgesic effects on CFA induced inflammatory pain (Figure 8B,D). Similarly, EA stimulation also efficiently alleviated CFA induced mechanical allodynia, as indicated by the results of CFA + i.pl. vehicle versus CFA + 100 Hz + i.pl. vehicle (Figure 8B) and CFA + i.t. vehicle versus CFA + 100 Hz + i.t. vehicle (Figure 8D). Notably, co-treatment with EA and i.t. A317491 did not further increase the analgesia compared with EA alone (Figure 8D).

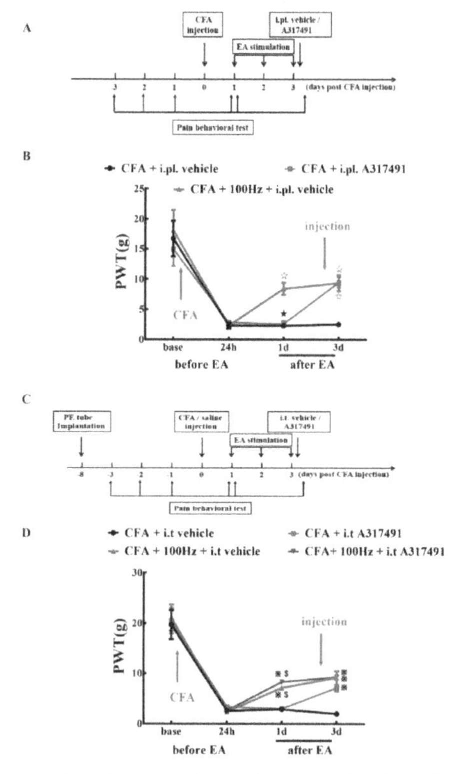


**Figure 8.** P2X3 inhibition contributed to the analgesic effects of 100 Hz EA on CFA-induced inflammatory pain. (**A**) Schematic flow diagram of the intraplantar injection of P2X3 antagonist A317491. (**B**) The effect of intraplantar injection of A317491 on the PWT. Data are presented as the mean ± SEM, $n = 6$. ☆ $P < 0.05$, compared with the CFA + i.pl. vehicle group; ★ $P < 0.05$, compared with the CFA + 100 Hz + i.pl. vehicle group. (**C**) Schematic flow diagram of the intrathecal injection of A317491. (**D**) The effect of intrathecal injection of A317491on the PWT. Data are presented as the mean ± SEM, $n = 6$. ※ $P < 0.05$, compared with the CFA + i.t vehicle group; $ $P < 0.05$, compared with the CFA + 100 Hz+ i.t vehicle group.

To further confirm that 100 Hz EA has an analgesic effect on chronic inflammatory pain by regulating P2X3, α β-me ATP was administered via i.pl. and i.t. to observe the changes in PWT in rats of each group. As shown in Figure 9, i.pl. and i.t. of α β-me ATP can reverse the analgesic effect of electroacupuncture, as indicated by the results of the CFA + 100 Hz + i.pl. vehicle group versus the CFA + 100 Hz + i.pl. α β-me ATP group (Figure 9B) and the CFA + 100 Hz + i.t. vehicle group versus the CFA + 100 Hz + i.t. α β-me ATP group (Figure 9D). This result provided evidence that EA reduced pain hypersensitivity through P2X3 modulation.

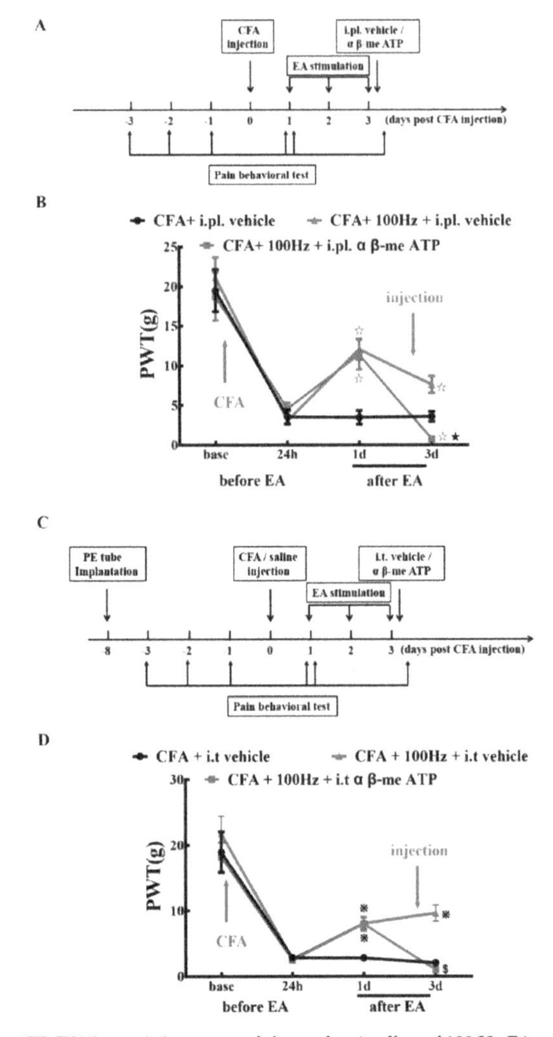


**Figure 9.** α β-me ATP (P2X3 agonist) attenuated the analgesic effect of 100 Hz EA on PWT in CFA rats. (**A**) Schematic flow diagram of the intraplantar injection of α β-me ATP. (**B**) The analgesic effect of 100Hz EA can be attenuated by intraplantar injection of α β-me ATP. Data are presented as the mean ± SEM, $n = 6$. ☆ $P < 0.05$, compared with the CFA + i.pl. vehicle group; ★ $P < 0.05$, compared with the CFA + 100Hz + i.pl. vehicle group. (**C**) Schematic flow diagram of the intrathecal injection of α β-me ATP. (**D**) Intrathecal injection of α β-me ATP may reduce the analgesic effect of 100 Hz EA. Data are presented as the mean ± SEM, $n = 6$. ※ $P < 0.05$, compared with the CFA + i.t vehicle group; $ $P < 0.05$, compared with the CFA + 100Hz + i.t vehicle group.

## 3. Discussion

The frequency of EA stimulation appears to be a determinant of the analgesic effect of EA [20]. The optimal frequency of EA treatment is not constant in different types of pathological pain [21]. While both 2 and 100 Hz EA treatment relieved type 2 diabetic neuropathic pain, 2 Hz exerted stronger analgesic effects 100 Hz [22]. However, we have demonstrated that the analgesic effect of EA was greater at 100 Hz than at 2 Hz in the scenario of inflammatory pain [7]. Similarly, 100 Hz, but not 2 Hz EA stimulation, could relieve post-incision pain [23]. In addition to the frequency, the number of stimulations may also significantly affect the analgesic effects of EA, indicative of the presence of a cumulative effect [24,25]. However, in the current study, we observed that daily EA stimulations for 3 days and 14 days provided comparable analgesia for persistent inflammatory pain. We concluded that a "ceiling" effect can occur during the application of EA stimulation. This finding may help determine the regimen of EA application for the treatment of chronic inflammatory pain.

Clinical studies and scientific studies have proved that EA has eminent analgesic effects, but the mechanism of EA analgesia is still an open question [26–28]. Cheng RS et al. reported that 4 Hz EA attenuated pain through the modulation of endorphins, whereas the analgesic effect of 200 Hz EA may be mediated through serotonin [29]. Wang Y et al. demonstrated that 100 Hz EA relieved inflammatory pain by increasing CXCL10, which chemoattracted opioid-containing macrophages and mediated the anti-nociceptive effect in the model of inflammatory pain [30]. Kim H W et al. proposed that 1 Hz electroacupuncture suppressed carrageenan-induced paw inflammation via sympathetic post-ganglionic neurons, while inflammation was restrained by 120 Hz EA in connection with the sympathoadrenal medullary axis [31]. Recently, increasing evidence has shown that the analgesic effect of EA is closely related to its regulation of ion channels in sensory neurons [32,33]. Particularly, increased attention has been paid to P2X3, which has been regarded as a potential target of inflammatory pain and neuropathic pain [34,35].

Inhibition of the P2X3 receptor through a selective antagonist, e.g., A317491 has been assessed as a potential approach for inflammatory pain management [36,37]. Using the model of CFA induced chronic inflammation, Qian Jiang et al. reported that the expression of P2X3 was markedly increased in DRG tissues [13]. Similar results were obtained in the current study. On the 3rd and 14th days after CFA injection, the expression of P2X3 was significantly increased in L4-6 DRG as indicated by immunofluorescence staining and western blotting. We confirmed that P2X3 was distributed in small- and medium-sized neurons, especially in the diameter range of 5–10 μm. Several studies shown that EA can be applied to treat different types of pain (such as neuropathic pain, inflammatory pain, and bone cancer pain) by down-regulating DRG P2X3 [34,38,39]. Consistently, we demonstrated that both short-term and long-term EA stimulation can effectively down-regulate the up-regulation of P2X3 in DRG induced by CFA. Few studies have focused on the expression of the P2X3 in the SCDH of CFA model. In this study, we found that the mean intensity of immunoreactivity and protein level of the P2X3 in the SCDH were significantly increased at three days and 14 days after CFA injection. In relation to DRG, elevation of P2X3 was also reversed by short-term and long-term EA in SCDH. Notably, that P2X3 is distributed in the presynaptic part of the spinal cord, i.e., the central terminal of nociceptors. Therefore, EA mediated inhibition of spinal P2X3 is relies on its effects on the DRG. The current findings suggested that EA might directly modulate pain signal transmission from first-order neurons (DRG) to second-order neurons (SCDH). Indeed, post-treatment with P2X3 inhibitor via intrathecal injection failed to further enhance the anti-nociceptive effects by EA, indicating that spinal P2X3 played essential roles underlying EA mediated analgesia. While overexpression of DRG and SCDH P2X3 in CFA rats nearly returned to normal levels after EA stimulation, but the hyperalgesia of CFA rats still existed despite relief. The reason for this phenomenon may be due to proinflammatory cytokines, NLRP, CB2, and other substances that are involved in CFA-induced chronic inflammatory pain. P2X3 is not the only determinant, even though it plays an important role in the occurrence and maintenance of chronic inflammatory pain [40–42].

α β-me ATP and A317491 were co-administered via i.pl. and i.t. to further prove that P2X3 is involved in the pathological process of chronic inflammatory pain. We found that α β-me ATP injection in normal rats can induce hyperalgesia, while A317491 injection in CFA rats can effectively reverse CFA-induced hyperalgesia, indicating that P2X3 is closely related to inflammatory pain induced by CFA. Then A317491 was administered, and the analgesic effect was consistent with 100 Hz EA, and effectively relieved CFA-induced mechanical hyperalgesia. However, α β-me ATP administered via i.pl. and i.t. effectively reduced the analgesic effect of 100 Hz EA. This finding reveals that P2X3 regulation may be a potential mechanism for the analgesic effect of EA.

We compared the ratio of change in the PWT and the P2X3 expression in DRG and SCDH between short-term and long-term EA stimulation in CFA rats. We discovered that the ratio of change in DRG P2X3 in long-term EA stimulation was slightly higher than that in short-term EA stimulation, but there was no significant difference. However, the ratio of change in SCDH P2X3 in short-term EA stimulation was slightly higher than that in long-term EA stimulation, and the difference was not significant. Therefore, we speculated that the analgesic effect of EA on chronic inflammatory pain may not be related to the term of stimulation.

## 4. Materials and Methods

### *4.1. Animals*

Male Sprague-Dawley (SD) rats (180–220 g) were purchased from the Experimental Animal Center of Zhejiang Chinese Medical University. All rats in this experiment were housed in a controlled environment (five rats per cage, temperature: 25 ± 2 °C, humidity: 55% ± 5%, and light: 12 h light/dark cycle) and were fed a standard rodent food and allowed distilled water ad libitum. All experimental procedures were approved by the Animal Care and Welfare Committee of Zhejiang Chinese Medical University, Zhejiang, China (ZSLL, 2015-022).

### *4.2. Experimental Design*

This study was divided into three parts.

In part one, the effects of short-term or long-term 100 Hz EA stimulation on CFA induced pain were monitored. For short-term EA stimulation experiments, rats were randomly assigned into four groups ($n$ = 9/group): (1) control group, (2) CFA group, (3) CFA + EA group, and (4) CFA + sham EA group. Rats were treated by acupuncture with (EA) or without (sham EA) 100 Hz electric current stimulation for 3 days after CFA injection. Pain behavioral tests were conducted according to the schedule (Figure 1A), i.e., on days -3, -2, -1, and zero before EA stimulation, and on days 1 and 3 after EA stimulation, respectively. After 3 days post EA stimulation, rats were sacrificed for tissue collection. The L4–6 DRG and lumbar spinal cord were removed for immunofluorescence staining or western blotting. For long-term EA stimulation experiments, rats were assigned to four groups as above, and administered with EA or sham EA for 14 days after CFA injection. Pain behavioral tests were conducted according to the schedule (Figure 1B), i.e., on days -3, -2, -1 and zero before EA stimulation, and on days 1, 3, 7 and 14 after EA stimulation. After fourteen days post EA stimulation, tissues were collected and applied as above.

In part two, the involvement of P2X3 in EA-mediated pain modulation was explored through the administration of a P2X3 antagonist (A317491) or agonist (α β-me ATP) via intraplantar injection (i.pl.) or intrathecal injection (i.t.). Rats were randomly assigned to eight groups ($n$ = 5/group), i.e., (1) control + i.pl. vehicle group, (2) control + i.pl. α β-me ATP group, (3) CFA + i.pl. A317491 group, (4) CFA + i.pl. vehicle group, (5) control + i.t. vehicle group, (6) control + i.t. α β-me ATP group, (7) CFA + i.t. A317491 group, and (8) CFA + i.t. vehicle group. For i.pl., antagonists or agonists were administered ipsilaterally on day three after CFA injection, and paw withdrawal thresholds (PWT) were recorded according to the schedule (Figure 7A). For i.t., antagonists or agonists were delivered via implanted PE tubes. PWT was recorded according to the schedule (Figure 7C).

In part three, to explore the role of P2X3 in DRG and SCDH on the analgesic effect of 100 Hz EA stimulation on CFA rats, we investigated whether peripheral subcutaneous injection or central intrathecal injection of P2X3 inhibitors could simulate the analgesic effect of 100 Hz EA (Figure 8). A group of 42 adult male rats ($n$ = 6/group) were divided into a (1) CFA + i.pl. vehicle group, (2) CFA

+ i.pl. A317491 group, (3) CFA + 100 Hz + i.pl. vehicle group, (4) CFA + i.t. vehicle group, (5) CFA + i.t. A317491 group, (6) CFA + 100 Hz + i.t. vehicle group, and a (7) CFA + 100 Hz + i.t. A317491 group. Then, we observed whether peripheral subcutaneous injection or central intrathecal injection of a P2X3 agonist (α β-me ATP) could reverse the analgesic effect of 100 Hz EA (Figure 9). We randomly assigned 36 adult male rats to six groups, including a (1) CFA + i.pl. vehicle group, (2) CFA + 100 Hz + i.pl. α β-me ATP group, (3) CFA + 100 Hz + i.pl. vehicle group, (4) CFA + i.t. vehicle group, (5) CFA + 100 Hz + i.t. α β-me ATP group, and a (6) CFA + 100 Hz + i.t. vehicle group.

### 4.3. Persistent Inflammatory Pain Model

Persistent inflammatory pain like responses were induced via i.pl. of 0.1mL CFA (Sigma-Aldrich, St. Louis, MO, USA) into the plantar surface of the right hind paws of SD rats. For the sham control, saline was applied via i.pl.

### 4.4. Paw Withdraw Threshold (PWT)

PWT was determined by the von Frey behavioral test, which was performed according to the up-down method described by Chaplan et al. [43]. Rats were placed in the individual testing cages for 30 min per day for three continuous days to adapt to the test environment. Before each test, rats were placed into the cage for at least 15 min to acclimate to the environment. The von Frey hairs (Stoelting Co, Thermo, Gilroy, CA, USA) were applied in a consecutive ascending order (0.4, 0.6, 1, 2, 4, 6, 8, 15, and 26 g) to the central surface of the hind paw and sustained for 5 s. The first hair applied corresponded to a force of 4 g. Brisk withdrawal or paw flinching was considered a positive response and marked as "X". A weaker stimulus was then applied. In the case of no responses, "O" was recorded, followed by a stronger stimulus. The interval between each stimulus was not less than 2 min. After the combination of "OX" or "XO" appeared, a series of four stimuli were applied and recorded as above. The 50% PWTs of the rats were calculated by the formula 50% PWTs (g) = 10 ^ (xf + k *δ- 4). "xf" is the logarithmic value of the last von Frey hair in the sequence, "k" is the corresponding value of the resulting sequence in the k-value table, and "δ" is the mean difference of each filament strength after logging (0.231 in the current cases). If a positive stimulus appeared five consecutive times, PWT was marked as 26 g. If five "X"s were recorded, PWT was marked as 0.4 g. If the value of 50% PWTs was greater than 26 g, 26 g was used as the maximum. And if the value of 50% PWTs was less than 0.4 g, 0.4 g was used as the minimum. The measuring time was fixed at 9:00–16:00, and the ambient temperature was 23 ± 2 °C.

### 4.5. EA Treatment

All rats in CFA + 100 Hz EA group were treated with the EA stimulus. Zusanli (ST36) and Kunlun (BL60) acupoints were taken from the bilateral legs of the rats. Acupuncture needles, 0.25 mm * 13 mm were used in this study. The needles were inserted into the acupoints at a depth of 5 mm and then stimulated by HANS Acupuncture Point Nerve Stimulator (HANS-200A Huawei Co., Ltd., Beijing, China). The parameters of the stimulator were as follows: 100 Hz, 0.5–1.5 mA (initial strength 0.5 mA, increased by 0.5 mA every 10 min) for a total of 30 min. The stimulus was conducted once daily in a period of 3 or 14 days. The rats in the CFA group were only given the same fixed time as the EA group. No treatment was performed in the control group. The CFA + sham EA group animals received needle insertion subcutaneously into ST36 and BL60 (1 mm in depth). The needles were connected to the electrodes without electrical stimulation. After finishing the EA or sham EA stimulation, the PWT was measured immediately.

### 4.6. Drug Treatment

α β-me ATP (P2X3 agonist) and A317491 (P2X3 antagonist) were purchased from Sigma-Aldrich (Sigma-Aldrich, Saint Louis, MO, USA), and dissolved in sterile 0.9 % saline solution to prepare stock solution (stored at −20 °C). They were diluted to the requested concentrations before each experiment.

For i.pl, α β-me ATP (600 nmol,10 μL) and A317491 (300 nmol,10 μL) were injected subcutaneously into the dorsal surface of the right hindpaw of rats. For i.t., α β-me ATP (300 nmol, 25 μL) and A317491 (100 nmol, 25 μL) were administered once on days 3 after CFA injection.

### 4.7. Immunofluorescence

Animals were sacrificed after the behavioral testing on days 3 and 14. Rats were deeply anesthetized using pentobarbital (80 mg/kg, i.p.) and transcardially perfused with 150 mL normal saline (4°C) and 400 mL 4% paraformaldehyde in 0.1 M phosphate-buffered saline (PBS) for prefixation. The L4–6 segments of the dorsal root ganglion (DRG) and lumbar spinal cord were removed and postfixed in 4% paraformaldehyde for 3 hours at 4 °C before transfer to 15% and 30% sucrose for dehydration. Tissues were embedded in Tissue-Tek O.C.T compound (SAKURA, Torrance, CA, USA). DRGs were cut at thickness of 14 μm, and frozen sections of the lumbar spinal cord were cut at thickness of 20 μm using a CryoStar (NX50 HOP, Thermo, Walldorf, Germany).

Sections were rinsed with TBST (0.1 % Tween-20) and blocked with 5% normal donkey serum for one hour at 37 °C. Sections were then incubated with rabbit anti-P2X3 (1:800 in 5 % normal donkey serum, Alomone, Jerusalem, Israel) overnight at 4 °C. The slides were then incubated in Alexa Fluor 647-conjugated AffiniPure donkey anti-rabbit IgG (H + L) (Jackson, West Grove, PA, USA) for one hour at 37 °C. Images were taken using the A1R confocal microscope (Nikon, Tokyo, Japan). We used NIS-Elements AR to calculate the mean intensity of P2X3-immunoreactivity (-ir) in the region-of-interest (ROI) in DRG and SCDH. The relative level of P2X3 in each group was normalized to the expression in the control group.

### 4.8. Western Blotting Analysis

Animals were sacrificed after the behavioral testing on days 3 and 14. Rats were deeply anesthetized using pentobarbital (80 mg/kg, i.p.) and transcardially perfused with 150 mL normal saline (4 °C).The L4–6 segments of the DRG and lumbar spinal cord were removed and stored at −80 °C. Tissues were homogenized in strong RIPA buffer (50 mM Tris (pH 7.5), 150 mM NaCl, 1% Triton X-100, 1% sodium deoxycholate, sodium orthovanadate, 0.1% Sodium dodecyl sulfate, Ethylene Diamine Tetraacetic Acid, sodium fluoride, leupeptin, and 1 nM PMSF). The homogenate was allowed to rest on ice for 30 min and centrifuged at 15,000 rpm for 15 min at 4 °C. The supernatant was then collected for further operations. The protein concentration of tissue lysates was determined with a BCA protein assay kit. Lysates were denatured and loaded (15 μg total protein per lane). Protein samples were separated on 5–10% Sodium dodecyl sulfate-polyacrylamide gelelectrophoresis gels and electrophoretically transferred to polyvinylidene difluoride (PVDF) membranes (Merck KGaA, Darmstadt, Germany). The membranes were blocked with 5% low-fat milk in TBST for one hour at room temperature. We used rabbit anti-P2X3 (1:1000 in 5% low-fat milk, Alomone, Jerusalem, Israel) as the primary antibody and horseradish peroxidase (HRP)-conjugated goat anti-rabbit IgG as the secondary antibody (1:10,000, CST, Danvers, MA, USA). Rabbit anti-GAPDH (HRP Conjugate) (1:1000, CST, Danvers, MA, USA) was used as the internal control. The membranes were developed with an ECL kit (Pierce, Rockford, lL , USA), and the signals were captured with an Image Quant LAS 4000 (GE, Pittsburgh, PA, USA). The density of each band was measured using Image Quant TL 7.0 analysis software (GE, Pittsburgh, PA, USA). The relative level of P2X3 in each group was normalized to the expression of the control group.

### 4.9. Statistical analysis

All data are expressed as the mean ± standard error of the mean (SEM). The PWTs among groups were compared by multi-factor analysis of variance (ANOVA), followed by Bonferroni's post hoc test to compare the significant difference between groups or between time points. All other data were analyzed by one-way ANOVA, followed by Bonferroni's post hoc tests. $P < 0.05$ was considered statistically significant.

## 5. Conclusions

In summary, we concluded that both short-term and long-term 100 Hz EA stimulation provided significant pain relief for chronic inflammatory pain, and this analgesic effect was related to the suppression of P2X3 in the DRG and SCDH.

**Author Contributions:** Methodology, S.W. and Y.X.; Software, X.L.; Formal Analysis, F.S.; Investigation, J.F.; Data Curation, W.W.; Writing—Original Draft Preparation, X.X.; Writing—Review & Editing, H.S.; Supervision, J.D.; Funding Acquisition, J.F.

**Acknowledgments:** This work was supported by the National Natural Science Foundation of China (8147377, 81603690, 81603692), the Major Scientific and Technological Project of Zhejiang Province (WKJ-ZJ-1419), the Zhejiang First-class Discipline (Chinese Medicine) Funding (Zhejiang Administration Letter (2016) No.6.), and the Talent Project of Zhejiang Association for Science and Technology (2017YCGC004).

**Conflicts of Interest:** The authors declare no conflict of interest.

## References

1. Luo, C.; Kuner, T.; Kuner, R. Synaptic plasticity in pathological pain. *Trends Neurosci.* **2014**, *37*, 343–355. [CrossRef] [PubMed]
2. Marret, E.; Kurdi, O.; Zufferey, P.; Bonnet, F. Effects of nonsteroidal antiinflammatory drugs on patient-controlled analgesia morphine side effects: Meta-analysis of randomized controlled trials. *Anesthesiology* **2005**, *102*, 1249–1260. [CrossRef] [PubMed]
3. Gao, F.; Xiang, H.C.; Li, H.P.; Jia, M.; Pan, X.L.; Pan, H.L.; Li, M. Electroacupuncture Inhibits NLRP3 Inflammasome Activation through CB2 Receptors in Inflammatory Pain. *Brain Behav. Immun.* **2017**, 67. [CrossRef] [PubMed]
4. Ooi Thye, C.; Critchley, H.O.D.; Horne, A.W.; Robert, E.; Erna, H.; Marie, F. The BMEA study: The impact of meridian balanced method electroacupuncture on women with chronic pelvic pain-a three-arm randomised controlled pilot study using a mixed-methods approach. *BMJ Open* **2015**, *5*, e008621.
5. Ulett, G.A.; Han, S.; Han, J.S. Electroacupuncture: Mechanisms and clinical application. *Biol. Psychiatry* **1998**, *44*, 129–138. [CrossRef]
6. Han, J.S. Acupuncture and endorphins. *Neurosci. Lett.* **2004**, *361*, 258–261. [CrossRef]
7. Fang, J.Q.; Du, J.Y.; Fang, J.F.; Xiao, T.; Le, X.Q.; Pan, N.F.; Yu, J.; Liu, B.Y. Parameter-specific analgesic effects of electroacupuncture mediated by degree of regulation TRPV1 and P2X3 in inflammatory pain in rats. *Life Sci.* **2018**, *200*, 69–80. [CrossRef]
8. Shiozaki, Y.; Sato, M.; Kimura, M.; Sato, T.; Tazaki, M.; Shibukawa, Y. Ionotropic P2X ATP Receptor Channels Mediate Purinergic Signaling in Mouse Odontoblasts. *Front. Physiol.* **2017**, *8*, 3. [CrossRef]
9. Bradbury, E.J.; Burnstock, G.; Mcmahon, S.B. The Expression of P2X 3 Purinoreceptors in Sensory Neurons: Effects of Axotomy and Glial-Derived Neurotrophic Factor. *Mol. Cell. Neurosci.* **1998**, *12*, 256–268. [CrossRef]
10. Liu, S.; Lv, Y.; Wan, X.X.; Song, Z.J.; Liu, Y.P.; Miao, S.; Wang, G.L.; Liu, G.J. Hedgehog signaling contributes to bone cancer pain by regulating sensory neuron excitability in rats. *Mol. Pain* **2018**, *14*, 1744806918767560. [CrossRef]
11. Wang, W.S.; Tu, W.Z.; Cheng, R.D.; He, R.; Ruan, L.H.; Zhang, L.; Gong, Y.S.; Fan, X.F.; Hu, J.; Cheng, B. Electroacupuncture and A-317491 depress the transmission of pain on primary afferent mediated by the P2X3 receptor in rats with chronic neuropathic pain states. *J. Neurosci. Res.* **2015**, *92*, 1703–1713. [CrossRef] [PubMed]
12. Meisner, J.G.; Reid, A.R.; Sawynok, J. Adrenergic regulation of P2X3 and TRPV1 receptors: Differential effects of spared nerve injury. *Neurosci. Lett.* **2008**, *444*, 172–175. [CrossRef] [PubMed]
13. Jiang, Q.; Li, W.X.; Sun, J.R.; Zhu, T.T.; Fan, J.; Yu, L.H.; Burnstock, G.; Yang, H.; Ma, B. Inhibitory effect of estrogen receptor beta on P2X3 receptors during inflammation in rats. *Purinergic Signal.* **2016**, *13*, 1–13. [CrossRef] [PubMed]
14. Brederson, J.D.; Jarvis, M.F. Homomeric and heteromeric P2X3 receptors in peripheral sensory neurons. *Curr. Opin. Investig. Drugs* **2008**, *9*, 716–725. [PubMed]

15. Vulchanova, L.; Riedl, M.S.; Shuster, S.J.; Stone, L.S.; Hargreaves, K.M.; Buell, G.; Surprenant, A.; North, R.A.; Elde, R. P2X3 is expressed by DRG neurons that terminate in inner lamina II. *Eur. J. Neurosci.* **1998**, *10*, 3470–3478. [CrossRef]
16. Kaan, T.K.Y.; Yip, P.K.; John, G.; Cefalu, J.S.; Nunn, P.A.; Ford, A.P.D.W.; Yu, Z.; Mcmahon, S.B. Endogenous purinergic control of bladder activity via presynaptic P2X3 and P2X2/3 receptors in the spinal cord. *J. Neurosci. Off. J. Soc. Neurosci.* **2010**, *30*, 4503. [CrossRef] [PubMed]
17. Zheng, X.B.; Zhang, Y.L.; Li, Q.; Liu, Y.G.; Wang, X.D.; Yang, B.L.; Zhu, G.C.; Zhou, C.F.; Gao, Y.; Liu, Z.X. Effects of 1,8-cineole on neuropathic pain mediated by P2X2 receptor in the spinal cord dorsal horn. *Sci. Rep.* **2019**, *9*, 7909. [CrossRef]
18. Zhang, R.X.; Wang, L.; Wang, X.; Ren, K.; Berman, B.M.; Lao, L. Electroacupuncture combined with MK-801 prolongs anti-hyperalgesia in rats with peripheral inflammation. *Pharmacol. Biochem. Behav.* **2005**, *81*, 146–151. [CrossRef]
19. Jianbo, Y.; Cong, Z.; Xiaoqin, L. The effects of electroacupuncture on the extracellular signal-regulated kinase 1/2/P2X3 signal pathway in the spinal cord of rats with chronic constriction injury. *Anesth. Analg.* **2013**, *116*, 239–246.
20. Han, J.S. Acupuncture: Neuropeptide release produced by electrical stimulation of different frequencies. *Trends Neurosci.* **2003**, *26*, 17–22. [CrossRef]
21. Lin, J.G.; Lo, M.W.; Wen, Y.R.; Hsieh, C.L.; Tsai, S.K.; Sun, W.Z. The effect of high and low frequency electroacupuncture in pain after lower abdominal surgery. *Pain* **2002**, *99*, 509–514. [CrossRef]
22. He, X.; Wei, J.; Shou, S.; Fang, J.; Jiang, Y. Effects of electroacupuncture at 2 and 100 Hz on rat type 2 diabetic neuropathic pain and hyperalgesia-related protein expression in the dorsal root ganglion. *J. Zhejiang Univ. Sci. B* **2017**, *18*, 239. [CrossRef] [PubMed]
23. Silva, M.L.; Silva, J.R.T.; Prado, W.A. 100-Hz Electroacupuncture but not 2-Hz Electroacupuncture is Preemptive Against Postincision Pain in Rats. *J. Acupunct. Meridian Stud.* **2016**, *9*, 200–206. [CrossRef] [PubMed]
24. Wang, J.; Gao, Y.; Chen, S.; Duanmu, C.; Zhang, J.; Feng, X.; Yan, Y.; Liu, J.; Litscher, G. The Effect of Repeated Electroacupuncture Analgesia on Neurotrophic and Cytokine Factors in Neuropathic Pain Rats. *Evid. -Based Complementray Altern. Med.* **2016**, *2016*, 1–11. [CrossRef] [PubMed]
25. Gao, Y.H.; Wang, J.Y.; Qiao, L.N.; Chen, S.P.; Tan, L.H.; Xu, Q.L.; Liu, J.L. NK cells mediate the cumulative analgesic effect of electroacupuncture in a rat model of neuropathic pain. *BMC Complementary Altern. Med.* **2014**, *14*, 1–9. [CrossRef] [PubMed]
26. Liu, B.; Liu, Y.; Qin, Z.; Zhou, K.; Xu, H.; He, L.; Li, N.; Su, T.; Sun, J.; Yue, Z.; et al. Electroacupuncture Versus Pelvic Floor Muscle Training Plus Solifenacin for Women With Mixed Urinary Incontinence: A Randomized Noninferiority Trial. *Mayo Clin. Proc.* **2019**, *94*, 54–65. [CrossRef]
27. Liu, Z.; Liu, Y.; Xu, H.; He, L.; Chen, Y.; Fu, L.; Li, N.; Lu, Y.; Su, T.; Sun, J.; et al. Effect of Electroacupuncture on Urinary Leakage Among Women With Stress Urinary Incontinence: A Randomized Clinical Trial. *JAMA* **2017**, *317*, 2493–2501. [CrossRef]
28. Alvarado-Sanchez, B.G.; Salgado-Ceballos, H.; Torres-Castillo, S.; Rodriguez-Silverio, J.; Lopez-Hernandez, M.E.; Quiroz-Gonzalez, S.; Sanchez-Torres, S.; Mondragon-Lozano, R.; Fabela-Sanchez, O. Electroacupuncture and Curcumin Promote Oxidative Balance and Motor Function Recovery in Rats Following Traumatic Spinal Cord Injury. *Neurochem. Res.* **2019**. [CrossRef]
29. Cheng, R.S.; Pomeranz, B. Electroacupuncture analgesia could be mediated by at least two pain-relieving mechanisms; endorphin and non-endorphin systems. *Life Sci.* **1979**, *25*, 1957–1962. [CrossRef]
30. Wang, Y.; Gehringer, R.; Mousa, S.A.; Hackel, D.; Brack, A.; Rittner, H.L. CXCL10 controls inflammatory pain via opioid peptide-containing macrophages in electroacupuncture. *PLoS ONE* **2014**, *9*, e94696. [CrossRef]
31. Kim, H.W.; Uh, D.K.; Yoon, S.Y.; Roh, D.H. Low-frequency electroacupuncture suppresses carrageenan-induced paw inflammation in mice via sympathetic post-ganglionic neurons, while high-frequency EA suppression is mediated by the sympathoadrenal medullary axis. *Brain Res. Bull.* **2008**, *75*, 698–705. [CrossRef] [PubMed]
32. Emery, E.C.; Berrocoso, E.M.; Chen, L.; Mcnaughton, P.A. HCN2 ion channels play a central role in inflammatory and neuropathic pain. *Science* **2011**, *333*, 1462–1466. [CrossRef] [PubMed]
33. Stevens, E.B.; Stephens, G.J. Recent advances in targeting ion channels to treat chronic pain. *Br. J. Pharmacol.* **2018**, *175*, 2133–2137. [CrossRef] [PubMed]

34. Zhou, Y.F.; Ying, X.M.; He, X.F.; Shou, S.Y.; Wei, J.J.; Tai, Z.X.; Shao, X.M.; Liang, Y.; Fang, F.; Fang, J.Q.; et al. Suppressing PKC-dependent membrane P2X3 receptor upregulation in dorsal root ganglia mediated electroacupuncture analgesia in rat painful diabetic neuropathy. *Purinergic Signal.* **2018**, *14*, 359–369. [CrossRef] [PubMed]
35. Nunez-Badinez, P.; Sepulveda, H.; Diaz, E.; Greffrath, W.; Treede, R.D.; Stehberg, J.; Montecino, M.; van Zundert, B. Variable transcriptional responsiveness of the P2X3 receptor gene during CFA-induced inflammatory hyperalgesia. *J. Cell Biochem.* **2018**, *119*, 3922–3935. [CrossRef] [PubMed]
36. Tariba, P.K.; Vukman, R.; Antonić, R.; Kovač, Z.; Uhač, I.; SimonićKocijan, S. The role of P2X3receptors in bilateral masseter muscle allodynia in rats. *Croat. Med. J.* **2016**, *57*, 530–539. [CrossRef] [PubMed]
37. Gu, Y.; Li, G.; Chen, Y.; Huang, L.Y. Epac-protein kinase C alpha signaling in purinergic P2X3R-mediated hyperalgesia after inflammation. *Pain* **2016**, *157*, 1541–1550. [CrossRef]
38. Cheng, R.D.; Tu, W.Z.; Wang, W.S.; Zou, E.M.; Cao, F.; Cheng, B.; Wang, J.Z.; Jiang, Y.X.; Jiang, S.H. Effect of electroacupuncture on the pathomorphology of the sciatic nerve and the sensitization of P2X(3) receptors in the dorsal root ganglion in rats with chronic constrictive injury. *Chin. J. Integr. Med.* **2013**, *19*, 374–379. [CrossRef]
39. Weng, Z.J.; Wu, L.Y.; Zhou, C.L.; Dou, C.Z.; Shi, Y.; Liu, H.R.; Wu, H.G. Effect of electroacupuncture on P2X3 receptor regulation in the peripheral and central nervous systems of rats with visceral pain caused by irritable bowel syndrome. *Purinergic Signal.* **2015**, *11*, 321–329. [CrossRef]
40. Li, M.H.; Suchland, K.L.; Ingram, S.L. Compensatory activation of cannabinoid cb2 receptor inhibition of gaba release in the rostral ventromedial medulla in inflammatory pain. *J. Neurosci.* **2017**, *37*, 626–636. [CrossRef]
41. Wang, K.; Wang, Z.; Cui, R.; Chu, H. Polysaccharopeptide from trametes versicolor blocks inflammatory osteoarthritis pain-morphine tolerance effects via activating cannabinoid type 2 receptor. *Int. J. Biol. Macromol.* **2019**, *126*, 805–810. [CrossRef] [PubMed]
42. Matsuoka, Y.; Yamashita, A.; Matsuda, M.; Kawai, K.; Sawa, T.; Amaya, F. The nlrp2 inflammasome in dorsal root ganglion as a novel molecular platform that produces inflammatory pain hypersensitivity. *Pain* **2019**. [CrossRef] [PubMed]
43. Chaplan, S.R.; Bach, F.W.; Pogrel, J.W.; Chung, J.M.; Yaksh, T.L. Quantitative assessment of tactile allodynia in the rat paw. *J. Neurosci. Methods* **1994**, *53*, 55–63. [CrossRef]

International Journal of
*Molecular Sciences*

*Article*

# RNA-Binding Proteins HuB, HuC, and HuD are Distinctly Regulated in Dorsal Root Ganglia Neurons from STZ-Sensitive Compared to STZ-Resistant Diabetic Mice

**Cosmin Cătălin Mustăciosu [1,†], Adela Banciu [2,†], Călin Mircea Rusu [1,3], Daniel Dumitru Banciu [2], Diana Savu [1], Mihai Radu [1,*] and Beatrice Mihaela Radu [3,4]**

1 Department of Life and Environmental Physics, 'Horia Hulubei' National Institute of Physics and Nuclear Engineering, Reactorului 30, 077125 Bucharest-Magurele, Romania; cosmin@nipne.ro (C.C.M.); calin.rusu@nipne.ro (C.M.R.); dsavu@nipne.ro (D.S.)
2 Department of Bioengineering and Biotechnology, Faculty of Medical Engineering, University Politehnica of Bucharest, Gheorghe Polizu Street 1-7, 011061 Bucharest, Romania; adela.banciu79@gmail.com (A.B.); danieldumitrubanciu@gmail.com (D.D.B.)
3 Department of Anatomy, Animal Physiology, and Biophysics, Faculty of Biology, University of Bucharest, Splaiul Independenţei 91-95, 050095 Bucharest, Romania; beatrice.radu@bio.unibuc.ro
4 Life, Environmental and Earth Sciences Division, Research Institute of the University of Bucharest (ICUB), Splaiul Independenţei 91-95, 050095 Bucharest, Romania
* Correspondence: mradu@nipne.ro
† These authors contributed equally to this work.

Received: 17 March 2019; Accepted: 19 April 2019; Published: 22 April 2019

**Abstract:** The neuron-specific *Elav*-like Hu RNA-binding proteins were described to play an important role in neuronal differentiation and plasticity by ensuring the post-transcriptional control of RNAs encoding for various proteins. Although *Elav*-like Hu proteins alterations were reported in diabetes or neuropathy, little is known about the regulation of neuron-specific *Elav*-like Hu RNA-binding proteins in sensory neurons of dorsal root ganglia (DRG) due to the diabetic condition. The goal of our study was to analyze the gene and protein expression of HuB, HuC, and HuD in DRG sensory neurons in diabetes. The diabetic condition was induced in CD-1 adult male mice with single-intraperitoneal injection of streptozotocin (STZ, 150 mg/kg), and 8-weeks (advanced diabetes) after induction was quantified the *Elav*-like proteins expression. Based on the glycemia values, we identified two types of responses to STZ, and mice were classified in STZ-resistant (diabetic resistant, glycemia $< 260$ mg/dL) and STZ-sensitive (diabetic, glycemia $> 260$ mg/dL). Body weight measurements indicated that 8-weeks after STZ-induction of diabetes, control mice have a higher increase in body weight compared to the diabetic and diabetic resistant mice. Moreover, after 8-weeks, diabetic mice ($19.52 \pm 3.52$ s) have longer paw withdrawal latencies in the hot-plate test than diabetic resistant ($11.36 \pm 1.92$ s) and control ($11.03 \pm 1.97$ s) mice, that correlates with the installation of warm hypoalgesia due to the diabetic condition. Further on, we evidenced the decrease of *Elav*-like gene expression in DRG neurons of diabetic mice (*Elavl2*, $0.68 \pm 0.05$ fold; *Elavl3*, $0.65 \pm 0.01$ fold; *Elavl4*, $0.53 \pm 0.07$ fold) and diabetic resistant mice (*Ealvl2*, $0.56 \pm 0.07$ fold; *Elavl3*, $0.32 \pm 0.09$ fold) compared to control mice. Interestingly, *Elav*-like genes have a more accentuated downregulation in diabetic resistant than in diabetic mice, although hypoalgesia was evidenced only in diabetic mice. The *Elav*-like gene expression changes do not always correlate with the Hu protein expression changes. To detail, HuB is upregulated and HuD is downregulated in diabetic mice, while HuB, HuC, and HuD are downregulated in diabetic resistant mice compared to control mice. To resume, we demonstrated HuD downregulation and HuB upregulation in DRG sensory neurons induced by diabetes, which might be correlated with altered post-transcriptional control of RNAs involved in the regulation of thermal hypoalgesia condition caused by the advanced diabetic neuropathy.

**Keywords:** *Elav*-like; Hu proteins; diabetes; streptozotocin; thermal response; hypoalgesia; dorsal root ganglia neurons

## 1. Introduction

Hu proteins are members of the RNA-binding proteins (RBP) superfamily and are encoded by Embryonic Lethal, Abnormal Vision, and Drosophila (*ELAV*) genes. The Hu proteins family has four members HuB (encoded by *ELAV-like 2* gene), HuC (encoded by *ELAV-like 3* gene), HuD (encoded by *ELAV-like 4* gene), and HuR or HuA (encoded by *ELAV-like 1* gene). Three of these proteins have identified as neuronal specific (i.e., HuB, HuC, and HuD), while the fourth is ubiquitary (HuR).

RBP are well known for the post-transcriptional control of RNAs encoding multiple proteins [1]. In particular, RBPs play essential roles in the nervous system, such as alternative splicing of neuronal proteins (i.e., neurotransmitters, membrane receptors, cell adhesion molecules, and components of signal transduction proteins), protection of the mRNAs for long-distance transport and guidance of the protein localization [2–5].

Neuronal-enriched *ELAV*-like (*nELAVL*) Hu proteins were described to play essential roles in neuronal development and plasticity [6,7] in the central and peripheral nervous system. *nELAVL* Hu proteins are binding to the adenylate-uridylate-rich (ARE) RNA elements in the 3′ untranslated regions (3′-UTR) of target proteins, including growth associated protein 43 (GAP-43) [8,9], c-myc and vascular endothelial growth factor (VEGF) [10], and neprilysin (a potent amyloid β degrading enzyme) [11] stabilizing them. Moreover, *nELAVL* Hu proteins autoregulate themselves [12] or interact/stabilize other neuronal RBPs, e.g., Musashi-1 [13] and NOVA1 [14]. In the central nervous system, *nELAVL* Hu-proteins have been involved in regulating neuronal excitability by controlling the glutamate synthesis pathway and their gene deletion induces spontaneous epileptic seizure activity [15], by binding to the mRNA encoding Kv1.1 voltage-gated potassium channels [16]. In the peripheral nervous system, *nELAVL* Hu proteins are localized in the dorsal root ganglia (DRG) neurons [17–20]. This anatomical localisation of *nELAVL* Hu-proteins is correlated with their functional role of binding mRNA encoding proteins (i.e., brain-derived neurotrophic factor, GAP-43) involved in peripheral nerve regeneration upon lesion [8,21,22], being upregulated in the early stages of nerve recovery.

The role of RBPs in diabetes and its complications was extensively documented [23]. To detail, it was described the regulation of beta-pancreatic cell function by various RBPs [24], including neuronal-enriched RBPs [25]. The altered regulatory function exerted by the ubiquitary HuR protein in diabetes was often described [25–28]. On the other hand, although the role of *nELAVL* Hu proteins was described in diabetes [29–31], yet no attention was paid so far to their expression changes in DRG sensory neurons associated with the diabetic condition.

We aimed to elucidate the role played by Hu proteins expressed by the DRG sensory neurons in diabetes. To this purpose, we have employed the streptozotocin (STZ)-induced model of diabetes in CD-1 adult male mice. We have explored the gene and protein expression for *nELAVL* Hu proteins in DRG neurons between diabetic and control mice and we correlated them with the changes in animal glycemia, body weight, or their response to hot thermal stimulation. We also analyzed the distinct changes between animals sensitive or resistant to the STZ-induction of diabetes.

## 2. Results

### 2.1. Diabetic Mice Have Changes in Glycemia and Body Weight Compared to Diabetic Resistant or Control Mice

We started our experimental protocol with two CD-1 mice groups: citrate buffer-injected group ($N = 20$) and STZ-injected group ($N = 20$). In the STZ-injected group, seven out of 20 animals died quickly. We measured the glycemia weekly for 7 weeks. Hyperglycemia was considered above 260 mg/dL, as previously described [32,33]. Considering the hyperglycemia threshold, at the end of 7 weeks we separated the surviving animals of the STZ-injected group ($N = 13$) into two subgroups: STZ-sensitive group ($N = 7$, glycemia > 260 mg/dL) and STZ-resistant group ($N = 6$, glycemia < 260 mg/dL), that will be further called diabetic group and diabetic resistant group, respectively. Then, we plotted the glycemia variation for the diabetic, diabetic resistant, and control mice groups (Figure 1).

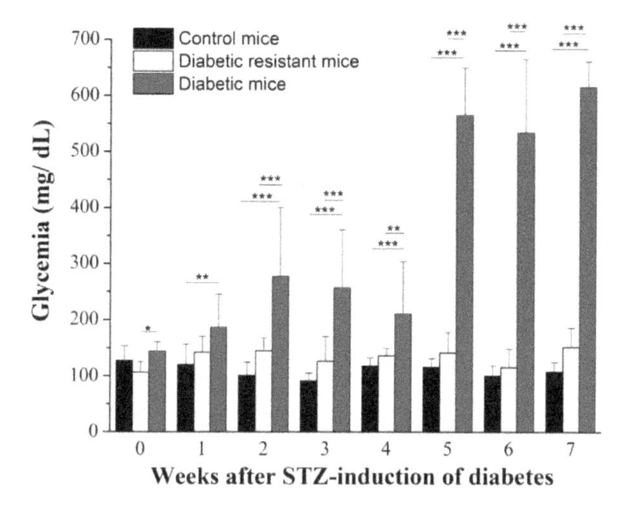


**Figure 1.** Blood glucose values (in mg/dL) were represented as mean ± SD for control, diabetic resistant, and diabetic mice. Statistical significance was indicated *** $p < 0.001$, ** $p < 0.01$, * $p < 0.05$.

Only the diabetic group had an increase in glycemia (from 144.16 ± 17.29 mg/dL to 615.50 ± 45.05 mg/dL, $N = 7$), while the diabetic resistant group (from 106.85 ± 18.73 mg/dL to 152.14 ± 33.55 mg/dL, $N = 6$) and the control group (from 127.89 ± 26.63 mg/dL to 107.89 ± 36.63 mg/dL, $N = 20$) had no significant changes. The two-way ANOVA analysis indicated statistical significance of the glycemia, for the diabetic condition, and for their interaction (Table S1). The one-way ANOVA weekly comparison between the animal groups indicated that diabetic mice had higher glycemia compared to diabetic resistant and control mice, starting from the first week after STZ-induction of diabetes and with this difference accentuating to fifth–seventh week (Table S2) and is indicated in Figure 1. The weekly comparison in the diabetic group showed the increase of glycemia up to the fifh week, followed by a plateau-like evolution up to the eighth week (Table S3). Meanwhile, the weekly comparison of the glycemia values for the control and diabetic groups did not indicate significant changes.

We have also measured the body weight for the diabetic, diabetic resistant and control CD-1 mice groups weekly, for 8 weeks, after the STZ-diabetes induction (Figure 2). All animal groups had an overall increase of the body weight, but the increase rate was higher for the control group (from 21.75 ± 2.75 g to 34.19 ± 2.66 g, $N = 20$) compared to the diabetic group (from 22.74 ± 2.68 g to 30.08 ± 2.66 g, $N = 7$) and diabetic-resistant group (from 22.74 ± 2.18 g to 28.65 ± 2.78 g, $N = 6$). The two-way ANOVA analysis indicated statistical significance of the body weight, of the diabetic condition, and of their interaction (Table S4). The one-way ANOVA comparison between the animal groups indicated that control mice were heavier than diabetic mice and diabetic resistant mice, starting from the fourth wk after STZ-induction of diabetes (Table S5). The one-way ANOVA weekly comparison in each animal group showed a continuous body weight increase, for the whole duration of the protocol (for 8 weeks), with statistical significance in all three animal groups (Table S6).

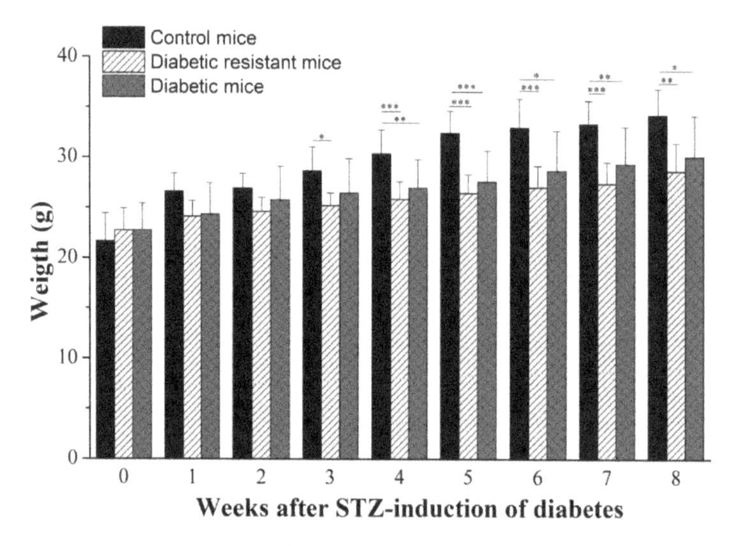


**Figure 2.** Body weight (in g) monitored for 8 weeks after streptozotocin (STZ)-induction of diabetes. Body weight values were represented as mean ± SD for control, diabetic resistant, and diabetic mice. Statistical significance is indicated * $p < 0.05$, ** $p < 0.01$, *** $p < 0.001$.

### *2.2. Diabetic Mice Have Longer Paw Withdrawal Latencies to Nociceptive Thermal Stimulation than Diabetic Resistant or Control Mice*

We employed the hot-plate test at the fixed temperature of 55 °C and we measured the paw withdrawal latency in order to evaluate the thermal response in diabetic, diabetic resistant, and control mice groups (Figure 3). The two-way ANOVA analysis indicated statistical significance of the latency, of the diabetic condition, and of their interaction (Table S7). We compared the initial paw withdrawal latency (L0) measured at the beginning of the protocol (before any treatment), with the final paw withdrawal latency (Lf) measured after 8 weeks of the STZ-induction of diabetes. In the diabetic group, the final latency (Lf = 19.52 ± 3.52 s, $N$ = 7) is significantly longer than the initial latency (L0 = 11.35 ± 1.69 s, unpaired $t$-test, $p < 0.001$). On the other hand an intergroup comparison of the final latency showed a statistical significant increase in the diabetic group (19.52 ± 3.52 s, $N$ = 7) compared to the diabetic resistant group (11.36 ± 1.92 s, $N$ = 6) or to the control group (11.03 ± 1.97 s, $N$ = 20) (Table S8).

### *2.3. Elav-Like Gene Expression in Mouse DRG Neurons Is Decreased in Diabetic and Diabetic Resistant Mice Compared to Control Mice*

The qRT-PCR analysis revealed the decrease of the *Elav*-like gene expression in mouse DRG neurons for all three *Elavl* (i.e., *Elavl2*, *Elavl3*, and *Elavl4*) in diabetic mice and diabetic resistant mice compared to control mice (Figure 4). In the DRG neurons of control mice, we obtained similar levels for *Elavl2*, *Elavl3*, and *Elavl4* genes (unshown data).

We demonstrated that *Elavl* genes expression is altered in diabetes, the two-way ANOVA analysis being statistical significant for the *Elavl* expression, for the diabetic condition and for their interaction (Table S9). *Elavl* genes were downregulated in the diabetic condition and strongly downregulated in the diabetic resistant condition in comparison with control. To detail, *Elavl2* expression decreased to 0.68 ± 0.05 fold in diabetic group and to 0.56 ± 0.07 fold in the diabetic resistant group compared to the control group. *Elavl3* expression decreased to 0.65 ± 0.01 fold in the diabetic group and to 0.32 ± 0.09 fold in the diabetic resistant group compared to control group. *Elavl4* expression decreased to 0.53 ± 0.07 fold in the diabetic resistant group compared to the control group, while in the diabetic group there was a tendency of expression increase without significance (Table S10).

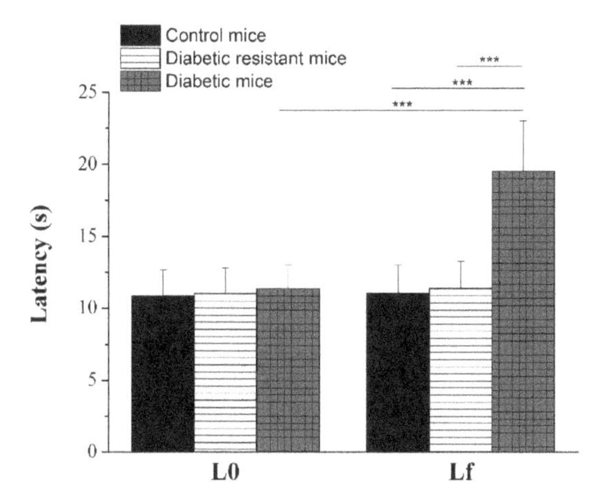


**Figure 3.** Paw withdrawal latencies (in s) in response to radiant heat (55 °C). Initial (L0) and final (Lf) withdrawal latencies were represented as mean ± SD for control, diabetic resistant, and diabetic mice. Statistical significance is indicated *** $p < 0.001$.

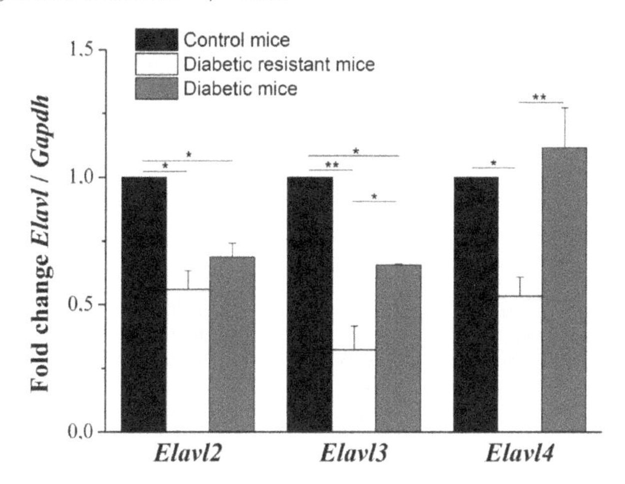


**Figure 4.** Fold change of *Elav*-like gene expression with respect to *Gapdh* expression in dorsal root ganglia (DRG) neurons of control, diabetic resistant, and diabetic mice. Statistical significance is indicated * $p < 0.05$, ** $p < 0.01$.

### 2.4. Hu Proteins Expression in Mouse DRG Neurons Is Decreased in Diabetic and Diabetic Resistant Mice Compared to Control Mice

The Hu protein expression in mouse DRG neurons was evaluated by immunofluorescence in control (Figure 5A–C), diabetic resistant (Figure 5D–F), and diabetic (Figure 5G–I) mice. We evidenced the expression of HuB, HuC, and HuD proteins in DRG neurons for all three CD-1 mice groups. We localized all three Hu proteins both in the soma and the neurites of the DRG neurons. Each Hu protein has a distinct distribution pattern in the soma and particularly HuC tends to organize in clusters. We observe a pronounced localisation of HuB and HuC in the neurites in diabetic conditions compared to diabetic resistant or control conditions.

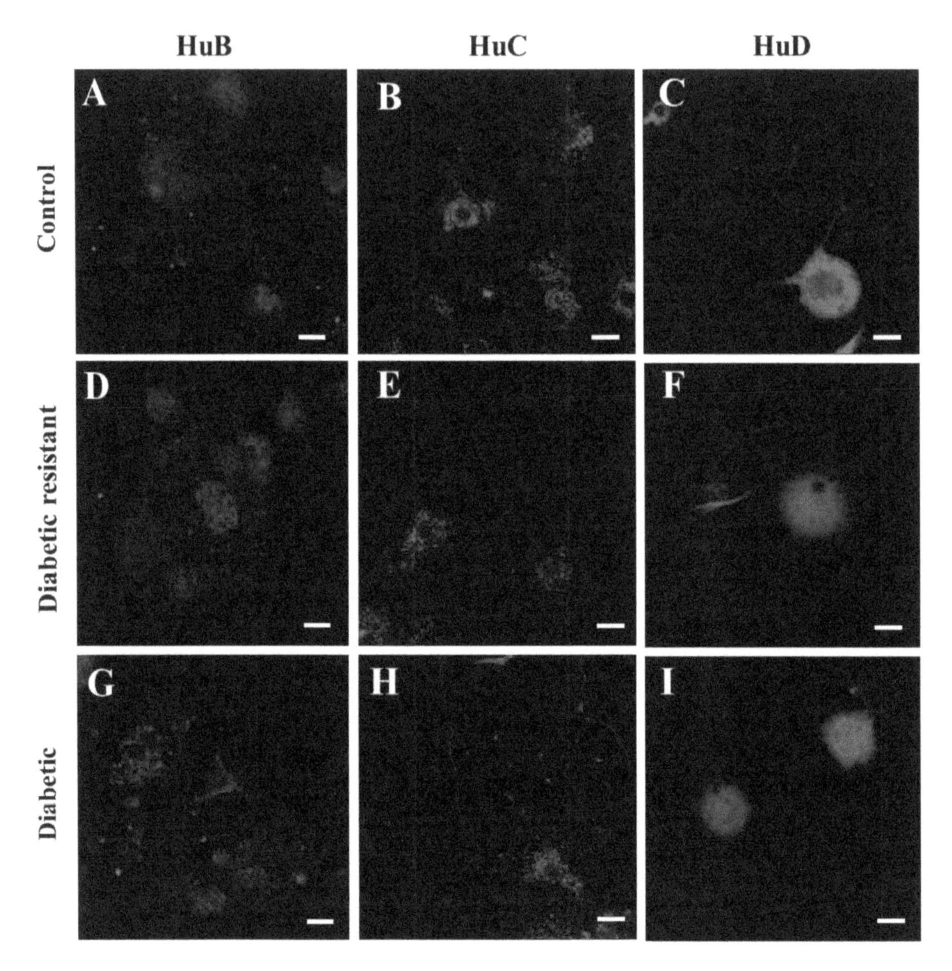


**Figure 5.** Hu proteins (HuB, HuC, and HuD ) expression in DRG neurons of control (**A–C**), diabetic resistant (**D–F**), and diabetic mice (**G–I**). The red labeling is obtained with rabbit polyclonal anti-ELAVL2, anti-ELAVL3, and anti-ELAVL4 antibodies, respectively, followed by the staining with donkey polyclonal anti-rabbit conjugated with Rhodamine Red X. Images are captured with an LSM 710 Zeiss laser scanning microscope using a 63× oil objective. Scale bar 10 μm.

Further on, we performed the quantitative analysis of the neuronal Hu proteins expression based on the mean fluorescence intensity (Figure 6) and we correlated these results with the *Elav*-like gene expression. In control mice, we obtained the following ranking for the protein expression HuD > HuC > HuB, and the one-way ANOVA analysis followed by post-hoc Bonferroni test indicated statistical significance between HuD and HuB expression ($p < 0.001$) and between HuD and HuC expression ($p < 0.001$). However, the distinct levels of HuB, HuC, and HuD expression in control mice are not in agreement with the *Elav*-like gene levels that are comparable. In diabetic and diabetic resistant conditions neuronal HuB, HuC, and HuD proteins were distinctly regulated compared to control conditions. The two-way ANOVA analysis indicated statistical significance of Hu proteins expression, for the diabetic condition and for their interaction (Table S11). In comparison with control mice, HuB protein was significanly downregulated in diabetic resistant mice and upregulated in diabetic mice, HuC protein was significantly downregulated in diabetic resistant mice, and HuD protein was significantly downregulated both in diabetic and diabetic resistant mice (Table S12).

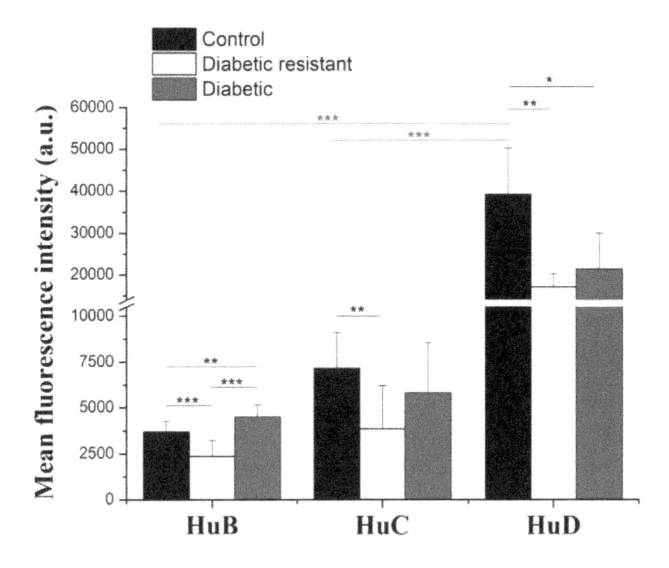


**Figure 6.** Hu protein expression based on mean fluorescence intensity analysis in DRG neurons of control, diabetic resistant, and diabetic mice. Data are expressed as mean ± SD in the captured images. Statistical significance is indicated * $p < 0.05$, ** $p < 0.01$, *** $p < 0.001$.

## 3. Discussion

In this study, we brought evidence that Hu proteins undergo expression changes that might be associated with the diabetic condition. First of all, it is necessary to discuss the model of diabetes that we employed in our study. Indeed, several mouse models for type 1 diabetes have been developed, the most employed being the STZ-induced diabetes, despite its variability, depending on the mice strain [34] or the development of diabetic neuropathy [35]. To detail, variable concentrations of STZ (single i.p. injection) were used in different mice strains to induce diabetes, i.e., ICR, ddY and BALB/c: 100–200 mg/kg and C57BL/6: 75–150 mg/kg [34]. Considering the different mice strain sensitivity to STZ-induction of diabetogenic state [35–37], in our study, we decided to induce diabetes in CD-1 adult mice with single i.p. STZ injection (150 mg/kg).

Despite its variability the STZ-induced model of diabetes is robust and used in multiple studies. However, researchers focus either on the mortality rate or on the resistance of the animal strain when injected with STZ, but in the STZ-"sensitive" animal strains little attention is paid to the rate of surviving animals that are resistant to the STZ-induction of diabetes. Some studies reported a subpopulation of mice [37] or rats [38] remains normoglycemic upon STZ-induction of diabetes, but do not explicitly consider these animals as "diabetic resistant". In our study, we are classifying STZ-resistant CD1-mice as "diabetic resistant" mice.

The resistance to STZ-induction of diabetes was previously described in different mice strains, including mice lacking phosphatase and a tensin homolog deleted from chromosome 10 [39] or nonobese diabetes-resistant mice [40,41]. On the other hand, the resistance to STZ-induction of diabetes in a certain percentage of animals belonging to the so-called 'sensitive' strains (e.g., CD-1 mice) is generally not discussed. For example, in CD-1 mice, the distinction between the induction of type 1 diabetes by single injection of high STZ dose (130 mg/kg or 150 mg/kg) and the induction of type 2 diabetes by multiple injections of low STZ dose (40 mg/kg) was reported [42], but the percentage of 100% reported in the text for the induction of diabetes animals by single injection of high STZ dose is not in agreement with the percentage of animals with hyperglycemia ≥600 mg/dL, one out of five animals (130 mg/kg STZ, 16% mortality) and three out of three (150 mg/kg STZ, 50% mortality) presented in the same study. In our study, we demonstrate the resistance to the induction of diabetes with a single injection of high STZ dose (150 mg/kg) in approximately 46% (CD-1 adult male mice, six out of 13 mice) of the surviving animals (37% mortality). We also evidence that diabetic-resistant mice have lower body weight values compared to control mice, but remain normoglycemic. Our data, indicating that diabetic mice have a lower body weight and a higher glycemia compared to control mice, are in agreement with previous protocols of STZ-induction of diabetes [37]. To resume, our study demonstrates that, in addition to the expected population of STZ-induced hyperglycemic CD-1 mice, a mice subpopulation develops resistance to STZ and we consider that a distinct analysis of the STZ-resistant normoglycemic mice should be done in each study, as the findings might be relevant in understanding the mechanisms of insulin resistance development in human patients.

In diabetic patients, diabetic peripheral neuropathy is gradually characterized by hyperalgesia and allodynia, followed by the development of hypoalgesia and finally the complete loss of sensation [43]. STZ-induced diabetes in mice or rat is associated with thermal hyperalgesia in early phases [43] and with thermal hypoalgesia in late stages of diabetes [44–46] in the absence of insulin therapy. Commonly thermal hypoalgesia precedes epidermal denervation in STZ-diabetic mice [47]. We confirm an increase in the paw withdrawal latency (thermal hypoalgesia) in diabetic mice 8-weeks after STZ-induction of diabetes, while diabetic resistant mice have similar paw withdrawal latencies compared to control mice. The absence of changes in the algesic profile of diabetic resistant mice is supported by previous reports showing that ~50% male Sprague-Dawley rats remain normoglycemic after STZ-injection, without significant changes in the algesic profile (no changes in the threshold or latency to heat noxious stimuli, or in the pressure pain threshold and frequency of withdrawal to brush and 20-g von Frey filament) compared to control rats [38].

Considering previous studies that reported the role played by Hu proteins from DRG neurons in hyperalgesia [19,48–52], we analyzed the expression changes of Hu proteins in diabetic and diabetic resistant mice compared to control mice and correlated these data with algesic profile to radiant heat exposure. To detail, HuR contributes to hyperalgesia either associated with experimental autoimmune encephalomyelitis [48] or with inflammation (exposure to bradykinin and interleukin-1) where stabilized cyclooxygenase-2 mRNA [49]. HuD is upregulated and contributes to pain hypersensitivity to mechanical and cold stimulation in antiretroviral-evoked painful neuropathy by regulating spinal ryanodine receptor-2 [50] or GAP43 [19,51] or contributes to thermal hot hyperalgesia in oxaliplatin-induced neuropathy by regulating GAP43 [52]. On the other hand, to the best of our knowledge, this is the first study documenting the role of Hu proteins in hypoalgesia associated with the diabetic condition, addressing the expression of Hu proteins in DRG neurons. Considering the previous reports regarding the role of *nELAVL* Hu-proteins in neuronal excitability by binding to the mRNAs encoding proteins from the glutamate synthesis pathway [15], or encoding Kv1.1 voltage-gated potassium channels [16], we might suppose that *nELAVL* Hu-proteins might also stabilize / regulate mRNA encoding other proteins (i.e., ion channels) involved in DRG neuronal excitability and being important players in the algesic profile and diabetes.

Our study brings evidence that *Elavl* genes and Hu proteins expression is distinctly regulated in DRG sensory neurons in diabetic, diabetic resistant, and control conditions, and we have tried to correlate these expression data with the final paw withdrawal latency in the hot plate test. Interestingly, the final paw withdrawal latency in diabetic-resistant mice has not significantly changed in comparison to control mice, which indicates that diabetic-resistance mice do not undergo changes in the algesic profile in radiant heat exposure after 8 weeks. In diabetic mice, *Elavl2* and *Elavl3* are downregulated, while HuB is upregulated and HuD is downregulated, compared to control mice. In diabetic resistant mice, both *Elavl* genes and Hu proteins are strongly downregulated, compared to control mice. It is very interesting to remark that, despite the lack of changes in the algesic profile of diabetic resistant mice, we reported significant *Elavl* gene and Hu protein expression changes in diabetic resistant mice compared to diabetic or control mice. Previous studies indicated HuD upregulation in thermal hyperalgesia [19,50–52] and our study brings evidence that HuD is downregulated in thermal hypoalgesia induced by the advanced diabetes status. Considering the role played by HuD upregulation in nerve regeneration upon lesion [21], a possible scenario in diabetes would be: (i) hyperalgesia (early phases of diabetes) is associated with HuD upregulation involved in nerve regeneration, (ii) hypoalgesia (late phases of diabetes) is associated with HuD downregulation, when its ability to regulate mRNA proteins involved in nerve recovery is overcome. However, HuD downregulation in late diabetes should be considered with caution as STZ-induced the same kind of expression changes in normoglycemic diabetic resistant mice. Extensive analysis of the algesic profile in diabetic and diabetic resistant mice in correlation with Hu proteins expression is necessary.

Our study also analyzed the immunolocalization of Hu proteins in correlation with the diabetic status. Previous immunostaining studies documented the expression of HuD [8,53], HuC/HuD (anti-16A11 antibody) [17], or all Hu proteins (anti-16A11 antibody) [18] in adult DRG neurons. The HuD immunopositivity in DRGs neurons was analyzed: (i) in the cell compartments, with distribution both in the soma and the axons [19] or (ii) in the subcellular structures (strong staining in the cytoplasm [8,18] and low staining in the nucleus, Golgi apparatus and mitochondria [18]). Our study indicates HuB, HuC, and HuD expression in the soma and neurites of the DRG neurons. However, our semi-quantitative analysis of Hu protein expression was limited to the soma of DRG neurons.

Although specific immunostaining was obtained for all three neuronal Hu proteins in different structures of the central nervous system or in the spinal cord [54], only HuD specific immunopositivity was analyzed in DRGs [8,53], but no specific targeting of HuB and HuC expression in DRGs was done. In our study, we bring evidence of the specific localisation of HuB and HuC in DRG neurons, and we also demonstrate that all three Hu proteins undergo expression changes in late diabetes.

Different neuronal types from hippocampus, cerebellum, olfactory cortex, neocortex, etc. were demonstrated to express from one to several Hu genes [54]. We might suppose that different subtypes of DRG neurons express various combinations of Hu genes, distinctly contributing to the regulation/stabilization of mRNA encoding proteins involved in the development of diabetic neuropathy and/or thermal hypoalgesia. To this purpose, subsequent colocalization studies of Hu proteins in DRG neurons might bring new insights.

To resume, our study analyzed the distinction between diabetic and diabetic resistant mice in the STZ-induction model and compares them with the control mice. We correlate the diabetic state with hyperglycemia, lower body weight, presence of late thermal hypoalgesia, *Elavl2* and *Elavl3* downregulation, HuB upregulation, and HuD downregulation in comparison to control conditions. Meanwhile, we correlate the diabetic resistant state with normoglycemia, slightly lower body weight, normal algesia, strong *Elavl2*, *Elavl3*, and *Elavl4* downregulation, HuB, HuC, and HuD downregulation compared to control conditions. In conclusion, we demonstrate the distinct expression regulation of *nELAVL* Hu proteins in diabetes and we consider that it is very important to understand if these Hu protein expression changes are also present in patients with peripheral diabetic neuropathy and if there is any correlation with the status of the disease.

## 4. Materials and Methods

### 4.1. Animals

Adult CD-1 male mice aged 6 weeks with a mean body weight of 20 g were acquired from the "Cantacuzino" Medico-Military National Institute of Research and Development. Animals ($N = 40$) were housed 3/cage in the animal husbandry of 'Horia Hulubei' National Institute of Physics and Nuclear Engineering, with food and water *ad libitum*. All procedures were in accordance with the European Guidelines on Laboratory Animal Care, and with the approval of the institutional Ethics Committee of the 'Horia Hulubei' National Institute of Physics and Nuclear Engineering (approval number 31/11.06.2015).

### 4.2. Streptozotocin-Induced Diabetes

Animals were divided into equal groups: 20 mice treated with citrate buffer solution and 20 mice treated with streptozotocin (STZ, #S0130, Sigma-Aldrich, St. Louis, MO, USA). Diabetes was induced with a single intraperitoneal injection of STZ, at a fixed volume of 300 μL/animal, at the final concentration of 150 mg/kg/body weight in 0.05 mol/L sodium citrate buffer, pH 4.5, as previously described [37,55,56]. Citrate buffer solution was also injected intraperitoneal at a fixed volume of 300 μL/animal. Upon data analysis, the surviving animals from the STZ-injected group were divided into two subgroups: STZ-sensitive group and STZ-resistant group (see Results Section 2.1). The timeline of the experimental protocol is presented in Figure 7.

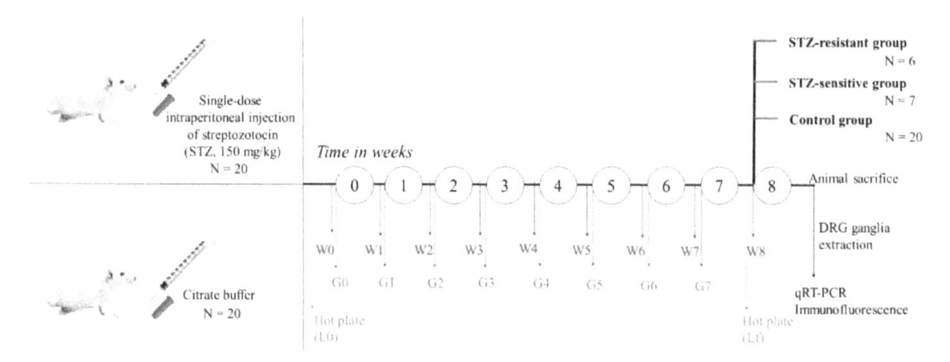


**Figure 7.** Timeline of the experimental protocol. Abbreviations W-weight, G-glycemia.

### 4.3. Body Weight Measurements

The body weight measurement was performed before the intraperitoneal injection (STZ or citrate buffer solution), and then repeated once per week (in the same day, at the beginning of the week, and at the same hour) for 8 weeks, as shown in Figure 2.

### 4.4. Glycemia Measurements

Blood glucose was measured from the tail vein blood by a glucometer (OneTouch, LifeScan, Milpitas, CA, USA). The blood glucose measurement was performed before the intraperitoneal injection (STZ or citrate buffer solution), and then repeated once per week (in the same day and at the same hour) for 7 weeks, as shown in Figure 1. Animals were fastened 12 h before the glycemia measurement. The body weight measurement was done alternately done before/after the glycemia measurement. In order to prevent any experimental bias, in the 8th week, the glycemia measurement was not performed, as animals were subjected to the hot-plate test.

### 4.5. Hot-Plate Test

In order to determine the occurrence of thermal hypersensitivity, the mice subgroups (control, STZ-sensitive and STZ-resistant) were subjected to the hot plate test. To detail, CD-1 mice were placed individually on a Hot Plate Analgesia Meter (Ugo Basile, Comerio, Varese, Italy), maintained at 55 °C and the latency to hind paw licking or flicking (whichever occurs first) was measured. The animals that did not respond within 30 s were removed from the hot plate to prevent paw damage. The hot plate test was performed before the intraperitoneal injection with STZ or citrate buffer solution (initial paw withdrawal latency, L0), and repeated after 8 weeks (final paw withdrawal latency, Lf), as shown in Figure 3.

### 4.6. Primary Cultures of Neurons from Dorsal Root Ganglia

Animals were sacrificed after 8 weeks from the intraperitoneal injection (at 16 weeks of age) and DRG neurons were obtained from all spinal levels of adult male CD1 mice as previously described [32,33]. The animals were exposed to $CO_2$ inhalation (1 min) followed by decapitation according to the European Guidelines on Laboratory Animal Care, with the approval of the institutional Ethics Committee of the 'Horia Hulubei' National Institute of Physics and Nuclear Engineering (approval number, 31/11.06.2015). DRGs were removed under sterile conditions and were immediately transferred into IncMix solution (in mM, NaCl 155, $K_2HPO4$ 1.5, HEPES 5.6, Na-HEPES 4.8, glucose 5). After cleaning the ganglia from the surrounding tissue and counting them, DRGs were incubated in a mixture of 1 mg/mL Collagenase from *Clostridium histolyticum*, type 1A (#C9891, Sigma-Aldrich, St. Louis, MO, USA) and 1 mg/mL Dispase from *Bacillus polymyxa* (#17105041, GIBCO, Invitrogen, Carlsbad, CA, USA) in IncMix solution for 1 h at 37 °C. Following enzyme treatment, the ganglia were washed once in Dulbecco's modified Eagle's medium Ham's F-12 (DMEM F-12, #D8900, Sigma-Aldrich, St. Louis, MO, USA) with 10% horse serum (#H1270, Sigma-Aldrich, St. Louis, MO, USA), before mechanical trituration in 0.5 mL DMEM F-12. The dissociated neurons were then washed by centrifugation (at 1000× *g* for 10 min, 25 °C) followed by resuspension in fresh medium. Following the final wash, the cell pellet was resuspended in DMEM F-12, containing 10% horse serum and 50 µg/mL gentamicin (#G1272, Sigma-Aldrich, St. Louis, MO, USA). Following a second trituration, neurons were seeded on 13-mm coverslips, previously coated with 1 mg/mL poly-D-lysine (#P0899, Sigma-Aldrich, St. Louis, MO, USA) for 1 h at 37 °C and after 24 h were further processed for the immunostaining protocol. In the case of qRT-PCR protocol, the extracted DRGs were directly subjected to the RNA extraction protocol.

### 4.7. RNA Isolation and Quantitative Real-Time PCR (qRT-PCR)

In order to quantify the expression levels of ELAV2, ELAV3, and ELAV4 in the DRG neurons from STZ-sensitive, STZ-resistant and control CD-1 mice subgroups, the total RNA was extracted from dissociated ganglia using the GenElute Mammalian Total RNA MiniPrep Kit (#RTN70, Sigma-Aldrich, St. Louis, MO, USA) according to the manufacturer's instructions. RNA concentrations were determined by spectrophotometric measurements at 260 and 280 nm (Beckman Coulter DU 730). In agreement with the manufacturer guidelines (Sigma-Aldrich, St. Louis, MO, USA) for the GenElute™ Mammalian Total RNA Miniprep Kit, in our experiments the $A_{260}$:$A_{280}$ ratio was 2.038 ± 0.07. Reverse transcription was performed using the High-Capacity cDNA Archieve Kit (Applied Biosystems, Foster City, California, USA). The relative abundance of ELAV transcripts was assessed by qRT-PCR using TaqMan methodology and the ABI Prism 7300 Sequence Detection System (Applied Biosystems). Reactions were carried out for 35 cycles in triplicate. *Elavl2* (#Mm00516015_m1), *Elavl3* (#Mm01151962_m1), *Elavl4* (#Mm01263580_mH) and the mouse control assay for glyceraldehydes-3-dehydrogenase (*Gapdh*, #Mm999999_g1) were obtained from Life Technologies (Carlsbad, CA, USA), and used in accordance with manufacturer's guidelines. From each animal group, 3 animals were sacrificed for the qRT-pCR analysis, and each gene was analyzed in triplicate. Quantitative RT-PCR data for each *Elavl* were normalized with *Gapdh* mRNA levels and relative amounts of mRNA were determined using the comparative cycle thresholds [57]

### 4.8. Immunofluorescence

DRG neurons in primary culture seeded on 13-mm coverslips were washed with PBS, fixed in 4% paraformaldehyde, permeabilized with 0.1% Triton X-100 and immunostained. Non-specific binding was blocked with donkey serum (#017-000-001, Jackson ImmunoResearch Laboratories, UK). DRG neurons were incubated with the primary antibodies overnight at 4 °C. We have used the following primary antibodies: rabbit polyclonal IgG anti-ELAVL2 antibody (1:100; #ab96471, Abcam, Cambridge, UK), rabbit polyclonal IgG anti-ELAVL3 antibody (1:100; #ab78467, Abcam), and rabbit polyclonal IgG anti-ELAVL4 antibody (1:100; #ab96474, Abcam), considering the target specificity of these antibodies described by previous studies [20,25,58,59]. Then, DRG neurons were incubated with the secondary antibody donkey polyclonal anti-rabbit IgG (H+L) conjugated with Rhodamine Red X (1:100; #111-295-003, Jackson ImmunoResearch Laboratories, UK) for 1h at room temperature. The primary antibody was omitted in negative-control samples. Images were captured using a confocal fluorescence microscope (LSM 710, Carl Zeiss, Oberkochen, Germany) equipped with a 63× oil objective. The following acquisition parameters were used: pinhole corresponding to 1 Airy Unit, 62 μm for the 488 nm laser, digital gain of 1.00, and 5% intensity of the laser, as we previously employed [60]. The acquisition parameter settings were kept fixed across all the image acquisition sessions. Image acquisition was carried out using Zeiss LSM Image Browser software (Carl Zeiss, Germany).

### 4.9. Digital Image Analysis

The fluorescence images were preprocessed using ImageJ. The outline of each neuronal soma was manually drawn and the quantitative analysis of the fluorescence signal was done (mean pixel intensity), as presviously described [61,62]. This quantification was carried out on both negative control (the primary antibody was omitted in the immunofluorescence protocol) and positive samples (full immunostaining protocol, including primary antibody). The average per group (control, diabetic and diabetic resistant) negative control mean intensity was subtracted from the mean intensity values of positive samples resulting the corrected mean pixel intensity. For each Hu protein, ~30 cells were scored from each mice group. Finally, we plotted the corrected mean pixel intensity calculated for each Hu protein (i.e., HuB, HuC, and HuD ) for the samples obtained from the control, diabetic and diabetic resistant mice group.

### 4.10. Statistical Analysis

Statistical analysis was performed using OriginPro 8 (OriginLab Corporation, Northampton, MA, USA).

Glycemia values were compared by: (i) by two-way ANOVA followed by post-hoc Bonferroni test (for testing the significance of glycemia change, diabetic condition contribution, and of their interaction), (ii) by one-way ANOVA followed by post-hoc Bonferroni test (for testing the inter-group significance of glycemia change for each week after the STZ-induction of diabetes), and (iii) by one-way ANOVA followed by post-hoc Bonferroni test (for testing the intra-group significance of glycemia change between different weeks after the STZ-induction of diabetes).

Body weight values were compared as follows: (i) by two-way ANOVA followed by post-hoc Bonferroni test (for testing the significance of body weight change, diabetic condition contribution and of their interaction), (ii) by one-way ANOVA followed by post-hoc Bonferroni test (for testing the inter-group significance of body weight change for each week after the STZ-induction of diabetes), and (iii) by one-way ANOVA followed by post-hoc Bonferroni test (for testing the intra-group significance of body weight change between different weeks after the STZ-induction of diabetes).

Paw withdrawal latency values obtained in the hot plate test were analyzed as follows: (i) by two-way ANOVA followed by post-hoc Bonferroni test (for testing the significance of final latency change, diabetic condition contribution and of their interaction), (ii) by one-way ANOVA followed by post-hoc Bonferroni test (for testing the inter-group significance of the final latency change), and (iii) by unpaired Student *t*-test (for testing the intra-group differences between the initial and final latency).

Quantitative RT-PCR data were analyzed as follows: (i) by two-way ANOVA followed by post-hoc Bonferroni test (for testing the significance of *Elavl* expression change, diabetic condition contribution and of their interaction) and (ii) by one-way ANOVA followed by post-hoc Bonferroni test (for testing the inter-group significance of *Elavl-2* expression, *Elavl-3* expression or *Elavl-4* expression change).

Mean grey levels obtained by immunofluorescence data analysis were compared as follows: (i) by two-way ANOVA followed by post-hoc Bonferroni test (for testing the significance of Hu protein expression change, diabetic condition contribution and of their interaction), (ii) by one-way ANOVA followed by post-hoc Bonferroni test (for testing the inter-group significance of HuB expression, HuC expression or HuD expression change and (iii) by one-way ANOVA followed by post-hoc Bonferroni test (for testing the intra-group significance of ELAV expression change in control mice).

Data were represented in OriginPro 8 (OriginLab Corporation, Northampton, MA, USA) as the mean ± SD. Differences were considered significant at $p < 0.05$. Statistical significance is indicated in figures as follows: * $p < 0.05$, ** $p < 0.01$, ** $p < 0.001$.

**Supplementary Materials:** Supplementary materials can be found at http://www.mdpi.com/1422-0067/20/8/1965/s1.

**Author Contributions:** C.C.M. (licence in veterinary medicine) designed and organized the animal protocol, performed the body weight, glycemia, and paw withdrawal latency measurements and did the animal sacrifice. A.B. and D.D.B. prepared the primary DRG neuronal cultures, performed the qRT-PCR experiments and analyzed the qRT-PCR data. D.S. performed the immunofluorescence experiments. C.M.R. did the digital analysis of the confocal microscopy images. B.M.R. and M.R. captured the confocal microscopy images, performed the statistical analysis of the experimental data, planned the experiments and wrote the paper.

**Funding:** This work was supported by the Romanian Ministry of Research and Innovation by means of the research grants PN 09 37 03 01/(2009-2015), PN 19 06 02 03/2019, and 36 PFE/2018. We are grateful to Dr. Adriana Georgescu from the Institute of Cellular Biology and Pathology "N. Simionescu" for helping us establish the protocol of STZ administration in mice and for her advices regarding the monitoring of the diabetes status. We also acknowledge Iulia Ghidu and Florentina Ioniţă for their participation to some of the immunofluorescence or qRT-PCR experiments during their master training. We also acknowledge Marzia Di Chio from the Department of Public Health and Community Medicine, University of Verona, Italy for helping us with the settings of the confocal microscope and the supervision of the image capture.

**Conflicts of Interest:** The authors declare no conflict of interest.

## References

1. Glisovic, T.; Bachorik, J.L.; Yong, J.; Dreyfuss, G. RNA-binding proteins and post-transcriptional gene regulation. *FEBS Lett.* **2008**, *582*, 1977–1986. [CrossRef] [PubMed]
2. Lee, C.J.; Irizarry, K. Alternative splicing in the nervous system: An emerging source of diversity and regulation. *Biol. Psychiatry.* **2003**, *54*, 771–776. [CrossRef]
3. Lipscombe, D. Neuronal proteins custom designed by alternative splicing. *Curr. Opin. Neurobiol.* **2005**, *15*, 358–363. [CrossRef]
4. Sutton, M.A.; Schuman, E.M. Local translational control in dendrites and its role in long-term synaptic plasticity. *J. Neurobiol.* **2005**, *64*, 116–131. [CrossRef]
5. Hengst, U.; Jaffrey, S.R. Function and translational regulation of mRNA in developing axons. *Semin. Cell. Dev. Biol.* **2007**, *18*, 209–215. [CrossRef]
6. Hinman, M.N.; Lou, H. Diverse molecular functions of Hu proteins. *Cell. Mol. Life Sci.* **2008**, *65*, 3168–3181. [CrossRef]
7. Perrone-Bizzozero, N.; Bird, C.W. Role of HuD in nervous system function and pathology. *Front. Biosci. (Schol. Ed.).* **2013**, *5*, 554–563. [CrossRef] [PubMed]
8. Anderson, K.D.; Merhege, M.A.; Morin, M.; Bolognani, F.; Perrone-Bizzozero, N.I. Increased expression and localization of the RNA-binding protein HuD and GAP-43 mRNA to cytoplasmic granules in DRG neurons during nerve regeneration. *Exp. Neurol.* **2003**, *183*, 100–108. [CrossRef]
9. Bolognani, F.; Tanner, D.C.; Merhege, M.; Deschênes-Furry, J.; Jasmin, B.; Perrone-Bizzozero, N.I. In vivo post-transcriptional regulation of GAP-43 mRNA by overexpression of the RNA-binding protein HuD. *J. Neurochem.* **2006**, *96*, 790–801. [CrossRef]
10. King, P.H. RNA-binding analyses of HuC and HuD with the VEGF and c-myc 3′-untranslated regions using a novel ELISA-based assay. *Nucleic Acids Res.* **2000**, *28*, E20. [CrossRef]
11. Lim, C.S.; Alkon, D.L. PKCε promotes HuD-mediated neprilysin mRNA stability and enhances neprilysin-induced Aβ degradation in brain neurons. *PLoS ONE* **2014**, *9*, e97756. [CrossRef]
12. Borgeson, C.D.; Samson, M.L. Shared RNA-binding sites for interacting members of the Drosophila ELAV family of neuronal proteins. *Nucleic Acids Res.* **2005**, *33*, 6372–6383. [CrossRef]
13. Ratti, A.; Fallini, C.; Cova, L.; Fantozzi, R.; Calzarossa, C.; Zennaro, E.; Pascale, A.; Quattrone, A.; Silani, V. A role for the ELAV RNA-binding proteins in neural stem cells: Stabilization of Msi1 mRNA. *J. Cell Sci.* **2006**, *119*, 1442–1452. [CrossRef]

14. Ratti, A.; Fallini, C.; Colombrita, C.; Pascale, A.; Laforenza, U.; Quattrone, A.; Silani, V. Post-transcriptional regulation of neuro-oncological ventral antigen 1 by the neuronal RNA-binding proteins ELAV. *J. Biol. Chem.* **2008**, *283*, 7531–7541. [CrossRef]
15. Ince-Dunn, G.; Okano, H.J.; Jensen, K.B.; Park, W.Y.; Zhong, R.; Ule, J.; Mele, A.; Fak, J.J.; Yang, C.; Zhang, C.; et al. Neuronal Elav-like (Hu) proteins regulate RNA splicing and abundance to control glutamate levels and neuronal excitability. *Neuron.* **2012**, *75*, 1067–1080. [CrossRef]
16. Sosanya, N.M.; Huang, P.P.; Cacheaux, L.P.; Chen, C.J.; Nguyen, K.; Perrone-Bizzozero, N.I.; Raab-Graham, K.F. Degradation of high affinity HuD targets releases Kv1.1 mRNA from miR-129 repression by mTORC1. *J. Cell Biol.* **2013**, *202*, 53–69. [CrossRef]
17. Fornaro, M.; Geuna, S. Confocal imaging of HuC/D RNA-binding proteins in adult rat primary sensory neurons. *Ann. Anat.* **2001**, *183*, 471–473. [CrossRef]
18. Fornaro, M.; Raimondo, S.; Lee, J.M.; Giacobini-Robecchi, M.G. Neuron-specific Hu proteins sub-cellular localization in primary sensory neurons. *Ann Anat.* **2007**, *189*, 223–228. [CrossRef]
19. Sanna, M.D.; Quattrone, A.; Mello, T.; Ghelardini, C.; Galeotti, N. The RNA-binding protein HuD promotes spinal GAP43 overexpression in antiretroviral-induced neuropathy. *Exp. Neurol.* **2014**, *261*, 343–353. [CrossRef]
20. Gomes, C.; Lee, S.J.; Gardiner, A.S.; Smith, T.; Sahoo, P.K.; Patel, P.; Thames, E.; Rodriguez, R.; Taylor, R.; Yoo, S.; et al. Axonal localization of neuritin/CPG15 mRNA is limited by competition for HuD binding. *J. Cell Sci.* **2017**, *130*, 3650–3662. [CrossRef]
21. Laedermann, C.J.; Pertin, M.; Suter, M.R.; Decosterd, I. Voltage-gated sodium channel expression in mouse DRG after SNI leads to re-evaluation of projections of injured fibers. *Mol. Pain.* **2014**, *10*, 19. [CrossRef]
22. Sanna, M.D.; Ghelardini, C.; Galeotti, N. HuD-mediated distinct BDNF regulatory pathways promote regeneration after nerve injury. *Brain Res.* **2017**, *1659*, 55–63. [CrossRef]
23. Nutter, C.A.; Kuyumcu-Martinez, M.N. Emerging roles of RNA-binding proteins in diabetes and their therapeutic potential in diabetic complications. *Wiley Interdiscip. Rev. RNA* **2018**, *9*. [CrossRef]
24. Magro, M.G.; Solimena, M. Regulation of β-cell function by RNA-binding proteins. *Mol. Metab.* **2013**, *2*, 348–355. [CrossRef]
25. Juan-Mateu, J.; Rech, T.H.; Villate, O.; Lizarraga-Mollinedo, E.; Wendt, A.; Turatsinze, J.V.; Brondani, L.A.; Nardelli, T.R.; Nogueira, T.C.; Esguerra, J.L.; et al. Neuron-enriched RNA-binding Proteins Regulate Pancreatic Beta Cell Function and Survival. *J. Biol. Chem.* **2017**, *292*, 3466–3480. [CrossRef]
26. Paukku, K.; Backlund, M.; De Boer, R.A.; Kalkkinen, N.; Kontula, K.K.; Lehtonen, J.Y. Regulation of AT1R expression through HuR by insulin. *Nucleic Acids Res.* **2012**, *40*, 5250–5261. [CrossRef]
27. Amadio, M.; Pascale, A.; Cupri, S.; Pignatello, R.; Osera, C.; D Agata, V.; D Amico, A.G.; Leggio, G.M.; Ruozi, B.; Govoni, S.; et al. Nanosystems based on siRNA silencing HuR expression counteract diabetic retinopathy in rat. *Pharmacol. Res.* **2016**, *111*, 713–720. [CrossRef]
28. Li, X.; Zeng, L.; Cao, C.; Lu, C.; Lian, W.; Han, J.; Zhang, X.; Zhang, J.; Tang, T.; Li, M. Long noncoding RNA MALAT1 regulates renal tubular epithelial pyroptosis by modulated miR-23c targeting of ELAVL1 in diabetic nephropathy. *Exp. Cell. Res.* **2017**, *350*, 327–335. [CrossRef]
29. Ishihara, E.; Nagahama, M.; Naruse, S.; Semba, R.; Miura, T.; Usami, M.; Narita, M. Neuropathological alteration of aquaporin 1 immunoreactive enteric neurons in the streptozotocin-induced diabetic rats. *Auton. Neurosci.* **2008**, *138*, 31–40. [CrossRef]
30. De Mello, S.T.; de Miranda Neto, M.H.; Zanoni, J.N.; Furlan, M.M. Effects of insulin treatment on HuC/HuD, NADH diaphorase, and nNOS-positive myoenteric neurons of the duodenum of adult rats with acute diabetes. *Dig. Dis. Sci.* **2009**, *54*, 731–737. [CrossRef] [PubMed]
31. Kim, C.; Lee, H.; Kang, H.; Shin, J.J.; Tak, H.; Kim, W.; Gorospe, M.; Lee, E.K. RNA-binding protein HuD reduces triglyceride production in pancreatic β cells by enhancing the expression of insulin-induced gene 1. *Biochim. Biophys. Acta.* **2016**, *1859*, 675–685. [CrossRef]
32. Radu, B.M.; Iancu, A.D.; Dumitrescu, D.I.; Flonta, M.L.; Radu, M. TRPV1 properties in thoracic dorsal root ganglia neurons are modulated by intraperitoneal capsaicin administration in the late phase of type 1 autoimmune diabetes. *Cell. Mol. Neurobiol.* **2013**, *33*, 187–196. [CrossRef]
33. Radu, B.M.; Dumitrescu, D.I.; Marin, A.; Banciu, D.D.; Iancu, A.D.; Selescu, T.; Radu, M. Advanced type 1 diabetes is associated with ASIC alterations in mouse lower thoracic dorsal root ganglia neurons. *Cell Biochem. Biophys.* **2014**, *68*, 9–23. [CrossRef] [PubMed]

34. Hayashi, K.; Kojima, R.; Ito, M. Strain differences in the diabetogenic activity of streptozotocin in mice. *Biol. Pharm. Bull.* **2006**, *29*, 1110–1119. [CrossRef] [PubMed]
35. Jolivalt, C.G.; Frizzi, K.E.; Guernsey, L.; Marquez, A.; Ochoa, J.; Rodriguez, M.; Calcutt, N.A. Peripheral Neuropathy in Mouse Models of Diabetes. *Curr. Protoc. Mouse Biol.* **2016**, *6*, 223–255.
36. Rossini, A.A.; Appel, M.C.; Williams, R.M.; Like, A.A. Genetic influence of the streptozotocin-induced insulitis and hyperglycemia. *Diabetes* **1977**, *26*, 916–920. [CrossRef]
37. Furman, B.L. Streptozotocin-Induced Diabetic Models in Mice and Rats. *Curr. Protoc. Pharmacol.* **2015**, *70*, 5.47.1–5.47.20. [PubMed]
38. Romanovsky, D.; Wang, J.; Al-Chaer, E.D.; Stimers, J.R.; Dobretsov, M. Comparison of metabolic and neuropathy profiles of rats with streptozotocin-induced overt and moderate insulinopenia. *Neuroscience* **2010**, *170*, 337–347. [CrossRef]
39. Kurlawalla-Martinez, C.; Stiles, B.; Wang, Y.; Devaskar, S.U.; Kahn, B.B.; Wu, H. Insulin hypersensitivity and resistance to streptozotocin-induced diabetes in mice lacking PTEN in adipose tissue. *Mol. Cell. Biol.* **2005**, *25*, 2498–2510. [CrossRef]
40. Kahraman, S.; Aydin, C.; Elpek, G.O.; Dirice, E.; Sanlioglu, A.D. Diabetes-resistant NOR mice are more severely affected by streptozotocin compared to the diabetes-prone NOD mice: Correlations with liver and kidney GLUT2 expressions. *J. Diabetes Res.* **2015**, *2015*, 450128. [CrossRef]
41. Tozzo, E.; Gnudi, L.; Kahn, B.B. Amelioration of insulin resistance in streptozotocin diabetic mice by transgenic overexpression of GLUT4 driven by an adipose-specific promoter. *Endocrinology* **1997**, *138*, 1604–1611. [CrossRef]
42. Ventura-Sobrevilla, J.; Boone-Villa, V.D.; Aguilar, C.N.; Román-Ramos, R.; Vega-Avila, E.; Campos-Sepúlveda, E.; Alarcón-Aguilar, F. Effect of varying dose and administration of streptozotocin on blood sugar in male CD1 mice. *Proc. West Pharmacol. Soc.* **2011**, *54*, 5–9.
43. Callaghan, B.C.; Cheng, H.T.; Stables, C.L.; Smith, A.L.; Feldman, E.L. Diabetic neuropathy: Clinical manifestations and current treatments. *Lancet Neurol.* **2012**, *11*, 521–534. [CrossRef]
44. Calcutt, N.A.; Freshwater, J.D.; Mizisin, A.P. Prevention of sensory disorders in diabetic Sprague-Dawley rats by aldose reductase inhibition or treatment with ciliary neurotrophic factor. *Diabetologia* **2004**, *47*, 718–724. [CrossRef] [PubMed]
45. Davidson, E.P.; Coppey, L.J.; Dake, B.; Yorek, M.A. Treatment of streptozotocin-induced diabetic rats with alogliptin: Effect on vascular and neural complications. *Exp Diabetes Res.* **2011**, *2011*, 810469. [CrossRef]
46. Murakami, T.; Iwanaga, T.; Ogawa, Y.; Fujita, Y.; Sato, E.; Yoshitomi, H.; Sunada, Y.; Nakamura, A. Development of sensory neuropathy in streptozotocin-induced diabetic mice. *Brain Behav.* **2013**, *3*, 35–41. [CrossRef]
47. Beiswenger, K.K.; Calcutt, N.A.; Mizisin, A.P. Dissociation of thermal hypoalgesia and epidermal denervation in streptozotocin-diabetic mice. *Neurosci. Lett.* **2008**, *442*, 267–272. [CrossRef]
48. Sanna, M.D.; Quattrone, A.; Galeotti, N. Silencing of the RNA-binding protein HuR attenuates hyperalgesia and motor disability in experimental autoimmune encephalomyelitis. *Neuropharmacology* **2017**, *123*, 116–125. [CrossRef]
49. Ohnishi, M.; Yukawa, R.; Akagi, M.; Ohsugi, Y.; Inoue, A. Bradykinin and interleukin-1β synergistically increase the expression of cyclooxygenase-2 through the RNA-binding protein HuR in rat dorsal root ganglion cells. *Neurosci. Lett.* **2019**, *694*, 215–219. [CrossRef]
50. Sanna, M.D.; Peroni, D.; Quattrone, A.; Ghelardini, C.; Galeotti, N. Spinal RyR2 pathway regulated by the RNA-binding protein HuD induces pain hypersensitivity in antiretroviral neuropathy. *Exp. Neurol.* **2015**, *267*, 53–63. [CrossRef] [PubMed]
51. Sanna, M.D.; Quattrone, A.; Ghelardini, C.; Galeotti, N. PKC-mediated HuD-GAP43 pathway activation in a mouse model of antiretroviral painful neuropathy. *Pharmacol. Res.* **2014**, *81*, 44–53. [CrossRef] [PubMed]
52. Sanna, M.D.; Ghelardini, C.; Galeotti, N. Altered Expression of Cytoskeletal and Axonal Proteins in Oxaliplatin-Induced Neuropathy. *Pharmacology* **2016**, *97*, 146–150. [CrossRef]
53. Clayton, G.H.; Perez, G.M.; Smith, R.L.; Owens, G.C. Expression of mRNA for the elav-like neural-specific RNA binding protein, HuD, during nervous system development. *Brain Res. Dev. Brain Res.* **1998**, *109*, 271–280. [CrossRef]
54. Okano, H.J.; Darnell, R.B. A hierarchy of Hu RNA binding proteins in developing and adult neurons. *J. Neurosci.* **1997**, *17*, 3024–3037. [CrossRef]

55. Georgescu, A.; Popov, D.; Dragan, E.; Dragomir, E.; Badila, E. Protective effects of nebivolol and reversal of endothelial dysfunction in diabetes associated with hypertension. *Eur. J. Pharmacol.* **2007**, *570*, 149–158. [CrossRef] [PubMed]
56. Tong, M.; Tuk, B.; Shang, P.; Hekking, I.M.; Fijneman, E.M.; Guijt, M.; Hovius, S.E.; van Neck, J.W. Diabetes-impaired wound healing is improved by matrix therapy with heparan sulfate glycosaminoglycan mimetic OTR4120 in rats. *Diabetes* **2012**, *61*, 2633–2641. [CrossRef] [PubMed]
57. Livak, K.J.; Schmittgen, T.D. Analysis of relative gene expression data using real-time quantitative PCR and the $2^{-\Delta\Delta Ct}$. *Methods* **2001**, *25*, 402–408. [CrossRef] [PubMed]
58. Tallafuss, A.; Kelly, M.; Gay, L.; Gibson, D.; Batzel, P.; Karfilis, K.V.; Eisen, J.; Stankunas, K.; Postlethwait, J.H.; Washbourne, P. Transcriptomes of post-mitotic neurons identify the usage of alternative pathways during adult and embryonic neuronal differentiation. *BMC Genom.* **2015**, *16*, 1100. [CrossRef] [PubMed]
59. Rodrigues, D.C.; Kim, D.S.; Yang, G.; Zaslavsky, K.; Ha, K.C.; Mok, R.S.; Ross, P.J.; Zhao, M.; Piekna, A.; Wei, W.; et al. MECP2 Is Post-transcriptionally Regulated during Human Neurodevelopment by Combinatorial Action of RNA-Binding Proteins and miRNAs. *Cell Rep.* **2016**, *17*, 720–734. [CrossRef] [PubMed]
60. Radu, B.M.; Osculati, A.M.M.; Suku, E.; Banciu, A.; Tsenov, G.; Merigo, F.; Di Chio, M.; Banciu, D.D.; Tognoli, C.; Kacer, P.; et al. All muscarinic acetylcholine receptors (M1-M5) are expressed in murine brain microvascular endothelium. *Sci. Rep.* **2017**, *7*, 5083. [CrossRef] [PubMed]
61. Dubový, P.; Jancálek, R.; Klusáková, I.; Svízenská, I.; Pejchalová, K. Intra- and extraneuronal changes of immunofluorescence staining for TNF-alpha and TNFR1 in the dorsal root ganglia of rat peripheral neuropathic pain models. *Cell. Mol. Neurobiol.* **2006**, *26*, 1205–1217. [CrossRef] [PubMed]
62. Tsunematsu, H.; Uyeda, A.; Yamamoto, N.; Sugo, N. Immunocytochemistry and fluorescence imaging efficiently identify individual neurons with CRISPR/Cas9-mediated gene disruption in primary cortical cultures. *BMC Neurosci.* **2017**, *18*, 55. [CrossRef] [PubMed]

International Journal of
*Molecular Sciences*

*Review*

# Nociceptor Signalling through ion Channel Regulation via GPCRs

**Isabella Salzer, Sutirtha Ray, Klaus Schicker and Stefan Boehm ***

Division of Neurophysiology and Neuropharmacology, Centre for Physiology and Pharmacology, Medical University of Vienna, Waehringerstrasse 13a, A-1090 Vienna, Austria; isabella.salzer@meduniwien.ac.at (I.S.); sutirtha.ray@meduniwien.ac.at (S.R.); klaus.schicker@meduniwien.ac.at (K.S.)
* Correspondence: stefan.boehm@meduniwien.ac.at; Tel.: +43-1-40160-31200

Received: 1 April 2019; Accepted: 13 May 2019; Published: 20 May 2019

**Abstract:** The prime task of nociceptors is the transformation of noxious stimuli into action potentials that are propagated along the neurites of nociceptive neurons from the periphery to the spinal cord. This function of nociceptors relies on the coordinated operation of a variety of ion channels. In this review, we summarize how members of nine different families of ion channels expressed in sensory neurons contribute to nociception. Furthermore, data on 35 different types of G protein coupled receptors are presented, activation of which controls the gating of the aforementioned ion channels. These receptors are not only targeted by more than 20 separate endogenous modulators, but can also be affected by pharmacotherapeutic agents. Thereby, this review provides information on how ion channel modulation via G protein coupled receptors in nociceptors can be exploited to provide improved analgesic therapy.

**Keywords:** nociceptor; inflammatory pain; G protein-coupled receptor; voltage-gated ion channel; TRP channel; $K_{2P}$ channel; $Ca^{2+}$-activated $Cl^-$ channel

## 1. Introduction

Nociception refers to *"neural processes of encoding and processing noxious stimuli"* as defined by the International Association for the Study of Pain. Noxious stimuli are *"actually or potentially tissue damaging events"* that need to act on nociceptors in order to cause pain. Accordingly, nociceptors are viewed as *"sensory receptors that are capable of transducing and encoding noxious stimuli"*. As such, nociceptors are peripheral nerve endings of first order nociceptive neurons; these are part of the peripheral nervous system with neuronal cell bodies located mostly in dorsal root ganglia and with central neurites projecting to second order nociceptive neurons located in the dorsal horn of the spinal cord [1].

Noxious stimuli that impinge on nociceptors comprise mechanical forces, temperature changes (heat and cold), and chemical agents (e.g., protons and plant-derived irritants such as capsaicin, menthol, or isothiocyanates). Apart from acting directly on nociceptors, such injurious impact may lead to inflammation, as do infections. This pathologic response is characterized by the release of a plethora of mediators from various types of cells including, amongst others, macrophages, mast cells, immune cells, platelets and the nociceptive neurons themselves [2]. Together, these mediators are called inflammatory soup and lead to an increased responsiveness of nociceptive neurons. This latter mechanism is known as sensitization and forms the pathophysiological basis of allodynia and hyperalgesia: pain in response to a non-nociceptive stimulus and increased pain sensitivity, respectively [1].

Constituents of the inflammatory soup comprise protons, nucleotides and nucleosides, enzymes (proteases), fatty acid derivatives (prostanglandins), biogenic amines (histamine, noradrenaline, and serotonin), cytokines, chemokines, neurotrophins and other peptides (bradykinin, endothelin, and tachykinins) [3]. These multifarious endogenous agents influence nociceptor signaling through a variety of different receptors:

- Protons act directly on ion channels that are members of either the TRP or the ASIC family [4,5].
- ATP as a prototypic nucleotide may activate a subset of ligand-gated ion channels known as P2X receptors [6].
- Cytokines such as various interleukins or tumor necrosis factors (TNFs) target different subtypes of cytokine receptors [7].
- Neurotrophins, in particular nerve growth factor, bind to high affinity tyrosine receptor kinases (trks) and to the low affinity receptor p75 [8].
- All others of the aforementioned inflammatory mediators and ATP elicit their actions on nociceptors via some type of G protein-coupled receptor (GPCR).

Hence, most of the influence of the inflammatory soup on nociceptors is mediated by GPCRs [3]. The common outcome of the separate actions of the single components contained in the inflammatory soup is sensitization of nociceptors, as mentioned above. The prime task of nociceptors is the transformation of noxious stimuli into action potentials that are propagated along the neurites of nociceptive neurons from the periphery to the spinal cord. Accordingly, sensitization means that this transformation of noxious stimuli into action potentials is facilitated, and this may occur through one of two possible mechanisms: reduction in the action potential threshold or increased responses to suprathreshold stimuli. In principle, these two pathophysiological alternatives underlie the clinical phenomena of allodynia and hyperalgesia, respectively [1]. Consequently, this review summarizes how activation of certain GPCRs can impinge on either of these two mechanisms underlying the sensitization of nociceptors.

Obviously, the transformation of noxious stimuli into action potentials relies on the coordinated operation of a variety of ion channels. Therefore, inflammatory mediators must ultimately act on the function of these ion channels to be able to sensitize nociceptors. In this regard, the present review summarizes signaling mechanisms that link an activation of GPCRs to changes in ion channel function in nociceptors.

When dealing with GPCRs expressed in peripheral nociceptive neurons, one must take into account that not all of them subserve stimulatory actions that in the end lead to sensitization. Several of these GPCRs mediate inhibitory effects which rather diminish than enhance neuronal excitability. For the sake of comprehensiveness, such inhibitory receptors are considered as well.

## 2. Ion Channels as Targets of GPCR Signaling in Peripheral Nociceptive Neurons

In this section, ion channel families are described in light of their roles in nociceptive neurons. In this respect, one can discern between ion channels that are directly involved in the sensation of noxious stimuli and those that are rather responsible for the ensuing generation and propagation of action potentials. The former group comprises TRP channels, ASICs, and mechanosensitive $K^+$ and Piezo channels, whereas voltage-activated $Na^+$ and $Ca^{2+}$ channels as well as various types of $K^+$ channels belong to the latter. This basic characterization of each of these ion channel families is followed by a description of the mechanisms that link activation of various GPCRs to changes in functions of these ion channels.

### *2.1. TRP Channels Involved in Pain Sensation*

Transient receptor potential (TRP) channels are expressed in a variety of tissues throughout the body, such as skin, kidney, bladder, vascular smooth muscle cells and the nervous system [9]. The TRP channel family consists of six sub-families: TRPA (Ankyrin), TRPC (canonical), TRPM (melastatin), TRPML (mucolipin), TRPP (polycystin), and TRPV (vanniloid), encoded by a total of 28 genes [10]. The latter five can be further divided into subtypes: TRPC1-7, TRPM1-8, TRPML1-3,

TRPP1-3, and TRPV1-6 [11]. The broad variety of TRP channels allows sensing both noxious and innocuous signals [12]. Thus far, TRPV1-4 [12] and TRPM3 [13] channels have been implicated in the sensation of noxious heat. TRPA1, TRPC5 and TRPM8 channels have been suggested to detect noxious cold temperatures [14]. In addition, both TRPA1 and TRPV4 subtypes are thought to be involved in the detection of noxious mechanical stimuli [12], while TRPA1, TRPV1, TRPV3, TRPV4, TRPM8, and TRPC3 may contribute to the sensation of itch [15,16].

The aforementioned TRP channel subtypes are expressed in different types of peripheral sensory neurons such as dorsal root ganglion (DRG) and trigeminal ganglion (TG) neurons. TRPV1 channels and TRPM8 channels are mainly expressed on separate sets of neurons. TRPV1 channels can be found in C-fibers, whereas TRPM8 channels can be found on both fiber types transferring noxious signals (A$\delta$- and C-fibers) [17]. Nevertheless, coexpression of TRPM8 and TRPV1 in one DRG neuron has been reported as well [18,19]. TRPA1 and TRPV1, in contrast, are mostly coexpressed in sensory neurons [17,19]. TRPV2 channels have been detected in A$\delta$-fibers [12]. TRPV3 channels can be found on keratinocytes rather than sensory neurons [20], and TRPV4 channels are expressed in variety of tissues including the CNS and peripheral neurons [21]. TRPM3 channels are expressed in a large subset of DRG and trigeminal ganglion neurons. TRPM3 mRNA can be detected in approximately 80% of these sensory neurons at a level that is comparable to that of TRPA1 and TRPV1. However, only small-diameter neurons produce currents mediated by TRPM3 [22].

The role of TRPV1 channels as sensors of noxious heat is well established [23]. In general, TRP channels resemble the structure of voltage-gated $K^+$ ($K_V$) channels. Each channel is made up of four subunits, each having six membrane spanning domains. Similar to $K_V$ channels, transmembrane domains 5 and 6 comprise the channel pore, whereas transmembrane domains 1–4 resemble a voltage sensor. Both N- and C-termini are found on the intracellular side [14] and harbor a number of regulatory domains. Channel trafficking and assembly is regulated by six so-called ankyrin repeats located at the N-terminus [24,25]. TRPV1 channels are not just activated by noxious heat, but also by voltage, binding of vanilloids, such as capsaicin, or high concentrations of $H^+$ ions [10]. As compared to $K_V$ channels, TRPV1 channels display a rather weak voltage sensitivity [26], which can be explained by the fact that the voltage-sensing transmembrane domains 1–4 remain fairly static during activation [26–28]. In addition, transmembrane domain 4 of TRPV1 channels contains a lower number of positively charged amino acids as compared to $K_V$ channels. Hence, an additional voltage-sensing segment might be required for TRP channels [14]. The gating in response to heat is regulated by the so-called TRP domain. However, this process remains incompletely understood [29]. The TRP domain spans 25 amino acid residues and is located immediately adjacent to transmembrane domain 6. It contains the TRP box, a stretch of conserved amino acid residues (WKFQR), which is a hallmark of TRP channels. The TRP domain is thought to be involved in a number of processes, like $PIP_2$ binding or channel assembly, but the exact mechanism still needs to be fully elucidated [24]. As mentioned before, $PIP_2$ is thought to regulate TRPV1 channel function, however it is still under debate if $PIP_2$ is a positive or a negative regulator [30]. Cryo-EM studies in nanodiscs revealed the position of $PIP_2$ in proximity to the vanilloid binding site. Binding of a vanilloid displaces a part of the $PIP_2$ molecule, which reaches into the vanilloid binding pocket. The removal of the phosphoinositide is thought to lead to channel gating [26]. Such an effect would rather point towards a negative regulatory effect of $PIP_2$.

The threshold value for classifying a thermal response as noxious was determined to be 43 °C [2]. TRPV1 channels activate at temperatures that exceed 43 °C, TRPV2 activate at even higher temperatures (>52 °C), whereas TRPV3 and TRPV4 channels gate in a temperature range between 26 °C and 34 °C [20]. Similar to TRPV1 channels, heterologously expressed TRPM3 channels activate at a temperature exceeding 40 °C [22]. Interestingly, mice lacking TRPV1 channels display a delayed nocifensive response only at temperatures exceeding 50 °C [31,32]. However, as compared to TRPV1 channels, the role of the other TRPV channels linked to the detection of noxious signals remains incompletely understood [12]. The role of TRPV2 to TRPV4 channels in detecting noxious signals remains debated [13], since both TRPV2 knock-out [33] and TRPV3/TRPV4 double knock-out [34] animals retain normal thermal and mechanical sensation. The nocifensive response times of TRPM3 knock-out mice is prolonged at temperatures exceeding 52 °C [22]. Mice lacking both TRPV1 and TRPM3 show a significantly increased nocifensive response time already at 45 °C. However, some sensory neurons still produce currents in response to heat. Only a triple knock-out of TRPV1, TRPM3 and, interestingly, TRPA1 leads to a complete heat-insensitivity of sensory neurons. Furthermore, these mice were completely heat insensitive in behavioral tests [35].

The detection of both noxious and innocuous cold signals is suggested to involve TRPM8 and TRPA1 channels [23]. TRPM8 channels gate at temperatures below 25 °C [20]. Knock-out animals of TRPM8 channels lose the ability to detect cool temperatures, but retain the ability to detect noxious cold signals below 15 °C [13]. Hence, the role of TRPM8 channels as cold sensors is well established, but an additional set of ion channels needs to be involved in detecting noxious cold temperatures. TRPA1 channels are thought to be involved, but their role remains controversial [20]. Rodent TRPA1 channels were found to be gated by noxious cold temperatures, however, that function is lost in primate TRPA1 channels [36]. By contrast, human TRPA1 channels, reconstituted in lipid bilayers, were found to be activated by noxious cold temperatures [37]. In addition to these conflicting results, TRPA1 channels are usually expressed on the same set of neurons as TRPV1 channels, which appears counterintuitive [17]. Furthermore, animal studies involving TRPA1 knock-out mice point towards an insignificant role in the detection of noxious cold temperatures [20]. While their role in the detection of mechanical stimuli remains controversial as well, their contribution to the detection of noxious chemical signals is well established [12]. A large number of structurally unrelated electrophilic compounds can gate TRPA1 channels [13]. These compounds covalently modify one or more of the 31 cysteine residues, which causes channel opening [38].

GPCR Regulation of TRP Channels

TRPV1 channels have been studied extensively for their modulation by GPCRs. Currents through TRPV1 channels are increased in response to inflammation, which mediates an enhanced depolarization and increased excitability [2]. The sensitization of TRPV1 channels can be mediated by both $G\alpha_q$- and $G\alpha_s$-coupled receptors. Stimulation of a $G\alpha_q$-coupled receptor leads to activation of phospholipase C (PLC), which hydrolyzes membrane bound phosphatitylinositol 1,4, bisphosphate ($PIP_2$) into soluble inositol 1,4,5 trisphosphate ($IP_3$) and membrane bound diacylglycerol (DAG, Figure 1). Subsequently, $IP_3$ binds to $IP_3$ receptors located at the membrane of the endoplasmic reticulum, which triggers the release of $Ca^{2+}$. DAG in turn activates protein kinase C, which phosphorylates target proteins. Every step of this cascade can interfere with the function of TRPV1 channels [39]. Presence of $PIP_2$ in the membrane is thought to decrease TRPV1 channel function by interfering with agonist binding [26]. If $PIP_2$ is depleted from the membrane in response to PLC activation, TRPV1 activity may increase [30]. The exact role of $PIP_2$ remains debated, as it was also shown to activate TRPV1 channels [30]. Activated PKC was shown to phosphorylate two serine residues at the C-terminus, which is thought to mediate sensitization [40]. A rise in cytosolic $Ca^{2+}$ is not considered to contribute to sensitization as it usually leads to rapid channel desensitization in response to prolonged activation [39]. A large number of inflammatory modulators was shown to increase TRPV1 channels via one of these mechanisms (Table 1).

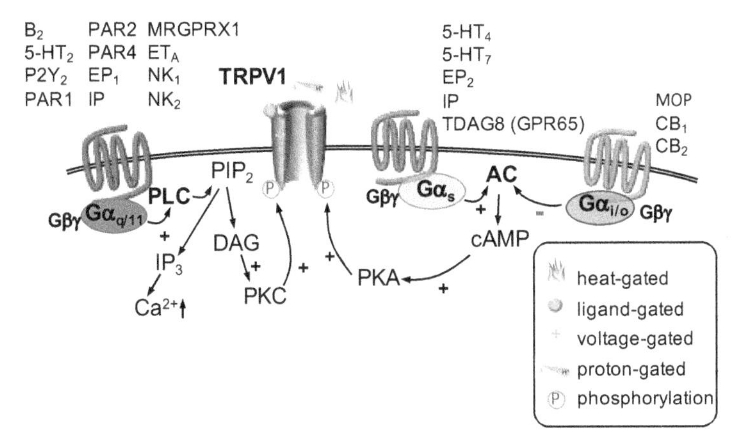


**Figure 1.** TRPV1 channels can be gated by different mechanisms (as indicated). Three major G-protein-dependent pathways modulate the function of TRPV1 channels. Activation of $G\alpha_{q/11}$-coupled receptors (**left**) leads to activation of phospholipase C (PLC), which hydrolyzes phosphatitylinositol 1,4, bisphosphate ($PIP_2$) into inositiol 1,4,5 trisphosphate ($IP_3$) and diacylglycerol (DAG). DAG activates protein kinase C (PKC), which phosphorylates TRPV1 channels, thereby increasing their function. Activation of a $G\alpha_s$-coupled receptor (**center**) stimulates adenylyl cyclase (AC), which produces cyclic adenosine monophosphate (cAMP). Subsequent activation of protein kinase A (PKA) leads to phosphorylation of TRPV1 channels and an increase in current. Stimulation of a $G\alpha_{i/o}$-coupled receptor (**right**) decreases AC activity. Therefore, less cAMP is formed, PKA is less active and hence TRPV1 channels are not phosphorylated which decreases their activity.

**Table 1.** GPCRs modulating TRPV1 function.

| GPCR Ligand | Involved GPCR | Pathway | Effect on TRPV1 | Reference |
|---|---|---|---|---|
| Bradykinin | $B_2$ | $G\alpha_q$-DAG-PKC | increased current | [2,41] |
| Serotonin | 5-$HT_2$ | $G\alpha_q$-DAG-PKC | increased current | [42–44] |
| | 5-$HT_4$ | $G\alpha_s$-AC-PKA | increased current | [43] |
| | 5-$HT_7$ | $G\alpha_s$-AC-PKA | increased current | [42] |
| UTP | $P2Y_2$ | $G\alpha_q$-DAG-PKC | increased current | [45–47] |
| BAM 8-22 | MRGPRX1 | $G\alpha_q$-DAG-PKC-$PIP_2$ | increased current | [48] |
| Proteases | PAR2 | $G\alpha_q$-PKC | increased current | [49,50] |
| | PAR1 | $G\alpha_q$-PKC | increased current | [51] |
| | PAR4 | $G\alpha_q$-PKC | increased current | [51] |
| $PGE_2$ | $EP_1$ | $G\alpha_q$-PKC | increased current | [52] |
| | $EP_2$ | $G\alpha_s$-PKA | increased current | [52–54] |
| $PGI_2$ | IP | $G\alpha_q$-PKC | increased current | [52] |
| | IP | $G\alpha_s$-PKA | increased current | [52] |
| Endothelin-1 | $ET_A$ | $G\alpha_q$ -PKC | increased current | [55,56] |
| Substance P | $NK_1$ | $G\alpha_q$ -PKC$\epsilon$ | increased current | [57] |
| | $NK_2$ | $G\alpha_q$-PKC | increased current | [58] |
| $H^+$ | TDAG8 (GPR65) | $G\alpha_s$-PKA | increased current | [59] |
| Morphine | MOP | $G\alpha_i$-reduced AC | decreased current | [60,61] |
| Endocannabinoids | $CB_1$ | $G\alpha_i$-reduced AC | decreased current | [62] |
| | $CB_2$ | $G\alpha_i$-reduced AC | decreased current | [63] |

AC, adenylyl cylcase; DAG, diacylglycerol; PKA, proteinkinase A; PKC, protein kinase C; BAM 8–22, bovine adrenal medulla peptide 8–22.

Activation of a $G\alpha_s$-coupled receptor stimulates the activity of adenylyl cyclase, which produces cyclic adenosine monophosphate (cAMP). This nucleotide is needed to activate protein kinase A (PKA), which then phosphorylates its target proteins. PKA-mediated phosphorylation of TRPV1 channels increases their sensitivity towards their agonists and reduces $Ca^{2+}$-mediated desensitization [54]. Several inflammatory mediators were found to sensitize TRPV1 channels utilizing this pathway (Table 1).

By contrast, activation of a $G\alpha_i$-coupled receptor decreases the activity of adenylyl cyclase which reduces the abundance of cAMP and subsequent activation of PKA (Table 1). Indeed, activation of $G\alpha_i$-coupled cannabinoid [62,63] and $\mu$-opioid (MOP) [60,61] receptors was shown to reduce currents of TRPV1 receptors, which is thought to contribute to the peripheral analgesic action of opioids [60,61] and cannabinoids [62,63].

In addition to the three major GPCR pathways, TRPV1 channels were shown to be sensitized by nerve growth factor (NGF), which requires the early activation of PI3 kinase and the presence of PKC and CamKII ($Ca^{2+}$/calmodulin dependent protein kinase II) [64]. The inflammatory mediator histamine sensitizes TRPV1 channels via $G\alpha_q$-coupled $H_1$ receptors. Instead of utilizing the signaling cascade described above, histamine-mediated sensitization requires activation of phospholipase A2 and lipoxigenases [65–67].

In sensory neurons, a variety of $G\alpha_{i/o}$-coupled receptors were found to inhibit currents through TRPM3 channels: including $GABA_B$ receptors [68–70], $\mu$-opioid receptors [68,69], somatostatin receptors [68,70], $CB_1$- [69], as well as, $CB_2$ receptors [68], and neuropeptide Y receptors [69]. Likewise, low concentrations of noradrenaline reduce TRPM3 activity, hinting towards $\alpha_2$ as mediating receptor. However, the adrenergic receptor involved was not further characterized [68]. Whether $\delta$-opioid receptors also mediate a TRPM3 inhibition remains controversial: while deltorphin, a $\delta$-selective peptide was able to reduce TRPM3 function [68], the small-molecule $\delta$-selective agonist SB205607 was not [69]. Likewise, activation of $G\alpha_{i/o}$-coupled metabotropic glutamate receptors ($mGluR_{4/6/7/8}$) did not reduce TRPM3 function [69]. In a heterologous system, $G\alpha_{q/11}$-coupled $M_1$ receptors were found to inhibit currents through TRPM3 channels [70]. However, in sensory neurons, activation of $G\alpha_{q/11}$-coupled $mGluR_5$ only weakly inhibited TRPM3 [68]. The inhibition of TRPM3 in sensory neurons involves activation pertussis toxin (PTX)-sensitive $G\alpha_{i/o}$-coupled receptors [68–70]. The effect did not require signaling downstream of $G\alpha_{i/o}$ activation, but relied on a direct interaction with the $\beta\gamma$ dimer [68–70].

TRPA1 channels are sensitized by $G\alpha_s$- and $G\alpha_q$-coupled receptors in sensory neurons. A $G\alpha_i$-mediated interaction has not been reported for sensory neurons. Activation of $G\alpha_q$-coupled PAR2 receptors increased currents mediated by TRPA1 receptors in dorsal root ganglion neurons. This interaction required the activation of PLC but none of the downstream products. Consequently, depletion of $PIP_2$ from the membrane was shown to be sufficient for this interaction [12]. The inflammatory mediator bradykinin was shown to increase currents through TRPA1 channels in dorsal root ganglion neurons. This effect was mediated by $G\alpha_q$-coupled $B_2$ receptors and required the activation of PLC. Interestingly, activation of PKA was further required for the interaction of $B_2$ receptors and TRPA1 channels [71]. An interaction of $G\alpha_q$-coupled bradykinin $B_1$ receptors with TRPA1 channels was reported in behavioral experiments. This interaction relied on activation of PLC and PKC [72]. Histamine was shown to cause nocifensive behavior in a TRPA1 dependent manner. It is thought to involve $G\alpha_q$-coupled $H_1$ receptors and activation of PLC [73]. Adenosine, another component of the inflammatory soup, was found to sensitize esophageal C-fibers and increase TRPA1 currents via $G\alpha_s$-coupled $A_{2A}$ receptors. Activation of PKA is necessary for this interaction [74]. Electrophilic metabolites of prostaglandins, however, were demonstrated to activate TRPA1 channels directly [75,76].

The inflammatory mediators prostaglandin E2 ($PGE_2$), bradykinin and histamine were tested for their influence on TRPM8 channel function. As opposed to the previously described members of the TRP channel family, the activity of the cool sensor TRPM8 is reduced in the presence of bradykinin and $PGE_2$. However, application of histamine did not interfere with TRPM8 channel function [77]. The action of bradykinin required the mobilization of PKC [77,78], whereas $PGE_2$ involved activation of PKA [77]. Bradykinin is assumed to act via $G\alpha_q$-coupled $B_2$ receptors [79] and several modes of action have been suggested: it was found that depletion of $PIP_2$ from the plasma membrane reduced heterologously expressed TRPM8 channel function [80]. However, these experiments were performed in absence of GPCRs and it remains to be established if $PIP_2$ depletion is also sufficient to reduce TRPM8 channel function in a native cell system. $PIP_2$ is hydrolyzed to form $IP_3$ and DAG, which is required to activate PKC. PKC is thought to activate protein phosphatase I (PPI), which is suggested to dephosphorylate TRPM8 channels. The dephosphorylation of TRPM8 channels is proposed to finally inhibit TRPM8 channel function [79]. More recently, both bradykinin and histamine were found to inhibit TRPM8 channels via a direct interaction of $G\alpha_q$ subunits with the channels. The inhibitory effect of both mediators did not require activation of PLC or any of the subsequent steps in the signaling cascade [81]. The receptor via which $PGE_2$ exerts its effect has not been determined, however it was found that activation of a $G\alpha_s$-coupled receptor and subsequent PKA stimulation was required. Furthermore, the exact mechanism how PKA modulates TRPM8 channel function remains unknown [79].

### 2.2. *Acid-Sensing Ion Channels*

Acid-sensing ion channels (ASIC) represent one of many ion channel families that detect noxious chemical stimuli. Additional chemical sensors are TRP channels, namely TRPA1 and TRPV1, as well as ATP-gated P2X receptors [82]. As the name suggests, acid sensing ion channels are activated in low pH conditions [83]. Such acidic conditions occur during an inflammatory response, ischemia, or fatiguing exercise [82]. ASICs can be divided into three subtypes ASIC1 to ASIC3. ASIC1 and ASIC2 can even be further subclassified into two splice variants each (ASIC1a, ASIC1b; ASIC2a, ASIC2b) [84]. A fourth analog, sometimes referred to as ASIC4 [85], rather affects expression levels of ASIC1a and ASIC3, instead of producing proton-gated currents [84]. The $EC_{50}$ for proton-mediated currents via ASIC1 and ASIC3 channels ranges between a pH of 6.2 to 6.8, whereas ASIC2 channels have an $EC_{50}$ between pH 4.1 and 5 [84]. All forms of ASICs can be detected in somata and peripheral ends of sensory neurons [85]. ASIC1a and ASIC3 channels are preferentially expressed in small diameter DRG neurons which also express TRPV1 and most likely subserve nociceptive function [86]. A functional channel is composed of three subunits and all but one subunit can participate in both homo- and heteromeric channels. Only ASIC2b does not form functional homomeric channels [87]. One subunit consists of two transmembrane domains, having both the N- and C-termini at the intracellular side [88,89]. Most of the protein is located at the extracellular side forming the large extracellular domain (termed ECD). The structure of the extracellular domain was compared to a hand holding a ball, which explains the peculiar terminology for parts of the ECD, such as palm, knuckle, thumb, finger and $\beta$-ball [90]. The ECD contains a number of regulatory domains: for example, an acidic pocket is formed by acidic amino acid residues at the subunit–subunit interface, which is involved in binding of $H^+$ ions and subsequent gating [89]. ASIC channels follow a three-state kinetic model, from a closed to an activated to an inactivated state. The recovery from inactivation can only be achieved in high pH conditions [88] and this desensitized state is thought to be regulated by the thumb domain within the ECD [90].

Genetic studies have suggested that ASICs play a role in sensing mechanical signals, but the exact gating mechanism is unknown, and their role remains heavily debated [91].

#### GPCR Regulation of ASICs

A few components of the inflammatory soup have been tested for their modulatory effect on acid-sensing ion channels: histamine was shown to selectively potentiate heterologously expressed ASIC1a channels. This process involved a direct action of histamine and did not require the presence of a GPCR [92]. The nucleotides UTP and ATP were shown to increase acid-induced currents (Figure 2) in rat dorsal root ganglion neurons as well as acid-induced membrane excitability. In this respect,

UTP was found to act via $G\alpha_q$-coupled $P2Y_2$ receptors and required the activation of PLC, subsequent stimulation of PKC and the presence of the anchoring protein PICK-1 (protein interacting with C-kinase 1) [93]. A similar effect can be observed when serotonin is applied: both ASIC-mediated currents and neuronal excitability are increased [94]. Serotonin was found to act via $G\alpha_q$-coupled 5-$HT_2$ receptors [94] and required activation of PKC [94,95]. Two phosphorylation sites, one at the N-terminus and one at the C-terminus of ASIC3, need to be phosphorylated for the full effect. Again, PICK-1 is necessary for PKC-mediated phosphorylation of ASIC channels. This scaffold protein is thought to bind to ASIC2b subunits in heterotrimeric channels and to link PKC to the channel and to enable phosphorylation [95]. Another $G\alpha_q$-coupled receptor, PAR2, was found to increase ASIC-mediated currents in rat pulmonary sensory neurons. Interestingly, neither PLC nor PKC were required for the PAR2 mediated current increase but the pathway involved was not studied further [96]. Depending on the activation mechanism of PAR2, one may observe an increase of cytosolic $Ca^{2+}$ following the $G\alpha_q$-dependent activation of PLC and formation of $IP_3$. On the other hand, PAR2 activation was shown to signal also via $G\alpha_{12/13}$ proteins, which activate Rho kinase and lead to ERK phosphorylation. Additionally, PAR2 activation may lead to $\beta$-arrestin recruitment. Whether PAR2 activation may also decrease or increase cAMP levels remains controversial [97].

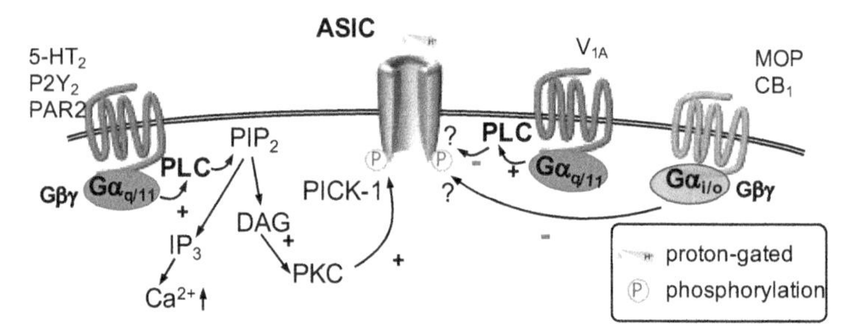


**Figure 2.** Acid sensing ion channels (ASIC) are gated by increasing concentrations of $H^+$. Two G-protein pathways modulate the function of ASICs. Activation of $G\alpha_{q/11}$-coupled receptors (**left**) leads to activation of phospholipase C (PLC) which hydrolyzes phosphatitylinositol 1,4, bisphosphate ($PIP_2$) into inositiol 1,4,5 trisphosphate ($IP_3$) and diacylglycerol (DAG). DAG activates protein kinase C (PKC) which phosphorylates ASICs increasing their function. The interaction of PKC with ASICs requires the presence of PICK-1 (protein interacting with C-kinase 1). By contrast, activation of $G\alpha_{q/11}$-coupled $V_{1A}$ receptors (**center**) were shown to decrease ASIC-mediated currents via an unknown mechanism. Stimulation of $G\alpha_{i/o}$-coupled receptors (**right**) was shown to decrease currents through ASICs involving an undetermined mechanism.

By contrast, activation of $G\alpha_{i/o}$-coupled receptors was shown to decrease currents through ASIC channels. First, stimulation of cannabinoid $CB_1$ receptors was found to reduce nocifensive behavior triggered by local acidosis which relies on an interaction between $CB_1$ receptors and ASIC channels [98]. Second, activation of $\mu$-opioid receptors was shown to decrease ASIC-mediated currents, neuronal excitability in dorsal root ganglion neurons and nocifensive behavior induced by local acidification [99]. In addition, nocifensive behavior provoked by mechanical stimuli [100] or local acidification [101] is reduced by local application of oxytocin. Oxytocin reduces ASIC-mediated currents in dorsal root ganglion neurons via activation of vasopressin $V_{1A}$ receptors. This effect was found to be dependent on $G\alpha_q$ activation, but further steps of the cascade were not tested [101]. It remains to be determined if the observed differences in $G\alpha_q$-mediated effects on ASIC channels, for example, depend on the recruitment of PICK-1.

### 2.3. Mechanosensitive Channels in Pain Sensation

The variety of mechanical stimuli detected by so-called mechanosensors in the sensory nervous system ranges from light to noxious mechanical stimuli. These specialized neurons express mechanotransducer channels [102]. To sense noxious mechanical stimuli, high-threshold mechanosensors are required, which should express ion channels that open in response to strong mechanical stimuli and lead to a depolarization of these neurons [103]. Acid sensing ion channels (as described above) were suggested to act as mechanotransducer channels in *C. elegans*. However, heterologously expressed mammalian ASICs do not gate in response to mechanical stimuli. Hence, such depolarizing mechanotransducer channels in nociceptive neurons remain to be identified [103].

One family of ion channels that contributes to the sensation of noxious mechanical stimuli is the family of two-pore $K^+$ channels ($K_{2P}$) [103]. The family of $K_{2P}$ channels consist of 15 members, which usually provide so-called background or leak currents, which are the major contributors to the resting membrane potential [11]. Three members of the $K_{2P}$ family were found to be involved in sensing noxious mechanical stimuli: $K_{2P}$2.1 (TREK1), $K_{2P}$4.1 (TRAAK), and $K_{2P}$10.1 (TREK2) [104]. A functional $K_{2P}$ channel is formed by two subunits consisting of four transmembrane segments each (TMS1–4). Both the N- and C-termini are located in the intracellular space and the linker regions between TMS1 and -2, as well as TMS3 and -4 are located inside the plasma membrane to form one selectivity filter each [105]. Accordingly, a total of eight transmembrane domains and four selectivity filter regions line the ion conduction pore. This structure of the pore is highly homologous to that of voltage-gated $K^+$ channels [104]. In closed conformation, the ion conduction pathway is blocked by lipid acyl side chains and membrane stretch directly gates $K_{2P}$ channels [104]. $K_{2P}$2.1 and $K_{2P}$4.1 channels are strongly expressed in small-diameter DRG neurons and only weakly expressed in medium- and large-diameter DRG neurons, whereas $K_{2P}$10.1 channels are exclusively expressed in small-diameter neurons [106]. Interestingly, knock-out of these channels leads to an increased nocifensive response to mechanical stimuli [107]. Since all these channels are selective $K^+$ channels, opening of $K_{2P}$ channels leads to an efflux of $K^+$ ions and subsequent hyperpolarization. It is thought that stretch-activation of $K_{2P}$ channels counteracts the activation of depolarizing mechanotransducer channels and thereby finetunes the mechanically induced nociceptive signal which is transferred to the brain [102]. Hence, it is clear that these three members of the $K_{2P}$ family contribute to the perception of noxious mechanical stimuli, but they cannot represent the primary depolarizing mechanotransducer channel [103]. Such a functional entity is rather provided by Piezo channels, in particular Piezo2, which contributes to mechanically activated currents in DRG neurons [108]. However, deletion of Piezo2 impairs touch, but sensitizes mechanical pain in mice [109]. Therefore, additional sensors of mechanical pain remain to be identified.

#### 2.3.1. GPCR Regulation of Mechanosensitive Potassium Channels

The function of mechanosensitive $K_{2P}$ channels can be adjusted by a number of modulators like arachidonic acid, polyunsaturated fatty acids, glutamate, noradrenaline, acetylcholine, TRH [105] or serotonin [107] (Figure 3). Arachidonic acid and polyunsaturated fatty acids activate these channels directly [105]. The other modulators influence $K_{2P}$ channel function via activation of GPCR pathways. All three major GPCR pathways affect $K_{2P}$2.1 and $K_{2P}$10.1 channel activity: phosphorylation of two different C-terminally located serine residues either by cAMP activated PKA or PKC leads to an inhibition of both subtypes. Phosphorylation of yet another serine residue by protein kinase G (PKG) on the other hand activates $K_{2P}$2.1 and $K_{2P}$10.1 channels. PKG is activated by an increase of cyclic guanosine monophosphate (cGMP) which in turn is formed by soluble guanylyl cyclase. Soluble guanylyl cyclase is directly activated by nitric oxide and does not involve activation of a GPCR [105]. In dorsal root ganglion neurons, prostaglandin $F_{2a}$ ($PGF_{2a}$) was shown to decrease $K_{2P}$ mediated currents [106]. The exact coupling mechanism was not elucidated, however, $PGF_{2a}$ is the endogenous ligand for $G\alpha_q$-coupled FP prostanoid receptors [110] and it is likely that PKC activation is involved in this process. In addition, prostaglandin $E_2$ ($PGE_2$)-induced nocifensive behavior was reduced in $K_{2P}10.1$ knock-out mice, but the mechanism of action was not investigated [111].

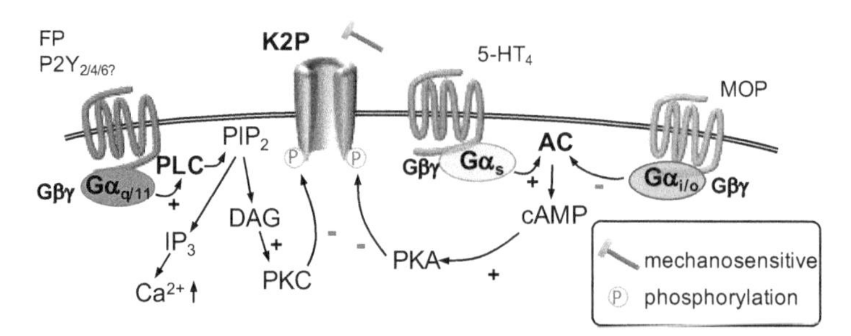


**Figure 3.** $K_{2P}$ channels are opened in response to mechanical stimuli. Stimulation of a $G\alpha_{q/11}$-coupled receptor (**left**) activates phospholipase C (PLC). Hydrolyzation of phosphatitylinositol 1,4, bisphosphate ($PIP_2$) forms inositiol 1,4,5 trisphosphate ($IP_3$) and diacylglycerol (DAG). DAG activates protein kinase C (PKC) which phosphorylates $K_{2P}$ channels decreasing their function. Activation of $G\alpha_s$-coupled receptors (**center**) leads to activation of adenylyl cyclase (AC) which produces cyclic adenosine monophosphate (cAMP). Thereafter, protein kinase A (PKA) is activated which phosphorylates $K_{2P}$ channels and decreases their currents. Stimulation of $G\alpha_{i/o}$-coupled receptors (**right**) decreases AC activity. Therefore, less cAMP is formed, PKA is less active and subsequently $K_{2P}$ channels are not phosphorylated, which increases their activity.

Activation of a $K^+$ permeable ion channel leads to an efflux of $K^+$ ions and a subsequent hyperpolarization. $K_{2P}$ channels are active at resting conditions and contribute to the formation of the resting membrane potential. Inhibition of these channels leads to a depolarization and a subsequent increase in excitability [107]. Other components of the inflammatory soup were examined for their effects on $K_{2P}$ channels: for example, serotonin was also found to inhibit $K_{2P}2.1$ and $K_{2P}10.1$ channels via activation of 5-$HT_4$ receptors in a heterologous cell system [112]. These GPCRs are $G\alpha_s$-coupled and activate PKA, which is thought to mediate this effect [107]. Application of UTP, which may act through $G\alpha_q$-coupled $P2Y_2$, $P2Y_4$ or $P2Y_6$ receptors, leads to an inhibition of $K_{2P}$ channels in mammary epithelial cells [113]. It remains to be determined if these inflammatory mediators also interact with mechanosensitive $K_{2P}$ channels in dorsal root ganglion neurons.

On the other hand, activation $\mu$-opioid receptors were found to increase $K_{2P}2.1$ currents in hippocampal astrocytes [114], in a heterologous cell system [115], and in substantia gelatinosa neurons of the spinal cord. The latter mechanism is thought to be involved in the antinociceptive actions of opioids [116]. All opioid receptors are coupled to $G\alpha_{i/o}$ G-proteins, which reduce the activity of adenylyl cyclase [110]. Subsequently, less cAMP is formed, which in turn leads to reduced PKA activity and less PKA-mediated phosphorylation of $K_{2P}$ channels [105]. Since opioid receptors are also expressed in peripheral sensory neurons [117], such an interaction between $\mu$-opioid receptors and $K_{2P}$ channels might also exist in peripheral sensory neurons and contribute to opioid-mediated antinociception.

#### 2.3.2. GPCR Regulation of Piezo Channels

Mechanically activated and rapidly adapting currents in DRG neurons are carried by Piezo2 channels [108] and get sensitized by the activation of $B_2$ bradykinin receptors, an effect that appears to involve PKA as well as PKC [118]. Akin currents are enhanced in the presence of ATP and UTP which act most likely through an activation of $P2Y_2$ receptors [119]. Likewise, mechanically activated and rapidly adapting currents in DRG neurons as well as currents in cells expressing recombinant Piezo2 channels are enhanced by intracellular GTP or GTP$\gamma$S which both lead to the activation of G proteins [120]. This confirms that the function of mechanosensitive Piezo2 channels can be enhanced by inflammatory mediators acting on GPCRs. Interestingly, Piezo2 channels can be inhibited by a depletion of membrane $PIP_2$, and this mechanism is believed to underlie the reduction of mechanically activated rapidly adapting currents in DRG neurons in response to an activation of TRPV1 by capsaicin [121]. Why such an effect cannot be observed in DRG neurons during the activation of $G\alpha q$-linked GPCRs, such as $B_2$ bradykinin and $P2Y_2$ receptors [118,119], remains open for future investigation.

### 2.4. Calcium-Activated Chloride Channels in Pain Sensation

Calcium-activated chloride (CaCC) channels occur in a variety of tissues. They are widely expressed in the nervous system but also other tissues like vascular smooth muscle cells. In addition, CaCCs are used as a marker for stroma tumors in the gastrointestinal tract and were initially termed DOG1 [122]. Functionally described CaCCs were linked to transmembrane proteins of unknown function 16A and B (TMEM16) [123–125]. Thereafter, the term anoctamin was introduced to account for its predicted eight (lat. octo) membrane spanning domains and its function as an anion channel [122,126]. The family of TMEM16 proteins consists of ten members termed TMEM16A to TMEM16K (I is left out), which correspond to anoctamin 1 to anoctamin 10 (ANO1 to ANO10) [127]. Only ANO1 and ANO2 were unequivocally identified as mediating calcium-activated anion currents, the other members were either identified as calcium-activated lipid scramblases or as dual-function scramblases/ ion channels [128,129]. The subtypes ANO1, ANO2 and ANO3 were found to be expressed in sensory neurons [130,131]. ANO1 was only detected in TRPV1 positive neurons [132], whereas about 50% of ANO3 positive neurons were also TRPV1 positive [131]. Both ANO1 [133] and ANO3 [131] are involved in nociceptive behavior in mouse models of inflammatory pain. Cyro-EM- [134–136] and X-ray studies [137] revealed that TMEM16 analogs actually consist of ten transmembrane domains. TMEM16A forms homodimers [138,139], where transmembrane domains 3–7 of both subunits form one separate ion conduction pathway creating a two-pore anion channel [134–136]. Calcium, needed for gating, binds to two regions of negatively charged amino acid residues in transmembrane domains 6–8. Subsequently, transmembrane domain 6 is displaced, which ultimately opens the channel [134,135]. In the absence of $Ca^{2+}$, CaCC function is not altered by changes in membrane voltage within physiological limits. Only voltages exceeding 100 mV may gate CaCCs directly if $Ca^{2+}$ is absent [140], presence of cytosolic $Ca^{2+}$ merely shifts the voltage-dependence to more physiological levels [141]. In sensory neurons, $Ca^{2+}$ can either rise in response to activation of a $G\alpha_q$-coupled receptor and subsequent release from intracellular stores [142], or due to an influx of extracellular $Ca^{2+}$ via $Ca^{2+}$-permeable ion channels like TRPV1 channels [143], or to a lesser extend voltage-gated $Ca^{2+}$ channels [142].

Mice lacking TRPV1 receptors, the canonical heat sensor, retain some sensitivity to noxious heat and CaCCs were suggested to fulfill that task [122]. Indeed, heterologously expressed TMEM16A channels produced currents at temperatures that exceeded 44 °C [144] and tissue specific knock-out reduced mechanically and thermally induced nocifensive behavior [132,133].

#### GPCR Regulation of Calcium-Activated Chloride Channels

As described above, there are three possible sources for $Ca^{2+}$ to activate CaCCs. One of these possibilities is to activate a $G\alpha_q$-coupled receptor and the subsequent signaling cascade (Figure 4). In dorsal root ganglion neurons, the inflammatory mediator bradykinin was demonstrated to increase neuronal excitability via gating of CaCCs which leads to a depolarizing $Cl^-$ efflux due to comparably high intracellular $Cl^-$ concentrations. The induction of $Cl^-$ currents through CaCCs by bradykinin required activation of PLC, formation of $IP_3$ and an increase of cytosolic $Ca^{2+}$ levels [145]. In addition, it was also shown that activation of proteinase-activated receptor 2 (PAR2) induced currents through CaCCs in dorsal root ganglion neurons. This action is dependent on increasing levels of cytosolic $Ca^{2+}$ and a close proximity of $IP_3$ receptors, which are located on the membrane of the endoplasmic reticulum, to ANO1 channels, which are located at the plasma membrane [142]. Furthermore, expression of both TMEM16A (ANO1) and PAR2 is induced in a model of neuropathic pain and both proteins are co-expressed in the same set of dorsal root ganglion neurons [146]. The inflammatory mediator serotonin was shown to induce currents mediated via CaCCs in dorsal root ganglion neurons. Even though all three types of 5-$HT_2$ receptors are expressed on small-diameter dorsal root ganglion neurons, only activation of 5-$HT_{2C}$ receptors was able to induce such currents. Furthermore, the according increase in excitability also required activation of TRPV1 channels, which provides an additional $Ca^{2+}$ source [44]. Other possible mediators for sensitization include nucleotides: Both, UTP and ATP were found to interact with CaCCs via activation of $G\alpha_q$-coupled $P2Y_1$ and $P2Y_2$ receptors. However, this interaction was only investigated in kidney [147] and pancreatic cells [148]. It remains to be established if such an interaction also contributes to nucleotide-mediated sensitization of sensory neurons.

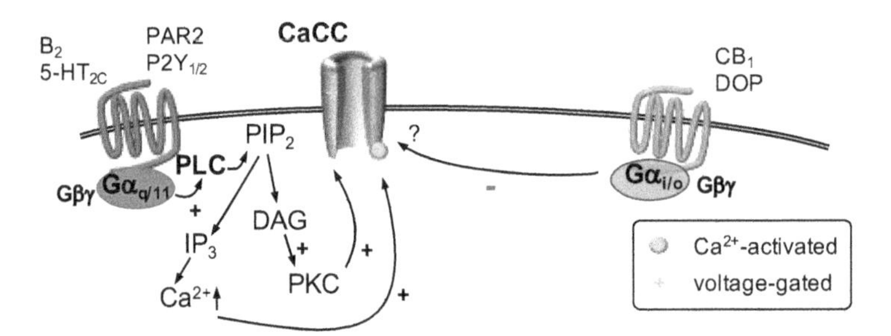


**Figure 4.** Calcium-activated $Cl^-$ channels (CaCC) are gated by increasing concentrations of cytosolic $Ca^{2+}$ and influenced by membrane voltage. The source of $Ca^{2+}$ may be $Ca^{2+}$-permeable ion channels located in proximity to CaCCs (not shown) or $Ca^{2+}$ released from intracellular stores. Stimulation of $G\alpha_{q/11}$-coupled receptors (**left**) activates phospholipase C (PLC). The subsequent hydrolysis of phosphatitylinositol 1,4, bisphosphate ($PIP_2$) forms inositiol 1,4,5 trisphosphate ($IP_3$) and diacylglycerol (DAG). DAG activates protein kinase C (PKC) which influences CaCC function. $IP_3$ binds to $IP_3$ receptors located in the membrane of the endoplasmic reticulum (ER) which causes $Ca^{2+}$ release from the ER. Activation of $G\alpha_{i/o}$-coupled receptors (**right**) was shown to decrease CaCC currents via an unknown mechanism.

In animal experiments, it was determined that activation of $G\alpha_i$ coupled cannabinoid $CB_1$ receptors may contribute to peripheral antinociception via an interaction with CaCCs [149]. In addition, central antinociception via activation of $G\alpha_i$-coupled $\delta$-opioid (DOP) receptors may involve an interaction with CaCCs [150]. However, it remains unclear how $G\alpha_i$-coupled receptors interfere with CaCC function.

### 2.5. Voltage-Gated $Na^+$ Channels

Voltage-gated $Na^+$ channels ($Na_V$) are crucial not just for excitable cells like central or peripheral neurons, skeletal or cardiac muscle cells, but also occur in immune cells, which are considered non-excitable [151]. In excitable cells, the principal role of $Na_V$ channels is generating action potentials. Action potentials are generated, if a sufficient number of $Na_V$ channels are activated in response to a local depolarization. As opposed to local depolarizations which can only spread over a few millimeters, action potentials can travel along several meters and thus transfer the information to the central nervous system. The stronger a local depolarization is, for example in response to a noxious stimulus, the more action potentials are triggered [2]. If $Na_V$ channels are rendered non-functional, action potentials cannot be evoked and information transfer to the central nervous system is stopped [152–155]. This mechanism is highlighted by patients carrying loss-of-function mutations in their $Na_V1.7$ or $Na_V1.9$ genes, who experience insensitivity towards painful stimuli [156].

To date, nine pore-forming $\alpha$-subunits of $Na_V$ channels are described, designated $Na_V1.1$ to $Na_V1.9$ [11]. Only $Na_V1.6$ to $Na_V1.9$ channels can be found in nociceptive neurons. A functional voltage-gated $Na^+$ channel is formed by a pore-forming $\alpha$-subunit and one additional auxiliary $\beta$-subunit, of which four ($\beta1$ to $\beta4$) have been described. An $\alpha$-subunit is composed of 24 transmembrane segments, which can be grouped into four domains (DI–DIV) of six transmembrane segments each (S1–S6) [151]. The transmembrane segments S1–S4 of each domain form the voltage-sensor, whereas all four S5 and S6 segments contribute to the channel pore. The S4 segments are of particular interest as they harbor a number of positively charged amino acid residues. These residues move the entire S4 segment upwards upon depolarization and lead to the gating of these ion channels. This basic principle of activation is conserved among all members of voltage-gated ion channels [157]. Voltage-gated $Na^+$ channels activate within a fraction of a millisecond and subsequently enter a fast-inactivated state. This inactivation is mediated by the intracellularly located DIII-DIV linker [151]. The subunits $Na_V1.5$, $Na_V1.8$ and $Na_V1.9$ have a low affinity for tetrodotoxin (TTX) as it ranges from 10 to 100 µM. The affinity for all other subtypes ranges between 1 to 10 nM [11]. The former subtypes are therefore described as TTX-insensitive and this represents a simple experimental tool to distinguish between $Na_V$s relevant for nociception and those that are not relevant for nociception [151]. In addition to the previously described channelopathies leading to pain-insensitive patients, the importance of voltage-gated $Na^+$ channels for nociception is highlighted by the fact that the most widely used local anesthetic drug, lidocaine, leads to a use-dependent block of these channels, which prevents the propagation of painful stimuli [151].

### 2.6. GPCR Regulation of Voltage-Gated $Na^+$ Channels

The activity of TTX-resistant voltage-gated $Na^+$ channels is affected by activation of $G\alpha_s$ coupled receptors (Figure 5). Direct activation of adenylyl cyclase by forskolin increases TTX-resistant $Na^+$ currents in dorsal root ganglion neurons. Activation of adenylyl cyclase increases the formation of cAMP, which wash shown to be involved in this process [158]. In sensory neurons, the inflammatory mediators serotonin [158], $PGE_2$ [159,160] and CGRP [161] were found increase TTX-resistant $Na^+$ currents via a mechanism involving $G\alpha_s$-coupled 5-$HT_4$ [162,163], $EP_4$ [164], and CGRP [161] receptors, respectively. Application of substance P leads to an increase in neuronal excitability in small diameter dorsal root ganglion neurons [165]. Substance P activates neurokinin 1 ($NK_1$) receptors, amongst others [110]. These $G\alpha_s$- and $G\alpha_q$-coupled receptors were found to increase TTX-resistant $Na^+$ currents in dorsal root ganglion neurons in a PKC$\epsilon$ dependent manner [166]. In addition, ATP, another component of the inflammatory soup, was also found to increase TTX-resistant $Na^+$ currents in sensory neurons. By contrast, TTX-sensitive currents were not affected by application of ATP [167]. The underlying signaling cascade was not elucidated and hence it remains to be determined if, for example, $G\alpha_s$-coupled $P2Y_{11}$ receptors could be involved.

On the other hand, activation of protease activated receptor 2 (PAR2) does not affect TTX-resistant $Na^+$ currents in nociceptive neurons [168]. These GPCRs are coupled to heterotrimeric $G\alpha_q$-proteins and lead to release of $Ca^{2+}$ from intracellular stores as well as activation of PKC [142].

With respect to $G\alpha_{i/o}$-coupled receptors, only activation of $\mu$-opioid receptors by the agonist DAMGO was investigated. In sensory neurons, application of DAMGO was able to prevent the $PGE_2$-mediated increase of TTX-resistant $Na^+$ channels [169]. The involved pathway was not investigated, but is seems reasonable to assume an interference of the $G\alpha_{i/o}$-pathway with the $G\alpha_s$-mediated potentiation of these NaV channels.

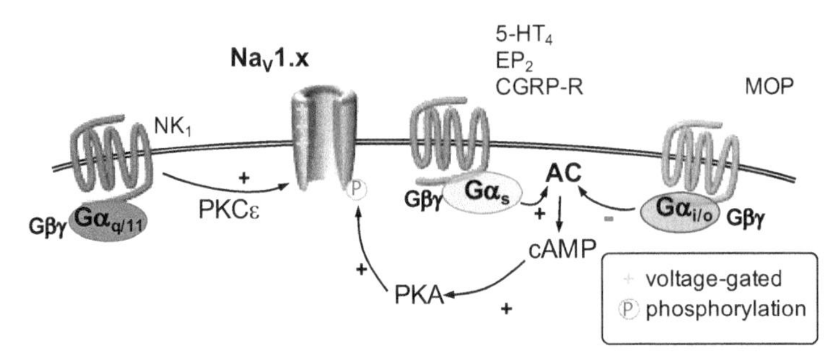


**Figure 5.** Na$_V$1.x channels are activated by depolarizing voltages (as indicated). Activation of G$\alpha_s$-coupled receptors stimulates adenylyl cyclase (AC) activity (**center**). Subsequently, cyclic adenosine monophosphate (cAMP) is formed, which activates protein kinase A (PKA). PKA-mediated phosphorylation of voltage-gated Na$^+$ channels increases their currents. By contrast, activation of G$\alpha_{i/o}$-coupled receptors (**right**) decreases AC activity and counteracts G$\alpha_s$-mediated current increases. Activation of a G$\alpha_{q/11}$-receptor (**left**)was shown to increase Na$_V$-mediated currents involving protein kinase C $\epsilon$ (PKC$\epsilon$).

## 2.7. *Voltage-Gated $Ca^{2+}$ Channels*

Neuronal calcium channels are protein complexes formed by a pore-forming $\alpha$1 subunit, one $\beta$ and one $\alpha 2\delta$ subunit, the latter regulating membrane trafficking and voltage dependence [170]. There are ten different $\alpha$1 subunits that can be divided into three families (Ca$_V$1.x–3.x). Each of these families consists of several subtypes that differ in biophysical parameters, expression pattern and physiological function. In addition, there is a clear distinction in the voltage dependence of Ca$_V$1.x and Ca$_V$2.x when compared to Ca$_V$3.x. While the latter activates at quite hyperpolarized potentials ($<-50$ mV), thus being termed low voltage activated channel (LVA), the former need much higher depolarizations to be opened and are called high voltage activated channels (HVA). In DRG neurons, all three families of Ca$_V$s are found [171] and all can be modulated by GPCRs to different extents. Depending on the type of GPCR, the channels are modulated via different pathways.

### GPCR Regulation of Voltage-Gated $Ca^{2+}$ Channels

The most prominent and by far best studied mechanism is the so-called voltage-dependent inhibition (Figure 6). This is found in Ca$_V$2.x channels, where the G-protein G$\beta\gamma$ subunits can directly bind to the calcium channel $\alpha$1 subunit. This binding event leads to a shift in the gating mode from *"wiilling"* to a *"reluctant"* one that manifests itself mainly in a marked slowing of activation [172–174]. The term "voltage dependent" refers to the fact that depolarization can relief the channels from inhibition and restore normal gating. As this type of inhibition is mostly exerted by G$\alpha_{i/o}$-coupled receptors it can be abolished by treating the cells with pertussis toxin (PTX) that ribosilates G$\alpha_{i/o}$-proteins and renders them inactive.

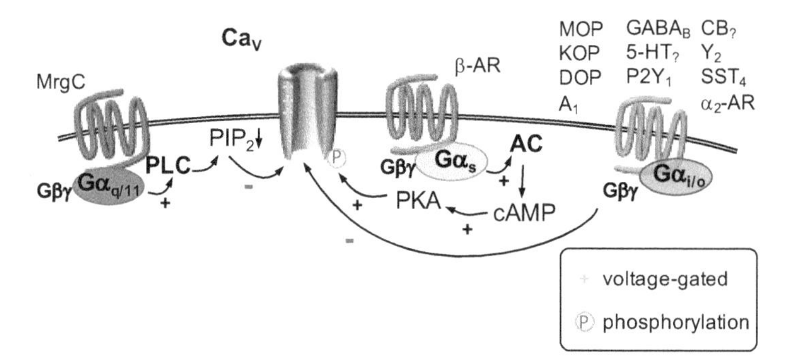


**Figure 6.** $Ca_V$ channels are activated by depolarizing voltages (as indicated). Activation of a $G\alpha_{i/o}$-coupled receptor (**right**) leads to a phenomenon called "voltage-dependent inhibition". This involves direct binding of the $G\beta\gamma$ dimer to $Ca_V2.x$ channels. Activation of $G\alpha_s$-coupled receptors (**center**) leads to activation of adenylyl cyclase (AC), which forms cAMP (cyclic adenosine monophosphate) to activate protein kinase A (PKA). PKA-mediated phosphorylation of $Ca_V1.x$ channels was shown to increase their currents. Activation of a $G\alpha_{q/11}$-coupled receptor (**left**) stimulates phospholipase C (PLC), which hydrolyzes phosphatitylinositol 4,5 bisphosphate ($PIP_2$). Depletion of $PIP_2$ is sufficient to decrease currents through $Ca_V$ channels.

One of the best studied examples for voltage dependent calcium channel inhibition is the action of opioid receptors. All three types of opioid receptors, $\mu$ (MOP), $\kappa$ (KOP) and $\delta$ (DOP), are found in DRG neurons, with the exact expression pattern depending on the cell type [175,176]. Initially, it was found that exposure of nociceptive neurons to opioids leads to a shortening of action potential durations [177–181]. It was demonstrated that application of [D-Ala2]-enkephalin (DADLE), an unspecific opioid receptor agonist, not only reduced action potential duration, but also diminished substance P release in these neurons [177]. As became clear later on, both effects were caused mainly by the inhibition of $Ca^{2+}$ channels [172,182]. The major target for opioid modulation are $Ca_V2.x$ channels [183–188]. These channels are found at the presynapse and govern neurotransmitter release [170], thus inhibition of $Ca_V2.x$ channels leads to reduced $Ca^{2+}$ influx and concomitantly reduced transmitter release from peripheral nociceptive neurons onto second-order neurons of the pain pathway located in the spinal dorsal horn. In line with the fact that opioid receptors couple predominantly to $G\alpha_{i/o}$ G proteins [110], opioid induced calcium channel inhibition was found to be PTX sensitive [188], and voltage dependent [172]. These findings were corroborated by intracellular administration of $G\alpha_o$ antiserum, that strongly reduced opioid receptor mediated $I_{Ca}$ inhibition [189], unequivocally demonstrating the mechanism of action.

Besides opioid receptors, many other GPCRs expressed in DRG neurons were found to lead to a similar kind of calcium channel modulation. For example $GABA_B$ [190], adenosine $A_1$ [191], 5-HT [163,192–195], P2Y [196], cannabinoid [197], neuropeptide Y $Y_2$ receptors [198–200], somatostatin $SST_4$ [201] or $\alpha_2$ adrenoceptors [202].

Besides the well studied $G\beta\gamma$ mediated voltage dependent inhibition, other mechanisms have also been described [174]. Most prominently, phospholipase C (PLC) mediated $PIP_2$ depletion can lead to inhibition of calcium channels [203–205]. Similar to $K_V7$ and $K_{ir}3$ channels (see below), $PIP_2$ stabilizes the open state of calcium channels. Thus, a reduction in membrane $PIP_2$ leads to a voltage independent decrease in channel open probability.

While several reports about this kind of inhibition exist from sympathetic neurons [174], there are few data from nociceptive neurons. However, given the similarity of receptor and channel expression between sympathetic and sensory neurons, it is reasonable to assume that these findings will also hold true in DRG neurons. Only recently, the Mas-related G protein coupled receptor type C (MrgC) was found to inhibit high voltage-gated calcium channels in a PLC dependent manner [206].

Recently, a potentially novel form of calcium channel inhibition has been described. Huang et al. [207] found that $GABA_B$ receptors not only inhibit HVA channels but also LVA channels. Inhibition of both channel types was PTX sensitive, however the LVA inhibition was strongly reduced by application of DTT, pointing to a novel mechanism that involves redox processes.

GPCRs cannot only inhibit $Ca^{2+}$ channels but they are also known to be able to facilitate their function. A classic example would be PKA which phosphorylates $Ca_V1.x$ channels and increases their currents [208]. In line with this, a broadening of the action potential upon application of noradrenaline to DRG neurons was reported. This increase in action potential duration could finally be attributed to an increase in $Ca_V1.x$ currents [202].

### 2.8. *Voltage-Gated $K^+$ Channels*

#### 2.8.1. $K_V7$ Channels

Amongst potassium channels, $K_V$ channels constitute the most diverse group with 12 known families [209]. Literature abounds on how various $K_V$ channels modulate nociception at different levels of the pain pathway [210]. The prototypical structural assembly of $K_V$ channels involves six transmembrane segments of which the first four (S1-S4) constitute the voltage-sensing domain (VSD) while the S5 and S6 segments constitute the pore through which $K^+$ ions pass [211,212]. Amongst others, the $K_V7$ channel family has received immense attention in lieu of its amenability by GPCR modulation [213,214], with five known members ($K_V7.1$– $K_V7.5$) encoded by KCNQ1-5 genes [209]. Four monomers come together in a homo- or heterotetrameric configuration in a subunit-specific way to yield a functional $K_V7$ channel [215]. The electrophysiological correlate of $K_V7$ channel activity is a slowly deactivating, non-inactivating current that has an activation threshold below −60 mV. This conductance is also known as M current as it was originally described as a current that is suppressed by an activation of muscarinic acetylcholine receptors [216]. In nociceptors, these channels contribute majorly to the resting membrane potential [217]. $K_V7$ channels are expressed in all functional parts of a first-order neuron which include free nerve endings, nodes of Ranvier, and the somata of dorsal root ganglion (DRG) neurons [218]. They regulate action potential (AP) firing, which is the basis for encoding of noxious stimuli in the pain pathway [219]. Enhancing $K_V7$ currents exerts analgesic effects since hyperpolarizing the resting membrane potential of nociceptors decreases neuronal excitability [220,221]. Similarly, inhibiting $K_V7$ currents is proalgesic since concomitant depolarization of the resting membrane potential enhances neuronal firing [222,223].

#### 2.8.2. GPCR Regulation of $K_V7$ Channels

A plethora of neurotransmitters and neuropeptides modulate $K_V7$ channel function via GPCR signaling (Figure 7), specifically of the $G\alpha_{q/11}$ class [224]. One of the early reports was of the nociception-relevant neuropeptide Substance P (SP) that inhibited $K_V7$ currents in bullfrog DRG neurons [225]. Subsequently it was revealed that neurokinin A (NKA inhibited currents through $K_V7$ channels in bullfrog DRG neuronsvia $NK_1$ receptors which were coupled to PTX-insensitive G proteins [226], even though this receptor may impinge on the functions of $K_V7$ channels through G protein-independent mechanisms as well [227]. The activation of $G\alpha_{q/11}$-coupled receptors leads phospholipase C$\beta$ (PLC$\beta$) to hydrolyze the membrane phospholipid phosphatidylinositol 4,5-bisphosphate ($PIP_2$) into inositol-1,4,5-trisphosphate ($IP_3$) and diacylglycerol (DAG) [228]. The function of $K_V7$ channels is governed by the presence of sufficient membrane $PIP_2$ pools and depletion of membrane $PIP_2$ levels leads $K_V7$ channels to close [229]. The proalgesic mediator bradykinin mediates inhibition of $K_V7$ currents via its actions on the $B_2$ receptor, a $G\alpha_{q/11}$-coupled receptor [230]. The active nociception mediated by bradykinin, consequent to enhanced neuronal excitability can be attenuated by prior application of the $K_V7$ channel opener retigabine [145]. In addition to membrane $PIP_2$ depletion, the inhibition of $K_V7$ channels via $IP_3$-mediated increase in intracellular $Ca^{2+}$ levels and subsequent binding to calmodulin is well known [231,232]. In sympathetic neurons, the governing factor for substrate ($PIP_2$)- versus product ($Ca^{2+}$)-mediated inhibition consequent to application of bradykinin is contingent upon $Ca^{2+}$ availability and rate of $PIP_2$ synthesis [233]. $B_2$ receptors are closely opposed to the endoplasmic reticulum (ER) where $IP_3$ can diffuse and consequently mobilize $Ca^{2+}$ reserves [216]. In DRG neurons, bradykinin primarily employs the $Ca^{2+}$ axis to inhibit $K_V7$ currents as evidenced by the fact that inhibition of $Ca^{2+}$ release from $IP_3$-sensitive stores with pharmacological tools as well as chelation of intracellular $Ca^{2+}$ prevents bradykinin-mediated inhibition of $K_V7$ channels, akin to direct activation of PLC [145]. One of the targets of the inflammatory soup is the protease-activated receptor 2 (PAR2), a $G\alpha_{q/11}$-coupled receptor expressed in nociceptors [49,234]. Activation of these receptors has an inhibitory impact on $K_V7$ currents leading to nociception which requires concurrent increase in cytosolic $Ca^{2+}$ levels in addition to depletion of $PIP_2$ levels [235]. Another example is the modulation of excitability in nociceptors by nucleotides. The $P2Y_1$ and $P2Y_2$ receptors are $G\alpha_{q/11}$-coupled receptors [236]. Activation of these receptors by the nucleotides adenosine diphosphate (ADP), and 2-thio-uridine triphosphate (2-thio-UTP), respectively, leads to the inhibition of currents through $K_V7$ channels [47]. Moreover, the observed effects were prevented by the application of U73122, a PLC inhibitor, inhibition of $Ca^{2+}$ATPases by thapsigargin, and chelation of intracellular $Ca^{2+}$ levels by BAPTA-AM [47].

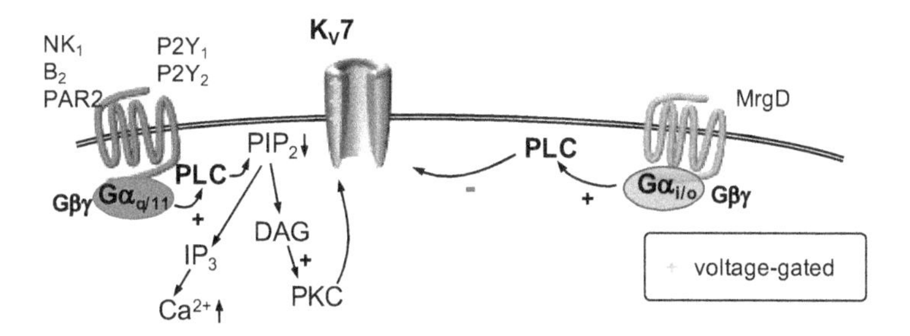


**Figure 7.** $K_V7$ channels are activated by depolarizing voltages (as indicated). $G\alpha_{q/11}$-coupled receptors (**left**) activate phospholipase C (PLC), which cleaves phosphatitylinositol 4,5, bisphosphate ($PIP_2$) into inositol 1,4,5 trisphosphate ($IP_3$) and diacylglycerol (DAG). Depletion of $PIP_2$ from the plasma membrane is sufficient to decrease currents through $K_V7$ channels. In addition, $Ca^{2+}$ is released from the endoplasmic reticulum subsequent to formation of $IP_3$. $Ca^{2+}$ decreases $K_V7$ currents via an interaction with calmodulin. Activation of a $G\alpha_{i/o}$-coupled receptor (**right**) was shown to decrease $K_V7$ currents in a PLC-dependent manner.

The GPCRs encoded by Mas-related genes (Mrgs) are a more recently identified subset of GPCRs that are widely expressed in sensory neurons and implicated in the modulation of nociceptive information [237,238]. Specifically, the MrgD isoform is expressed in DRG neurons especially in non-peptidergic, small-diameter IB4-postive C-fiber somata [239,240]. Activation of endogenous MrgD with the agonist alanine results in the inhibition of $K_V7$ currents in DRG neurons, mainly employing a pertussis toxin-sensitive pathway implicating the involvement of $G\alpha_{i/o}$. Such an inhibition translates into enhanced neuronal firing in phasic DRG neurons, which classically shoot single single APs [241]. Recombinant cell-lines coexpressing MrgD receptors and $K_V7.2/7.3$ heteromers exhibit an inhibition of $K_V7$ currents upon stimulation with alanine, an effect that could be reversed partially by pharmacologically blocking $G\alpha_{i/o}$ and reversed completely by PLC inhibition [241].

### 2.8.3. $K_V1.4$, $K_V3.4$ and $K_V4$ Channels

The $K_V$ channels $K_V1.4$, $K_V3.4$ and $K_V4$ members contribute to the so-called transient A-current ($I_A$) [242,243], which plays a key role in regulating AP firing in DRG neurons [210,244]. The somata, axons and central terminals of DRG neurons that abut the spinal dorsal horn are enriched with $K_V3.4$ channels. The expression of $K_V4.3$ channels, on the other hand is restricted to the somata of non-peptidergic DRG neurons [245]. The rapidly inactivating $K_V3.4$ channel is a key player in AP repolarization in DRG neurons [246,247]. Activating PKC through physiological or pharmacological means leads to a decline in fast N-type inactivation in endogenously expressed $K_V3.4$ channels. This directly impacts the biophysical properties of the nociceptor: the AP gets narrowed while the AP repolarization is accelerated [248]. Specific siRNAs that selectively target $K_V3.4$ channel expression abolish the changes in AP waveform mediated by PKC activation [248]. In a rat model of cervical spinal cord injury (SCI), the surface expression of $K_V3.4$ channels was shown to be impaired; such dysregulation was associated with the failure of PKC to shorten the AP duration in DRG neurons [249]. Similarly, the phosphatase calcineurin (CaN) antagonizes PKC activity as revealed by a reduction in the inactivation of $K_V3.4$ channels upon pharmacological inhibition of the former [250].

2.8.4. GPCR Regulation of $K_V1.4$, $K_V3.4$ and $K_V4$ Channels

Neuromedin U (NMU) is a neuropeptide that decreases neuronal excitability in DRG neurons via the enhancement of $I_A$ currents through its actions on the NMU type 1 receptor (NMUR1, Figure 8) [251]. NMUR1 couples to $G\alpha_o$ proteins, PKA and the ERK pathway in a sequential manner [251]. On the other hand, the cyclic undecapeptide urotensin-II activates the urotensin-II receptor (UTR), which couples to $G\alpha_{q/11}$ [252,253]. This leads to a reduction in $I_A$ in a dose-dependent fashion in trigeminal ganglion neurons, mediated via the activation of PKC [254]. The concomitant recruitment of ERK signaling cascade culminates in an enhanced excitability of TG neurons [254]. In rat TG neurons, the $K_V1.4$, $K_V3.4$, $K_V4.2$ and $K_V4.3$ channels are co-expressed with $P2Y_2$ receptors [255], which in turn couple to different G proteins [256]. The application of UTP, an agonist at the $P2Y_2$ receptor inhibits $I_A$ currents via the ERK pathway and enhances excitability in these neurons, an effect that can be reversed by the $P2Y_2$ receptor antagonist suramin [255].

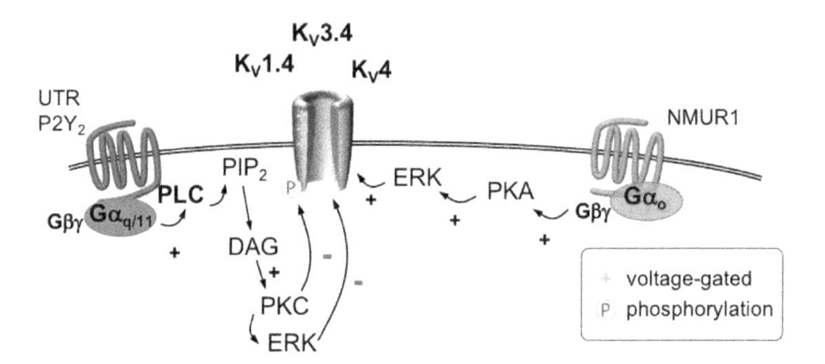


**Figure 8.** The so-called A-current is mediated by voltage-activated $K_V1.4$, $K_V3.4$, and $K_V4$ channels. Activation of a $G\alpha_o$-coupled receptor (**right**) increases A-type currents via a mechanism involving the $G\beta\gamma$ dimer, protein kinase A (PKA) and extracellular signal-regulated kinase (ERK). Stimulation of $G\alpha_{q/11}$-coupled receptors (**left**) activates phospholipase C (PLC) leading to hydrolysis of phosphatityl 4,5 bisphosphate ($PIP_2$) to diacylglycerol (DAG) and $IP_3$. DAG activates protein kinase C (PKC), which phosphorylates A-type channels and thus inhibits these channels. ERK is activated in parallel which also phosphorylates A-type channels and thereby decreases their function.

### 2.9. G-Protein Activated, Inwardly Rectifying Potassium Channels

G protein activated, inwardly rectifying potassium channels (GIRK) are homo- or heterotetrameric channels formed from four different subunits ($K_{ir}3.1$–3.4) encoded by the genes KCNJ3, KCNJ6, KCNJ9 and KCNJ5, respectively [257,258]. They are activated by pertussis toxin sensitive G proteins via binding of $\beta\gamma$ dimers to the channel [257]. In addition, it has also been demonstrated that the G$\alpha$ subunit can directly bind to the channels modulating their basal activity in the absence of GPCR activation [259]. Besides their G protein-mediated modulation, it has been demonstrated several times that $K_{ir}3$ channels bind $PIP_2$ and that this is necessary for their function [260,261].

GPCR Modulation of Girk Channels

Compared to their role in the CNS, data on their physiological role in DRG neurons is scarce [258]. In rat DRG neurons, co-localization of GIRK channels and $\mu$-opioid receptors has been found [262]. The picture is complicated by the fact that, while all four subunits are expressed in rat and human DRG neurons [117,263], only low mRNA levels and a lack of immunostaining have been reported in mice [117]. This is corroborated by the fact that local application of DAMGO (Figure 9), a $\mu$-opioid receptor agonist, is ineffective in inflammatory pain mouse models [117]. This lack of effect, however, could be overcome by expressing $K_{ir}3.2$ in $Na_V1.8$ expressing nociceptive neurons [117], demonstrating the importance of $K_{ir}3$ channels for peripheral analgesia. These findings are contrasted by the fact that GIRK currents could be induced in a small number of about 15–20% of mouse DRG neurons by application of DAMGO [264], rendering the interpretation of mouse data difficult.

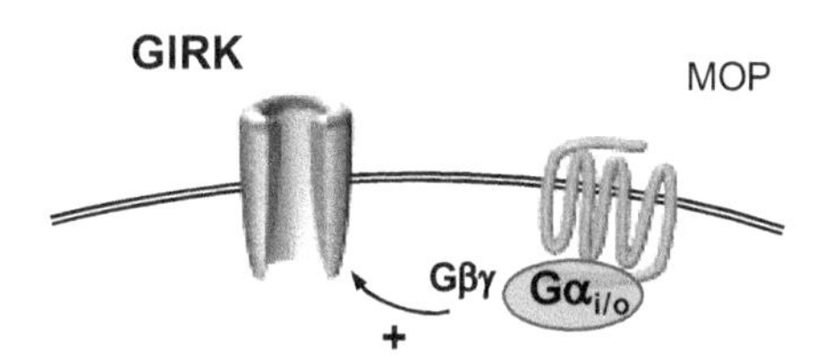


**Figure 9.** G-protein activated, inwardly rectifying $K^+$ channels (GIRK) are activated subsequent to stimulation of G$\alpha_{i/o}$-coupled receptors. The dissociated G$\beta\gamma$ dimer binds directly to GIRK channels.

## 3. Conclusions

In this review, members of ten different ion channel families that are expressed in sensory neurons are described with respect to their contribution to nociception. Appropriate gating of these channels is subject to modulation by at least 35 different types of GPCRs, which are targeted by more than 20 separate endogenous modulators (Table 2). Thereby, GPCRs provide the largest superfamily of receptors, activation of which can mediate pro- as well as antinociceptive effects . Accordingly, these GPCRs are relevant as potential targets for analgesic drugs. However, only a few of them are currently exploited in analgesic therapy, such as opioid, cannabinoid, and CGRP receptors or prostanoid receptors as indirect targets of cyclooxygenase inhibitors. Therefore, this review should be viewed as incitement to further investigate how the modulation of ion channels via GPCRs might be tackled to provide novel pharmacotherapeutic agents for improved analgesic therapy.

**Table 2.** Endogenous ligands for GPCRs modulating ion channel function.

| Endogenous Ligand | GPCR | ASIC | CaCC | $Ca_V$ | GIRK | $K_2P$ | $K_V1.4$<br>$K_V3.4$<br>$K_V4$ | $K_V7$ | $Na_V$ | TRPA1<br>TRPM3<br>TRPM8<br>TRPV1 | Piezo |
|---|---|---|---|---|---|---|---|---|---|---|---|
| Adenosine | $A_1$ | | | $Ca_V$ | | | | | | | |
| | $A_{2A}$ | | | | | | | | | TRPA1 | |
| Alanine | MrgD | | | | | | | $K_V7$ | | | |
| BAM 8-22 | MRGPRX1 | | | | | | | | | TRPV1 | |
| Bradykinin | $B_1$ | | | | | | | | | TRPA1 | |
| | $B_2$ | | CaCC | | | | | $K_V7$ | | TRPA1 | Piezo2 |
| | | | | | | | | | | TRPM8 | |
| | | | | | | | | | | TRPV1 | |
| CGRP | CGRP-R | | | | | | | | $Na_V$ | | |
| Endo- | $CB_1$ | ASIC | CaCC | | | | | | | TRPV1 | |
| cannabinoids | $CB_2$ | | | | | | | | | TRPV1 | |
| | $CB_?$ | | | $Ca_V$ | | | | | | | |
| Endothelin 1 | $ET_A$ | | | | | | | | | TRPV1 | |
| GABA | $GABA_B$ | | | $Ca_V$ | | | | | | | |
| $H^+$ | TDAG8/ | | | | | | | | | TRPV1 | |
| | (GPR65) | | | | | | | | | | |
| Histamine | $H_1$ | | | | | | | | | TRPA1 | |
| Neuromedin U | NMUR1 | | | | | | $K_V1.4$ | | | | |
| | | | | | | | $K_V3.4$ | | | | |
| | | | | | | | $K_V4$ | | | | |
| Neuropeptide Y | $Y_2$ | | | $Ca_V$ | | | | | | | |
| Noradrenaline | $\alpha_2$ | | | $Ca_V$ | | | | | | | |
| | $\beta$ | | | $Ca_V$ | | | | | | | |
| Nucleotides | $P2Y_1$ | | CaCC | $Ca_V$ | | | | $K_V7$ | | | |
| | $P2Y_2$ | ASIC | CaCC | | | | $K_V1.4$ | $K_V7$ | | | Piezo2 |
| | | | | | | | $K_V3.4$ | | | | |
| | | | | | | | $K_V4$ | | | | |
| | $P2Y_{2/4/6?}$ | | | | | $K_2P$ | | | | | |
| Opioids | MOP | ASIC | | $Ca_V$ | GIRK | $K_2P$ | | | $Na_V$ | TRPV1 | |
| | DOP | | CaCC | $Ca_V$ | | | | | | | |
| | KOP | | | $Ca_V$ | | | | | | | |
| Prostaglandins | $EP_1$ | | | | | | | | | TRPV1 | |
| | $EP_2$ | | | | | | | | $Na_V$ | TRPV1 | |
| | IP | | | | | | | | | TRPV1 | |
| | FP | | | | | $K_2P$ | | | | | |
| Proteases | PAR1 | | | | | | | | | TRPV1 | |
| | PAR2 | ASIC | CaCC | | | | | | | TRPV1 | |
| | PAR4 | | | | | | | | | TRPV1 | |
| Serotonin | $5\text{-}HT_2$ | ASIC | CaCC | | | | | | | TRPV1 | |
| | $5\text{-}HT_4$ | | | | | $K_2P$ | | | $Na_V$ | TRPV1 | |
| | $5\text{-}HT_7$ | | | | | | | | | TRPV1 | |
| | $5\text{-}HT_?$ | | | $Ca_V$ | | | | | | | |
| Somatostatin | $SST_4$ | | | $Ca_V$ | | | | | | | |
| Substance P/ | $NK_1$ | | | | | | | $K_V7$ | $Na_V$ | TRPV1 | |
| neurokinin A | $NK_2$ | | | | | | | | | TRPV1 | |
| Urotensin | UTR | | | | | | $K_V1.4$ | | | | |
| | | | | | | | $K_V3.4$ | | | | |
| | | | | | | | $K_V4$ | | | | |
| ? | MrgC | | | $Ca_V$ | | | | | | | |

ASIC, acid sensing ion channel; CaCC, $Ca^{2+}$-activated $Cl^-$ channel; $Ca_V$, voltage-gated $Ca^{2+}$ channel; GIRK, G-protein activated; inwardly rectifying $K^+$ channel; $K_2P$, two-pore $K^+$ channel; $K_V$, voltage-gated $K^+$ channel; $Na_V$, voltage-gated $Na^+$ channel; TRP, transient receptor potential channel; TRPA, ankyrin family; TRPM, melastatin family; TRPV, vanilloid family; BAM 8-22, bovine adrenal medulla peptide 8-22; CGRP, calcidonin-gene related peptide; CGRP-R, CGRP receptor; GABA, $\gamma$-amino butyric acid; ?, unknown.

**Author Contributions:** I.S., S.R., K.S. and S.B. wrote the manuscript; I.S. performed artwork.

**Funding:** Work in the authors' laboratory was supported by the doctoral program CCHD funded by the Austrian Science Fund (FWF, W1205) and the Medical University of Vienna as well as by the inter-university cluster project "Novel scaffolds for improved antiepileptic drugs" financed by the University of Vienna and the Medical University of Vienna.

**Conflicts of Interest:** The authors declare no conflict of interest.

## Abbreviations

The following abbreviations are used in this manuscript:

| | |
|---|---|
| $\alpha_2$ AR | $\alpha_2$ adrenoceptor |
| $\beta$ AR | $\beta$ adrenoceptor |
| $\mu$-OR | $\mu$ opioid receptor |
| 5-$HT_{2/4}$ | 5-hydroxytryptamine receptor 2 or 4 |
| $A_1$ | adenosine $A_1$ receptor |
| AC | adenylyl cyclase |
| ADP | adenosine diphosphate |
| ANO1-10 | anoctamin 1 to 10 |
| AP | action potential |
| ASIC | acid sensing ion channel |
| ATP | adenosine triphosphate |
| $B_2$ | bradykinin $B_2$ receptor |
| BAPTA-AM | 1,2-Bis(2-aminophenoxy)ethane-*N*,*N*,*N*′,*N*′-tetraacetic acid tetrakis(acetoxymethyl ester) |
| $Ca_V$ | voltage-gated $Ca^{2+}$ channel |
| CaCC | $Ca^{2+}$-activated $Cl^-$ channel |
| CamKII | $Ca^{2+}$/calmodulin dependent protein kinase II |
| cAMP | cyclic adenosine monophosphate |
| CaN | calcineurin |
| $CB_{1/2}$ | cannabinoid $CB_1$ or $CB_2$ receptor |
| cGMP | cyclic guanosine monophosphate |
| CGRP | calcitonin gene-related peptide |
| CGRP-R | calcitonin gene-related peptide receptor |
| CNS | central nervous system |
| Cryo-EM | cryogenic electron microscopy |
| DADLE | [D-Ala2]-enkephalin |
| DAG | diacylglycerol |
| DAMGO | [D-$Ala^2$, NMe-$Phe^4$, Gly-$ol^5$]-enkephalin |
| DI-IV | domain I to IV of $Ca_V$ and $Na_V$ channels |
| DOG1 | discovered on GIST 1 |
| DOP | $\delta$ opioid receptor |
| DRG | dorsal root ganglion |
| ECD | extracellular domain |
| $EP_{2/4}$ | prostanoid $EP_2$ or $EP_4$ receptor |
| ER | endoplasmic reticulum |
| ERK | extracellular signal-regulated kinase |
| $ET_A$ | endothelin $ET_A$ receptor |
| FP | prostanoid FP receptor |
| $G\alpha_{i/o}$ | $G\alpha_{i/o}$ protein |
| $G\alpha_{q/11}$ | $G\alpha_{q/11}$ protein |
| $G\alpha_s$ | $G\alpha_s$ protein |
| $G\beta\gamma$ | $G\beta\gamma$ dimer |
| $GABA_B$ | $\gamma$-aminobutyric acid receptor B |
| GIRK | G-protein activated, inwardly rectifying potassium channels |
| GIST | gastrointestinal stroma tumor |
| GPCR | G-protein coupled receptor |

| | |
|---|---|
| HVA | high voltage activated $Ca^{2+}$ channel |
| $I_{+}A$ | A-type $K^{+}$ current |
| IP | prostanoid IP receptor |
| $IP_3$ | inositol 1,4,5 trisphosphate |
| $K_{2P}$ | two-pore $K^{+}$ channel |
| $K_{ir}$ | inwardly rectifying potassium channel |
| $K_V$ | voltage-gated $K^{+}$*channel* |
| KCNJ | gene name potassium voltage-gated channel subfamily J |
| KCNQ | gene name potassium voltage-gated channel subfamily Q |
| KOP | $\kappa$ opioid receptor |
| LVA | low voltage activated $Ca^{2+}$ channel |
| MOP | $\mu$ opioid receptor |
| MrgC | mas-related G protein coupled receptor type C |
| MrgD | mas-related G protein coupled receptor type D |
| mRNA | messenger ribonucleic acid |
| $Na_V$ | voltage-gated $Na^{+}$ channel |
| NGF | nerve growth factor |
| $NK_{1/2}$ | tachynin $NK_1$ or $NK_2$ receptor |
| NKA | neurokinin A |
| NMU | neuromedin U |
| NMUR | NMU receptor |
| P2X | purinergic P2X receptor |
| P2Y | purinergic P2Y receptor |
| PAR2/ 4 | protease-activated receptor type 2 or type 4 |
| $PGE_2$ | prostaglandin $E_2$ |
| $PGF_{2a}$ | prostaglandin $F_{2a}$ |
| PICK-1 | protein interacting with C-kinase 1 |
| $PIP_2$ | phosphatitylinositol 4,5 bisphosphate |
| PKA | protein kinase A |
| PKC | protein kinase C |
| PKG | protein kinase G |
| PLA−2 | phospholipase $A_2$ |
| PLC | phospholipase C |
| PPI | protein phosphatase I |
| PTX | pertussis toxin |
| S1–4 | transmembrane segment 1–4 of voltage-gated channels |
| SCI | spinal cord injury |
| siRNA | small interfering RNA |
| SP | substance P |
| $SST_4$ | somatostatin $SST_4$ receptor |
| TG | trigeminal ganglion |
| TMEM16 | transmembrane protein of unknown function family 16 |
| TMS | transmembrane segment |
| TNF | tumor necrosis factor |
| TRAAK | TWIK-related arachidonic acid activated $K^{+}$ channel = $K_{2P}4.1$ |
| TREK1/2 | TWIK-related $K^{+}$ channel 1 ($K_{2P}2.1$) or TWIK-related $K^{+}$ channel 2 ($K_{2P}10.1$) |
| TRH | thyrotropin-releasing hormon |
| trk | tyrosine receptor kinase |
| TRPA1 | transient receptor potential channel ankyrin family |
| TRPM8 | transient receptor potential channel melastatin family |
| TRPV | transient receptor potential channel vanilloid family |
| TTX | tetrodotoxin |
| TWIK | tandem of pore domains in a weak inward rectifying $K^{+}$ channel |
| UTP | uridine triphosphate |
| UTR | urotensin-II receptor |
| $V_{1A}$ | vasopressin $V_{1A}$ receptor |
| $Y_2$ | neuropeptide Y $Y_2$ receptor |

## References


1. Loeser, J.D.; Treede, R.D. The Kyoto protocol of IASP Basic Pain Terminology. *Pain* **2008**, *137*, 473–477. [CrossRef]
2. Basbaum, A.I.; Bautista, D.M.; Scherrer, G.; Julius, D. Cellular and molecular mechanisms of pain. *Cell* **2009**, *139*, 267–284. [CrossRef]
3. Hucho, T.; Levine, J.D. Signaling pathways in sensitization: Toward a nociceptor cell biology. *Neuron* **2007**, *55*, 365–376. [CrossRef] [PubMed]
4. Venkatachalam, K.; Montell, C. TRP channels. *Annu. Rev. Biochem.* **2007**, *76*, 387–417. [CrossRef] [PubMed]
5. Baron, A.; Lingueglia, E. Pharmacology of acid-sensing ion channels—Physiological and therapeutical perspectives. *Neuropharmacology* **2015**, *94*, 19–35. [CrossRef]
6. North, R.A.; Jarvis, M.F. P2X receptors as drug targets. *Mol. Pharmacol.* **2013**, *83*, 759–769. [CrossRef]
7. Cook, A.D.; Christensen, A.D.; Tewari, D.; McMahon, S.B.; Hamilton, J.A. Immune Cytokines and Their Receptors in Inflammatory Pain. *Trends Immunol.* **2018**, *39*, 240–255. [CrossRef]
8. Denk, F.; Bennett, D.L.; McMahon, S.B. Nerve Growth Factor and Pain Mechanisms. *Annu. Rev. Neurosci.* **2017**, *40*, 307–325. [CrossRef]
9. Kaneko, Y.; Szallasi, A. Transient receptor potential (TRP) channels: A clinical perspective. *Br. J. Pharmacol.* **2014**, *171*, 2474–2507. [CrossRef]
10. Basso, L.; Altier, C. Transient Receptor Potential Channels in neuropathic pain. *Curr. Opin. Pharmacol.* **2017**, *32*, 9–15. [CrossRef] [PubMed]
11. Alexander, S.P.; Striessnig, J.; Kelly, E.; Marrion, N.V.; Peters, J.A.; Faccenda, E.; Harding, S.D.; Pawson, A.J.; Sharman, J.L.; Southan, C.; et al. THE CONCISE GUIDE TO PHARMACOLOGY 2017/18: Voltage-gated ion channels. *Br. J. Pharmacol.* **2017**, *174*, S160–S194. [CrossRef]
12. Dai, Y. TRPs and pain. *Semin. Immunopathol.* **2016**, *38*, 277–291. [CrossRef]
13. Mickle, A.D.; Shepherd, A.J.; Mohapatra, D.P. Sensory TRP channels: The key transducers of nociception and pain. *Prog. Mol. Biol. Transl. Sci.* **2015**, *131*, 73–118.
14. Steinberg, X.; Lespay-Rebolledo, C.; Brauchi, S. A structural view of ligand-dependent activation in thermoTRP channels. *Front. Physiol.* **2014**, *5*, 171. [CrossRef]
15. Wang, G.; Wang, K. The Ca2+-Permeable Cation Transient Receptor Potential TRPV3 Channel: An Emerging Pivotal Target for Itch and Skin Diseases. *Mol. Pharmacol.* **2017**, *92*, 193–200. [CrossRef] [PubMed]
16. Zhang, X. Targeting TRP ion channels for itch relief. *Naunyn-Schmiedeberg's Arch. Pharmacol.* **2015**, *388*, 389–399. [CrossRef] [PubMed]
17. Kobayashi, K.; Fukuoka, T.; Obata, K.; Yamanaka, H.; Dai, Y.; Tokunaga, A.; Noguchi, K. Distinct expression of TRPM8, TRPA1, and TRPV1 mRNAs in rat primary afferent neurons with adelta/c-fibers and colocalization with trk receptors. *J. Comp. Neurol.* **2005**, *493*, 596–606. [CrossRef] [PubMed]
18. Okazawa, M.; Inoue, W.; Hori, A.; Hosokawa, H.; Matsumura, K.; Kobayashi, S. Noxious heat receptors present in cold-sensory cells in rats. *Neurosci. Lett.* **2004**, *359*, 33–36. [CrossRef]
19. Usoskin, D.; Furlan, A.; Islam, S.; Abdo, H.; Lönnerberg, P.; Lou, D.; Hjerling-Leffler, J.; Haeggström, J.; Kharchenko, O.; Kharchenko, P.V.; et al. Unbiased classification of sensory neuron types by large-scale single-cell RNA sequencing. *Nat. Neurosci.* **2015**, *18*, 145–153. [CrossRef]
20. Palkar, R.; Lippoldt, E.K.; McKemy, D.D. The molecular and cellular basis of thermosensation in mammals. *Curr. Opin. Neurobiol.* **2015**, *34*, 14–19. [CrossRef]
21. White, J.P.M.; Cibelli, M.; Urban, L.; Nilius, B.; McGeown, J.G.; Nagy, I. TRPV4: Molecular Conductor of a Diverse Orchestra. *Physiol. Rev.* **2016**, *96*, 911–973. [CrossRef]
22. Vriens, J.; Owsianik, G.; Hofmann, T.; Philipp, S.E.; Stab, J.; Chen, X.; Benoit, M.; Xue, F.; Janssens, A.; Kerselaers, S.; et al. TRPM3 is a nociceptor channel involved in the detection of noxious heat. *Neuron* **2011**, *70*, 482–494. [CrossRef]
23. Laing, R.J.; Dhaka, A. ThermoTRPs and Pain. *Neurosci. A Rev. J. Bringing Neurobiol. Neurol. Psychiatry* **2016**, *22*, 171–187. [CrossRef]
24. Latorre, R.; Zaelzer, C.; Brauchi, S. Structure-functional intimacies of transient receptor potential channels. *Q. Rev. Biophys.* **2009**, *42*, 201–246. [CrossRef]
25. Jardín, I.; López, J.J.; Diez, R.; Sánchez-Collado, J.; Cantonero, C.; Albarrán, L.; Woodard, G.E.; Redondo, P.C.; Salido, G.M.; Smani, T.; et al. TRPs in Pain Sensation. *Front. Physiol.* **2017**, *8*, 392. [CrossRef]

26. Gao, Y.; Cao, E.; Julius, D.; Cheng, Y. TRPV1 structures in nanodiscs reveal mechanisms of ligand and lipid action. *Nature* **2016**, *534*, 347–351. [CrossRef]
27. Cao, E.; Liao, M.; Cheng, Y.; Julius, D. TRPV1 structures in distinct conformations reveal activation mechanisms. *Nature* **2013**, *504*, 113–118. [CrossRef]
28. Liao, M.; Cao, E.; Julius, D.; Cheng, Y. Structure of the TRPV1 ion channel determined by electron cryo-microscopy. *Nature* **2013**, *504*, 107–112. [CrossRef]
29. Gregorio-Teruel, L.; Valente, P.; Liu, B.; Fernández-Ballester, G.; Qin, F.; Ferrer-Montiel, A. The Integrity of the TRP Domain Is Pivotal for Correct TRPV1 Channel Gating. *Biophys. J.* **2015**, *109*, 529–541. [CrossRef]
30. Rohacs, T. Phosphoinositide regulation of TRPV1 revisited. *Pflugers Arch. Eur. J. Physiol.* **2015**, *467*, 1851–1869. [CrossRef]
31. Caterina, M.J.; Leffler, A.; Malmberg, A.B.; Martin, W.J.; Trafton, J.; Petersen-Zeitz, K.R.; Koltzenburg, M.; Basbaum, A.I.; Julius, D. Impaired nociception and pain sensation in mice lacking the capsaicin receptor. *Science* **2000**, *288*, 306–313. [CrossRef]
32. Davis, J.B.; Gray, J.; Gunthorpe, M.J.; Hatcher, J.P.; Davey, P.T.; Overend, P.; Harries, M.H.; Latcham, J.; Clapham, C.; Atkinson, K.; et al. Vanilloid receptor-1 is essential for inflammatory thermal hyperalgesia. *Nature* **2000**, *405*, 183–187. [CrossRef]
33. Park, U.; Vastani, N.; Guan, Y.; Raja, S.N.; Koltzenburg, M.; Caterina, M.J. TRP vanilloid 2 knock-out mice are susceptible to perinatal lethality but display normal thermal and mechanical nociception. *J. Neurosci. Off. J. Soc. Neurosci.* **2011**, *31*, 11425–11436. [CrossRef]
34. Huang, S.M.; Li, X.; Yu, Y.; Wang, J.; Caterina, M.J. TRPV3 and TRPV4 ion channels are not major contributors to mouse heat sensation. *Mol. Pain* **2011**, *7*, 37. [CrossRef]
35. Vandewauw, I.; De Clercq, K.; Mulier, M.; Held, K.; Pinto, S.; Van Ranst, N.; Segal, A.; Voet, T.; Vennekens, R.; Zimmermann, K.; et al. A TRP channel trio mediates acute noxious heat sensing. *Nature* **2018**, *555*, 662–666. [CrossRef]
36. Chen, J.; Kang, D.; Xu, J.; Lake, M.; Hogan, J.O.; Sun, C.; Walter, K.; Yao, B.; Kim, D. Species differences and molecular determinant of TRPA1 cold sensitivity. *Nat. Commun.* **2013**, *4*, 2501. [CrossRef]
37. Moparthi, L.; Survery, S.; Kreir, M.; Simonsen, C.; Kjellbom, P.; Högestätt, E.D.; Johanson, U.; Zygmunt, P.M. Human TRPA1 is intrinsically cold- and chemosensitive with and without its N-terminal ankyrin repeat domain. *Proc. Natl. Acad. Sci. USA* **2014**, *111*, 16901–16906. [CrossRef]
38. Nilius, B.; Appendino, G.; Owsianik, G. The transient receptor potential channel TRPA1: From gene to pathophysiology. *Pflugers Arch. Eur. J. Physiol.* **2012**, *464*, 425–458. [CrossRef]
39. Rohacs, T. Regulation of transient receptor potential channels by the phospholipase C pathway. *Adv. Biol. Regul.* **2013**, *53*, 341–355. [CrossRef] [PubMed]
40. Julius, D. TRP channels and pain. *Annu. Rev. Cell Dev. Biol.* **2013**, *29*, 355–384. [CrossRef]
41. Tang, H.B.; Inoue, A.; Oshita, K.; Nakata, Y. Sensitization of vanilloid receptor 1 induced by bradykinin via the activation of second messenger signaling cascades in rat primary afferent neurons. *Eur. J. Pharmacol.* **2004**, *498*, 37–43. [CrossRef]
42. Ohta, T.; Ikemi, Y.; Murakami, M.; Imagawa, T.; Otsuguro, K.I.; Ito, S. Potentiation of transient receptor potential V1 functions by the activation of metabotropic 5-HT receptors in rat primary sensory neurons. *J. Physiol.* **2006**, *576*, 809–822. [CrossRef]
43. Sugiuar, T.; Bielefeldt, K.; Gebhart, G.F. TRPV1 function in mouse colon sensory neurons is enhanced by metabotropic 5-hydroxytryptamine receptor activation. *J. Neurosci. Off. J. Soc. Neurosci.* **2004**, *24*, 9521–9530. [CrossRef]
44. Salzer, I.; Gantumur, E.; Yousuf, A.; Boehm, S. Control of sensory neuron excitability by serotonin involves 5HT2C receptors and Ca(2+)-activated chloride channels. *Neuropharmacology* **2016**, *110*, 277–286. [CrossRef]
45. Moriyama, T.; Iida, T.; Kobayashi, K.; Higashi, T.; Fukuoka, T.; Tsumura, H.; Leon, C.; Suzuki, N.; Inoue, K.; Gachet, C.; et al. Possible involvement of P2Y2 metabotropic receptors in ATP-induced transient receptor potential vanilloid receptor 1-mediated thermal hypersensitivity. *J. Neurosci. Off. J. Soc. Neurosci.* **2003**, *23*, 6058–6062. [CrossRef]
46. Malin, S.A.; Davis, B.M.; Koerber, H.R.; Reynolds, I.J.; Albers, K.M.; Molliver, D.C. Thermal nociception and TRPV1 function are attenuated in mice lacking the nucleotide receptor P2Y2. *Pain* **2008**, *138*, 484–496. [CrossRef]

47. Yousuf, A.; Klinger, F.; Schicker, K.; Boehm, S. Nucleotides control the excitability of sensory neurons via two P2Y receptors and a bifurcated signaling cascade. *Pain* **2011**, *152*, 1899–1908. [CrossRef] [PubMed]
48. Solinski, H.J.; Zierler, S.; Gudermann, T.; Breit, A. Human sensory neuron-specific Mas-related G protein-coupled receptors-X1 sensitize and directly activate transient receptor potential cation channel V1 via distinct signaling pathways. *J. Biol. Chem.* **2012**, *287*, 40956–40971. [CrossRef]
49. Amadesi, S.; Nie, J.; Vergnolle, N.; Cottrell, G.S.; Grady, E.F.; Trevisani, M.; Manni, C.; Geppetti, P.; McRoberts, J.A.; Ennes, H.; et al. Protease-activated receptor 2 sensitizes the capsaicin receptor transient receptor potential vanilloid receptor 1 to induce hyperalgesia. *J. Neurosci. Off. J. Soc. Neurosci.* **2004**, *24*, 4300–4312. [CrossRef] [PubMed]
50. Dai, Y.; Moriyama, T.; Higashi, T.; Togashi, K.; Kobayashi, K.; Yamanaka, H.; Tominaga, M.; Noguchi, K. Proteinase-activated receptor 2-mediated potentiation of transient receptor potential vanilloid subfamily 1 activity reveals a mechanism for proteinase-induced inflammatory pain. *J. Neurosci. Off. J. Soc. Neurosci.* **2004**, *24*, 4293–4299. [CrossRef] [PubMed]
51. Vellani, V.; Kinsey, A.M.; Prandini, M.; Hechtfischer, S.C.; Reeh, P.; Magherini, P.C.; Giacomoni, C.; McNaughton, P.A. Protease activated receptors 1 and 4 sensitize TRPV1 in nociceptive neurones. *Mol. Pain* **2010**, *6*, 61. [CrossRef]
52. Moriyama, T.; Higashi, T.; Togashi, K.; Iida, T.; Segi, E.; Sugimoto, Y.; Tominaga, T.; Narumiya, S.; Tominaga, M. Sensitization of TRPV1 by EP1 and IP reveals peripheral nociceptive mechanism of prostaglandins. *Mol. Pain* **2005**, *1*, 3. [CrossRef]
53. Schnizler, K.; Shutov, L.P.; Van Kanegan, M.J.; Merrill, M.A.; Nichols, B.; McKnight, G.S.; Strack, S.; Hell, J.W.; Usachev, Y.M. Protein kinase A anchoring via AKAP150 is essential for TRPV1 modulation by forskolin and prostaglandin E2 in mouse sensory neurons. *J. Neurosci. Off. J. Soc. Neurosci.* **2008**, *28*, 4904–4917. [CrossRef] [PubMed]
54. Bao, Y.; Gao, Y.; Yang, L.; Kong, X.; Yu, J.; Hou, W.; Hua, B. The mechanism of $\mu$-opioid receptor (MOR)-TRPV1 crosstalk in TRPV1 activation involves morphine anti-nociception, tolerance and dependence. *Channels* **2015**, *9*, 235–243. [CrossRef]
55. Plant, T.D.; Zöllner, C.; Mousa, S.A.; Oksche, A. Endothelin-1 potentiates capsaicin-induced TRPV1 currents via the endothelin A receptor. *Exp. Biol. Med.* **2006**, *231*, 1161–1164.
56. Plant, T.D.; Zöllner, C.; Kepura, F.; Mousa, S.S.; Eichhorst, J.; Schaefer, M.; Furkert, J.; Stein, C.; Oksche, A. Endothelin potentiates TRPV1 via ETA receptor-mediated activation of protein kinase C. *Mol. Pain* **2007**, *3*, 35. [CrossRef] [PubMed]
57. Zhang, H.; Cang, C.L.; Kawasaki, Y.; Liang, L.L.; Zhang, Y.Q.; Ji, R.R.; Zhao, Z.Q. Neurokinin-1 receptor enhances TRPV1 activity in primary sensory neurons via PKCepsilon: A novel pathway for heat hyperalgesia. *J. Neurosci. Off. J. Soc. Neurosci.* **2007**, *27*, 12067–12077. [CrossRef] [PubMed]
58. Sculptoreanu, A.; Aura Kullmann, F.; de Groat, W.C. Neurokinin 2 receptor-mediated activation of protein kinase C modulates capsaicin responses in DRG neurons from adult rats. *Eur. J. Neurosci.* **2008**, *27*, 3171–3181. [CrossRef]
59. Chen, Y.J.; Huang, C.W.; Lin, C.S.; Chang, W.H.; Sun, W.H. Expression and function of proton-sensing G-protein-coupled receptors in inflammatory pain. *Mol. Pain* **2009**, *5*, 39. [CrossRef]
60. Vetter, I.; Wyse, B.D.; Monteith, G.R.; Roberts-Thomson, S.J.; Cabot, P.J. The mu opioid agonist morphine modulates potentiation of capsaicin-evoked TRPV1 responses through a cyclic AMP-dependent protein kinase A pathway. *Mol. Pain* **2006**, *2*, 22. [CrossRef]
61. Endres-Becker, J.; Heppenstall, P.A.; Mousa, S.A.; Labuz, D.; Oksche, A.; Schäfer, M.; Stein, C.; Zöllner, C. Mu-opioid receptor activation modulates transient receptor potential vanilloid 1 (TRPV1) currents in sensory neurons in a model of inflammatory pain. *Mol. Pharmacol.* **2007**, *71*, 12–18. [CrossRef]
62. Anand, U.; Otto, W.R.; Sanchez-Herrera, D.; Facer, P.; Yiangou, Y.; Korchev, Y.; Birch, R.; Benham, C.; Bountra, C.; Chessell, I.P.; et al. Cannabinoid receptor CB2 localisation and agonist-mediated inhibition of capsaicin responses in human sensory neurons. *Pain* **2008**, *138*, 667–680. [CrossRef]
63. Sántha, P.; Jenes, A.; Somogyi, C.; Nagy, I. The endogenous cannabinoid anandamide inhibits transient receptor potential vanilloid type 1 receptor-mediated currents in rat cultured primary sensory neurons. *Acta Physiol. Hung.* **2010**, *97*, 149–158. [CrossRef]
64. Bonnington, J.K.; McNaughton, P.A. Signalling pathways involved in the sensitisation of mouse nociceptive neurones by nerve growth factor. *J. Physiol.* **2003**, *551*, 433–446. [CrossRef]

65. Kim, B.M.; Lee, S.H.; Shim, W.S.; Oh, U. Histamine-induced Ca(2+) influx via the PLA(2)/lipoxygenase/ TRPV1 pathway in rat sensory neurons. *Neurosci. Lett.* **2004**, *361*, 159–162. [CrossRef]
66. Shim, W.S.; Tak, M.H.; Lee, M.H.; Kim, M.; Kim, M.; Koo, J.Y.; Lee, C.H.; Kim, M.; Oh, U. TRPV1 mediates histamine-induced itching via the activation of phospholipase A2 and 12-lipoxygenase. *J. Neurosci. Off. J. Soc. Neurosci.* **2007**, *27*, 2331–2337. [CrossRef]
67. Kajihara, Y.; Murakami, M.; Imagawa, T.; Otsuguro, K.; Ito, S.; Ohta, T. Histamine potentiates acid-induced responses mediating transient receptor potential V1 in mouse primary sensory neurons. *Neuroscience* **2010**, *166*, 292–304. [CrossRef]
68. Dembla, S.; Behrendt, M.; Mohr, F.; Goecke, C.; Sondermann, J.; Schneider, F.M.; Schmidt, M.; Stab, J.; Enzeroth, R.; Leitner, M.G.; et al. Anti-nociceptive action of peripheral mu-opioid receptors by G-beta-gamma protein-mediated inhibition of TRPM3 channels. *eLife* **2017**, *6*, 1–32. [CrossRef]
69. Quallo, T.; Alkhatib, O.; Gentry, C.; Andersson, D.A.; Bevan, S. G protein $\beta\gamma$ subunits inhibit TRPM3 ion channels in sensory neurons. *eLife* **2017**, *6*, 1–22. [CrossRef]
70. Badheka, D.; Yudin, Y.; Borbiro, I.; Hartle, C.M.; Yazici, A.; Mirshahi, T.; Rohacs, T. Inhibition of Transient Receptor Potential Melastatin 3 ion channels by G-protein $\beta\gamma$ subunits. *eLife* **2017**, *6*, 1–21. [CrossRef]
71. Wang, S.; Dai, Y.; Fukuoka, T.; Yamanaka, H.; Kobayashi, K.; Obata, K.; Cui, X.; Tominaga, M.; Noguchi, K. Phospholipase C and protein kinase A mediate bradykinin sensitization of TRPA1: A molecular mechanism of inflammatory pain. *Brain A J. Neurol.* **2008**, *131*, 1241–1251. [CrossRef]
72. Meotti, F.C.; Figueiredo, C.P.; Manjavachi, M.; Calixto, J.B. The transient receptor potential ankyrin-1 mediates mechanical hyperalgesia induced by the activation of B1 receptor in mice. *Biochem. Pharmacol.* **2017**, *125*, 75–83. [CrossRef]
73. Andrade, E.L.; Luiz, A.P.; Ferreira, J.; Calixto, J.B. Pronociceptive response elicited by TRPA1 receptor activation in mice. *Neuroscience* **2008**, *152*, 511–520. [CrossRef]
74. Brozmanova, M.; Mazurova, L.; Ru, F.; Tatar, M.; Hu, Y.; Yu, S.; Kollarik, M. Mechanisms of the adenosine A2A receptor-induced sensitization of esophageal C fibers. *Am. J. Physiol. Gastrointest. Liver Physiol.* **2016**, *310*, G215–223. [CrossRef]
75. Taylor-Clark, T.E.; Undem, B.J.; Macglashan, D.W.; Ghatta, S.; Carr, M.J.; McAlexander, M.A. Prostaglandin-induced activation of nociceptive neurons via direct interaction with transient receptor potential A1 (TRPA1). *Mol. Pharmacol.* **2008**, *73*, 274–281. [CrossRef]
76. Weng, Y.; Batista-Schepman, P.A.; Barabas, M.E.; Harris, E.Q.; Dinsmore, T.B.; Kossyreva, E.A.; Foshage, A.M.; Wang, M.H.; Schwab, M.J.; Wang, V.M.; et al. Prostaglandin metabolite induces inhibition of TRPA1 and channel-dependent nociception. *Mol. Pain* **2012**, *8*, 75. [CrossRef]
77. Linte, R.M.; Ciobanu, C.; Reid, G.; Babes, A. Desensitization of cold- and menthol-sensitive rat dorsal root ganglion neurones by inflammatory mediators. *Exp. Brain Res.* **2007**, *178*, 89–98. [CrossRef]
78. Premkumar, L.S.; Raisinghani, M.; Pingle, S.C.; Long, C.; Pimentel, F. Downregulation of transient receptor potential melastatin 8 by protein kinase C-mediated dephosphorylation. *J. Neurosci. Off. J. Soc. Neurosci.* **2005**, *25*, 11322–11329. [CrossRef]
79. Babes, A.; Ciobanu, A.C.; Neacsu, C.; Babes, R.M. TRPM8, a sensor for mild cooling in mammalian sensory nerve endings. *Curr. Pharm. Biotechnol.* **2011**, *12*, 78–88. [CrossRef]
80. Yudin, Y.; Lukacs, V.; Cao, C.; Rohacs, T. Decrease in phosphatidylinositol 4,5-bisphosphate levels mediates desensitization of the cold sensor TRPM8 channels. *J. Physiol.* **2011**, *589*, 6007–6027. [CrossRef]
81. Zhang, X.; Mak, S.; Li, L.; Parra, A.; Denlinger, B.; Belmonte, C.; McNaughton, P.A. Direct inhibition of the cold-activated TRPM8 ion channel by G$\alpha$q. *Nat. Cell Biol.* **2012**, *14*, 851–858. [CrossRef]
82. Abdelhamid, R.E.; Sluka, K.A. ASICs Mediate Pain and Inflammation in Musculoskeletal Diseases. *Physiology* **2015**, *30*, 449–459. [CrossRef]
83. Hwang, S.W.; Oh, U. Current concepts of nociception: Nociceptive molecular sensors in sensory neurons. *Curr. Opin. Anaesthesiol.* **2007**, *20*, 427–434. [CrossRef]
84. Alexander, S.P.; Peters, J.A.; Kelly, E.; Marrion, N.V.; Faccenda, E.; Harding, S.D.; Pawson, A.J.; Sharman, J.L.; Southan, C.; Davies, J.A.; et al. THE CONCISE GUIDE TO PHARMACOLOGY 2017/18: Ligand-gated ion channels. *Br. J. Pharmacol.* **2017**, *174*, S130–S159. [CrossRef]
85. Deval, E.; Lingueglia, E. Acid-Sensing Ion Channels and nociception in the peripheral and central nervous systems. *Neuropharmacology* **2015**, *94*, 49–57. [CrossRef]

86. Ugawa, S.; Ueda, T.; Yamamura, H.; Shimada, S. In situ hybridization evidence for the coexistence of ASIC and TRPV1 within rat single sensory neurons. *Brain Research. Mol. Brain Res.* **2005**, *136*, 125–133. [CrossRef]
87. Wemmie, J.A.; Taugher, R.J.; Kreple, C.J. Acid-sensing ion channels in pain and disease. *Nat. Rev. Neurosci.* **2013**, *14*, 461–471. [CrossRef]
88. Yoder, N.; Gouaux, E. Divalent cation and chloride ion sites of chicken acid sensing ion channel 1a elucidated by X-ray crystallography. *PLoS ONE* **2018**, *13*, e0202134.
89. Rash, L.D. Acid-Sensing Ion Channel Pharmacology, Past, Present, and Future .... *Adv. Pharmacol.* **2017**, *79*, 35–66.
90. Krauson, A.J.; Carattino, M.D. The Thumb Domain Mediates Acid-sensing Ion Channel Desensitization. *J. Biol. Chem.* **2016**, *291*, 11407–11419. [CrossRef]
91. Cheng, Y.R.; Jiang, B.Y.; Chen, C.C. Acid-sensing ion channels: Dual function proteins for chemo-sensing and mechano-sensing. *J. Biomed. Sci.* **2018**, *25*, 46. [CrossRef] [PubMed]
92. Nagaeva, E.I.; Tikhonova, T.B.; Magazanik, L.G.; Tikhonov, D.B. Histamine selectively potentiates acid-sensing ion channel 1a. *Neurosci. Lett.* **2016**, *632*, 136–140. [CrossRef]
93. Ren, C.; Gan, X.; Wu, J.; Qiu, C.Y.; Hu, W.P. Enhancement of acid-sensing ion channel activity by metabotropic P2Y UTP receptors in primary sensory neurons. *Purinergic Signal.* **2016**, *12*, 69–78. [CrossRef]
94. Qiu, F.; Qiu, C.Y.; Liu, Y.Q.; Wu, D.; Li, J.D.; Hu, W.P. Potentiation of acid-sensing ion channel activity by the activation of 5-HT2 receptors in rat dorsal root ganglion neurons. *Neuropharmacology* **2012**, *63*, 494–500. [CrossRef]
95. Deval, E.; Salinas, M.; Baron, A.; Lingueglia, E.; Lazdunski, M. ASIC2b-dependent regulation of ASIC3, an essential acid-sensing ion channel subunit in sensory neurons via the partner protein PICK-1. *J. Biol. Chem.* **2004**, *279*, 19531–19539. [CrossRef]
96. Gu, Q.; Lee, L.Y. Effect of protease-activated receptor 2 activation on single TRPV1 channel activities in rat vagal pulmonary sensory neurons. *Exp. Physiol.* **2009**, *94*, 928–936. [CrossRef]
97. Zhao, P.; Metcalf, M.; Bunnett, N.W. Biased signaling of protease-activated receptors. *Front. Endocrinol.* **2014**, *5*, 67. [CrossRef]
98. Liu, Y.Q.; Qiu, F.; Qiu, C.Y.; Cai, Q.; Zou, P.; Wu, H.; Hu, W.P. Cannabinoids inhibit acid-sensing ion channel currents in rat dorsal root ganglion neurons. *PLoS ONE* **2012**, *7*, e45531. [CrossRef]
99. Cai, Q.; Qiu, C.Y.; Qiu, F.; Liu, T.T.; Qu, Z.W.; Liu, Y.M.; Hu, W.P. Morphine inhibits acid-sensing ion channel currents in rat dorsal root ganglion neurons. *Brain Res.* **2014**, *1554*, 12–20. [CrossRef]
100. Kubo, A.; Shinoda, M.; Katagiri, A.; Takeda, M.; Suzuki, T.; Asaka, J.; Yeomans, D.C.; Iwata, K. Oxytocin alleviates orofacial mechanical hypersensitivity associated with infraorbital nerve injury through vasopressin-1A receptors of the rat trigeminal ganglia. *Pain* **2017**, *158*, 649–659. [CrossRef]
101. Qiu, F.; Qiu, C.Y.; Cai, H.; Liu, T.T.; Qu, Z.W.; Yang, Z.; Li, J.D.; Zhou, Q.Y.; Hu, W.P. Oxytocin inhibits the activity of acid-sensing ion channels through the vasopressin, V1A receptor in primary sensory neurons. *Br. J. Pharmacol.* **2014**, *171*, 3065–3076. [CrossRef] [PubMed]
102. Delmas, P.; Hao, J.; Rodat-Despoix, L. Molecular mechanisms of mechanotransduction in mammalian sensory neurons. *Nat. Rev. Neurosci.* **2011**, *12*, 139–153. [CrossRef]
103. Ranade, S.S.; Syeda, R.; Patapoutian, A. Mechanically Activated Ion Channels. *Neuron* **2015**, *87*, 1162–1179. [CrossRef] [PubMed]
104. Brohawn, S.G. How ion channels sense mechanical force: Insights from mechanosensitive K2P channels TRAAK, TREK1, and TREK2. *Ann. N. Y. Acad. Sci.* **2015**, *1352*, 20–32. [CrossRef]
105. Enyedi, P.; Czirják, G. Molecular background of leak K+ currents: Two-pore domain potassium channels. *Physiol. Rev.* **2010**, *90*, 559–605. [CrossRef] [PubMed]
106. Viatchenko-Karpinski, V.; Ling, J.; Gu, J.G. Characterization of temperature-sensitive leak K+ currents and expression of TRAAK, TREK-1, and TREK2 channels in dorsal root ganglion neurons of rats. *Mol. Brain* **2018**, *11*, 40. [CrossRef]
107. Li, X.Y.; Toyoda, H. Role of leak potassium channels in pain signaling. *Brain Res. Bull.* **2015**, *119*, 73–79. [CrossRef] [PubMed]
108. Coste, B.; Mathur, J.; Schmidt, M.; Earley, T.J.; Ranade, S.; Petrus, M.J.; Dubin, A.E.; Patapoutian, A. Piezo1 and Piezo2 are essential components of distinct mechanically activated cation channels. *Science* **2010**, *330*, 55–60. [CrossRef]

109. Zhang, M.; Wang, Y.; Geng, J.; Zhou, S.; Xiao, B. Mechanically Activated Piezo Channels Mediate Touch and Suppress Acute Mechanical Pain Response in Mice. *Cell Rep.* **2019**, *26*, 1419–1431.e4. [CrossRef] [PubMed]
110. Alexander, S.P.; Christopoulos, A.; Davenport, A.P.; Kelly, E.; Marrion, N.V.; Peters, J.A.; Faccenda, E.; Harding, S.D.; Pawson, A.J.; Sharman, J.L.; et al. THE CONCISE GUIDE TO PHARMACOLOGY 2017/18: G protein-coupled receptors. *Br. J. Pharmacol.* **2017**, *174*, S17–S129. [CrossRef] [PubMed]
111. Pereira, V.; Busserolles, J.; Christin, M.; Devilliers, M.; Poupon, L.; Legha, W.; Alloui, A.; Aissouni, Y.; Bourinet, E.; Lesage, F.; et al. Role of the TREK2 potassium channel in cold and warm thermosensation and in pain perception. *Pain* **2014**, *155*, 2534–2544. [CrossRef] [PubMed]
112. Patel, A.J.; Honoré, E.; Maingret, F.; Lesage, F.; Fink, M.; Duprat, F.; Lazdunski, M. A mammalian two pore domain mechano-gated S-like K+ channel. *EMBO J.* **1998**, *17*, 4283–4290. [CrossRef] [PubMed]
113. Srisomboon, Y.; Zaidman, N.A.; Maniak, P.J.; Deachapunya, C.; O'Grady, S.M. P2Y receptor regulation of K2P channels that facilitate K+ secretion by human mammary epithelial cells. *Am. J. Physiol. Cell Physiol.* **2018**, *314*, C627–C639. [CrossRef]
114. Woo, D.H.; Bae, J.Y.; Nam, M.H.; An, H.; Ju, Y.H.; Won, J.; Choi, J.H.; Hwang, E.M.; Han, K.S.; Bae, Y.C.; et al. Activation of Astrocytic μ-opioid Receptor Elicits Fast Glutamate Release Through TREK-1-Containing K2P Channel in Hippocampal Astrocytes. *Front. Cell. Neurosci.* **2018**, *12*, 319. [CrossRef]
115. Devilliers, M.; Busserolles, J.; Lolignier, S.; Deval, E.; Pereira, V.; Alloui, A.; Christin, M.; Mazet, B.; Delmas, P.; Noel, J.; et al. Activation of TREK-1 by morphine results in analgesia without adverse side effects. *Nat. Commun.* **2013**, *4*, 2941. [CrossRef]
116. Cho, P.S.; Lee, H.K.; Lee, S.H.; Im, J.Z.; Jung, S.J. DAMGO modulates two-pore domain K(+) channels in the substantia gelatinosa neurons of rat spinal cord. *Korean J. Physiol. Pharmacol. Off. J. Korean Physiol. Soc. Korean Soc. Pharmacol.* **2016**, *20*, 525–531. [CrossRef] [PubMed]
117. Nockemann, D.; Rouault, M.; Labuz, D.; Hublitz, P.; McKnelly, K.; Reis, F.C.; Stein, C.; Heppenstall, P.A. The K(+) channel GIRK2 is both necessary and sufficient for peripheral opioid-mediated analgesia. *EMBO Mol. Med.* **2013**, *5*, 1263–1277. [CrossRef]
118. Dubin, A.E.; Schmidt, M.; Mathur, J.; Petrus, M.J.; Xiao, B.; Coste, B.; Patapoutian, A. Inflammatory signals enhance piezo2-mediated mechanosensitive currents. *Cell Rep.* **2012**, *2*, 511–517. [CrossRef]
119. Lechner, S.G.; Lewin, G.R. Peripheral sensitisation of nociceptors via G-protein-dependent potentiation of mechanotransduction currents. *J. Physiol.* **2009**, *587*, 3493–3503. [CrossRef]
120. Jia, Z.; Ikeda, R.; Ling, J.; Gu, J.G. GTP-dependent run-up of Piezo2-type mechanically activated currents in rat dorsal root ganglion neurons. *Mol. Brain* **2013**, *6*, 57. [CrossRef]
121. Borbiro, I.; Badheka, D.; Rohacs, T. Activation of TRPV1 channels inhibits mechanosensitive Piezo channel activity by depleting membrane phosphoinositides. *Sci. Signal.* **2015**, *8*, ra15. [CrossRef]
122. Oh, U.; Jung, J. Cellular functions of TMEM16/anoctamin. *Pflugers Arch. Eur. J. Physiol.* **2016**, *468*, 443–453. [CrossRef]
123. Schroeder, B.C.; Cheng, T.; Jan, Y.N.; Jan, L.Y. Expression cloning of TMEM16A as a calcium-activated chloride channel subunit. *Cell* **2008**, *134*, 1019–1029. [CrossRef] [PubMed]
124. Caputo, A.; Caci, E.; Ferrera, L.; Pedemonte, N.; Barsanti, C.; Sondo, E.; Pfeffer, U.; Ravazzolo, R.; Zegarra-Moran, O.; Galietta, L.J.V. TMEM16A, a membrane protein associated with calcium-dependent chloride channel activity. *Science* **2008**, *322*, 590–594. [CrossRef]
125. Yang, Y.D.; Cho, H.; Koo, J.Y.; Tak, M.H.; Cho, Y.; Shim, W.S.; Park, S.P.; Lee, J.; Lee, B.; Kim, B.M.; et al. TMEM16A confers receptor-activated calcium-dependent chloride conductance. *Nature* **2008**, *455*, 1210–1215. [CrossRef] [PubMed]
126. Falzone, M.E.; Malvezzi, M.; Lee, B.C.; Accardi, A. Known structures and unknown mechanisms of TMEM16 scramblases and channels. *J. Gen. Physiol.* **2018**, *150*, 933–947. [CrossRef]
127. Huang, F.; Wong, X.; Jan, L.Y. International Union of Basic and Clinical Pharmacology. LXXXV: calcium-activated chloride channels. *Pharmacol. Rev.* **2012**, *64*, 1–15. [CrossRef]
128. Suzuki, J.; Fujii, T.; Imao, T.; Ishihara, K.; Kuba, H.; Nagata, S. Calcium-dependent phospholipid scramblase activity of TMEM16 protein family members. *J. Biol. Chem.* **2013**, *288*, 13305–13316. [CrossRef] [PubMed]
129. Scudieri, P.; Caci, E.; Venturini, A.; Sondo, E.; Pianigiani, G.; Marchetti, C.; Ravazzolo, R.; Pagani, F.; Galietta, L.J.V. Ion channel and lipid scramblase activity associated with expression of TMEM16F/ANO6 isoforms. *J. Physiol.* **2015**, *593*, 3829–3848. [CrossRef] [PubMed]

130. Zhao, L.; Li, L.I.; Ma, K.T.; Wang, Y.; Li, J.; Shi, W.Y.; Zhu, H.E.; Zhang, Z.S.; Si, J.Q. NSAIDs modulate GABA-activated currents via Ca2+-activated Cl- channels in rat dorsal root ganglion neurons. *Exp. Ther. Med.* **2016**, *11*, 1755–1761. [CrossRef]
131. Huang, F.; Wang, X.; Ostertag, E.M.; Nuwal, T.; Huang, B.; Jan, Y.N.; Basbaum, A.I.; Jan, L.Y. TMEM16C facilitates Na(+)-activated K+ currents in rat sensory neurons and regulates pain processing. *Nat. Neurosci.* **2013**, *16*, 1284–1290. [CrossRef]
132. Cho, H.; Yang, Y.D.; Lee, J.; Lee, B.; Kim, T.; Jang, Y.; Back, S.K.; Na, H.S.; Harfe, B.D.; Wang, F.; et al. The calcium-activated chloride channel anoctamin 1 acts as a heat sensor in nociceptive neurons. *Nat. Neurosci.* **2012**, *15*, 1015–1021. [CrossRef]
133. Lee, B.; Cho, H.; Jung, J.; Yang, Y.D.; Yang, D.J.; Oh, U. Anoctamin 1 contributes to inflammatory and nerve-injury induced hypersensitivity. *Mol. Pain* **2014**, *10*, 5. [CrossRef]
134. Dang, S.; Feng, S.; Tien, J.; Peters, C.J.; Bulkley, D.; Lolicato, M.; Zhao, J.; Zuberbühler, K.; Ye, W.; Qi, L.; et al. Cryo-EM structures of the TMEM16A calcium-activated chloride channel. *Nature* **2017**, *552*, 426–429. [CrossRef]
135. Paulino, C.; Kalienkova, V.; Lam, A.K.M.; Neldner, Y.; Dutzler, R. Activation mechanism of the calcium-activated chloride channel TMEM16A revealed by cryo-EM. *Nature* **2017**, *552*, 421–425. [CrossRef]
136. Paulino, C.; Neldner, Y.; Lam, A.K.; Kalienkova, V.; Brunner, J.D.; Schenck, S.; Dutzler, R. Structural basis for anion conduction in the calcium-activated chloride channel TMEM16A. *eLife* **2017**, *6*, 1–23. [CrossRef]
137. Brunner, J.D.; Lim, N.K.; Schenck, S.; Duerst, A.; Dutzler, R. X-ray structure of a calcium-activated TMEM16 lipid scramblase. *Nature* **2014**, *516*, 207–212. [CrossRef]
138. Fallah, G.; Römer, T.; Detro-Dassen, S.; Braam, U.; Markwardt, F.; Schmalzing, G. TMEM16A(a)/anoctamin-1 shares a homodimeric architecture with CLC chloride channels. *Mol. Cell. Proteom. MCP* **2011**, *10*, M110.004697. [CrossRef]
139. Sheridan, J.T.; Worthington, E.N.; Yu, K.; Gabriel, S.E.; Hartzell, H.C.; Tarran, R. Characterization of the oligomeric structure of the Ca(2+)-activated Cl- channel Ano1/TMEM16A. *J. Biol. Chem.* **2011**, *286*, 1381–1388. [CrossRef]
140. Xiao, Q.; Yu, K.; Perez-Cornejo, P.; Cui, Y.; Arreola, J.; Hartzell, H.C. Voltage- and calcium-dependent gating of TMEM16A/Ano1 chloride channels are physically coupled by the first intracellular loop. *Proc. Natl. Acad. Sci. USA* **2011**, *108*, 8891–8896. [CrossRef]
141. Ma, K.; Wang, H.; Yu, J.; Wei, M.; Xiao, Q. New Insights on the Regulation of Ca2+ -Activated Chloride Channel TMEM16A. *J. Cell. Physiol.* **2017**, *232*, 707–716. [CrossRef]
142. Jin, X.; Shah, S.; Liu, Y.; Zhang, H.; Lees, M.; Fu, Z.; Lippiat, J.D.; Beech, D.J.; Sivaprasadarao, A.; Baldwin, S.A.; et al. Activation of the Cl- channel ANO1 by localized calcium signals in nociceptive sensory neurons requires coupling with the IP3 receptor. *Sci. Signal.* **2013**, *6*, ra73. [CrossRef]
143. Takayama, Y.; Uta, D.; Furue, H.; Tominaga, M. Pain-enhancing mechanism through interaction between TRPV1 and anoctamin 1 in sensory neurons. *Proc. Natl. Acad. Sci. USA* **2015**, *112*, 5213–5218. [CrossRef]
144. Cho, H.; Oh, U. Anoctamin 1 mediates thermal pain as a heat sensor. *Curr. Neuropharmacol.* **2013**, *11*, 641–651. [CrossRef]
145. Liu, B.; Linley, J.E.; Du, X.; Zhang, X.; Ooi, L.; Zhang, H.; Gamper, N. The acute nociceptive signals induced by bradykinin in rat sensory neurons are mediated by inhibition of M-type K+ channels and activation of Ca2+-activated Cl- channels. *J. Clin. Investig.* **2010**, *120*, 1240–1252. [CrossRef]
146. Zhang, Y.; Wang, H.; Ke, J.; Wei, Y.; Ji, H.; Qian, Z.; Liu, L.; Tao, J. Inhibition of A-Type K+ Channels by Urotensin-II Induces Sensory Neuronal Hyperexcitability Through the PKCα-ERK Pathway. *Endocrinology* **2018**, *159*, 2253–2263. [CrossRef]
147. Rajagopal, M.; Kathpalia, P.P.; Thomas, S.V.; Pao, A.C. Activation of P2Y1 and P2Y2 receptors induces chloride secretion via calcium-activated chloride channels in kidney inner medullary collecting duct cells. *Am. J. Physiology. Ren. Physiol.* **2011**, *301*, F544–553. [CrossRef]
148. Wang, J.; Haanes, K.A.; Novak, I. Purinergic regulation of CFTR and Ca(2+)-activated Cl(-) channels and K(+) channels in human pancreatic duct epithelium. *Am. J. Physiology. Cell Physiol.* **2013**, *304*, C673–684. [CrossRef]
149. Romero, T.R.L.; Pacheco, D.D.F.; Duarte, I.D.G. Probable involvement of Ca(2+)-activated Cl(-) channels (CaCCs) in the activation of CB1 cannabinoid receptors. *Life Sci.* **2013**, *92*, 815–820. [CrossRef]

150. Pacheco, D.d.F.; Pacheco, C.M.D.F.; Duarte, I.D.G. δ-Opioid receptor agonist SNC80 induces central antinociception mediated by Ca2+ -activated Cl- channels. *J. Pharm. Pharmacol.* **2012**, *64*, 1084–1089. [CrossRef]
151. Cardoso, F.C.; Lewis, R.J. Sodium channels and pain: From toxins to therapies. *Br. J. Pharmacol.* **2018**, *175*, 2138–2157. [CrossRef]
152. Hodgkin, A.L.; Huxley, A.F. The components of membrane conductance in the giant axon of Loligo. *J. Physiol.* **1952**, *116*, 473–496. [CrossRef] [PubMed]
153. Hodgkin, A.L.; Huxley, A.F. The dual effect of membrane potential on sodium conductance in the giant axon of Loligo. *J. Physiol.* **1952**, *116*, 497–506. [CrossRef] [PubMed]
154. Hodgkin, A.L.; Huxley, A.F. Currents carried by sodium and potassium ions through the membrane of the giant axon of Loligo. *J. Physiol.* **1952**, *116*, 449–472. [CrossRef] [PubMed]
155. Hodgkin, A.L.; Huxley, A.F.; Katz, B. Measurement of current-voltage relations in the membrane of the giant axon of Loligo. *J. Physiol.* **1952**, *116*, 424–448. [CrossRef] [PubMed]
156. Brouwer, B.A.; Merkies, I.S.J.; Gerrits, M.M.; Waxman, S.G.; Hoeijmakers, J.G.J.; Faber, C.G. Painful neuropathies: The emerging role of sodium channelopathies. *J. Peripher. Nerv. Syst. JPNS* **2014**, *19*, 53–65. [CrossRef]
157. Bezanilla, F. How membrane proteins sense voltage. *Nat. Rev. Mol. Cell Biol.* **2008**, *9*, 323–332. [CrossRef]
158. Cardenas, L.M.; Cardenas, C.G.; Scroggs, R.S. 5HT increases excitability of nociceptor-like rat dorsal root ganglion neurons via cAMP-coupled TTX-resistant Na(+) channels. *J. Neurophysiol.* **2001**, *86*, 241–248. [CrossRef]
159. Gold, M.S.; Reichling, D.B.; Shuster, M.J.; Levine, J.D. Hyperalgesic agents increase a tetrodotoxin-resistant Na+ current in nociceptors. *Proc. Natl. Acad. Sci. USA* **1996**, *93*, 1108–1112. [CrossRef]
160. England, S.; Bevan, S.; Docherty, R.J. PGE2 modulates the tetrodotoxin-resistant sodium current in neonatal rat dorsal root ganglion neurones via the cyclic AMP-protein kinase A cascade. *J. Physiol.* **1996**, *495*, 429–440. [CrossRef]
161. Natura, G.; von Banchet, G.S.; Schaible, H.G. Calcitonin gene-related peptide enhances TTX-resistant sodium currents in cultured dorsal root ganglion neurons from adult rats. *Pain* **2005**, *116*, 194–204. [CrossRef] [PubMed]
162. Scroggs, R.S. Serotonin upregulates low- and high-threshold tetrodotoxin-resistant sodium channels in the same subpopulation of rat nociceptors. *Neuroscience* **2010**, *165*, 1293–1300. [CrossRef] [PubMed]
163. Cardenas, C.G.; Del Mar, L.P.; Scroggs, R.S. Two parallel signaling pathways couple 5HT1A receptors to N- and L-type calcium channels in C-like rat dorsal root ganglion cells. *J. Neurophysiol.* **1997**, *77*, 3284–3296. [CrossRef] [PubMed]
164. Matsumoto, S.; Ikeda, M.; Yoshida, S.; Tanimoto, T.; Takeda, M.; Nasu, M. Prostaglandin E2-induced modification of tetrodotoxin-resistant Na+ currents involves activation of both EP2 and EP4 receptors in neonatal rat nodose ganglion neurones. *Br. J. Pharmacol.* **2005**, *145*, 503–513. [CrossRef] [PubMed]
165. Moraes, E.R.; Kushmerick, C.; Naves, L.A. Characteristics of dorsal root ganglia neurons sensitive to Substance P. *Mol. Pain* **2014**, *10*, 73. [CrossRef] [PubMed]
166. Cang, C.L.; Zhang, H.; Zhang, Y.Q.; Zhao, Z.Q. PKCepsilon-dependent potentiation of TTX-resistant Nav1.8 current by neurokinin-1 receptor activation in rat dorsal root ganglion neurons. *Mol. Pain* **2009**, *5*, 33. [CrossRef]
167. Song, J.H.; Shin, Y.K.; Lee, C.S.; Bang, H.; Park, M. Effects of ATP on TTX-sensitive and TTX-resistant sodium currents in rat sensory neurons. *Neuroreport* **2001**, *12*, 3659–3662. [CrossRef]
168. Kayssi, A.; Amadesi, S.; Bautista, F.; Bunnett, N.W.; Vanner, S. Mechanisms of protease-activated receptor 2-evoked hyperexcitability of nociceptive neurons innervating the mouse colon. *J. Physiol.* **2007**, *580*, 977–991. [CrossRef]
169. Gold, M.S.; Levine, J.D. DAMGO inhibits prostaglandin E2-induced potentiation of a TTX-resistant Na+ current in rat sensory neurons in vitro. *Neurosci. Lett.* **1996**, *212*, 83–86. [CrossRef]
170. Catterall, W.A. Voltage-gated calcium channels. *Cold Spring Harb. Perspect. Biol.* **2011**, *3*, a003947. [CrossRef]
171. Acosta, C.G.; López, H.S. delta opioid receptor modulation of several voltage-dependent Ca(2+) currents in rat sensory neurons. *J. Neurosci. Off. J. Soc. Neurosci.* **1999**, *19*, 8337–8348. [CrossRef]
172. Bean, B.P. Neurotransmitter inhibition of neuronal calcium currents by changes in channel voltage dependence. *Nature* **1989**, *340*, 153–156. [CrossRef]

173. Dolphin, A.C. Beta subunits of voltage-gated calcium channels. *J. Bioenerg. Biomembr.* **2003**, *35*, 599–620. [CrossRef]
174. Proft, J.; Weiss, N. G protein regulation of neuronal calcium channels: Back to the future. *Mol. Pharmacol.* **2015**, *87*, 890–906. [CrossRef]
175. Scherrer, G.; Imamachi, N.; Cao, Y.Q.; Contet, C.; Mennicken, F.; O'Donnell, D.; Kieffer, B.L.; Basbaum, A.I. Dissociation of the opioid receptor mechanisms that control mechanical and heat pain. *Cell* **2009**, *137*, 1148–1159. [CrossRef]
176. Snyder, L.M.; Chiang, M.C.; Loeza-Alcocer, E.; Omori, Y.; Hachisuka, J.; Sheahan, T.D.; Gale, J.R.; Adelman, P.C.; Sypek, E.I.; Fulton, S.A.; et al. Kappa Opioid Receptor Distribution and Function in Primary Afferents. *Neuron* **2018**, *99*, 1274–1288.e6. [CrossRef]
177. Mudge, A.W.; Leeman, S.E.; Fischbach, G.D. Enkephalin inhibits release of substance P from sensory neurons in culture and decreases action potential duration. *Proc. Natl. Acad. Sci. USA* **1979**, *76*, 526–530. [CrossRef]
178. Werz, M.A.; Grega, D.S.; MacDonald, R.L. Actions of mu, delta and kappa opioid agonists and antagonists on mouse primary afferent neurons in culture. *J. Pharmacol. Exp. Ther.* **1987**, *243*, 258–263.
179. Werz, M.A.; Macdonald, R.L. Heterogeneous sensitivity of cultured dorsal root ganglion neurones to opioid peptides selective for mu- and delta-opiate receptors. *Nature* **1982**, *299*, 730–733. [CrossRef]
180. Werz, M.A.; Macdonald, R.L. Dynorphin reduces calcium-dependent action potential duration by decreasing voltage-dependent calcium conductance. *Neurosci. Lett.* **1984**, *46*, 185–190. [CrossRef]
181. Werz, M.A.; MacDonald, R.L. Opioid peptides selective for mu- and delta-opiate receptors reduce calcium-dependent action potential duration by increasing potassium conductance. *Neurosci. Lett.* **1983**, *42*, 173–178. [CrossRef]
182. Macdonald, R.L.; Werz, M.A. Dynorphin A decreases voltage-dependent calcium conductance of mouse dorsal root ganglion neurones. *J. Physiol.* **1986**, *377*, 237–249. [CrossRef]
183. Andrade, A.; Denome, S.; Jiang, Y.Q.; Marangoudakis, S.; Lipscombe, D. Opioid inhibition of N-type Ca2+ channels and spinal analgesia couple to alternative splicing. *Nat. Neurosci.* **2010**, *13*, 1249–1256. [CrossRef]
184. Gross, R.A.; Macdonald, R.L. Dynorphin A selectively reduces a large transient (N-type) calcium current of mouse dorsal root ganglion neurons in cell culture. *Proc. Natl. Acad. Sci. USA* **1987**, *84*, 5469–5473. [CrossRef]
185. Heinke, B.; Gingl, E.; Sandkühler, J. Multiple targets of $\mu$-opioid receptor-mediated presynaptic inhibition at primary afferent A$\delta$- and C-fibers. *J. Neurosci. Off. J. Soc. Neurosci.* **2011**, *31*, 1313–1322. [CrossRef]
186. King, A.P.; Hall, K.E.; Macdonald, R.L. kappa- and mu-Opioid inhibition of N-type calcium currents is attenuated by 4beta-phorbol 12-myristate 13-acetate and protein kinase C in rat dorsal root ganglion neurons. *J. Pharmacol. Exp. Ther.* **1999**, *289*, 312–320.
187. Moises, H.C.; Rusin, K.I.; Macdonald, R.L. Mu- and kappa-opioid receptors selectively reduce the same transient components of high-threshold calcium current in rat dorsal root ganglion sensory neurons. *J. Neurosci. Off. J. Soc. Neurosci.* **1994**, *14*, 5903–5916. [CrossRef]
188. Su, X.; Wachtel, R.E.; Gebhart, G.F. Inhibition of calcium currents in rat colon sensory neurons by K- but not mu- or delta-opioids. *J. Neurophysiol.* **1998**, *80*, 3112–3119. [CrossRef]
189. Wiley, J.W.; Moises, H.C.; Gross, R.A.; MacDonald, R.L. Dynorphin A-mediated reduction in multiple calcium currents involves a G(o) alpha-subtype G protein in rat primary afferent neurons. *J. Neurophysiol.* **1997**, *77*, 1338–1348. [CrossRef]
190. Dolphin, A.C.; McGuirk, S.M.; Scott, R.H. An investigation into the mechanisms of inhibition of calcium channel currents in cultured sensory neurones of the rat by guanine nucleotide analogues and ( )-baclofen. *Br. J. Pharmacol.* **1989**, *97*, 263–273. [CrossRef]
191. Dolphin, A.C.; Forda, S.R.; Scott, R.H. Calcium-dependent currents in cultured rat dorsal root ganglion neurones are inhibited by an adenosine analogue. *J. Physiol.* **1986**, *373*, 47–61. [CrossRef]
192. Cardenas, C.G.; Del Mar, L.P.; Scroggs, R.S. Variation in serotonergic inhibition of calcium channel currents in four types of rat sensory neurons differentiated by membrane properties. *J. Neurophysiol.* **1995**, *74*, 1870–1879. [CrossRef] [PubMed]
193. Del Mar, L.P.; Cardenas, C.G.; Scroggs, R.S. Serotonin inhibits high-threshold Ca2+ channel currents in capsaicin-sensitive acutely isolated adult rat DRG neurons. *J. Neurophysiol.* **1994**, *72*, 2551–2554. [CrossRef] [PubMed]

194. Dunlap, K.; Fischbach, G.D. Neurotransmitters decrease the calcium conductance activated by depolarization of embryonic chick sensory neurones. *J. Physiol.* **1981**, *317*, 519–535. [CrossRef]
195. Holz, G.G.; Shefner, S.A.; Anderson, E.G. Serotonin decreases the duration of action potentials recorded from tetraethylammonium-treated bullfrog dorsal root ganglion cells. *J. Neurosci. Off. J. Soc. Neurosci.* **1986**, *6*, 620–626. [CrossRef]
196. Gerevich, Z.; Borvendeg, S.J.; Schröder, W.; Franke, H.; Wirkner, K.; Nörenberg, W.; Fürst, S.; Gillen, C.; Illes, P. Inhibition of N-type voltage-activated calcium channels in rat dorsal root ganglion neurons by P2Y receptors is a possible mechanism of ADP-induced analgesia. *J. Neurosci. Off. J. Soc. Neurosci.* **2004**, *24*, 797–807. [CrossRef] [PubMed]
197. Ross, R.A.; Coutts, A.A.; McFarlane, S.M.; Anavi-Goffer, S.; Irving, A.J.; Pertwee, R.G.; MacEwan, D.J.; Scott, R.H. Actions of cannabinoid receptor ligands on rat cultured sensory neurones: Implications for antinociception. *Neuropharmacology* **2001**, *40*, 221–232. [CrossRef]
198. Bleakman, D.; Colmers, W.F.; Fournier, A.; Miller, R.J. Neuropeptide Y inhibits Ca2+ influx into cultured dorsal root ganglion neurones of the rat via a Y2 receptor. *Br. J. Pharmacol.* **1991**, *103*, 1781–1789. [CrossRef] [PubMed]
199. Ewald, D.A.; Matthies, H.J.; Perney, T.M.; Walker, M.W.; Miller, R.J. The effect of down regulation of protein kinase C on the inhibitory modulation of dorsal root ganglion neuron Ca2+ currents by neuropeptide Y. *J. Neurosci. Off. J. Soc. Neurosci.* **1988**, *8*, 2447–2451. [CrossRef]
200. Walker, M.W.; Ewald, D.A.; Perney, T.M.; Miller, R.J. Neuropeptide Y modulates neurotransmitter release and Ca2+ currents in rat sensory neurons. *J. Neurosci. Off. J. Soc. Neurosci.* **1988**, *8*, 2438–2446. [CrossRef]
201. Gorham, L.; Just, S.; Doods, H. Somatostatin 4 receptor activation modulates G-protein coupled inward rectifying potassium channels and voltage stimulated calcium signals in dorsal root ganglion neurons. *Eur. J. Pharmacol.* **2014**, *736*, 101–106. [CrossRef]
202. Abdulla, F.A.; Smith, P.A. Ectopic alpha2-adrenoceptors couple to N-type Ca2+ channels in axotomized rat sensory neurons. *J. Neurosci. Off. J. Soc. Neurosci.* **1997**, *17*, 1633–1641. [CrossRef]
203. Gamper, N.; Reznikov, V.; Yamada, Y.; Yang, J.; Shapiro, M.S. Phosphatidylinositol [correction] 4,5-bisphosphate signals underlie receptor-specific Gq/11-mediated modulation of N-type Ca2+ channels. *J. Neurosci. Off. J. Soc. Neurosci.* **2004**, *24*, 10980–10992. [CrossRef] [PubMed]
204. Lechner, S.G.; Hussl, S.; Schicker, K.W.; Drobny, H.; Boehm, S. Presynaptic inhibition via a phospholipase C- and phosphatidylinositol bisphosphate-dependent regulation of neuronal Ca2+ channels. *Mol. Pharmacol.* **2005**, *68*, 1387–1396. [CrossRef]
205. Wu, L.; Bauer, C.S.; Zhen, X.G.; Xie, C.; Yang, J. Dual regulation of voltage-gated calcium channels by PtdIns(4,5)P2. *Nature* **2002**, *419*, 947–952. [CrossRef]
206. Li, Z.; He, S.Q.; Tseng, P.Y.; Xu, Q.; Tiwari, V.; Yang, F.; Shu, B.; Zhang, T.; Tang, Z.; Raja, S.N.; et al. The inhibition of high-voltage-activated calcium current by activation of MrgC11 involves phospholipase C-dependent mechanisms. *Neuroscience* **2015**, *300*, 393–403. [CrossRef]
207. Huang, D.; Huang, S.; Peers, C.; Du, X.; Zhang, H.; Gamper, N. GABAB receptors inhibit low-voltage activated and high-voltage activated Ca(2+) channels in sensory neurons via distinct mechanisms. *Biochem. Biophys. Res. Commun.* **2015**, *465*, 188–193. [CrossRef] [PubMed]
208. Hofmann, F.; Flockerzi, V.; Kahl, S.; Wegener, J.W. L-type CaV1.2 calcium channels: From in vitro findings to in vivo function. *Physiol. Rev.* **2014**, *94*, 303–326. [CrossRef]
209. Gutman, G.A.; Chandy, K.G.; Adelman, J.P.; Aiyar, J.; Bayliss, D.A.; Clapham, D.E.; Covarriubias, M.; Desir, G.V.; Furuichi, K.; Ganetzky, B.; et al. International Union of Pharmacology. XLI. Compendium of voltage-gated ion channels: Potassium channels. *Pharmacol. Rev.* **2003**, *55*, 583–586. [CrossRef]
210. Tsantoulas, C.; McMahon, S.B. Opening paths to novel analgesics: The role of potassium channels in chronic pain. *Trends Neurosci.* **2014**, *37*, 146–158. [CrossRef]
211. Swartz, K.J. Towards a structural view of gating in potassium channels. *Nat. Reviews. Neurosci.* **2004**, *5*, 905–916. [CrossRef]
212. Kuang, Q.; Purhonen, P.; Hebert, H. Structure of potassium channels. *Cell. Mol. Life Sci. CMLS* **2015**, *72*, 3677–3693. [CrossRef]
213. Brown, D.A.; Passmore, G.M. Neural KCNQ (Kv7) channels. *Br. J. Pharmacol.* **2009**, *156*, 1185–1195. [CrossRef]

214. Du, X.; Gao, H.; Jaffe, D.; Zhang, H.; Gamper, N. M-type K+ channels in peripheral nociceptive pathways. *Br. J. Pharmacol.* **2018**, *175*, 2158–2172. [CrossRef]
215. Howard, R.J.; Clark, K.A.; Holton, J.M.; Minor, D.L. Structural insight into KCNQ (Kv7) channel assembly and channelopathy. *Neuron* **2007**, *53*, 663–675. [CrossRef]
216. Delmas, P.; Brown, D.A. Pathways modulating neural KCNQ/M (Kv7) potassium channels. *Nat. Rev. Neurosci.* **2005**, *6*, 850–862. [CrossRef]
217. Du, X.; Hao, H.; Gigout, S.; Huang, D.; Yang, Y.; Li, L.; Wang, C.; Sundt, D.; Jaffe, D.B.; Zhang, H.; Gamper, N. Control of somatic membrane potential in nociceptive neurons and its implications for peripheral nociceptive transmission. *Pain* **2014**, *155*, 2306–2322. [CrossRef]
218. Rivera-Arconada, I.; Roza, C.; Lopez-Garcia, J.A. Enhancing m currents: A way out for neuropathic pain? *Front. Mol. Neurosci.* **2009**, *2*, 10. [CrossRef]
219. Passmore, G.M.; Selyanko, A.A.; Mistry, M.; Al-Qatari, M.; Marsh, S.J.; Matthews, E.A.; Dickenson, A.H.; Brown, T.A.; Burbidge, S.A.; Main, M.; et al. KCNQ/M currents in sensory neurons: Significance for pain therapy. *J. Neurosci. Off. J. Soc. Neurosci.* **2003**, *23*, 7227–7236. [CrossRef]
220. Zheng, Y.; Xu, H.; Zhan, L.; Zhou, X.; Chen, X.; Gao, Z. Activation of peripheral KCNQ channels relieves gout pain. *Pain* **2015**, *156*, 1025–1035. [CrossRef]
221. Ray, S.; Salzer, I.; Kronschläger, M.T.; Boehm, S. The paracetamol metabolite N-acetylp-benzoquinone imine reduces excitability in first- and second-order neurons of the pain pathway through actions on KV7 channels. *Pain* **2019**, *160*, 954–964. [CrossRef]
222. Zheng, Q.; Fang, D.; Liu, M.; Cai, J.; Wan, Y.; Han, J.S.; Xing, G.G. Suppression of KCNQ/M (Kv7) potassium channels in dorsal root ganglion neurons contributes to the development of bone cancer pain in a rat model. *Pain* **2013**, *154*, 434–448. [CrossRef]
223. Yu, T.; Li, L.; Liu, H.; Li, H.; Liu, Z.; Li, Z. KCNQ2/3/5 channels in dorsal root ganglion neurons can be therapeutic targets of neuropathic pain in diabetic rats. *Mol. Pain* **2018**, *14*, 1744806918793229. [CrossRef]
224. Brown, B.S.; Yu, S.P. Modulation and genetic identification of the M channel. *Prog. Biophys. Mol. Biol.* **2000**, *73*, 135–166. [CrossRef]
225. Ishimatsu, M. Substance P produces an inward current by suppressing voltage-dependent and -independent K+ currents in bullfrog primary afferent neurons. *Neurosci. Res.* **1994**, *19*, 9–20. [CrossRef]
226. Akasu, T.; Ishimatsu, M.; Yamada, K. Tachykinins cause inward current through NK1 receptors in bullfrog sensory neurons. *Brain Res.* **1996**, *713*, 160–167. [CrossRef]
227. Lin, C.C.J.; Chen, W.N.; Chen, C.J.; Lin, Y.W.; Zimmer, A.; Chen, C.C. An antinociceptive role for substance P in acid-induced chronic muscle pain. *Proc. Natl. Acad. Sci. USA* **2012**, *109*, E76–83. [CrossRef]
228. Neves, S.R.; Ram, P.T.; Iyengar, R. G protein pathways. *Science* **2002**, *296*, 1636–1639. [CrossRef]
229. Suh, B.C.; Hille, B. Recovery from muscarinic modulation of M current channels requires phosphatidylinositol 4,5-bisphosphate synthesis. *Neuron* **2002**, *35*, 507–520. [CrossRef]
230. Higashida, H.; Brown, D.A. Two polyphosphatidylinositide metabolites control two K+ currents in a neuronal cell. *Nature* **1986**, *323*, 333–335. [CrossRef]
231. Selyanko, A.A.; Brown, D.A. Intracellular calcium directly inhibits potassium M channels in excised membrane patches from rat sympathetic neurons. *Neuron* **1996**, *16*, 151–162. [CrossRef]
232. Gamper, N.; Shapiro, M.S. Calmodulin mediates Ca2+-dependent modulation of M-type K+ channels. *J. Gen. Physiol.* **2003**, *122*, 17–31. [CrossRef]
233. Brown, D.A.; Hughes, S.A.; Marsh, S.J.; Tinker, A. Regulation of M(Kv7.2/7.3) channels in neurons by PIP(2) and products of PIP(2) hydrolysis: Significance for receptor-mediated inhibition. *J. Physiol.* **2007**, *582*, 917–925. [CrossRef]
234. Soh, U.J.; Dores, M.R.; Chen, B.; Trejo, J. Signal transduction by protease-activated receptors. *Br. J. Pharmacol.* **2010**, *160*, 191–203. [CrossRef]
235. Linley, J.E.; Rose, K.; Patil, M.; Robertson, B.; Akopian, A.N.; Gamper, N. Inhibition of M current in sensory neurons by exogenous proteases: A signaling pathway mediating inflammatory nociception. *J. Neurosci. Off. J. Soc. Neurosci.* **2008**, *28*, 11240–11249. [CrossRef]
236. von Kügelgen, I. Pharmacological profiles of cloned mammalian P2Y-receptor subtypes. *Pharmacol. Ther.* **2006**, *110*, 415–432. [CrossRef]
237. Dong, X.; Han, S.; Zylka, M.J.; Simon, M.I.; Anderson, D.J. A diverse family of GPCRs expressed in specific subsets of nociceptive sensory neurons. *Cell* **2001**, *106*, 619–632. [CrossRef]

238. Choi, S.S.; Lahn, B.T. Adaptive evolution of MRG, a neuron-specific gene family implicated in nociception. *Genome Res.* **2003**, *13*, 2252–2259. [CrossRef]
239. Zhang, L.; Taylor, N.; Xie, Y.; Ford, R.; Johnson, J.; Paulsen, J.E.; Bates, B. Cloning and expression of MRG receptors in macaque, mouse, and human. *Brain Res. Mol. Brain Res.* **2005**, *133*, 187–197. [CrossRef]
240. Zylka, M.J.; Rice, F.L.; Anderson, D.J. Topographically distinct epidermal nociceptive circuits revealed by axonal tracers targeted to Mrgprd. *Neuron* **2005**, *45*, 17–25. [CrossRef]
241. Crozier, R.A.; Ajit, S.K.; Kaftan, E.J.; Pausch, M.H. MrgD activation inhibits KCNQ/M-currents and contributes to enhanced neuronal excitability. *J. Neurosci. Off. J. Soc. Neurosci.* **2007**, *27*, 4492–4496. [CrossRef]
242. Baldwin, T.J.; Tsaur, M.L.; Lopez, G.A.; Jan, Y.N.; Jan, L.Y. Characterization of a mammalian cDNA for an inactivating voltage-sensitive K+ channel. *Neuron* **1991**, *7*, 471–483. [CrossRef]
243. Shibata, R.; Nakahira, K.; Shibasaki, K.; Wakazono, Y.; Imoto, K.; Ikenaka, K. A-type K+ current mediated by the Kv4 channel regulates the generation of action potential in developing cerebellar granule cells. *J. Neurosci. Off. J. Soc. Neurosci.* **2000**, *20*, 4145–4155. [CrossRef]
244. Vydyanathan, A.; Wu, Z.Z.; Chen, S.R.; Pan, H.L. A-type voltage-gated K+ currents influence firing properties of isolectin B4-positive but not isolectin B4-negative primary sensory neurons. *J. Neurophysiol.* **2005**, *93*, 3401–3409. [CrossRef]
245. Chien, L.Y.; Cheng, J.K.; Chu, D.; Cheng, C.F.; Tsaur, M.L. Reduced expression of A-type potassium channels in primary sensory neurons induces mechanical hypersensitivity. *J. Neurosci. Off. J. Soc. Neurosci.* **2007**, *27*, 9855–9865. [CrossRef]
246. Liu, P.W.; Blair, N.T.; Bean, B.P. Action Potential Broadening in Capsaicin-Sensitive DRG Neurons from Frequency-Dependent Reduction of Kv3 Current. *J. Neurosci. Off. J. Soc. Neurosci.* **2017**, *37*, 9705–9714. [CrossRef]
247. Zemel, B.M.; Ritter, D.M.; Covarrubias, M.; Muqeem, T. A-Type KV Channels in Dorsal Root Ganglion Neurons: Diversity, Function, and Dysfunction. *Front. Mol. Neurosci.* **2018**, *11*, 253. [CrossRef]
248. Ritter, D.M.; Ho, C.; O'Leary, M.E.; Covarrubias, M. Modulation of Kv3.4 channel N-type inactivation by protein kinase C shapes the action potential in dorsal root ganglion neurons. *J. Physiol.* **2012**, *590*, 145–161. [CrossRef]
249. Ritter, D.M.; Zemel, B.M.; Hala, T.J.; O'Leary, M.E.; Lepore, A.C.; Covarrubias, M. Dysregulation of Kv3.4 channels in dorsal root ganglia following spinal cord injury. *J. Neurosci. Off. J. Soc. Neurosci.* **2015**, *35*, 1260–1273. [CrossRef]
250. Zemel, B.M.; Muqeem, T.; Brown, E.V.; Goulão, M.; Urban, M.W.; Tymanskyj, S.R.; Lepore, A.C.; Covarrubias, M. Calcineurin Dysregulation Underlies Spinal Cord Injury-Induced K+ Channel Dysfunction in DRG Neurons. *J. Neurosci. Off. J. Soc. Neurosci.* **2017**, *37*, 8256–8272. [CrossRef]
251. Zhang, Y.; Jiang, D.; Zhang, Y.; Jiang, X.; Wang, F.; Tao, J. Neuromedin U type 1 receptor stimulation of A-type K+ current requires the $\beta\gamma$ subunits of Go protein, protein kinase A, and extracellular signal-regulated kinase 1/2 (ERK1/2) in sensory neurons. *J. Biol. Chem.* **2012**, *287*, 18562–18572. [CrossRef]
252. Saetrum Opgaard, O.; Nothacker, H.; Ehlert, F.J.; Krause, D.N. Human urotensin II mediates vasoconstriction via an increase in inositol phosphates. *Eur. J. Pharmacol.* **2000**, *406*, 265–271. [CrossRef]
253. Maguire, J.J.; Davenport, A.P. Is urotensin-II the new endothelin? *Br. J. Pharmacol.* **2002**, *137*, 579–588. [CrossRef]
254. Zhang, M.; Gao, C.X.; Wang, Y.P.; Ma, K.T.; Li, L.; Yin, J.W.; Dai, Z.G.; Wang, S.; Si, J.Q. The association between the expression of PAR2 and TMEM16A and neuropathic pain. *Mol. Med. Rep.* **2018**, *17*, 3744–3750. [CrossRef]
255. Li, N.; Lu, Z.Y.; Yu, L.H.; Burnstock, G.; Deng, X.M.; Ma, B. Inhibition of G protein-coupled P2Y2 receptor induced analgesia in a rat model of trigeminal neuropathic pain. *Mol. Pain* **2014**, *10*, 21. [CrossRef]
256. Jacobson, K.A.; Ivanov, A.A.; de Castro, S.; Harden, T.K.; Ko, H. Development of selective agonists and antagonists of P2Y receptors. *Purinergic Signal.* **2009**, *5*, 75–89. [CrossRef]
257. Glaaser, I.W.; Slesinger, P.A. Structural Insights into GIRK Channel Function. *Int. Rev. Neurobiol.* **2015**, *123*, 117–160.
258. Nagi, K.; Pineyro, G. Kir3 channel signaling complexes: Focus on opioid receptor signaling. *Front. Cell. Neurosci.* **2014**, *8*, 186. [CrossRef]

259. Peleg, S.; Varon, D.; Ivanina, T.; Dessauer, C.W.; Dascal, N. G(alpha)(i) controls the gating of the G protein-activated K(+) channel, GIRK. *Neuron* **2002**, *33*, 87–99. [CrossRef]
260. Huang, C.L.; Feng, S.; Hilgemann, D.W. Direct activation of inward rectifier potassium channels by PIP2 and its stabilization by Gbetagamma. *Nature* **1998**, *391*, 803–806. [CrossRef]
261. Sui, J.L.; Petit-Jacques, J.; Logothetis, D.E. Activation of the atrial KACh channel by the betagamma subunits of G proteins or intracellular Na+ ions depends on the presence of phosphatidylinositol phosphates. *Proc. Natl. Acad. Sci. USA* **1998**, *95*, 1307–1312. [CrossRef]
262. Khodorova, A.; Navarro, B.; Jouaville, L.S.; Murphy, J.E.; Rice, F.L.; Mazurkiewicz, J.E.; Long-Woodward, D.; Stoffel, M.; Strichartz, G.R.; Yukhananov, R.; et al. Endothelin-B receptor activation triggers an endogenous analgesic cascade at sites of peripheral injury. *Nat. Med.* **2003**, *9*, 1055–1061. [CrossRef]
263. Gao, X.F.; Zhang, H.L.; You, Z.D.; Lu, C.L.; He, C. G protein-coupled inwardly rectifying potassium channels in dorsal root ganglion neurons. *Acta Pharmacol. Sin.* **2007**, *28*, 185–190. [CrossRef]
264. Stötzner, P.; Spahn, V.; Celik, M.Ö.; Labuz, D.; Machelska, H. Mu-Opioid Receptor Agonist Induces Kir3 Currents in Mouse Peripheral Sensory Neurons - Effects of Nerve Injury. *Front. Pharmacol.* **2018**, *9*, 1478. [CrossRef]

International Journal of *Molecular Sciences*

*Review*

# Ion Channels Involved in Tooth Pain

**Kihwan Lee [1,†], Byeong-Min Lee [2,†], Chul-Kyu Park [1], Yong Ho Kim [1,*] and Gehoon Chung [2,3,*]**

1 Gachon Pain Center and Department of Physiology, College of Medicine, Gachon University, Incheon 406-799, Korea; key1479@gmail.com (K.L.); pck0708@gachon.ac.kr (C.-K.P.)
2 Department of Oral Physiology and Program in Neurobiology, School of Dentistry, Seoul National University, Seoul 08826, Korea; bmjj88@snu.ac.kr
3 Dental Research Institute, Seoul National University, Seoul 03080, Korea
* Correspondence: euro16@gachon.ac.kr (Y.H.K.); gehoon@snu.ac.kr (G.C.); Tel.: +82-2-880-2332 (G.C.)
† These authors equally contributed to this work.

Received: 31 March 2019; Accepted: 3 May 2019; Published: 8 May 2019

**Abstract:** The tooth has an unusual sensory system that converts external stimuli predominantly into pain, yet its sensory afferents in teeth demonstrate cytochemical properties of non-nociceptive neurons. This review summarizes the recent knowledge underlying this paradoxical nociception, with a focus on the ion channels involved in tooth pain. The expression of temperature-sensitive ion channels has been extensively investigated because thermal stimulation often evokes tooth pain. However, temperature-sensitive ion channels cannot explain the sudden intense tooth pain evoked by innocuous temperatures or light air puffs, leading to the hydrodynamic theory emphasizing the microfluidic movement within the dentinal tubules for detection by mechanosensitive ion channels. Several mechanosensitive ion channels expressed in dental sensory systems have been suggested as key players in the hydrodynamic theory, and TRPM7, which is abundant in the odontoblasts, and recently discovered PIEZO receptors are promising candidates. Several ligand-gated ion channels and voltage-gated ion channels expressed in dental primary afferent neurons have been discussed in relation to their potential contribution to tooth pain. In addition, in recent years, there has been growing interest in the potential sensory role of odontoblasts; thus, the expression of ion channels in odontoblasts and their potential relation to tooth pain is also reviewed.

**Keywords:** tooth pain; TRP channels; odontoblasts; piezo; purinergic; trigeminal ganglion

## 1. Introduction

The tooth is a unique sensory system that senses external stimuli predominantly as nociception. Most of the nerves innervating tooth pulp have been presumed to be nociceptors since most axons in tooth pulp are unmyelinated or small fibers that are myelinated [1]. However, this belief was challenged by multiple observations that pulpal nerves possess physical and chemical properties of large myelinated Aβ fibers. Due to these paradoxical findings, a new concept of "algoneurons" was introduced [2,3].

The structure of the tooth is comprised of densely vascularized and innervated tooth pulp covered by two layers of hard tissue—the dentin and enamel [3,4]. The dentin and enamel are distinguished by their microstructure and mineral content. The outermost enamel layer is the hardest tissue in the body, with minerals forming 97% of its weight. The dentin layer lies between the tooth pulp and the enamel layer and has an intermediate hardness with a mineral content slightly higher than that of bone, providing resilience to the enamel. The most notable property of dentin is its microstructure. Dentin is made of thousands of microtubules—dentinal tubules—filled with dentin tubular fluid. Odontoblasts are the cells that deposit the calcium matrix to form dentin and constitute a cellular single layer at the inter-surface of the dentin and the tooth pulp. Each odontoblast possesses a process that protrudes into the dentinal tubules (Figure 1).

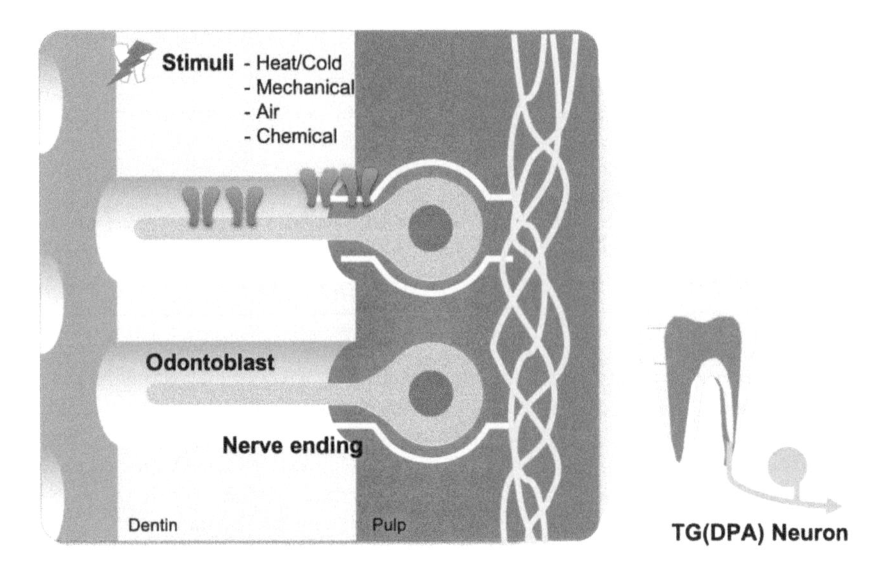


**Figure 1.** Anatomical features of the dental pain sensory system. Odontoblasts comprise the outermost cell layer of dental pulp tissue, which is advantageous to odontoblasts playing the role of a sensory transducer. Some nerve endings of dental primary afferents (DPAs) spread into the dentinal tubule. This structural nature establishes a distinctive sensory mechanism for the tooth.

The structure of teeth results in a unique pattern of nociception. One example is a special condition known as dentin hypersensitivity—the exaggerated nociception in teeth caused by non-noxious mechanical, chemical, or thermal stimuli without the pulpal inflammation predisposed or the nerve damage in the adjacent tissue [5–8]. While the molecular mechanisms underlying dentin hypersensitivity have not been fully elucidated, one promising hypothesis—the hydrodynamic theory—states that external stimuli cause the movement of the dentin tubular fluid to, ultimately, excite nerve fibers in the pulp to initiate pain. This provides the most plausible explanation for dental cold hypersensitivity of all the hypotheses that have been proposed, although not without controversy [9–20]. Another example is the pulsating nature of tooth pain often described by chronic pulpitis patients. This phenomenon is presumed to be caused by hydrostatic pressure applied to the edematous tooth pulp in the restricted space within the dentin and enamel. Both the pulsating pain associated with pulpal inflammation and the hydrodynamic theory of dental hypersensitivity require a mechanosensitive receptor as a key molecule. However, understanding such a receptor and its associated mechanism of action only began not long ago. This review summarizes the most recent advances in the understanding of the molecular and cellular mechanisms of mechanotransduction in the context of tooth pain.

The tooth is exposed to drastic temperature changes of the oral cavity. Although the harsh thermal conditions from food consumption hardly induce tooth pain under normal circumstances because of the excellent thermal insulating of the enamel tissue [21–24], mild temperature changes can induce intense pain with exposed dentin or pulpal inflammation. For example, noxious cold induces sharp and transient pain while noxious heat induces dull and lasting pain [25,26]. To elucidate the molecular mechanisms associated with temperature-driven tooth pain, the expression and physiology of molecular thermosensor candidates, such as the transient receptor potential (TRP) channel superfamily, have been investigated. A large variety of temperature receptors that may play critical roles in the transduction of tooth pain are expressed in dental primary afferent nerves [22,27] and odontoblasts [5,6,8–10,18].

In addition, voltage-gated and ligand-gated ion channels take important roles in tooth pain. Not only are various types of voltage-gated ion channels expressed in the trigeminal sensory nerve on common nerve cells, but they are also expressed in odontoblast cells [22,28–32]. Previous studies have indicated that small molecules, such as adenosine 5′-triphosphate (ATP), and their ionotropic receptors, the P2X family, play an important role in the sensory system for tooth pain [19,33–35]. In this review, we summarize the research on temperature-sensitive, mechanosensitive, ligand-gated, and voltage-gated ion channels and their role in the sensory system for tooth pain.

## 2. Thermo-Sensitive Ion Channels

Since the temperature-gated TRPV1 ion channel was first cloned from a subset of trigeminal and dorsal root ganglia (DRG) neurons [36,37], several members of the TRP superfamily have been discovered and proposed as potential molecular temperature sensors [38]. These TRP channels have been hypothesized to be key contributors for the keen sense of temperature in teeth, and the functional expression of TRP channels in dental primary afferent neurons and in odontoblasts has been massively investigated (Figure 2) [24,26].

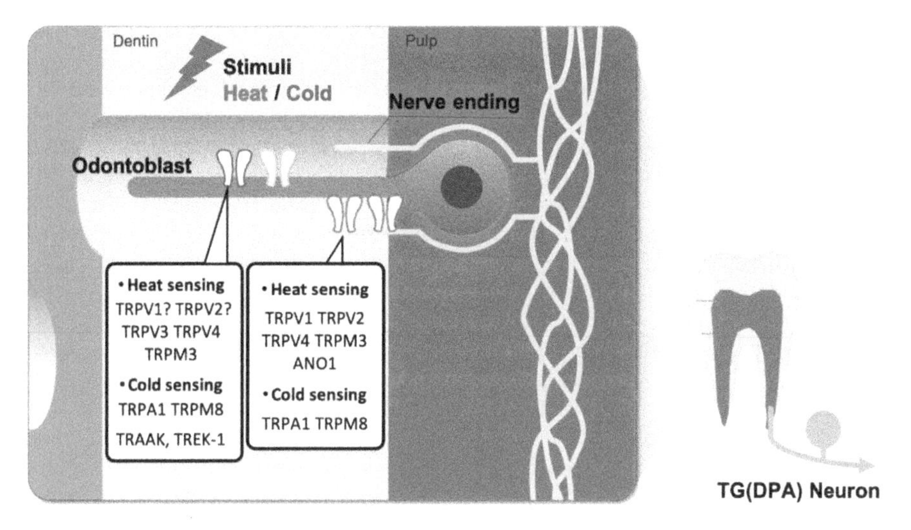


**Figure 2.** Thermosensitive ion channels in the dental sensory system. External heat or cold stimuli cause activation of thermosensitive ion channels in dental primary afferent (DPA) nerve ending or odontoblast cells, therefore dental pain transduces from thermal stimuli.

### 2.1. *Thermo-Sensing Ion Channels in the Trigeminal Nerve*

TRPV1 is a polymodal receptor activated by high temperatures over 43 °C or irritant chemicals including capsaicin and proton. TRPV1 is believed to play a central role in nociception because it is primarily expressed in small- to medium-peptidergic nociceptive neurons, and its activation is modulated by various inflammatory and nerve-damage-inducing mediators. Immunohistochemical investigation demonstrated TRPV1 expression in 20% of rat trigeminal ganglion (TG) cells, mostly in small- to medium-sized, as expected [39]. Interestingly, a retrograde labeling study revealed that only 8% of tooth pulpal neurons were TRPV1-positive, whereas 26% of TG neurons innervating facial skin were TRPV1-positive, which was contrary to the previous speculation that most nerves innervating the tooth pulp are nociceptive [40–42]. Conversely, the functional analysis of retrograde-labeled dental primary afferent neurons showed the opposite results; the neuron response to capsaicin application was more abundant for dental primary afferents than for TG neurons in calcium imaging studies [43] and in whole-cell patch clamp experiments [44]. Single-cell RT-PCR analysis revealed that most dental primary afferents are TRPV1-positive, whereas two other immunohistochemical analyses reported that only 17–34% of pulp-innervating neurons were TRPV1-positive [45,46]. The reason for this discrepancy is not clear. Of note, lipopolysaccharides (LPS) from Gram (−) bacteria upregulated TRPV1 expression in TG [47], and Complete Freund's Adjuvant (CFA) upregulated TRPV1 in TG neurons innervating adjacent teeth [48], suggesting the potential contribution of TRPV1 to tooth pain under the pulpitis condition. Interestingly, estrogen is also upregulated TRPV1 and anoctamin-1 (ANO1)—a potential heat-sensing ion channel—in female rat TG neurons and induced an increased pain response to TRPV1 agonists [49]. The physiological meaning of estrogen-induced upregulation for heat sensing ion channels is not clear and needs to be considered when designing pain studies.

TRPV2 is an ion channel homolog to TRPV1 with a higher threshold (>52 °C). TRPV2 is different from TRPV1 in that it does not respond to capsaicin nor acid and is preferentially expressed in medium- to large-sized myelinated neurons [50]. The immunohistochemical analysis of retrograde-labeled TG neurons revealed a TRPV2 expression pattern quite the opposite to that of TRPV1 [51]. While 14% of the TG cells showed immunoreactivity to TRPV2 mostly in medium- to large-sized, 37% of neurons innervating tooth pulp was TRPV2-positive, whereas only 9% of neurons to the facial skin were positive. Another immunohistochemical analysis using a double-labeling technique confirmed the mutually exclusive expression of TRPV1 and TRPV2 in pulpal neurons, with 32–51% TRPV2 positive cells [45,46]. These findings are consistent with previous reports that pulpal neurons are mostly medium- to large-sized myelinated neurons that lose their myelination upon entering tooth pulp [2,4,52–56], suggesting that teeth might have a distinct nociceptive system.

In addition, the expression of TRPV4 and TRPM3 was observed in retrogradely labeled dental afferent neurons [57–59]. Because TRPV4 activates at innocuously warm temperature between 27 and 35 °C, it is believed to play a role in the maintenance of body temperature, rather than in nociception [60]. On the other hand, TRPM3, or long TRPC3 as previously known, was recently discovered to have an activation threshold of 40 °C and became a prominent candidate of noxious heat detector [61].

Since cold stimuli induce tooth pain more frequently than hot, cold-sensitive TRP channels might play a role in the transduction of tooth pain. TRPA1 and TRPM8 are cold-sensitive TRP channel subtypes activated at temperatures below 17 and 25 °C, respectively [62,63]. Calcium imaging experiments with cold stimuli under 18 °C revealed that cold-sensitive neurons are more abundant in the TG than in the DRG (15% and 7%, respectively) [43]. TRPA1 upregulation in a tooth injury rat model proposes the importance of TRPA1 in tooth pain [64]. A subsequent study combining electrophysiological recording with single-cell RT-PCR and immunocytochemistry revealed the functional expression of TRPA1 and TRPM8 in rodent dental primary afferent neurons [43]. Interestingly, the expression of TRPA1 and TRPM8 channels was lower than that of TRPV1 in dental primary afferent (DPA) neurons. TRPA1 and TRPM8 were, moreover, co-expressed in some of the TRPV1-positive DPA neurons, suggesting an ambiguity between cold and hot stimuli-induced tooth pain. A recent study suggested that acute heat sensation requires any of functional TRPV1, TRPA1, and TRPM3 ion channels, and only triple knock-out mice showed a lack of acute withdrawal response to noxious heat compared to the intact normal response to cold stimuli, which suggests a redundant mechanism for heat detection [65]. Whether dental sensory systems utilize a similar mechanism is unclear.

### 2.2. Thermo-Sensing Ion Channels in Odontoblast Cells

Odontoblasts deposit calcium matrix at the outer surface of tooth pulp to form the dentin layer. Due to this anatomical location, the potential secondary role of odontoblasts as a member of the sensory system has been continuously proposed [3,21,22,66–74]. The expression of temperature-sensing TRP channels in odontoblasts has been investigated by several researchers, but the results have been diverse. While calcium imaging, immunohistochemical detection, and single cell RT-PCR all revealed the negative expression of heat-sensing TRPV1 and TRPV2 channels in acutely isolated odontoblasts from adult rat incisors [71], calcium imaging and electrophysiological recording of the odontoblasts cultured from neonatal rat pulpal slices showed positive responses to TRPV1, TRPV2, TRPV3, TRPV4, and TRPM3 [69,72]. It was not clear whether TRPV1 or TRPV2 channel-expressing odontoblasts were damaged or lost during acute isolation, whether the odontoblasts cultured from pulpal slices did not faithfully reflect the naïve odontoblasts, or whether it was from the age difference. Results from cold-sensing TRPA1 and TRPM8 investigation are more perplexing. While TRPA1 and TRPM8 were not detected in both acutely isolated odontoblasts and in pulpal slice-derived odontoblasts [71,72], another study showed both TRPA1 and TRPM8 in rat odontoblasts cultured from pulpal slices [70].

The results from human odontoblasts are less diverse. TRPV1-4 and TRPM8 have been detected by functional, immunohistochemical, western-blotting and electron microscopic tests [66,67,75]. TRPA1, however, showed controversial results. While in one study [67] immunohistochemical analysis of decalcified healthy human molar sections detected TRPA1 expression, another study did not [68]. Further clarification is required to determine the expression of TRPA1 in human odontoblasts. Nonetheless, it is very probable that odontoblasts functionally express temperature-sensing TRP channels and that these channels might confer odontoblasts with the ability to detect hot and cold temperatures. Many questions remained to be answered, including whether odontoblasts, if activated, can transfer these signals to pulpal neurons.

### 2.3. *Other Aspects of the Thermo-Sensing Ion Channels in the Dental Sensory System*

TRAAK and TREK-1 channels are also considered as potent thermosensitive ion channels [32,38]. Noël and his colleagues demonstrated that TRAAK and TREK-1 participate in the heat and cold sensing functions of TRP channels [76]. Their expression in odontoblast cells was demonstrated in a rodent model and in human pulp tissue [8,77]. Many other ion channels, including voltage-gated $Na^+$ ($Na_V$) channels (VGSCs), have been thoroughly studied as molecular thermosensors [78,79]. Recently, other types of dental cells, such as human tooth pulp fibroblasts and periodontal ligament (PDL) cells, were also shown to express temperature sensitive TRP channels [80,81]. These findings suggest that apart from odontoblasts, other cell types, such as pulp fibroblast cells or PDL cells, might contribute to the response to noxious thermal stimuli. Further studies are needed to elucidate the thermosensing mechanisms of various cell types surrounding tooth tissue. Alternatively, some efforts to characterize the dental sensory system by Next Generation Sequencing (NGS) studies have also been performed [82,83]. Combining these results with new emerging experimental methodologies, such as NGS or multi-omics studies of dental sensory systems, understanding of the temperature-induced tooth pain perception mechanisms may prove to be a significant scientific breakthrough.

## 3. Mechanosensitive Channels in Tooth Pain

It is difficult to explain tooth pain strictly by transduction of noxious temperatures by thermo-TRP channels. Temperature transduction cannot explain the sudden and intense tooth pain elicited by innocuous stimuli, such as an air puff, water spray, or sweet substances, or the pulsating pain often described by chronic pulpitis patients. Evidence from clinical studies suggests that the movement of dentin tubular fluid by temperature change might cause the sudden intense tooth pain from an air puff or spray of water. The sudden intense pain can also be generated in the micro-movement of cracked tooth parts during mastication. In addition, tooth structure can be mechanically deformed in response to thermal changes [3,13–16]. Pulsating pain in chronic pulpitis results from hydrostatic blood pressure applied to inflamed and swollen pulp tissue contained within the hard dentin structures [84,85]. All of these are suggested molecular transducers of mechanical force or stretch expressed in the dental sensory system, that are activated upon mechanical stimulation from movement of dentinal fluid, or deformation of microstructure (Figure 3) [13,18].

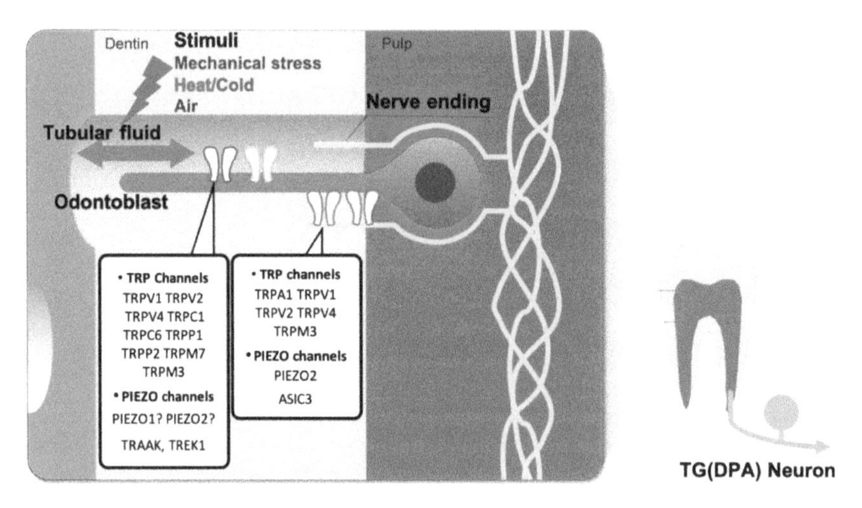


**Figure 3.** Mechanosensitive ion channels in the dental sensory system. According to the hydrodynamic theory of dental nociception, movement of the dentine tubular fluid generated by external stimuli, such as thermal or mechanical stress, activates mechanosensitive ion channels in odontoblasts or dental primary afferent (DPA) nerve ending extend into the dentinal tubule. Thus, mechanosensitive ion channels are regarded as major players in dental nociception. These ion channels can also be activated with directly applied mechanical stress.

### 3.1. TRP Channels

Several TRP channel superfamily members that exhibit mechanosensitivity include TRPC1, TRPC6, TRPV1, TRPV2, TRPV4, TRPM3, TRPM4, TRPM7, TRPA1, and TRPP2 [86]. Of these channels, the expression of TRPV1, TRPV2, TRPV4, TRPM3, TRPM7, and TRPA1 was reported in TG neurons [59,87,88], while TRPV1, TRPV2, TRPV4, TRPM3, and TRPA1 were shown in dental afferent neurons with retrograde labelling [43,57–59,72,89,90].

TRPV1, although this is still in debate, has been proposed to have mechanosensitivity. Bladder and urothelial epithelial cells from TRPV1-deleted mice showed markedly diminished responses to stretch [91]. The expression of TRPV1 in TG neurons innervating tooth pulp or in odontoblasts is also controversial, as elaborated in the previous section. The mechanosensitivity of TRPA1 is similar. While TRPA1-deleted mice showed a higher threshold and reduced response to mechanical stimuli [92], another TRPA1-null mouse line reported no difference in mechanical threshold compared to wild-type mice [89]. Ex vivo skin-nerve recordings from TRPA1-null mice showed deficits in mechanical sensitivity [93]. Although the role of TRPA1 as a cellular mechanical transducer is unclear, it suggests that TRPA1 may be implicated in mechanical hyperalgesia under pathological conditions. A recent report on the upregulation of TRPA1 in an experimental tooth injury model suggests that TRPA1 is still a promising candidate transducer in teeth [94].

TRPV4 is expressed in many cell types and tissues where mechanosensitivity is critical, such as hair cells of the inner ear, vibrissae Merkel cells, sensory ganglia, chondrocytes, osteoclasts, osteoblasts, and keratinocytes, as well as cutaneous A- and C-fiber terminals [95]. Studies conducted in TRPV4-null mice revealed that TRPV4 is related to the development of acute inflammatory mechanical hyperalgesia [95–97]. TRPV4-deleted mice showed reduced C-fiber sensitization for mechanical and hypotonic stimuli [98], suggesting TRPV4 involvement in osmotic mechanical hyperalgesia and nociceptor sensitization [98,99]. Recently, one study showed TRPV4 expression in the nerves of human tooth pulps and that TRPV4 expression was upregulated in human tooth pulp nerves of symptomatic teeth associated with pulpitis [100].

The investigation of non-neuronal cells revealed the expression of TRPC1, TRPC6, TRPV4, TRPM3, TRPM7, TRPP1, and TRPP2 in rodent odontoblasts [72,90,101,102] and TRPV1, TRPV2, TRPV4, and TRPM3 in pulp cells from neonate rats after in vitro differentiation into odontoblasts [72]; this suggests that these channels might function as molecular mechanotransducers that possibly confer mechanosensitivity to odontoblasts. TRPM7 is a unique ion channel with mechanosensitivity attached to a kinase, as shown by a touch-unresponsive zebrafish mutant [103]. Interestingly, TRPM7 expression was detected in most odontoblasts, predominantly in the odontoblastic process region [101], and TRPM7-specific inhibitor blocked mechanically-evoked calcium responses in odontoblasts [101], suggesting that TRPM7 might mediate mechanical sensitivity in odontoblasts. TRPP1 and TRPP2, which act together as a mechanical receptor, are present on the surface of odontoblasts and appear to be located at the base of the primary cilium [104].

Recent publications strongly suggest that IB4-positive non-peptidergic afferents play an important role transducing mechanical stimuli in the skin [105,106]. Chung and his colleagues showed a non-peptidergic mechanosensitive subpopulation in TG neurons that might be responsible for the detection of dentin tubular fluid [107]. However, the mechanical transducer molecule responsible for tooth pain in non-peptidergic polymodal nociceptors remains to be elucidated by future research.

### 3.2. PIEZO Channels

Since PIEZO family ion channels were cloned in mammals, the PIEZO gene family have been considered as putative mechanosensitive ion channel proteins [108–114]. PIEZO1 and PIEZO2 were identified by efforts to elucidate mammalian mechanosensing mechanisms which could not be clearly understood by TRP channels. While PIEZO channels are broadly expressed in a wide range of mammalian mechanosensitive cell types, PIEZO2 channels are identified as low-threshold mechanoreceptors in sensory DRG neurons and Merkel cells [115–117]. Moreover, the depletion of PIEZO2 in sensory DRG neurons and Merkel cells resulted in the dramatic reduction of rapidly adapting mechanically induced currents, suggesting a critical role of PIEZO2 as low threshold mechanoreceptor [118]. These findings have great implications for tooth pain research because low threshold mechanoreceptors are regarded as major players in tooth pain sensory systems, considering that mild mechanical stimuli could cause severe tooth pain. Moreover, several studies revealed that major populations of dental primary afferent neurons consist of A-fibers regarded as low-threshold mechanoreceptors [2,53]. Recently, many groups have examined the functional expression of PIEZO2 in the dental sensory system. Won et al. demonstrated PIEZO2 expression in murine dental primary afferent neurons by single-cell RT-PCR and in situ hybridization and function by recording rapidly adapting inward current induced by direct pocking [119]. Interestingly, PIEZO2 positive dental primary afferent neurons were medium-to-large sized and co-expressed with TRPV1, Nav1.8, and CGRP, which are regarded as nociceptive neuronal marker genes. These results indicate that PIEZO2 positive low-threshold mechanoreceptor neurons innervating teeth are 'algoneurons' that also, paradoxically, act as nociceptors.

To verify the role of these low-threshold mechanoreceptors in the odontoblast cells, other studies have been performed to verify PIEZO expression in odontoblast cells. An electrophysiological study with odontoblast cells co-cultured with IB4-negative medium-sized TG neurons elucidated the role of odontoblasts as mechanosensitive transducer cells [120]. Inward currents were detected from TG neurons when mechanical stimulation was applied to neighboring odontoblast cells. Interestingly, this odontoblast-induced inward current from TG neurons was antagonized with a PIEZO1 selective blocker. Three-dimensional imaging with focused ion beam-scanning electron microscopy revealed that PIEZO2 is expressed in nearly all rodent matured odontoblast cells and is absent in immature cells [121]. PIEZO2 proteins were detected selectively in odontoblastic processes that protrude into dentinal tubules. In another study, however, odontoblastic response to mechanical stimulation was inhibited by a specific antagonist of PIEZO1 [120]. These controversial results indicate the essential role of PIEZO ion channels in dental sensory systems as putative mechanosensors but also suggest that more research is needed to comprehensively understand complex dental mechanosensing systems.

*3.3. ASIC Channels*

Acid-sensing ion channels (ASICs) were initially implicated in mechanotransduction because their phylogenetic homologs in *Caenorhabditis elegans*—the mechanosensory (MEC) channel subunits—are essential for the perception of touch. Three members of the ASIC family (ASIC1-3) are expressed in peripheral mechanoreceptors and nociceptors in mammals. Six ASIC proteins encoded by four genes have been identified, ASIC1a, ASIC1b, ASIC2a, ASIC2b, ASIC3, and ASIC4, which differ in their kinetics, external pH sensitivity, tissue distribution, and pharmacological properties [122]. ASIC-2 mRNA is expressed in both small-diameter and large-diameter neurons and colocalized within single sensory neurons in the TG [123]. One-third of TG neurons that project towards the tooth pulp are immunoreactive (IR) to ASIC3 [124]. A single-cell RT-PCR study revealed that the ASIC3 mRNA is expressed in 67% of pulpal afferent neurons [58,59]. Human odontoblasts display immunoreactivity for ASIC2 as well as the ENaC-β and ENaC-γ, but not the ENaC-α, subunits [125]. These findings suggest a role for ASIC3 in the mechanotransduction of tooth sensitivity.

*3.4. TREK-1 Potassium Channels*

The primary function of the two-pore potassium ($K_{2P}$) channels is to mediate $K^+$-selective leak currents that regulate cell excitability through a hyperpolarized resting membrane potential [126]. Several members of the $K_{2P}$ channel family including TRESK, TRAAK/KCNK4, TASK, TREK, and THIK are intrinsically mechanosensitive, and all are expressed in the DRG [127,128]. $K_{2P}$ channels are well established regulators of primary afferent fibers excitability. Two kinds of high conductance $Ca^{2+}$-activated $K^+$ ($K_{Ca}$) channels and TREK-1 channels (TWIK-related $K^+$ channels) have been identified as putative mechanotransduction channels [19,46,102,113,114,129–131]. Investigation of $K_{2P}$ ion channels in the mammalian tooth pulp and in the odontoblast membrane revealed TREK1 mRNA expression in human odontoblasts [58,77]. Consequently, TREK-1 channels when stretch-activated may participate in the signal transduction to afferent nerve endings.

## 4. Ligand-Gated Channels

*ATP: Purinergic Receptors*

ATP acts as an extracellular signaling molecule that affects numerous downstream factors and signaling cascades. Signaling involving a purine nucleotide or nucleoside, such as ATP, is called purinergic signaling and is associated with multiple levels of nociception and immune responses in the oral system [132]. For example, P2X receptors (P2XRs) are expressed in the nociceptive TG cells [133,134] as well as in tooth pulp cells [35,135,136]. P2X positive nerve fibers have been detected in the subodontoblastic plexus close to odontoblasts [33,135,137]. P2XR2 and P2XR3 receptors have been found in both pulp nerves and a subpopulation of rat TG neurons [134,138–141]. In addition, a study showed that the presence of the P2X3 receptor and possibly the heteromeric P2X2/3 receptor in the trigeminal subnucleus caudalis (Vc) initiates and maintains the central sensitization in rat tooth pulp nociceptive neurons [142].

Recent studies suggest that P2X3 receptor activation by ATP induces tooth nociception in rat tooth pulp [33,139]. Importantly, an ATP derivative is sufficient to elicit behavioral pain sensation in tooth pulp [143] and odontoblasts contribute to the sensory function of teeth by releasing ATP in response to physical stimuli [19,66,129,144,145]. Furthermore, odontoblasts themselves express different P2XR subtypes (Figure 4) [34,146]. Since blocking extracellular ATP release results in the inhibition of interodontoblastic communication, ATP might regulate the physiology of odontoblasts via autocrine or paracrine mechanisms [19]. G-protein coupled P2Y ATP receptors are also present in pulp cells [136], TG neurons [147,148], trigeminal satellite glial cells [149], and odontoblasts [129,136].

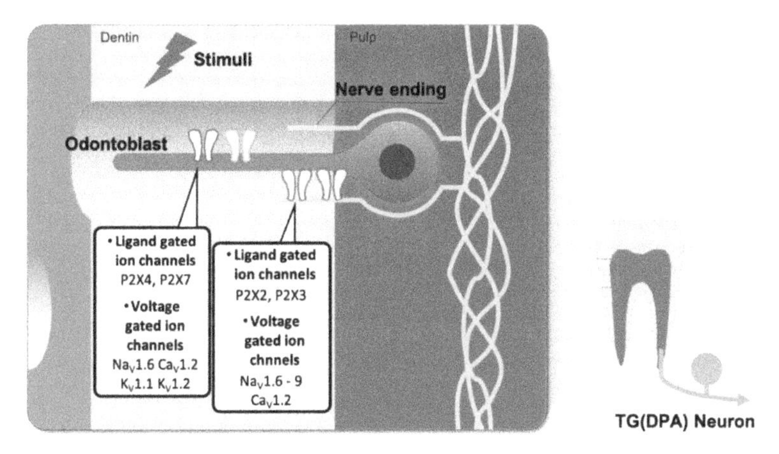


**Figure 4.** Other types of ion channels such as ligand gated ion channels and voltage gated ion channels expressed in the dental sensory system. ATP molecules released by adjacent odontoblast or fibroblast cells in pulp by external stimuli and they induce activation of purinergic receptors in odontoblasts or DPA neurons. Various types of voltage gated ion channels are also expressed in dental pain sensory cells but their functions are not clearly revealed.

## 5. Voltage-Gated Ion Channels

### *5.1. Voltage-Gated Sodium Channels*

VGSCs are responsible for action potential generation and excitability of the cell membrane. Nine different VGSC isoforms have been discovered in the mammalian nervous system, with $Na_V1.6$ and $Na_V1.7$ being the most abundant in the peripheral nervous system [150] and nociceptive sensory neurons [151], respectively. Immunohistochemical analysis of pulp tissue taken from pulpitis patients revealed expression of $Na_V1.7$ and $Na_V1.8$ with greater immunoreactivity in the pulp from patients with painful pulpitis [152–154]. Closer investigation of tooth pulp from pulpitis patients showed an increased expression of $Na_V1.7$ in the nerve bundles at intact and demyelinating nodes of Ranvier compared with healthy tooth pulp [155], while no significant difference for $Na_V1.6$ expression was observed [156]; together, this suggests that $Na_V1.7$ might play a role in inflammatory tooth pain.

Since expression of VGSCs is an important property of excitable cells, the demonstration of $Na_V1.6$ expression in non-neuronal pulpal cells, such as pulpal immune cells, dendritic pulpal cells, and odontoblasts [29], has gathered a robust interest. In addition, electrophysiology, immunohistochemistry, RT-PCR, and in situ hybridization of odontoblasts differentiated from human dental pulp explants has revealed the expression and functionality of $Na_V1.1$, $Na_V1.2$, and $Na_V1.3$ [28]. Interestingly, patch-clamp recording of the cultured human tooth pulp cells revealed rapidly inactivating TTX-sensitive $Na^+$ currents and membrane properties similar to neuronal satellite cells but not to odontoblasts [157]. The molecular and cellular identity of such pulpal cells is still unknown, and whether odontoblasts or other pulpal cells are indeed excitable and, if so, what their function would be, is unclear.

$Na_V1.9$ is the VGSC most recently identified [158]. $Na_V1.9$ is preferentially expressed in small-diameter DRG neurons, TG neurons, and myenteric neurons [159–162] $Na_V1.9$ is activated at voltages near the resting membrane potential and generates a relatively persistent current [159]. $Na_V1.9$ channels may also have a role in inflammatory pain, but not in neuropathic pain [158,163]. In addition, an investigation of $Na_V1.9$ in rats revealed the innervation of $Na_V1.9$-IR fibers in lip skin and in the tooth pulp of non-painful teeth, suggesting a role of this VGSC isoform in orofacial pain [164]. Recently, a study found that $Na_V1.9$ was increased in the axons of symptomatic pulpitis of permanent painful human teeth compared to the tooth pulp of permanent non-painful teeth (Figure 4) [165].

### *5.2. Voltage-Gated Calcium Channels*

Several lines of evidence have shown that DRG and spinal cord neurons express $Ca_V1.2$ [166], while L-type $Ca_V$ channels are broadly expressed in skeletal and cardiac muscle, neurons, auditory hair cells, pancreatic cells, and the retina [167]. Electrophysiological examination of the DRG showed that L-type $Ca_V$ channels are present largely in small and large neurons, although these channels are regulated during chronic pain [168]. One study using RT-PCR showed that L-type $Ca_V$ channels are downregulated in DRG upon chronic constriction injury (CCI) and sciatic nerve axotomy in rats [169]. suggesting that decreases in $Ca_V1.2$ and $Ca_V1.3$ in DRG could contribute to the hyperexcitability of neuropathic pain by modulating $Ca^{2+}$-dependent inactivation or facilitation as negative feedback [170]. Inversely, $Ca_V1.2$ is upregulated in the spinal cord in a spinal nerve ligation (SNL) model. One study reported that Cav1.2 functions as a key factor for the differentiation of tooth pulp stem cells [171]. In addition, several lines of evidence indicate that $Ca_V1.2$ may have a central role in odontoblast behavior at both the physiological and pathological levels [31,172–174].

### *5.3. Voltage-Gated Potassium Channels*

Patch-clamp recording revealed the presence of the voltage-gated potassium channel ($K_V$ channel) in cultured human dental pulp cells [175] and in human odontoblasts [73]. Calcium-activated potassium ($K_{Ca}$) channels that display mechanosensitivity are also present in odontoblast cells [30,31,176], and their concentration at the apical pole of odontoblasts could have relevance in the sensory transduction process of teeth [73].

## 6. Conclusions

Tooth pain greatly undermines patient quality of life. Tooth pain arises from distinct mechanisms from other pain types because of the unique neurochemical properties and anatomical structure of dense innervation and vascularization under hard tissue. The physiology of tooth pain involves the complex orchestration of ion channels introduced in this review (Table 1). Still, the present understanding is vague. Many questions remain, such as how mechanosensitive ion channels involved in tooth pain are molecular identified, whether odontoblasts function as primary sensory cells, and, if so, how they provide signals to underlying nerves. Elucidating these questions will provide the basis for understanding tooth pain and can lead to the development of therapeutics specifically targeting tooth pain.

**Table 1.** Tabular summary of ion channels expressed in dental pain sensory system and their functions.

| Ion Channel Type | Cell Type | Expressed Ion Channels | Remarks |
| --- | --- | --- | --- |
| Thermo-sensitive | Odontoblast | Heat sensing ion channels TRPV1? TRPV2? TRPV3 TRPV4 TRPM3 | Heat-induced dental pain in healthy or pathological state (Odontoblast transducer theory) |
| | | Cold sensing ion channels TRPA1 TRPM8 | Cold-induced dental pain in healthy or pathological state (odontoblast transducer theory) |
| | | TRAAK, TREK-1 | K2p channels may play a role as thermos-sensors (Neural theory) |
| | DPA neurons | Heat sensing ion channels TRPV1 TRPV2 TRPV4 TRPM3 ANO1 | Heat induced dental pain in healthy or pathol gical state (Neural theory) |
| | | Cold sensing ion channels TRPA1 TRPM8 | Cold induced dental pain in healthy or pathological state (Neural theory) |
| | Others PDL cells/Fibroblast | Thermosesing TRP channels | Function in dental pain sensing mechanism is not clear |

**Table 1.** *Cont.*

| Ion Channel Type | Cell Type | Expressed Ion Channels | Remarks |
|---|---|---|---|
| Mechano-sensitive | Odontoblast | TRP channels<br>TRPV1 TRPV2 TRPV4 TRPC1 TRPC6<br>TRPP1 TRPP2 TRPM7 TRPM3 | Sensing movement of dentine tubular fluid (Hydrodynamic theory) or microdeformation of tooth structure |
| | | Piezo channels<br>PIEZO1? PIEZO2? | |
| | | TREK1 | |
| | DPA neurons | TRP channels<br>TRPA1 TRPV1 TRPV2 TRPV4<br>TRPM3 TRPM7 | |
| | | Piezo channels<br>PIEZO2 | |
| | | ASIC3 | |
| Ligand-gated | Odontoblast | Purinergic receptors<br>P2X4, P2X7 | Paracrine or autocrine signaling molecule |
| | DPA neurons | Ligand gated ion channels<br>P2X2, P2X3 | Paracrine or autocrine signaling molecule |
| Voltage-gated | Odontoblast | Voltage gated ion channels<br>$Na_V$1.6 $Ca_V$1.2 $K_V$1.1 $K_V$1.2 | Role of voltage gated ion channels in odontoblasts is not clear |
| | DPA neurons | Voltage gated ion channels<br>$Na_V$1.6-9 $Ca_V$1.2 $K_V$1.1 $K_V$1.2 | Function in transmission of nociceptive information |

**Funding:** This work was supported by the National Research Foundation of Korea (NRF) grant (NRF-2018R1D1A1B07049067, NRF-2017M3C7A1025602 and NRF-2015R1A1A1A05027503) funded by the Korean government.

**Conflicts of Interest:** The authors declare no conflict of interest. All authors approved the final manuscript.

## References

1. Närhi, M.; Jyväsjärvi, E.; Virtanen, A.; Huopaniemi, T.; Ngassapa, D.; Hirvonen, T. Role of intradental A- and C-type nerve fibres in dental pain mechanisms. *Proc. Fin. Dent. Soc.* **1992**, *88*, 507–516.
2. Fried, K.; Sessle, B.J.; Devor, M. The paradox of pain from tooth pulp: Low-threshold "algoneurons"? *Pain* **2011**, *152*, 2685–2689. [CrossRef]
3. Hossain, M.Z.; Bakri, M.M.; Yahya, F.; Ando, H.; Unno, S.; Kitagawa, J. The role of transient receptor potential (Trp) channels in the transduction of dental pain. *Int. J. Mol. Sci.* **2019**, *20*, 526. [CrossRef]
4. Byers, M.R.; Närhi, M.V. Dental injury models: Experimental tools for understanding neuroinflammatory interactions and polymodal nociceptor functions. *Crit. Rev. Oral Biol. Med.* **1999**, *10*, 4–39. [CrossRef]
5. Dababneh, R.H.; Khouri, A.T.; Addy, M. Dentine hypersensitivity—An enigma? A review of terminology, mechanisms, aetiology and management. *Br. Dent. J.* **1999**, *187*, 606–611. [CrossRef] [PubMed]
6. Dowell, P.; Addy, M.; Dummer, P. Dentine hypersensitivity: Aetiology, differential diagnosis and management. *Br. Dent. J.* **1985**, *158*, 92–96. [CrossRef]
7. Mitchell, D.A.; Mitchell, L. *Oxford Handbook of Clinical Dentistry*, 4th ed.; Oxford University Press: Oxford, UK, 2005; Volume 260.
8. Sole-Magdalena, A.; Martinez-Alonso, M.; Coronado, C.A.; Junquera, L.M.; Cobo, J.; Vega, J.A. Molecular basis of dental sensitivity: The odontoblasts are multisensory cells and express multifunctional ion channels. *Ann. Anat.* **2018**, *215*, 20–29. [CrossRef]
9. Brannstrom, M.; Astrom, A. The hydrodynamics of the dentine; its possible relationship to dentinal pain. *Int. Dent. J.* **1972**, *22*, 219–227.
10. Brannstrom, M.; Linden, L.A.; Astrom, A. The hydrodynamics of the dental tubule and of pulp fluid. A discussion of its significance in relation to dentinal sensitivity. *Caries Res.* **1967**, *1*, 310–317. [CrossRef] [PubMed]
11. Chidchuangchai, W.; Vongsavan, N.; Matthews, B. Sensory transduction mechanisms responsible for pain caused by cold stimulation of dentine in man. *Arch. Oral Biol.* **2007**, *52*, 154–160. [CrossRef]

12. Rahim, Z.H.; Bakri, M.M.; Zakir, H.M.; Ahmed, I.A.; Zulkifli, N.A. High fluoride and low ph level have been detected in popular flavoured beverages in malaysia. *Pak. J. Med. Sci.* **2014**, *30*, 404–408. [CrossRef]
13. Horiuchi, H.; Matthews, B. In-vitro observations on fluid flow through human dentine caused by pain-producing stimuli. *Arch. Oral Biol.* **1973**, *18*, 275–294. [CrossRef]
14. Jacobs, H.R.; Thompson, R.E.; Brown, W.S. Heat transfer in teeth. *J. Dent. Res.* **1973**, *52*, 248–252. [CrossRef]
15. Linsuwanont, P.; Palamara, J.E.A.; Messer, H.H. An investigation of thermal stimulation in intact teeth. *Arch. Oral Biol.* **2007**, *52*, 218–227. [CrossRef]
16. Linsuwanont, P.; Versluis, A.; Palamara, J.E.; Messer, H.H. Thermal stimulation causes tooth deformation: A possible alternative to the hydrodynamic theory? *Arch. Oral Biol.* **2008**, *53*, 261–272. [CrossRef]
17. Lloyd, B.A.; Mcginley, M.B.; Brown, W.S. Thermal stress in teeth. *J. Dent. Res.* **1978**, *57*, 571–582. [CrossRef]
18. Sessle, B.J. The neurobiology of facial and dental pain: present knowledge, future directions. *J. Dent. Res.* **1987**, *66*, 962–981. [CrossRef]
19. Shibukawa, Y.; Sato, M.; Kimura, M.; Sobhan, U.; Shimada, M.; Nishiyama, A.; Kawaguchi, A.; Soya, M.; Kuroda, H.; Katakura, A.; et al. Odontoblasts as sensory receptors: Transient receptor potential channels, pannexin-1, and ionotropic ATP receptors mediate intercellular odontoblast-neuron signal transduction. *Pflug. Arch.* **2015**, *467*, 843–863. [CrossRef]
20. Trowbridge, H.O.; Franks, M.; Korostoff, E.; Emling, R. Sensory response to thermal stimulation in human teeth. *J. Endod.* **1980**, *6*, 405–412. [CrossRef]
21. Bleicher, F. Odontoblast physiology. *Exp. Cell Res.* **2014**, *325*, 65–71. [CrossRef]
22. Chung, G.; Jung, S.J.; Oh, S.B. Cellular and molecular mechanisms of dental nociception. *J. Dent. Res.* **2013**, *92*, 948–955. [CrossRef]
23. Renton, T. Dental (odontogenic) pain. *Rev. Pain* **2011**, *5*, 2–7. [CrossRef]
24. Sessle, B.J. Peripheral and central mechanisms of orofacial inflammatory pain. *Int. Rev. Neurobiol.* **2011**, *97*, 179–206.
25. Ahn, D.K.; Doutova, E.A.; Mcnaughton, K.; Light, A.R.; Narhi, M.; Maixner, W. Functional properties of tooth pulp neurons responding to thermal stimulation. *J. Dent. Res.* **2012**, *91*, 401–406. [CrossRef]
26. Henry, M.A.; Hargreaves, K.M. Peripheral mechanisms of odontogenic pain. *Dent. Clin. N. Am.* **2007**, *51*, 19–44. [CrossRef]
27. Jain, N.; Gupta, A.N.M. An insight into neurophysiology of pulpal pain: Facts and hypotheses. *Korean J. Pain* **2013**, *26*, 347–355. [CrossRef]
28. Allard, B.; Magloire, H.; Couble, M.L.; Maurin, J.C.; Bleicher, F. Voltage-gated sodium channels confer excitability to human odontoblasts: Possible role in tooth pain transmission. *J. Biol. Chem.* **2006**, *281*, 29002–29010. [CrossRef]
29. Byers, M.R.; Rafie, M.M.; Westenbroek, R.E. Dexamethasone effects on Na(V)1.6 in tooth pulp, dental nerves, and alveolar osteoclasts of adult rats. *Cell Tissue Res.* **2009**, *338*, 217–226. [CrossRef]
30. Ichikawa, H.; Kim, H.J.; Shuprisha, A.; Shikano, T.; Tsumura, M.; Shibukawa, Y.; Tazaki, M. Voltage-dependent sodium channels and calcium-activated potassium channels in human odontoblasts in vitro. *J. Endod.* **2012**, *38*, 1355–1362. [CrossRef]
31. Lundgren, T.; Linde, A. Voltage-gated calcium channels and nonvoltage-gated calcium uptake pathways in the rat incisor odontoblast plasma membrane. *Calcif Tissue Int.* **1997**, *60*, 79–85. [CrossRef]
32. Chung, G. Trp channels in dental pain. *Open Pain J.* **2013**, *6*, 31–36. [CrossRef]
33. Cook, S.P.; Vulchanova, L.; Hargreaves, K.M.; Elde, R.; Mccleskey, E.W. Distinct ATP receptors on pain-sensing and stretch-sensing neurons. *Nature* **1997**, *387*, 505–508. [CrossRef]
34. Lee, B.M.; Jo, H.; Park, G.; Kim, Y.H.; Park, C.K.; Jung, S.J.; Chung, G.; Oh, S.B. Extracellular ATP induces calcium signaling in odontoblasts. *J. Dent. Res.* **2017**, *96*, 200–207. [CrossRef]
35. Renton, T.; Yiangou, Y.; Baecker, P.A.; Ford, A.P.; Anand, P. Capsaicin receptor VR1 And ATP purinoceptor P2X3 in painful and nonpainful human tooth pulp. *J. Orofac. Pain* **2003**, *17*, 245–250.
36. Caterina, M.J.; Leffler, A.; Malmberg, A.B.; Martin, W.J.; Trafton, J.; Petersen-Zeitz, K.R.; Koltzenburg, M.; Basbaum, A.I.; Julius, D. Impaired nociception and pain sensation in mice lacking the capsaicin receptor. *Science* **2000**, *288*, 306–313. [CrossRef] [PubMed]
37. Caterina, M.J.; Schumacher, M.A.; Tominaga, M.; Rosen, T.A.; Levine, J.D.; Julius, D. The capsaicin receptor: A heat-activated ion channel in the pain pathway. *Nature* **1997**, *389*, 816–824. [CrossRef] [PubMed]

38. Vriens, J.; Nilius, B.; Voets, T. Peripheral Thermosensation In Mammals. *Nat. Rev. Neurosci.* **2014**, *15*, 573–589. [CrossRef] [PubMed]
39. Ichikawa, H.; Sugimoto, T. VR1-immunoreactive primary sensory neurons in the rat trigeminal ganglion. *Brain Res.* **2001**, *890*, 184–188. [CrossRef]
40. Byers, M.R.; Suzuki, H.; Maeda, T. Dental neuroplasticity, neuro-pulpal interactions, and nerve regeneration. *Microsc. Res. Tech.* **2003**, *60*, 503–515. [CrossRef]
41. Fried, K.; Aldskogius, H.; Hildebrand, C. Proportion of unmyelinated axons in rat molar and incisor tooth pulps following neonatal capsaicin treatment and/or sympathectomy. *Brain Res.* **1988**, *463*, 118–123. [CrossRef]
42. Hildebrand, C.; Fried, K.; Tuisku, F.; Johansson, C.S. Teeth and tooth nerves. *Prog. Neurobiol.* **1995**, *45*, 165–222. [CrossRef]
43. Park, C.-K.; Kim, M.S.; Fang, Z.; Li, H.Y.; Jung, S.J.; Choi, S.-Y.; Lee, S.J.; Park, K.; Kim, J.S.; Oh, S.B. Functional expression of thermo-transient receptor potential channels in dental primary afferent neurons: Implication for tooth pain. *J. Biol. Chem.* **2006**, *281*, 17304–17311. [CrossRef]
44. Kim, H.Y.; Chung, G.; Jo, H.J.; Kim, Y.S.; Bae, Y.C.; Jung, S.J.; Kim, J.S.; Oh, S.B. Characterization of dental nociceptive neurons. *J. Dent. Res.* **2011**, *90*, 771–776. [CrossRef]
45. Stenholm, E.; Bongenhielm, U.; Ahlquist, M.; Fried, K. Vrl- and Vrl-L-like immunoreactivity in normal and injured trigeminal dental primary sensory neurons of the rat. *Acta Odontol. Scand.* **2002**, *60*, 72–79. [CrossRef]
46. Gibbs, J.L.; Melnyk, J.L.; Basbaum, A.I. Differential Trpv1 And Trpv2 channel expression in dental pulp. *J. Dent. Res.* **2011**, *90*, 765–770. [CrossRef]
47. Chung, M.K.; Lee, J.; Duraes, G.; Ro, J.Y. Lipopolysaccharide-induced pulpitis up-regulates Trpv1 in trigeminal ganglia. *J. Dent. Res.* **2011**, *90*, 1103–1107. [CrossRef]
48. Matsuura, S.; Shimizu, K.; Shinoda, M.; Ohara, K.; Ogiso, B.; Honda, K.; Katagiri, A.; Sessle, B.J.; Urata, K.; Iwata, K. Mechanisms underlying ectopic persistent tooth-pulp pain following pulpal inflammation. *PLoS ONE* **2013**, *8*, e52840. [CrossRef]
49. Yamagata, K.; Sugimura, M.; Yoshida, M.; Sekine, S.; Kawano, A.; Oyamaguchi, A.; Maegawa, H.; Niwa, H. Estrogens exacerbate nociceptive pain via up-regulation of TRPV1 and ANO1 in trigeminal primary neurons of female rats. *Endocrinology* **2016**, *157*, 4309–4317. [CrossRef]
50. Caterina, M.J.; Rosen, T.A.; Tominaga, M.; Brake, A.J.; Julius, D. A capsaicin-receptor homologue with a high threshold for noxious heat. *Nature* **1999**, *398*, 436–441. [CrossRef]
51. Ichikawa, H.; Sugimoto, T. Vanilloid receptor 1-like receptor-immunoreactive primary sensory neurons in the rat trigeminal nervous system. *NSC* **2000**, *101*, 719–725. [CrossRef]
52. Lisney, S.J. Some anatomical and electrophysiological properties of tooth-pulp afferents in the cat. *J. Physiol.* **1978**, *284*, 19–36. [CrossRef]
53. Paik, S.K.; Park, K.P.; Lee, S.K.; Ma, S.K.; Cho, Y.S.; Kim, Y.K.; Rhyu, I.J.; Ahn, D.K.; Yoshida, A.; Bae, Y.C. Light and electron microscopic analysis of the somata and parent axons innervating the rat upper molar and lower incisor pulp. *Neuroscience* **2009**, *162*, 1279–1286. [CrossRef]
54. Fried, K.; Arvidsson, J.; Robertson, B.; Brodin, E.; Theodorsson, E. Combined retrograde tracing and enzyme/immunohistochemistry of trigeminal ganglion cell bodies innervating tooth pulps in the rat. *NSC* **1989**, *33*, 101–109. [CrossRef]
55. Fried, K.; Hildebrand, C. Axon number and size distribution in the developing feline inferior alveolar nerve. *J. Neurol. Sci.* **1982**, *53*, 169–180. [CrossRef]
56. Henry, M.A.; Luo, S.; Levinson, S.R. Unmyelinated nerve fibers in the human dental pulp express markers for myelinated fibers and show sodium channel accumulations. *BMC Neurosci.* **2012**, *13*, 29. [CrossRef]
57. Flegel, C.; Schobel, N.; Altmuller, J.; Becker, C.; Tannapfel, A.; Hatt, H.; Gisselmann, G. RNA-seq analysis of human trigeminal and dorsal root ganglia with a focus on chemoreceptors. *PLoS ONE* **2015**, *10*, e0128951. [CrossRef]
58. Hermanstyne, T.O.; Markowitz, K.; Fan, L.; Gold, M.S. Mechanotransducers in rat pulpal afferents. *J. Dent. Res.* **2008**, *87*, 834–838. [CrossRef]
59. Vandewauw, I.; Owsianik, G.; Voets, T. Systematic and quantitative mrna expression analysis of trp channel genes at the single trigeminal and dorsal root ganglion level in mouse. *Bmc Neurosci.* **2013**, *14*, 1. [CrossRef]
60. Güler, A.D.; Lee, H.; Iida, T.; Shimizu, I.; Tominaga, M.; Caterina, M. Heat-evoked activation of the ion channel, Trpv4. *J. Neurosci.* **2002**, *22*, 6408–6414. [CrossRef] [PubMed]

61. Held, K.; Voets, T.; Vriens, J. Trpm3 in temperature sensing and beyond. *Temperature* **2015**, *2*, 201–213. [CrossRef]
62. Bandell, M.; Story, G.M.; Hwang, S.W.; Viswanath, V.; Eid, S.R.; Petrus, M.J.; Earley, T.J.; Patapoutian, A. Noxious cold ion channel trpa1 is activated by pungent compounds and bradykinin. *Neuron* **2004**, *41*, 849–857. [CrossRef]
63. Colburn, R.W.; Lubin, M.L.; Stone, D.J., Jr.; Wang, Y.; Lawrence, D.; D'andrea, M.R.; Brandt, M.R.; Liu, Y.; Flores, C.M.; Qin, N. Attenuated cold sensitivity in Trpm8 null mice. *Neuron* **2007**, *54*, 379–386. [CrossRef] [PubMed]
64. Haas, E.T.; Rowland, K.; Gautam, M. Tooth injury increases expression of the cold sensitive trp channel Trpa1 in trigeminal neurons. *Arch. Oral Biol.* **2011**, *56*, 1604–1609. [CrossRef] [PubMed]
65. Vandewauw, I.; De Clercq, K.; Mulier, M.; Held, K.; Pinto, S.; Van Ranst, N.; Segal, A.; Voet, T.; Vennekens, R.; Zimmermann, K.; et al. A trp channel trio mediates acute noxious heat sensing. *Nature* **2018**, *555*, 662–666. [CrossRef] [PubMed]
66. Egbuniwe, O.; Grover, S.; Duggal, A.K.; Mavroudis, A.; Yazdi, M.; Renton, T.; Di Silvio, L.; Grant, A.D. Trpa1 and Trpv4 activation in human odontoblasts stimulates atp release. *J. Dent. Res.* **2014**, *93*, 911–917. [CrossRef] [PubMed]
67. El Karim, I.A.; Linden, G.J.; Curtis, T.M.; About, I.; Mcgahon, M.K.; Irwin, C.R.; Lundy, F.T. Human odontoblasts express functional thermo-sensitive trp channels: Implications for dentin sensitivity. *Pain* **2011**, *152*, 2211–2223. [CrossRef] [PubMed]
68. Tazawa, K.; Ikeda, H.; Kawashima, N.; Okiji, T. Transient receptor potential melastatin (Trpm) 8 is expressed in freshly isolated native human odontoblasts. *Arch. Oral Biol.* **2017**, *75*, 55–61. [CrossRef] [PubMed]
69. Tsumura, M.; Sobhan, U.; Muramatsu, T.; Sato, M.; Ichikawa, H.; Sahara, Y.; Tazaki, M.; Shibukawa, Y. Trpv1-mediated calcium signal couples with cannabinoid receptors and sodium–calcium exchangers in rat odontoblasts. *Cell Calcium* **2012**, *52*, 124–136. [CrossRef]
70. Tsumura, M.; Sobhan, U.; Sato, M.; Shimada, M.; Nishiyama, A.; Kawaguchi, A.; Soya, M.; Kuroda, H.; Tazaki, M.; Shibukawa, Y. Functional expression of Trpm8 and Trpa1 channels in rat odontoblasts. *PLoS ONE* **2013**, *8*, e82233. [CrossRef]
71. Yeon, K.Y.; Chung, G.; Shin, M.S.; Jung, S.J.; Kim, J.S.; Oh, S.B. Adult rat odontoblasts lack noxious thermal sensitivity. *J. Dent. Res.* **2009**, *88*, 328–332. [CrossRef]
72. Son, A.R.; Yang, Y.M.; Hong, J.H.; Lee, S.I.; Shibukawa, Y.; Shin, D.M. Odontoblast trp channels and thermo/mechanical transmission. *J. Dent. Res.* **2009**, *88*, 1014–1019. [CrossRef] [PubMed]
73. Allard, B.; Couble, M.-L.; Magloire, H.; Bleicher, F. Characterization and gene expression of high conductance calcium-activated potassium channels displaying mechanosensitivity in human odontoblasts. *J. Biol. Chem.* **2000**, *275*, 25556–25561. [CrossRef]
74. Magloire, H.; Couble, M.-L.; Thivichon-Prince, B.; Maurin, J.-C.; Bleicher, F. Odontoblast: A mechano-sensory cell. *J. Exp. Zool. Part B* **2009**, *312*, 416–424. [CrossRef] [PubMed]
75. Wen, W.; Que, K.; Zang, C.; Wen, J.; Sun, G.; Zhao, Z.; Li, Y. Expression and distribution of three transient receptor potential vanilloid (Trpv) channel proteins in human odontoblast-like cells. *J. Mol. Histol.* **2017**, *48*, 367–377. [CrossRef] [PubMed]
76. Noel, J.; Zimmermann, K.; Busserolles, J.; Deval, E.; Alloui, A.; Diochot, S.; Guy, N.; Borsotto, M.; Reeh, P.; Eschalier, A.; et al. The mechano-activated K+ channels traak and Trek-1 control both warm and cold perception. *EMBO J.* **2009**, *28*, 1308–1318. [CrossRef] [PubMed]
77. Magloire, H.; Lesage, F.; Couble, M.L.; Lazdunski, M.; Bleicher, F. Expression and localization of Trek-1 K+ channels in human odontoblasts. *J. Dent. Res.* **2003**, *82*, 542–545. [CrossRef] [PubMed]
78. Luiz, A.P.; Macdonald, D.I.; Santana-Varela, S.; Millet, Q.; Sikandar, S.; Wood, J.N.; Emery, E.C. Cold sensing by Nav1.8-positive and Nav1.8-negative sensory neurons. *Proc. Natl. Acad. Sci. USA* **2019**, *116*, 3811–3816. [CrossRef] [PubMed]
79. Magloire, H. Odontoblast and dentin thermal sensitivity. *Pain* **2011**, *152*, 2191–2192. [CrossRef] [PubMed]
80. El Karim, I.A.; Linden, G.J.; Curtis, T.M.; About, I.; Mcgahon, M.K.; Irwin, C.R.; Killough, S.A.; Lundy, F.T. Human dental pulp fibroblasts express the "cold-sensing" transient receptor potential channels Trpa1 and Trpm8. *J. Endod.* **2011**, *37*, 473–478. [CrossRef] [PubMed]
81. Son, G.Y.; Hong, J.H.; Chang, I.; Shin, D.M. Induction of IL-6 And IL-8 by activation of thermosensitive Trp channels in human pdl cells. *Arch. Oral Biol.* **2015**, *60*, 526–532. [CrossRef]

82. Kogelman, L.J.A.; Christensen, R.E.; Pedersen, S.H.; Bertalan, M.; Hansen, T.F.; Jansen-Olesen, I.; Olesen, J. Whole transcriptome expression of trigeminal ganglia compared to dorsal root ganglia in rattus norvegicus. *Neuroscience* **2017**, *350*, 169–179. [CrossRef]
83. Nguyen, M.Q.; Wu, Y.; Bonilla, L.S.; Von Buchholtz, L.J.; Ryba, N.J.P. Diversity amongst trigeminal neurons revealed by high throughput single cell sequencing. *PLoS ONE* **2017**, *12*, e0185543. [CrossRef]
84. Heyeraas, K.J.; Berggreen, E. Interstitial fluid pressure in normal and inflamed pulp. *Crit. Rev. Oral Biol. Med.* **1999**, *10*, 328–336. [CrossRef]
85. Van Hassel, H.J. Physiology of the human dental pulp. *Oral Surg. Oral Med. Oral Pathol.* **1971**, *32*, 126–134. [CrossRef]
86. Kamkin, A.; Kiseleva, I. *Mechanosensitivity of the Nervous System*; Springer: Basel, Switzerland, 2009; Volume 2, pp. 23–49.
87. Vriens, J.; Owsianik, G.; Hofmann, T.; Philipp, S.E.; Stab, J.; Chen, X.; Benoit, M.; Xue, F.; Janssens, A.; Kerselaers, S.; et al. Trpm3 is a nociceptor channel involved in the detection of noxious heat. *Neuron* **2011**, *70*, 482–494. [CrossRef]
88. Wei, X.; Edelmayer, R.M.; Yan, J.; Dussor, G. Activation of Trpv4 on dural afferents produces headache-related behavior in a preclinical rat model. *Cephalalgia* **2011**, *31*, 1595–1600. [CrossRef]
89. Bautista, D.M.; Jordt, S.E.; Nikai, T.; Tsuruda, P.R.; Read, A.J.; Poblete, J.; Yamoah, E.N.; Basbaum, A.I.; Julius, D. Trpa1 mediates the inflammatory actions of environmental irritants and proalgesic agents. *Cell* **2006**, *124*, 1269–1282. [CrossRef]
90. Sato, M.; Sobhan, U.; Tsumura, M.; Kuroda, H.; Soya, M.; Masamura, A.; Nishiyama, A.; Katakura, A.; Ichinohe, T.; Tazaki, M.; et al. Hypotonic-induced stretching of plasma membrane activates transient receptor potential vanilloid channels and sodium-calcium exchangers in mouse odontoblasts. *J. Endod.* **2013**, *39*, 779–787. [CrossRef]
91. Birder, L.A.; Nakamura, Y.; Kiss, S.; Nealen, M.L.; Barrick, S.; Kanai, A.J.; Wang, E.; Ruiz, G.; De Groat, W.C.; Apodaca, G.; et al. Altered urinary bladder function in mice lacking the vanilloid receptor Trpv1. *Nat. Neurosci.* **2002**, *5*, 856–860. [CrossRef]
92. Kwan, K.Y.; Allchorne, A.J.; Vollrath, M.A.; Christensen, A.P.; Zhang, D.S.; Woolf, C.J.; Corey, D.P. Trpa1 contributes to cold, mechanical, and chemical nociception but is not essential for hair-cell transduction. *Neuron* **2006**, *50*, 277–289. [CrossRef]
93. Kwan, K.Y.; Glazer, J.M.; Corey, D.P.; Rice, F.L.; Stucky, C.L. Trpa1 modulates mechanotransduction in cutaneous sensory neurons. *J. Neurosci.* **2009**, *29*, 4808–4819. [CrossRef]
94. Shibukawa, Y.; Suzuki, T. A Voltage-dependent transient K(+) current in rat dental pulp cells. *Jpn. J. Physiol.* **2001**, *51*, 345–353. [CrossRef]
95. Levine, J.D.; Alessandri-Haber, N. Trp channels: Targets for the relief of pain. *Biochim. Biophys. Acta* **2007**, *1772*, 989–1003. [CrossRef]
96. Alessandri-Haber, N.; Dina, O.A.; Joseph, E.K.; Reichling, D.; Levine, J.D. A transient receptor potential vanilloid 4-dependent mechanism of hyperalgesia is engaged by concerted action of inflammatory mediators. *J. Neurosci.* **2006**, *26*, 3864–3874. [CrossRef]
97. Alessandri-Haber, N.; Yeh, J.J.; Boyd, A.E.; Parada, C.A.; Chen, X.; Reichling, D.B.; Levine, J.D. Hypotonicity induces Trpv4-mediated nociception in rat. *Neuron* **2003**, *39*, 497–511. [CrossRef]
98. Chen, X.; Alessandri-Haber, N.; Levine, J.D. Marked attenuation of inflammatory mediator-induced c-fiber sensitization for mechanical and hypotonic stimuli in Trpv4−/− Mice. *Mol. Pain* **2007**, *3*, 31. [CrossRef]
99. Alessandri-Haber, N.; Dina, O.A.; Chen, X.; Levine, J.D. Trpc1 and Trpc6 channels cooperate with Trpv4 to mediate mechanical hyperalgesia and nociceptor sensitization. *J. Neurosci.* **2009**, *29*, 6217–6228. [CrossRef]
100. Bakri, M.M.; Yahya, F.; Munawar, K.M.M.; Kitagawa, J.; Hossain, M.Z. Transient receptor potential vanilloid 4 (Trpv4) expression on the nerve fibers of human dental pulp is upregulated under inflammatory condition. *Arch. Oral Biol.* **2018**, *89*, 94–98. [CrossRef]
101. Won, J.; Vang, H.; Kim, J.H.; Lee, P.R.; Kang, Y.; Oh, S.B. Trpm7 mediates mechanosensitivity in adult rat odontoblasts. *J. Dent. Res.* **2018**, *97*, 1039–1046. [CrossRef]
102. Kwon, M.; Baek, S.H.; Park, C.K.; Chung, G.; Oh, S.B. Single-cell Rt-Pcr and immunocytochemical detection of mechanosensitive transient receptor potential channels in acutely isolated rat odontoblasts. *Arch. Oral Biol.* **2014**, *59*, 1266–1271. [CrossRef]

103. Low, S.E.; Amburgey, K.; Horstick, E.; Linsley, J.; Sprague, S.M.; Cui, W.W.; Zhou, W.; Hirata, H.; Saint-Amant, L.; Hume, R.I.; et al. Trpm7 is required within zebrafish sensory neurons for the activation of touch-evoked escape behaviors. *J. Neurosci.* **2011**, *31*, 11633–11644. [CrossRef]
104. Thivichon-Prince, B.; Couble, M.L.; Giamarchi, A.; Delmas, P.; Franco, B.; Romio, L.; Struys, T.; Lambrichts, I.; Ressnikoff, D.; Magloire, H.; et al. Primary cilia of odontoblasts: possible role in molar morphogenesis. *J. Dent. Res.* **2009**, *88*, 910–915. [CrossRef]
105. Abrahamsen, B.; Zhao, J.; Asante, C.O.; Cendan, C.M.; Marsh, S.; Martinez-Barbera, J.P.; Nassar, M.A.; Dickenson, A.H.; Wood, J.N. The cell and molecular basis of mechanical, cold, and inflammatory pain. *Science* **2008**, *321*, 702–705. [CrossRef]
106. Cavanaugh, D.J.; Lee, H.; Lo, L.; Shields, S.D.; Zylka, M.J.; Basbaum, A.I.; Anderson, D.J. Distinct subsets of unmyelinated primary sensory fibers mediate behavioral responses to noxious thermal and mechanical stimuli. *Proc. Natl. Acad. Sci. USA* **2009**, *106*, 9075–9080. [CrossRef]
107. Chung, M.K.; Jue, S.S.; Dong, X. Projection of non-peptidergic afferents to mouse tooth pulp. *J. Dent. Res.* **2012**, *91*, 777–782. [CrossRef]
108. Coste, B.; Xiao, B.; Santos, J.S.; Syeda, R.; Grandl, J.; Spencer, K.S.; Kim, S.E.; Schmidt, M.; Mathur, J.; Dubin, A.E.; et al. Piezo proteins are pore-forming subunits of mechanically activated channels. *Nature* **2012**, *483*, 176–181. [CrossRef]
109. Coste, B.; Mathur, J.; Schmidt, M.; Earley, T.J.; Ranade, S.; Petrus, M.J.; Dubin, A.E.; Patapoutian, A. Piezo1 and Piezo2 are essential components of distinct mechanically activated cation channels. *Science* **2010**, *330*, 55–60. [CrossRef]
110. Delmas, P.; Coste, B. Mechano-gated ion channels in sensory systems. *Cell* **2013**, *155*, 278–284. [CrossRef]
111. Lumpkin, E.A.; Caterina, M.J. Mechanisms of sensory transduction in the skin. *Nature* **2007**, *445*, 858–865. [CrossRef]
112. Lumpkin, E.A.; Marshall, K.L.; Nelson, A.M. The cell biology of touch. *J. Cell Biol.* **2010**, *191*, 237–248. [CrossRef]
113. Ranade, S.S.; Syeda, R.; Patapoutian, A. Mechanically activated ion channels. *Neuron* **2015**, *87*, 1162–1179. [CrossRef] [PubMed]
114. Sharif-Naeini, R. Contribution of mechanosensitive ion channels to somatosensation. *Prog. Mol. Biol. Transl. Sci.* **2015**, *131*, 53–71.
115. Ikeda, H.; Heinke, B.; Ruscheweyh, R.; Sandkuhler, J. Synaptic plasticity in spinal lamina I projection neurons that mediate hyperalgesia. *Science* **2003**, *299*, 1237–1240. [CrossRef] [PubMed]
116. Maksimovic, S.; Nakatani, M.; Baba, Y.; Nelson, A.M.; Marshall, K.L.; Wellnitz, S.A.; Firozi, P.; Woo, S.H.; Ranade, S.; Patapoutian, A.; et al. Epidermal merkel cells are mechanosensory cells that tune mammalian touch receptors. *Nature* **2014**, *509*, 617–621. [CrossRef]
117. Woo, S.H.; Ranade, S.; Weyer, A.D.; Dubin, A.E.; Baba, Y.; Qiu, Z.; Petrus, M.; Miyamoto, T.; Reddy, K.; Lumpkin, E.A.; et al. Piezo2 is required for merkel-cell mechanotransduction. *Nature* **2014**, *509*, 622–626. [CrossRef] [PubMed]
118. Ranade, S.S.; Woo, S.H.; Dubin, A.E.; Moshourab, R.A.; Wetzel, C.; Petrus, M.; Mathur, J.; Begay, V.; Coste, B.; Mainquist, J.; et al. Piezo2 is the major transducer of mechanical forces for touch sensation in mice. *Nature* **2014**, *516*, 121–125. [CrossRef] [PubMed]
119. Won, J.; Vang, H.; Lee, P.R.; Kim, Y.H.; Kim, H.W.; Kang, Y.; Oh, S.B. Piezo2 expression in mechanosensitive dental primary afferent neurons. *J. Dent. Res.* **2017**, *96*, 931–937. [CrossRef] [PubMed]
120. Sato, M.; Ogura, K.; Kimura, M.; Nishi, K.; Ando, M.; Tazaki, M.; Shibukawa, Y. Activation of mechanosensitive transient receptor potential/piezo channels in odontoblasts generates action potentials in cocultured isolectin B4-negative medium-sized trigeminal ganglion neurons. *J. Endod.* **2018**, *44*, 984–991. [CrossRef] [PubMed]
121. Khatibi Shahidi, M.; Krivanek, J.; Kaukua, N.; Ernfors, P.; Hladik, L.; Kostal, V.; Masich, S.; Hampl, A.; Chubanov, V.; Gudermann, T.; et al. Three-dimensional imaging reveals new compartments and structural adaptations in odontoblasts. *J. Dent. Res.* **2015**, *94*, 945–954. [CrossRef]
122. Krishtal, O. The Asics: Signaling Molecules? Modulators? *Trends Neurosci.* **2003**, *26*, 477–483. [CrossRef]
123. Ugawa, S.; Ueda, T.; Takahashi, E.; Hirabayashi, Y.; Yoneda, T.; Komai, S.; Shimada, S. Cloning and functional expression of Asic-β2, a splice variant of Asic-β. *Neuroreport* **2001**, *12*, 2865–2869. [CrossRef]
124. Ichikawa, H.; Sugimoto, T. The co-expression of Asic3 with calcitonin gene-related peptide and parvalbumin in the rat trigeminal ganglion. *Brain Res.* **2002**, *943*, 287–291. [CrossRef]

125. Sole-Magdalena, A.; Revuelta, E.G.; Menenez-Diaz, I.; Calavia, M.G.; Cobo, T.; Garcia-Suarez, O.; Perez-Pinera, P.; De Carlos, F.; Cobo, J.; Vega, J.A. Human odontoblasts express transient receptor protein and acid-sensing ion channel mechanosensor proteins. *Microsc. Res. Tech.* **2011**, *74*, 457–463. [CrossRef]
126. Honore, E. The Neuronal Background K2p Channels: Focus On Trek1. *Nat. Rev. Neurosci.* **2007**, *8*, 251–261. [CrossRef]
127. Kang, D.; Kim, D. Trek-2 (K2p10.1) and tresk (K2p18.1) are major background K+ channels in dorsal root ganglion neurons. *Am. J. Physiol. Cell Physiol.* **2006**, *291*, C138–C146. [CrossRef]
128. Marsh, B.; Acosta, C.; Djouhri, L.; Lawson, S.N. Leak $K^+$ channel mrnas in dorsal root ganglia: Relation to inflammation and spontaneous pain behaviour. *Mol. Cell Neurosci.* **2012**, *49*, 375–386. [CrossRef]
129. Sato, M.; Furuya, T.; Kimura, M.; Kojima, Y.; Tazaki, M.; Sato, T.; Shibukawa, Y. Intercellular odontoblast communication via ATP mediated by pannexin-1 channel and phospholipase C-coupled receptor activation. *Front. Physiol.* **2015**, *6*, 326. [CrossRef]
130. Nishiyama, A.; Sato, M.; Kimura, M.; Katakura, A.; Tazaki, M.; Shibukawa, Y. Intercellular signal communication among odontoblasts and trigeminal ganglion neurons via glutamate. *Cell Calcium* **2016**, *60*, 341–355. [CrossRef]
131. Lesage, F.; Lazdunski, M. Molecular and functional properties of two-pore-domain potassium channels. *Am. J. Physiol. Ren. Physiol.* **2000**, *279*, F793–F801. [CrossRef]
132. Lim, J.C.; Mitchell, C.H. Inflammation, pain, and pressure—Purinergic signaling in oral tissues. *J. Dent. Res.* **2012**, *91*, 1103–1109. [CrossRef]
133. Kim, Y.S.; Paik, S.K.; Cho, Y.S.; Shin, H.S.; Bae, J.Y.; Moritani, M.; Yoshida, A.; Ahn, D.K.; Valtschanoff, J.; Hwang, S.J.; et al. Expression of P2x3 receptor in the trigeminal sensory nuclei of the rat. *J. Comp. Neurol.* **2008**, *506*, 627–639. [CrossRef]
134. Staikopoulos, V.; Sessle, B.J.; Furness, J.B.; Jennings, E.A. Localization of P2x2 and P2x3 receptors in rat trigeminal ganglion neurons. *Neuroscience* **2007**, *144*, 208–216. [CrossRef]
135. Alavi, A.M.; Dubyak, G.R.; Burnstock, G. Immunohistochemical evidence for atp receptors in human dental pulp. *J. Dent. Res.* **2001**, *80*, 476–483. [CrossRef]
136. Wang, W.; Yi, X.; Ren, Y.; Xie, Q. Effects of adenosine triphosphate on proliferation and odontoblastic differentiation of human dental pulp cells. *J. Endod.* **2016**, *42*, 1483–1489. [CrossRef]
137. Sharma, C.G.; Pradeep, A.R. Gingival crevicular fluid osteopontin levels in periodontal health and disease. *J. Periodontol.* **2006**, *77*, 1674–1680. [CrossRef]
138. Chung, M.K.; Guler, A.D.; Caterina, M.J. Trpv1 shows dynamic ionic selectivity during agonist stimulation. *Nat. Neurosci.* **2008**, *11*, 555–564. [CrossRef]
139. Jiang, J.; Gu, J. Expression of adenosine triphosphate P2x3 receptors in rat molar pulp and trigeminal ganglia. *Oral Surg. Oral Med. Oral Pathol. Oral Radiol. Endod.* **2002**, *94*, 622–626. [CrossRef]
140. Kuroda, H.; Shibukawa, Y.; Soya, M.; Masamura, A.; Kasahara, M.; Tazaki, M.; Ichinohe, T. Expression of P2x(1) and P2x(4) receptors in rat trigeminal ganglion neurons. *Neuroreport* **2012**, *23*, 752–756. [CrossRef]
141. Matsuka, Y.; Neubert, J.K.; Maidment, N.T.; Spigelman, I. Concurrent release of ATP and substance P within guinea pig trigeminal ganglia in vivo. *Brain Res.* **2001**, *915*, 248–255. [CrossRef]
142. Hu, B.; Chiang, C.Y.; Hu, J.W.; Dostrovsky, J.O.; Sessle, B.J. P2x receptors in trigeminal subnucleus caudalis modulate central sensitization in trigeminal subnucleus oralis. *J. Neurophysiol.* **2002**, *88*, 1614–1624. [CrossRef]
143. Adachi, K.; Shimizu, K.; Hu, J.W.; Suzuki, I.; Sakagami, H.; Koshikawa, N.; Sessle, B.J.; Shinoda, M.; Miyamoto, M.; Honda, K.; et al. Purinergic receptors are involved in tooth-pulp evoked nocifensive behavior and brainstem neuronal activity. *Mol. Pain* **2010**, *6*, 59. [CrossRef]
144. Liu, X.; Wang, C.; Fujita, T.; Malmstrom, H.S.; Nedergaard, M.; Ren, Y.F.; Dirksen, R.T. External dentin stimulation induces ATP release in human teeth. *J. Dent. Res.* **2015**, *94*, 1259–1266. [CrossRef]
145. Liu, X.; Yu, L.; Wang, Q.; Pelletier, J.; Fausther, M.; Sevigny, J.; Malmstrom, H.S.; Dirksen, R.T.; Ren, Y.F. Expression of ecto-atpase Ntpdase2 in human dental pulp. *J. Dent. Res.* **2012**, *91*, 261–267. [CrossRef]
146. Shiozaki, Y.; Sato, M.; Kimura, M.; Sato, T.; Tazaki, M.; Shibukawa, Y. Ionotropic P2x ATP receptor channels mediate purinergic signaling in mouse odontoblasts. *Front. Physiol.* **2017**, *8*, 3. [CrossRef]
147. Kawaguchi, A.; Sato, M.; Kimura, M.; Ichinohe, T.; Tazaki, M.; Shibukawa, Y. Expression and function of purinergic P2y12 receptors in rat trigeminal ganglion neurons. *Neurosci. Res.* **2015**, *98*, 17–27. [CrossRef]
148. Li, N.; Lu, Z.Y.; Yu, L.H.; Burnstock, G.; Deng, X.M.; Ma, B. Inhibition of g protein-coupled P2y2 receptor induced analgesia in a rat model of trigeminal neuropathic pain. *Mol. Pain* **2014**, *10*, 21. [CrossRef]

149. Magni, G.; Merli, D.; Verderio, C.; Abbracchio, M.P.; Ceruti, S. P2y2 receptor antagonists as anti-allodynic agents in acute and sub-chronic trigeminal sensitization: Role of satellite glial cells. *Glia* **2015**, *63*, 1256–1269. [CrossRef]
150. Chen, L.; Huang, J.; Zhao, P.; Persson, A.K.; Dib-Hajj, F.B.; Cheng, X.; Tan, A.; Waxman, S.G.; Dib-Hajj, S.D. Conditional knockout of Nav1.6 in adult mice ameliorates neuropathic pain. *Sci. Rep.* **2018**, *8*, 3845. [CrossRef]
151. Nassar, M.A.; Stirling, L.C.; Forlani, G.; Baker, M.D.; Matthews, E.A.; Dickenson, A.H.; Wood, J.N. Nociceptor-specific gene deletion reveals a major role for Nav1.7 (PN1) in acute and inflammatory pain. *Proc. Natl. Acad. Sci. USA* **2004**, *101*, 12706–12711. [CrossRef]
152. Beneng, K.; Renton, T.; Yilmaz, Z.; Yiangou, Y.; Anand, P. Sodium channel Na V 1.7 immunoreactivity in painful human dental pulp and burning mouth syndrome. *BMC Neurosci.* **2010**, *11*, 71. [CrossRef]
153. Renton, T.; Yiangou, Y.; Plumpton, C.; Tate, S.; Bountra, C.; Anand, P. Sodium channel Nav1.8 immunoreactivity in painful human dental pulp. *BMC Oral Health* **2005**, *5*, 5. [CrossRef]
154. Warren, C.A.; Mok, L.; Gordon, S.; Fouad, A.F.; Gold, M.S. Quantification of neural protein in extirpated tooth pulp. *J. Endod.* **2008**, *34*, 7–10. [CrossRef]
155. Luo, S.; Perry, G.M.; Levinson, S.R.; Henry, M.A. Nav1.7 expression is increased in painful human dental pulp. *Mol. Pain* **2008**, *4*, 16. [CrossRef]
156. Luo, S.; Perry, G.M.; Levinson, S.R.; Henry, M.A. Pulpitis increases the proportion of atypical nodes of ranvier in human dental pulp axons without a change in Nav1.6 sodium channel expression. *Neuroscience* **2010**, *169*, 1881–1887. [CrossRef]
157. Davidson, R.M. Neural form of voltage-dependent sodium current in human cultured dental pulp cells. *Arch. Oral Biol.* **1994**, *39*, 613–620. [CrossRef]
158. Dib-Hajj, S.D.; Tyrrell, L.; Black, J.A.; Waxman, S.G.; Nan, A. Novel voltage-gated Na channel, is expressed preferentially in peripheral sensory neurons and down-regulated after axotomy. *Proc. Natl. Acad. Sci. USA* **1998**, *95*, 8963–8968. [CrossRef]
159. Dib-Hajj, S.; Black, J.A.; Cummins, T.R.; Waxman, S.G. Nan/Nav1.9: A sodium channel with unique properties. *Trends Neurosci.* **2002**, *25*, 253–259. [CrossRef]
160. Fang, X.; Djouhri, L.; Black, J.A.; Dib-Hajj, S.D.; Waxman, S.G.; Lawson, S.N. The presence and role of the tetrodotoxin-resistant sodium channel Na(v)1.9 (Nan) in nociceptive primary afferent neurons. *J. Neurosci.* **2002**, *22*, 7425–7433. [CrossRef]
161. Fang, X.; Djouhri, L.; Mcmullan, S.; Berry, C.; Waxman, S.G.; Okuse, K.; Lawson, S.N. Intense isolectin-b4 binding in rat dorsal root ganglion neurons distinguishes C-fiber nociceptors with broad action potentials and high Nav1.9 expression. *J. Neurosci.* **2006**, *26*, 7281–7292. [CrossRef]
162. Rugiero, F.; Mistry, M.; Sage, D.; Black, J.A.; Waxman, S.G.; Crest, M.; Clerc, N.; Delmas, P.; Gola, M. Selective expression of a persistent tetrodotoxin-resistant $Na^+$ current and Nav1.9 subunit in myenteric sensory neurons. *J. Neurosci.* **2003**, *23*, 2715–2725. [CrossRef]
163. Tate, S.; Benn, S.; Hick, C.; Trezise, D.; John, V.; Mannion, R.J.; Costigan, M.; Plumpton, C.; Grose, D.; Gladwell, Z.; et al. Two sodium channels contribute to the Ttx-r sodium current in primary sensory neurons. *Nat. Neurosci.* **1998**, *1*, 653–655. [CrossRef] [PubMed]
164. Padilla, F.; Couble, M.L.; Coste, B.; Maingret, F.; Clerc, N.; Crest, M.; Ritter, A.M.; Magloire, H.; Delmas, P. Expression and localization of the Nav1.9 sodium channel in enteric neurons and in trigeminal sensory endings: Implication for intestinal reflex function and orofacial pain. *Mol. Cell. Neurosci.* **2007**, *35*, 138–152. [CrossRef] [PubMed]
165. Wells, J.E.; Bingham, V.; Rowland, K.C.; Hatton, J. Expression of Nav1.9 channels in human dental pulp and trigeminal ganglion. *J. Endod.* **2007**, *33*, 1172–1176. [CrossRef] [PubMed]
166. Dobremez, E.; Bouali-Benazzouz, R.; Fossat, P.; Monteils, L.; Dulluc, J.; Nagy, F.; Landry, M. Distribution and regulation of L-type calcium channels in deep dorsal horn neurons after sciatic nerve injury in rats. *Eur. J. Neurosci.* **2005**, *21*, 3321–3333. [CrossRef] [PubMed]
167. Namkung, Y.; Skrypnyk, N.; Jeong, M.J.; Lee, T.; Lee, M.S.; Kim, H.L.; Chin, H.; Suh, P.G.; Kim, S.S.; Shin, H.S. Requirement for the L-Type $Ca^{2+}$ channel $\alpha$(1d) subunit in postnatal pancreatic $\beta$ cell generation. *J. Clin. Investig.* **2001**, *108*, 1015–1022. [CrossRef] [PubMed]
168. Scroggs, R.S.; Fox, A.P. Calcium current variation between acutely isolated adult rat dorsal root ganglion neurons of different size. *J. Physiol.* **1992**, *445*, 639–658. [CrossRef]

169. Kim, C.; Jun, K.; Lee, T.; Kim, S.S.; Mcenery, M.W.; Chin, H.; Kim, H.L.; Park, J.M.; Kim, D.K.; Jung, S.J.; et al. Altered nociceptive response in mice deficient in the α(1b) subunit of the voltage-dependent calcium channel. *Mol. Cell. Neurosci.* **2001**, *18*, 235–245. [CrossRef]
170. Tang, Q.; Bangaru, M.L.; Kostic, S.; Pan, B.; Wu, H.E.; Koopmeiners, A.S.; Yu, H.; Fischer, G.J.; Mccallum, J.B.; Kwok, W.M.; et al. $Ca^{2+}$-dependent regulation of $Ca^{2+}$ currents in rat primary afferent neurons: role of camkii and the effect of injury. *J. Neurosci.* **2012**, *32*, 11737–11749. [CrossRef]
171. Ju, Y.; Ge, J.; Ren, X.; Zhu, X.; Xue, Z.; Feng, Y.; Zhao, S. Cav1.2 of L-type calcium channel is a key factor for the differentiation of dental pulp stem cells. *J. Endod.* **2015**, *41*, 1048–1055. [CrossRef]
172. Seux, D.; Joffre, A.; Fosset, M.; Magloire, H. Immunohistochemical localization of L-type calcium channels in the developing first molar of the rat during odontoblast differentiation. *Arch. Oral Biol.* **1994**, *39*, 167–170. [CrossRef]
173. Shibukawa, Y.; Suzuki, T. $Ca^{2+}$ signaling mediated by Ip3-dependent $Ca^{2+}$ releasing and store-operated $Ca^{2+}$ channels in rat odontoblasts. *J. Bone Miner Res.* **2003**, *18*, 30–38. [CrossRef]
174. Westenbroek, R.E.; Anderson, N.L.; Byers, M.R. Altered localization of Cav1.2 (L-Type) calcium channels in nerve fibers, schwann cells, odontoblasts, and fibroblasts of tooth pulp after tooth injury. *J. Neurosci. Res.* **2004**, *75*, 371–383. [CrossRef]
175. Davidson, R.M. Potassium currents in cells derived from human dental pulp. *Arch. Oral Biol.* **1993**, *38*, 803–811. [CrossRef]
176. Lundgren, T.; Nannmark, U.; Linde, A. Calcium ion activity and ph in the odontoblast-predentin region: Ion-selective microelectrode measurements. *Calcif. Tissue Int.* **1992**, *50*, 134–136. [CrossRef]

International Journal of
*Molecular Sciences*

*Article*

# Hydrophobic Amines and Their Guanidine Analogues Modulate Activation and Desensitization of ASIC3

**Vasilii Y Shteinikov [1,*], Natalia N Potapieva [1], Valery E Gmiro [2] and Denis B Tikhonov [1]**

[1] I.M. Sechenov Institute of Evolutionary Physiology and Biochemistry RAS, St. Petersburg 194223, Russia; potapieva2004@mail.ru (N.N.P.); denistikhonov2002@yahoo.com (D.B.T.)
[2] Institute of Experimental Medicine, RAMS, St. Petersburg 197376, Russia; gmiro2119@gmail.com
* Correspondence: vasilii.shteinikov@gmail.com

Received: 15 March 2019; Accepted: 4 April 2019; Published: 6 April 2019

**Abstract:** Acid-sensing ion channel 3 (ASIC3) is an important member of the acid-sensing ion channels family, which is widely expressed in the peripheral nervous system and contributes to pain sensation. ASICs are targeted by various drugs and toxins. However, mechanisms and structural determinants of ligands' action on ASIC3 are not completely understood. In the present work we studied ASIC3 modulation by a series of "hydrophobic monoamines" and their guanidine analogs, which were previously characterized to affect other ASIC channels via multiple mechanisms. Electrophysiological analysis of action via whole-cell patch clamp method was performed using rat ASIC3 expressed in Chinese hamster ovary (CHO) cells. We found that the compounds studied inhibited ASIC3 activation by inducing acidic shift of proton sensitivity and slowed channel desensitization, which was accompanied by a decrease of the equilibrium desensitization level. The total effect of the drugs on the sustained ASIC3-mediated currents was the sum of these opposite effects. It is demonstrated that drugs' action on activation and desensitization differed in their structural requirements, kinetics of action, and concentration and state dependencies. Taken together, these findings suggest that effects on activation and desensitization are independent and are likely mediated by drugs binding to distinct sites in ASIC3.

**Keywords:** acid-sensing ion channel (ASIC); drug action; ligand-gated ion channel; pharmacology; small molecule; nociception; ASIC3

---

## 1. Introduction

Acid-sensing ion channels (ASICs) are cation channels from the degenerin/epithelial sodium channel (DEG/ENaC) superfamily. They are activated by fast acidification of the media, while prolonged exposure leads to their desensitization. There are five paralogous genes in this group, with the expression products of *ASIC1*, *2*, and *3* forming functional trimeric channels. *ASIC1* and *ASIC2* are predominantly expressed in the central nervous system, whereas *ASIC3* is more common in the peripheral nervous system [1]. The functions they fulfill also vary. ASIC1 and ASIC2 have been shown to contribute to the excitatory postsynaptic currents [2] and synaptic plasticity [3], and are also involved in the pathologic processes in stroke and ischemia [4,5]. On the other hand, ASIC3 is typically associated with peripheral nociception [6]. Another important difference is that ASIC3 channels, unlike other ASICs, do not fully desensitize during prolonged activation, supporting a significant sustained current [7].

The involvement of ASICs, in particular ASIC3, in the perception of pain has been firmly established in a number of studies (for review see [8]). The use of ASIC inhibitors in rats and humans was shown to alleviate cutaneous pain and hyperalgesia [9–11]. Surprisingly, knockout of the *ASIC3* gene in mice did not lead to a loss or significant decrease of their pain responses compared to wild type [12]. In fact, in the study of Kang et al. [13], triple knockout (for *ASIC1a, 2* and *3* genes) mice showed enhanced pain sensitivity. This phenomenon can potentially be explained by the different roles of the ASIC channels in different species or particular levels and by specific details of their expression [14]. Large acidification-evoked currents were also shown in cardiac afferents, where they propagated cardiac pain and angina [15]. Characteristics of those currents are closely matched by heteromeric ASIC3/ASIC2b channels [16]. Other pain-associated conditions are also mediated by ASIC3, such as migraines [17], osteoarthritis [18], and muscle inflammation [19].

Given the importance of their role and the potential of new functions' discovery, it is not surprising that ASIC pharmacology receives quite a lot of attention [20]. The pioneering paper by Waldmann et al. [21] described the action of amiloride, a common modulator of ENaC channels, which was found to be a low-affinity inhibitor of ASICs. Focusing on ASIC3 for the purposes of this work, there are several groups of drugs to be noted. The abovementioned amiloride inhibits peak currents of ASIC3 but does not affect the window current. Even more interestingly, in high concentrations it is capable of inducing said window current by itself, without acidification [22]. 2-Guanidine-4-methylquinazoline (GMQ) was also originally described as an ASIC3 modulator, although, unlike its predecessor, it has a potentiating effect. Like amiloride, GMQ can also evoke ASIC3 currents in neutral pH [23]. Later it was found that GMQ and its derivatives can also modulate ASIC1a [24].

Several endogenous compounds were shown to potentiate ASIC3 currents, including FMRFamides and related peptides [25], agmatine [26], and serotonin [27], with the last one only affecting the sustained component of the response. Agmatine was also able to activate the channels directly.

On the other hand, toxins mostly display inhibitory action on ASIC3. A number of sea anemone toxins, such as APETx2 [28] and Ugr 9-1 [29], inhibit both peak and window currents in ASIC3. MitTx [30], which locks the channel in the open state, also works on ASIC3 but in significantly higher concentrations than on the other subunits.

In our research [31] we focused our attention on a group of small-molecule ligands we collectively called *hydrophobic monoamines*. Despite their structural simplicity, further investigations revealed quite complex effects that they can induce on ASIC channels [32]. We found that they can block the channel pore, affect the activation curve in either direction, and shift the desensitization curve to more acidic values, often with several effects observed for a single compound. Additionally, through this line of investigation a potential physiological modulator of ASICs (i.e., histamine) was discovered [33]. Its effects were specific to ASIC1a homomers. However, outside of initial assessment [31], the action of monoamines on ASIC3 was never studied. Thus, in the present work we attempted to elucidate the mechanisms of action of hydrophobic monoamines and their guanidine analogs on ASIC3 channels. Other compounds that were found to affect ASIC1a and/or ASIC2a, such as some antidepressants [34] and histamine receptor agonists [35], were also included in the study.

## 2. Results

### 2.1. Drug Selection

Several groups of compounds were selected for the present study. The IEM line ofcompounds was originally designed as glutamate receptor agonists [36,37]. Their activity on ASIC channels was subsequently shown by our group [31,32]. Memantine [38] and 9-aminoacridine [39] also affect glutamate receptors as well as ASICs [31]. Other drugs included long-established antidepressants amitriptyline and tianeptine [40,41] and histamine receptor modulators imetit, dimaprit, and thioperamide [42,43]; their effects on ASICs were recently established in [34] and [35], respectively. It is important to note that in previous studies only the effects on ASIC1a and ASIC2a were examined, with ASIC3 covered very briefly in [31].

## 2.2. Estimation of Drug Activities

For the sixteen compounds presented in Figure 1 we estimated the effects on peak and sustained currents evoked by acidification from pH 7.4 to 6.85, which caused 10% ± 7% ($n$ = 11) of maximal peak response, and to pH 6.0, which caused 74% ± 16% ($n$ = 11) maximal peak response. The compounds were applied simultaneously with acidification at a concentration of 0.5 mM. These applications were repeated 3–7 times to reach the effect's equilibrium point and then 3–10 washout acidifications were done until complete recovery was achieved.

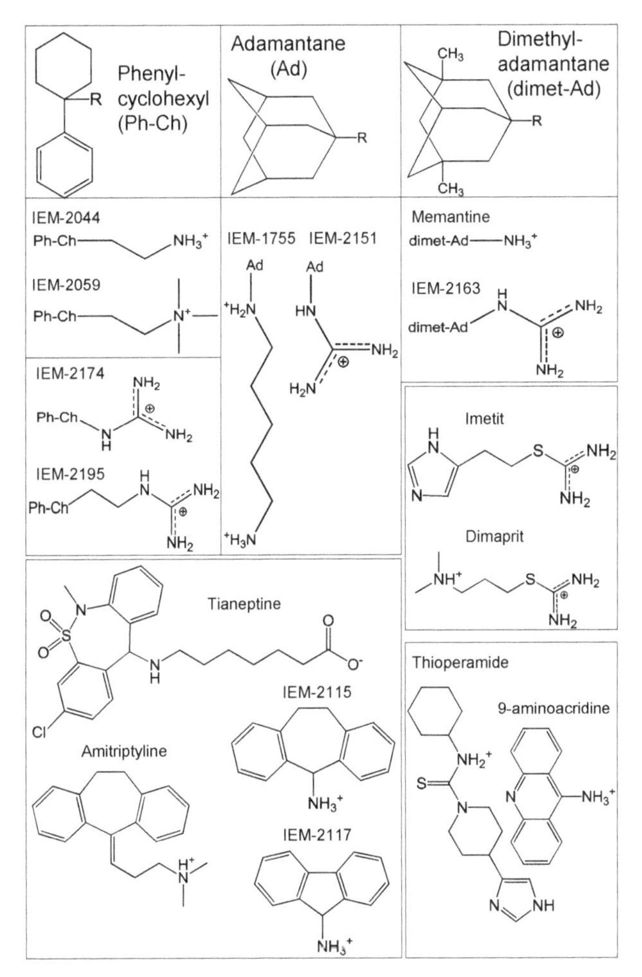


**Figure 1.** Chemical structure of the tested compounds. The first row represents common hydrophobic moieties (Ph-Ch, Ad, and dimet-Ad) of IEM compounds and memantine, with their terminal radicals (R) shown directly below.

The results are presented in Figure 2. At pH 6.85 (Figure 2A, with sample traces shown in Figure 2B,C) the peak component of the response was strongly inhibited by a number of compounds, the most potent being IEM-2195 at 85% ± 7% ($n$ = 6) inhibition, and only IEM-2117 slightly potentiated the peak response by 42% ± 21% ($n$ = 5). On the other hand, sustained current was typically potentiated, with the strongest effect by IEM-2117 at 382% ± 84% ($n$ = 5). IEM-2163 and IEM-2151 were the only compounds that reduced the sustained current by 42% ± 21% ($n$ = 9) and 29% ± 4% ($n$ = 5), respectively. At pH 6.0 (Figure 2D, with sample traces shown in Figure 2E,F) the drugs' effect on peak current disappeared, while for sustained current the overall picture stayed the same and the effects even somewhat increased in magnitude, with maximal potentiation by IEM-2117 reaching 498% ± 196% ($n$ = 5) and inhibition by IEM-2163 at 54% ± 23% ($n$ = 8). We can conclude from the data that (1) structural determinants of the effects on peak and sustained components of the response did not coincide and (2) only the effect on peak component demonstrated pronounced pH-dependence.

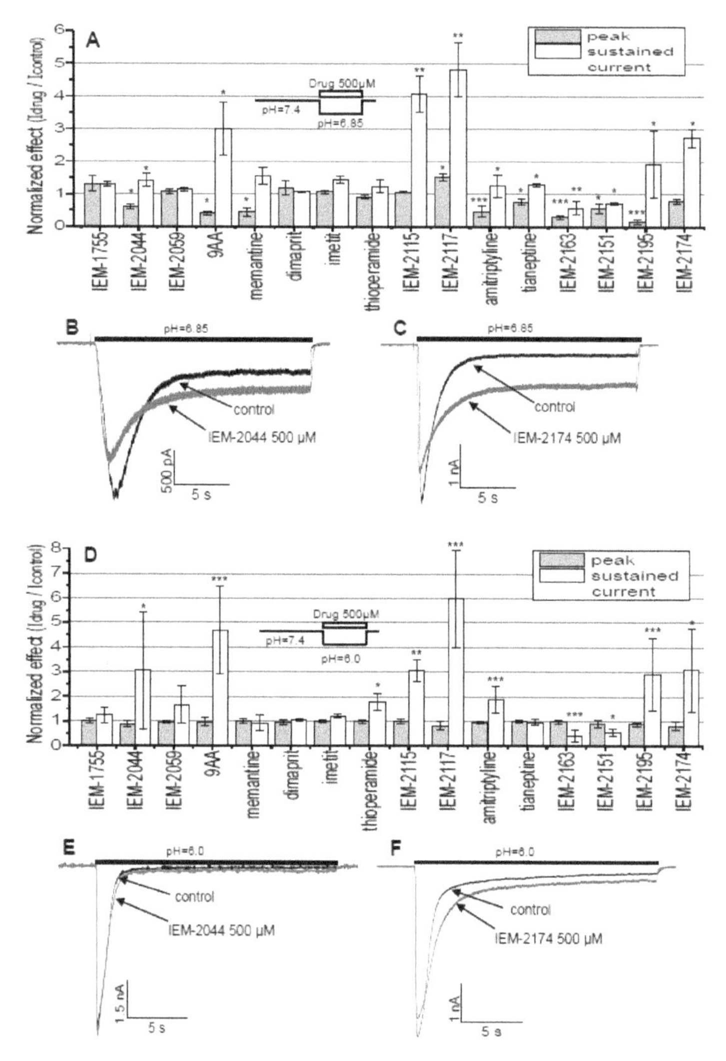


**Figure 2.** Estimation of the drug activities. Number of * denotes statistical significance at * $p < 0.05$, ** $p < 0.01$, or *** $p < 0.001$, respectively, $n \geq 5$. (**A**) Effects of 500 µM of the compounds studied on the ASIC3 activated by pH drop from 7.4 to 6.85. Compounds were applied simultaneously with activation. (**B**,**C**) Representative examples of ASIC3 responses in control and in the presence of IEM-2044 (**B**) and IEM-2174 (**C**) when activated by pH drop from 7.4 to 6.85. (**D**) effects of 500 µM of the compounds studied on the ASIC3 activated by pH drop from 7.4 to 6.0. Compounds were applied simultaneously with activation. (**E**,**F**) Representative examples of ASIC3 responses in control and in the presence of IEM-2044 (**E**) and IEM-2174 (**F**) when activated by pH drop from 7.4 to 6.0.

Typically, potentiation of the sustained current was accompanied by deceleration of the response decay, reflecting an effect on desensitization. In control experiments the decay time constant was 462 ± 143 ms ($n$ = 8). The most drastic increase was seen for IEM-2195, which changed the decay time constant of the response to 4891 ± 1996 ms ($n$ = 6). Inhibition of sustained current by IEM-2163 or IEM-2151 did not elicit significant changes of the response kinetics.

### 2.3. pH and Concentration Dependencies

For detailed analysis we selected IEM-2163 and IEM-2195, as they demonstrate opposite effects (inhibition and potentiation, respectively) on the sustained currents. First, we estimated the pH-dependence of action on peak currents (Figure 3A). Figure 3B demonstrates that both compounds caused a parallel shift of activation to more acidic values without affecting maximal response. IEM-2163 at 0.5 mM shifted the $pH_{50}$ value from 6.26 ± 0.02 in control to 6.17 ± 0.06. The shift caused by IEM-2195 was about equal, with the $pH_{50}$ of activation being 6.17 ± 0.04 in the presence of this drug.

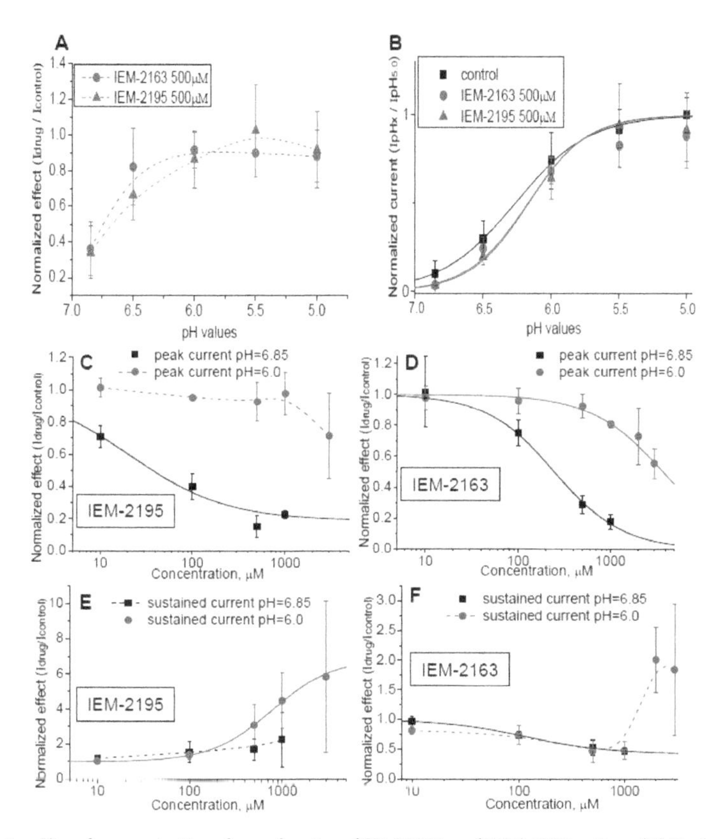


**Figure 3.** pH and concentration dependencies of IEM-2163 and IEM-2195 action. (**A**) Peak current inhibition was pH-dependent, the effects of both drugs disappeared under strong acidification. (**B**) IEM-2163 and IEM-2195 caused acidic shift of the ASIC3 activation curve. (**C–F**) Concentration dependencies of IEM-2163 (**D**,**F**) and IEM-2195 (**C**,**E**) action on peak (**C**,**D**) and sustained (**E**,**F**) currents. Fitting is shown in solid lines. Note that the concentration dependence of IEM-2163's action on sustained current was biphasic at pH 6.0 where the current inhibition was small. Low concentrations caused inhibition, but at high concentrations the effect was inverted.

We then studied the concentration dependencies of the actions of IEM-2163 and IEM-2195 on peak and sustained components of the response at pH 6.85 and 6.0 (Figure 3C–F). At pH 6.85, peak inhibition by IEM-2195 (Figure 3C)—which reflects its effect on activation—was well-fitted by the Hill equation, with optimal parameters nH = 0.82 ± 0.54, $IC_{50}$ = 21 ± 15 μM. At pH 6.0, no significant effect was detected for concentrations up to 1 mM (Figure 3C). Our attempts to further increase the concentration led to poor clamp stability, resulting in highly diverging data at higher concentrations. In contrast, potentiation of the sustained current was well established at pH 6.0 (Figure 3E), where inhibition of activation was absent. The fitting resulted in $EC_{50}$ = 784.2 ± 122.8 μM, nH = 1.31 ± 0.06, and maximal effect 588% ± 55% potentiation.

IEM-2163 also strongly inhibited peak current. The $IC_{50}$ at pH 6.85 (Figure 3D) was 245.64 ± 16.62 μM, nH = 1.17 ± 0.08. Similar to IEM-2195, peak inhibition at pH 6.0 was not significant. Sustained currents were significantly inhibited by IEM-2163 at pH 6.85 as well (Figure 3F), $IC_{50}$ = 117.3 ± 5.6 μM, nH = 1.12 ± 0.07. A peculiar concentration dependence was observed for IEM-2163's action at pH 6.0. Low concentrations caused progressive inhibition, but at around 1 mM the effect reached saturation at the level of 52% ± 16% of inhibition ($n$ = 5), and at 3 mM, despite large data diversity, we saw an apparent potentiation by 84% ± 111% ($n$ = 5). To ensure this was not an artifact of data variation, we performed additional experiments at 2 mM, which complied with the observed reversion of the effect resulting in 101% ± 56% ($n$ = 6) potentiation.

An explanation of such concentration dependencies could be that they reflect a mixture of two distinct effects: pH-dependent inhibition of activation, which is responsible for the inhibition of peak component and window component in low concentrations; and reduction of desensitization, which determines the potentiation of the sustained current at high concentrations. Thus, analysis of concentration dependencies provided arguments in favor of the independence of drug effects on activation and desensitization.

### 2.4. Dependence of Action on the Application Protocols

Next, we compared drug effects in different application protocols. In addition to the protocol of simultaneous application (see above) we applied the drugs continuously or during 30 s immediately before activation by pH drop. The results are shown in Figure 4. The peak response evoked by pH 6.0 was not strongly affected, regardless of the application protocol for both compounds. More interestingly, peak response evoked by pH 6.85 was inhibited only if the drug was present during acidification and not only before it (Figure 4A,E). This finding can be explained by two different mechanisms: (1) the compounds interact only with the open channels and/or (2) the kinetics of their action is very fast. The effects on the sustained currents also depended on the application protocol. Application of IEM-2163 before activation by pH 6.0 resulted in 143% ± 41% ($n$ = 5) potentiation, while under other conditions the drug caused inhibition (Figure 4B). For IEM-2195, in all protocols potentiation at pH 6.0 was higher than at pH 6.85 (two-way ANOVA for "protocol" and "pH" as factors, $F(1,36) = 25.838$, $p < 0.001$). This complex behavior is readily explained by the existence of two separate effects: pH-dependent inhibition of activation, which was also seen as peak inhibition; and pH-independent reduction of desensitization, with the total effect on the sustained currents being a sum of them.

As we diminished inhibition by the use of pH 6.0 for activation or by drug application only at neutral pH before activation, the anti-desensitizing effect on the sustained current increased (for IEM-2195) or became apparent (in the case of IEM-2163).

### 2.5. Kinetics of Action

Observation of the drug effects throughout the series of activations in the drug presence revealed an interesting tendency (Figure 5). Unlike the typical monotonic effect development, in experiments with IEM-2163, the sustained current was strongly inhibited during the first activation in the presence of the drug, but in subsequent activations the inhibition became less pronounced. The washout process was also non-monotonic—in the first activation it demonstrated a significant "overshoot"—the response decay was much slower than in control, resulting in the current at the end of activation being higher than the control one (Figure 5A). The control parameters were eventually reached after 5–7 activations. For IEM-2195, its potentiation developed monotonically but, similarly to IEM-2163, there was also a washout "overshoot"—in the first washout activation the current at the end of the response was even higher than in the last activation with the drug (Figure 5B).

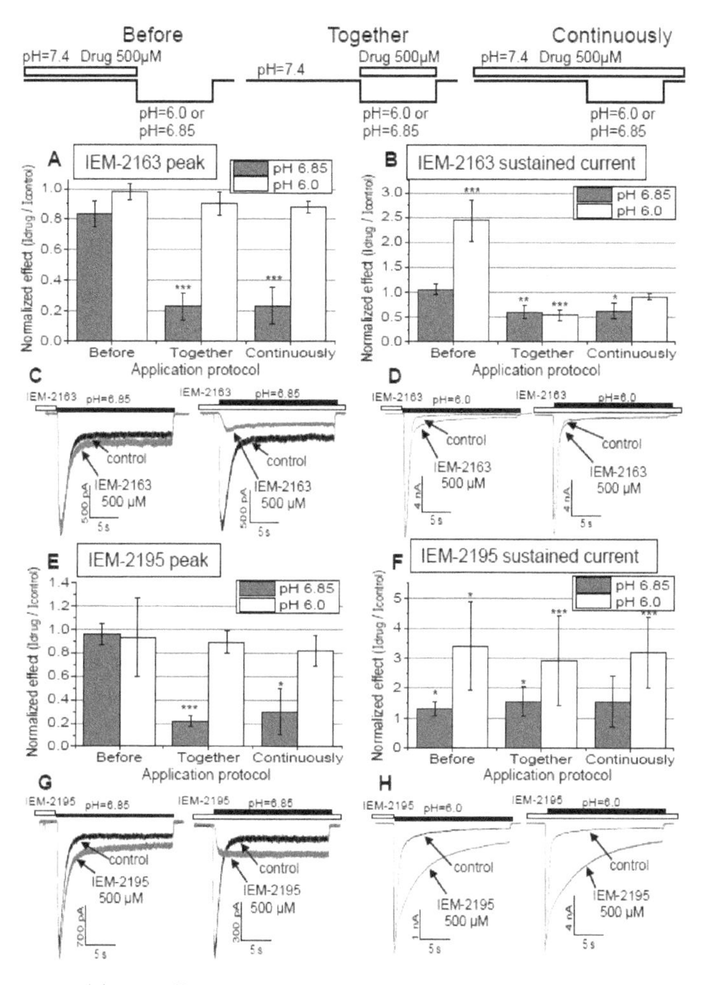


**Figure 4.** Effects of drug application protocol on peak and sustained currents of ASIC3 in different activating pH. Number of * denotes statistical significance at * $p < 0.05$, ** $p < 0.01$, or *** $p < 0.001$, respectively, $n \geq 5$. (**A**) Effects of IEM-2163 on peak current and (**B**) sustained current. (**C**,**D**) Representative examples of ASIC3 responses in different activating pH (6.85 in (**C**) and 6.0 in (**D**)) and application protocols (left panels: application before activation, right panels: continuous application). (**E**–**H**) Same for IEM-2195. IEM-2163 (**A**–**D**) had mostly similar effects regardless of the protocol used, inhibiting peak current at pH 6.85 and sustained current for both pH values, with two notable exceptions. When applied before activation with activating pH 6.85 it had no effect at all, and in the same protocol but with activating pH 6.0 it strongly potentiated sustained current. IEM-2195 (**E**–**H**) mostly had a similar profile, but it typically potentiated the sustained current, although this effect was significantly weaker at pH 6.85, essentially disappearing when applied before activation.

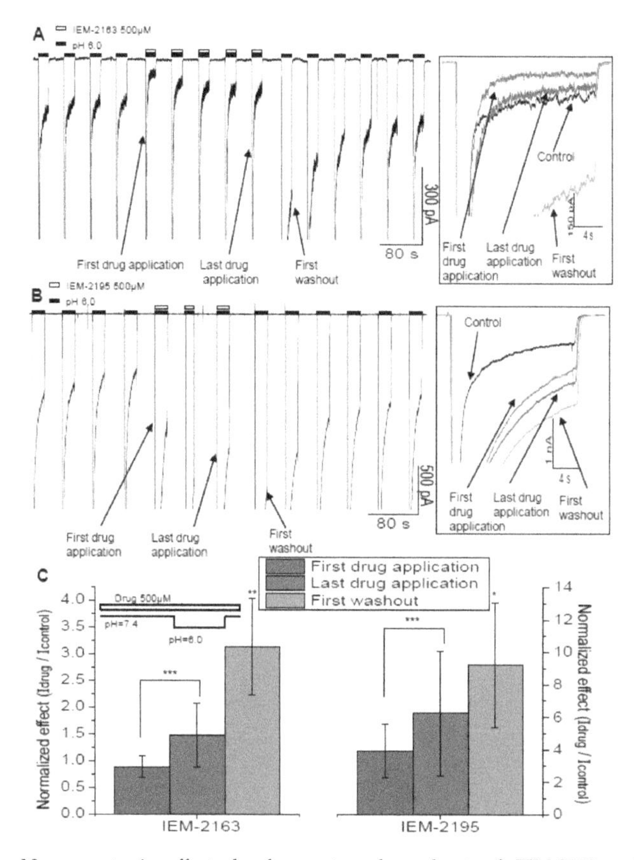


**Figure 5.** Non-monotonic effect development and washout of IEM-2163 and IEM-2195. (**A**,**B**) Representative recordings of the time course of experiments with IEM-2163 (**A**) and IEM-2195 (**B**). To the right are overlaid responses from the main panel. (**C**) The effect was the most pronounced in the protocol of continuous application. Number of * denotes statistical significance at * $p < 0.05$, ** $p < 0.01$, or *** $p < 0.001$, respectively, $n \geq 5$. The average values were calculated as the ratio of amplitudes for the first drug application, last drug application, and first washout to the last control response, respectively.

This phenomenon was the most pronounced with continuous drug application (Figure 5C). However, if the drugs were applied only before the channel activation, the overshoot effect disappeared. In this case both IEM-2163 and IEM-2195 caused potentiation, and recovery from it developed monotonically.

Our explanation of these effects is that inhibition of activation is fast while the effect on desensitization is much slower. Inhibition develops during the first activation in the presence of the drugs and is just as rapidly washed out during the first activation without them, while the reduction of desensitization requires several minutes to develop and wash out. This explanation agrees with the protocol dependence—fast peak inhibition required the drug's presence during activation, whereas a slow effect on desensitization could be obtained during long pre-application and remained even if the drug was absent from the solution during the activation. Thus, the effects on activation and desensitization differed not only in structural determinants (Figure 2) as well as concentration and pH dependence (Figure 3), but also in their kinetics.

## 2.6. Biphasic Drug Effects, when Applied Exclusively to the Sustained Current

The fact that ASIC3s do not desensitize completely and can mediate significant sustained current allows for one more type of experiment (Figure 6), which is helpful for the analysis of the mechanism of action. We activated the channels in the absence of a drug and only applied it when the current reached the sustained level. Washout was also performed during this prolonged activation, without returning to the neutral pH. Typical currents are presented in Figure 6A,B. Application of IEM-2195 (Figure 6A) caused fast transient "on current", which then slowly returned to the equilibrium level similar to the value of sustained current potentiation observed in the previous experiments. Removal of the drug resulted in a large transient "tail current" before returning to the control value. Application and removal of IEM-2163 caused similar "on" and "tail" transient currents, although they had smaller amplitude (Figure 6B). The main difference between the drugs was the direction of the change in the sustained current's amplitude at the equilibrium level, which was potentiated by IEM-2195 and inhibited by IEM-2163, respectively. We were especially careful to ensure that this unusual behavior was not an artifact of the solution exchange. Additionally, "tail" currents for both compounds demonstrated clear concentration dependence (Figure 6C), with fitting resulting in $EC_{50}$ = 269.06 ± 15.35 µM, nH = 1.76 ± 0.11 for IEM-2195 and $EC_{50}$ = 319.69 ± 44.08 µM, nH = 2.00 ± 0.33 for IEM-2163.

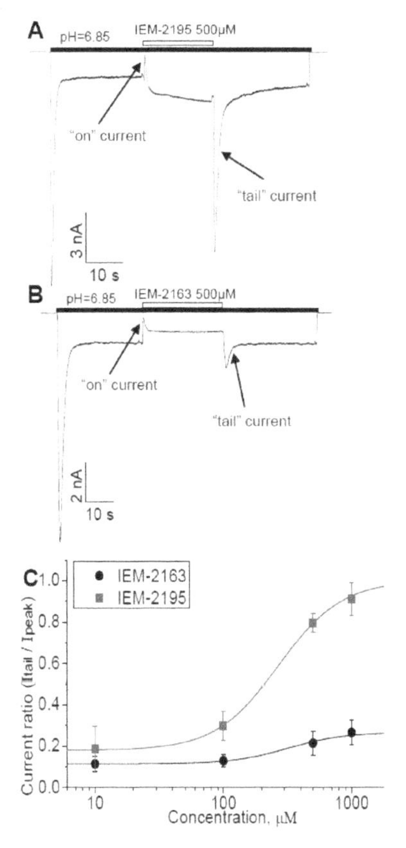


**Figure 6.** IEM-2163 and IEM-2195 cause transient currents when applied to sustained response. (**A**,**B**) Representative recordings. Fast drug application caused transient current decrease ("on" current), while washout induced transient increase ("tail" current). These transients reflect the presence of two opposite effects with different kinetics. (**C**) Concentration dependencies of "tail" currents.

We suggest that the observed "on" and "tail" currents reflect kinetics and complex mechanisms of drug action, which include the inhibition of activation and reduction of desensitization. We suggest that the "on" current appears because inhibitory action develops quickly, while slow modulation of desensitization is responsible for the subsequent equilibrium level of the sustained current. The change of this equilibrium effect (potentiation by IEM-2195 and inhibition by IEM-2163) may depend on the balance between these two opposite actions. Fast inhibition of activation would also be responsible for the "tail" currents, resulting from an acidic shift of activation (see Figure 3B). Purportedly, in this case the drug-bound channels would remain in the resting state even under conditions of acidic pH. Thus, fast removal of a drug would allow protons to bind and activate the channels.

To further check this suggestion, we performed analogous experiments with some other drugs (Figure 7). 9-Aminoacridine and IEM-2044, which inhibit peak and potentiate the sustained component of the response, also demonstrated pronounced "on" and "tail" currents. In contrast, for IEM-2059 and IEM-1755 these transient currents were absent. In analogous experiments with agmatine performed by Li et al. [26], no "on" or "tail" currents were shown, probably because agmatine leads to an alkaline shift of activation and has an overall potentiating effect on ASIC3 currents.

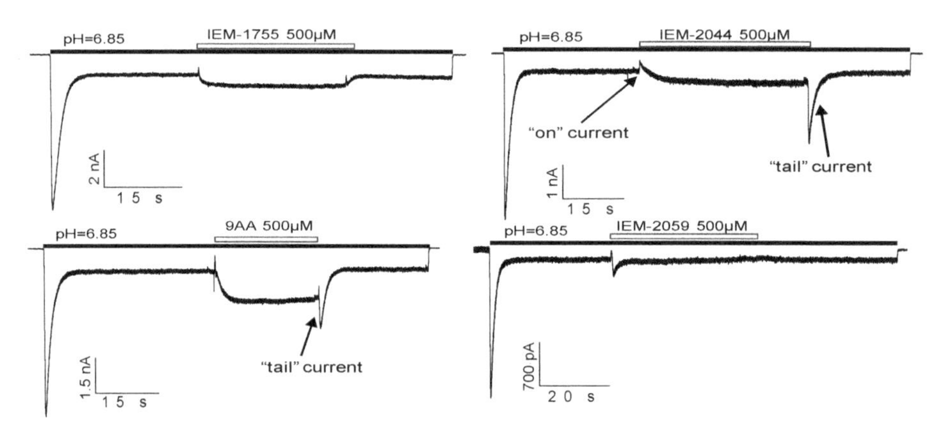


**Figure 7.** "On" and "tail" currents for different drugs. Transient currents were pronounced for the drugs which caused peak inhibition at pH 6.85 due to the activation shift (9AA, IEM-2044, see Figure 2A).

## 3. Discussion

In the present work we demonstrated that many hydrophobic monoamines and their guanidine analogs affected ASIC3 in submillimolar concentrations. Two of them, IEM-2163 and IEM-2195, were studied in detail, whereby we found that their effects are best explained by the existence of two distinct mechanisms. The first mechanism is the acidic shift of activation that results in fast pH-dependent peak inhibition. The second one is deceleration of the ASIC3 desensitization, which raises the equilibrium level of the sustained current, thus effectively increasing its amplitude. The total drug effect on the sustained current depended on the ratio of these two independent types of actions.

We found that in a large series of drugs there was no correlation between these two types of action. Effect on activation was found to be pH-dependent, whereas modulation of desensitization was similar at pH 6.85 and pH 6.0. Elucidation of these types of action in turn required different application protocols. Concentration dependencies were also apparently separated, with effect on activation developing at lower concentrations than the effect on desensitization. There was also a drastic difference in kinetics, as the effect on activation was much faster than on desensitization. Taken together, these data suggest that two distinct types of action are mediated by drugs binding to different sites.

Drug effects on ASIC3 have been the subject matter of numerous studies. For instance, a detailed examination of GMQ and a representative series of its derivatives [44] allows for comparison with our data. In particular, compounds containing two aromatic rings and a guanidine group used at a concentration of 1 mM induced an acidic shift in the activation curve of ASIC3, similarly to IEM-2163, IEM-2195, and some other compounds in our work (Figure 2). On the other hand, GMQ and a few other derivatives induced an alkaline shift of activation, while we found no such effect for our compounds. Amiloride, a known ASIC blocker, also causes an alkaline shift of activation in high concentrations (0.5–1 mM). [24]. A similar but much weaker effect was induced by agmatine [26].

Notably, to detect a shift of activation it is necessarily to study the ligands' effects with both weak and strong acidifications, and such data are not available for a number of other compounds.

Analysis of drugs' action on sustained current is more complex, as it can be mediated by the effects on both channel activation and desensitization. Additionally, in some experimental setups ASIC3 does not mediate such currents in control, complicating quantitative estimations of effects. Various drugs, including GMQ [23], agmatine [26], and amiloride [22], induce or potentiate sustained ASIC3-mediated currents evoked by modest acidifications. Serotonin [27] and FMRFamide [25] potentiate the sustained current, while simultaneously slowing down desensitization kinetics under conditions of strong acidification. Note that these two compounds did not affect the peak component. In the present work we showed that drugs reduced the speed of response decay and increased final equilibrium level of sustained current amplitude under conditions of modest acidifications. We also experimentally separated this effect from their influence on activation, allowing us to detect such an effect for IEM-2163 despite its total inhibitory action. We are not aware of the proven examples of the compounds inhibiting the sustained current via modulation of desensitization.

In this regard it is interesting to compare drugs' effect on ASIC1a and ASIC3. In our previous paper [32] we demonstrated that many monoamines and their guanidine analogs affect the steady-state desensitization of ASIC1a by shifting its pH dependence to more acidic values, although this effect does not lead to the appearance of sustained current. The opposite effect, alkaline shift of the steady-state desensitization, was not revealed for small molecules but only for psalmotoxin [45]. Thus, if we assume that a similar process underlies desensitization in both ASIC1a and ASIC3, there is an apparent commonality in the direction of drug action, although it manifests differently, according to the channel type. In contrast, the drug action on activation properties is notably diverse. For instance, IEM-2044 and amitriptyline have opposite effects on different channels, inducing an alkaline shift of activation on ASIC1a [32,34] and an acidic one on ASIC3, while 9AA shifts the activation to more acidic values in both cases [32]. Histamine only enhanced the activation of ASIC1a and was inactive against ASIC3 [33]. Thus, we do not see a correlation for action on activation of ASIC1a and ASIC3. Similarly, in [44], GMQ and its derivatives also demonstrated varying effects on ASIC1a and ASIC3, with some compounds acting differently on different channels and others having the same effect regardless of the target.

The problem of the binding site(s) of ASIC ligands in the extracellular domain is intensely debated. According to recent structural data, channel "activation involves 'closure' of the thumb domain into the acidic pocket, expansion of the lower palm domain and an iris-like opening of the channel gate. The linkers between the upper and lower palm domains serve as a molecular 'clutch', and undergo a simple rearrangement to permit rapid desensitization" [46]. Another study [47] suggests that the protonable residues in the acidic pocket affect ASIC pH dependence, but in the palm domain they are responsible for the regulation of desensitization kinetics as well as prevention of the sustained currents. Thus, different regions participate in complex allosteric interactions which contribute to activation and desensitization, which in turn significantly complicates estimation of ligands' binding site(s). In addition, particular mutations can unequally affect different modes of ligands' action. One such example is Glu-79 in the palm domain of ASIC3 [48]. While it has been shown to be a crucial element for direct opening of the channel by GMQ, its mutation did not elicit any changes in GMQ's effects on activation, but instead altered GMQ's influence on the channel inactivation. The effect of mutations on ligands' binding and action can also be either direct or allosteric. These data, together with the complex structure–activity relationships revealed in the present and other studies, raise the possibility that low-weight drugs can bind to more than one site in the extracellular domain of ASICs. For instance, binding to the acidic pocket could control effects on activation, whereas binding to the palm domain could be responsible for desensitization effects.

In our work we have focused our attention primarily on the low-to-moderate (pH 6.85–6.0) acidification range. While the more powerful acidification (pH < 5.0) that is frequently used in other studies can indeed occur [49] in vivo, it typically accompanies severe conditions such as tumors and open fractures. However, physiological processes and less-drastic pathologies usually stay in the less-acidic pH range [50–53]. Additionally, the research of Salinas et al. [54] indicates that ASIC3 activations by different levels of acidification are facilitated by distinct mechanisms. If one were to assume that those mechanisms in turn mediate specific physiological responses, then our work shows potential for the development of state-dependent drugs, which would affect only the specific response, without influencing other channel functions.

## 4. Materials and Methods

### 4.1. Chemicals and Synthesis

The synthesis of IEM compounds was performed at the Institute of Experimental Medicine, Saint-Petersburg, Russia as described in References [55–58]. The rest of the drugs were obtained from Tocris Bioscience and Sigma Aldrich.

### 4.2. Cell Culture and Transfection

Chinese hamster ovary (CHO) cells, purchased from Evrogen company (Evrogen, Moscow, Russia), were cultured in a humidified atmosphere of 5% $CO_2$ at 37 °C. Standard culture conditions were used for cell maintenance (Dulbecco's modified Eagle's medium (DMEM), 10% fetal bovine serum, 5% gentamicin). Transfection of plasmid encoding rat ASIC3 subunit was done using Lipofectamine 2000 (Invitrogen, Carlsbad, CA, USA) following the manufacturer's protocol. We received expression vectors encoding rat ASIC3 as a gift from A. Staruschenko [59]. Those vectors were described in Reference [60]. Cells were transfected with 0.5 mg rASIC3 cDNA + 0.5 mg eGFP per 35-mm dish to achieve the expression of homomeric channels.

### 4.3. Drugs and Solutions

Pipette solution was prepared as 100 mM CsF, 40 mM CsCl, 5 mM NaCl, 0.5 mM $CaCl_2$, 10 mM HEPES, and 5 mM EGTA, and its pH was adjusted to 7.35 with CsOH. For cells' perfusion, an extracellular solution with 143 mM NaCl, 5 mM KCl, 2.5 mM $CaCl_2$, 2 mM $MgCl_2$, 18 mM D-glucose, 10 mM HEPES, and 10 mM MES was used, with its pH adjusted to 7.4. Drug-containing solutions were prepared from extracellular solution and their pH was adjusted again if necessary.

### 4.4. Electrophysiology

Electrophysiological experiments were performed 48–72 h after transfection. Green fluorescence detected with a Leica DMIL microscope was used to identify transfected cells. Current recordings were acquired with an EPC-8 (HEKA Elektronik, Lambrecht, Germany) patch clamp amplifier in whole-cell voltage-clamp mode at a membrane potential of −80 mV. The data were stored on a personal computer via Patchmaster software (HEKA Elektronik, Lambrecht, Germany). The recordings where access resistance and capacitance changed by more than 10% over the course of the experiment were excluded from the analysis.

### 4.5. Experimental Protocol

A standard experiment included a 30 s application of conditioning solution, followed by 20 s of activating solution. Then, the process was repeated until at least three responses in a row were differing from each other by less than 10% of their amplitude. In experiments where the drug was applied during conditioning period, we had to reduce its potential effect on the open channels. To achieve this, channel activation was preceded by a 3 s flush of drug-free conditioning solution.

When the drug was applied only to the sustained current, both activating solution and the drug were applied until sustained current stabilized, with no other time constraints.

To account for response variability during the experiment, the control responses were averaged before drug application and after washout.

### 4.6. Data Analysis and Statistics

The values in the text are given as mean ± standard deviation (SD) with $n \geq 5$. To test for effects' significance, paired *t*-tests (drug versus control) or ANOVA were used, as appropriate, via the IBM SPSS Statistics software package (IBM, Armonk, NY, USA). OriginPro 8.1 (OriginLab Corporation, Northampton, MA, USA) was used for fitting of the data.

**Author Contributions:** Conceptualization, V.Y.S. and D.B.T.; Formal analysis, V.Y.S.; Funding acquisition, D.B.T.; Investigation, V.Y.S.; Methodology, N.N.P. and D.B.T.; Project administration, N.N.P.; Resources, N.N.P. and V.E.G.; Supervision, D.B.T.; Visualization, V.Y.S.; Writing—original draft, V.Y.S. and D.B.T.; Writing—review and editing, V.Y.S., V.E.G., and D.B.T.

**Funding:** This research received no external funding.

**Acknowledgments:** Authors would like to thank A. Staruschenko (Medical College of Wisconsin, WI, USA) for ASIC3 plasmid.

**Conflicts of Interest:** The authors declare no conflict of interest.

## Abbreviations

| | |
|---|---|
| ASIC | Acid-sensing ion channel |
| CHO | Chinese hamster ovary (cells) |
| DEG/ENaC | Degenerin/Epithelial sodium channels |
| GMQ | 2-Guanidine-4-methylquinazoline |

## References

1. Deval, E.; Lingueglia, E. Acid-Sensing Ion Channels and nociception in the peripheral and central nervous systems. *Neuropharmacology* **2015**, *94*, 49–57. [CrossRef]
2. Kreple, C.J.; Lu, Y.; Taugher, R.J.; Schwager-Gutman, A.L.; Du, J.; Stump, M.; Wang, Y.; Ghobbeh, A.; Fan, R.; Cosme, C.V.; et al. Acid-sensing ion channels contribute to synaptic transmission and inhibit cocaine-evoked plasticity. *Nat. Neurosci.* **2014**, *17*, 1083–1091. [CrossRef] [PubMed]
3. Huang, Y.; Jiang, N.; Li, J.; Ji, Y.-H.; Xiong, Z.-G.; Zha, X. Two aspects of ASIC function: Synaptic plasticity and neuronal injury. *Neuropharmacology* **2015**, *94*, 42–48. [CrossRef]
4. Kweon, H.-J.; Suh, B.-C. Acid-sensing ion channels (ASICs): therapeutic targets for neurological diseases and their regulation. *BMB Rep.* **2013**, *46*, 295–304. [CrossRef] [PubMed]
5. Xiong, Z.-G.; Xu, T.-L. The role of ASICS in cerebral ischemia. *Wiley Interdiscip. Rev. Membr. Transp. Signal* **2012**, *1*, 655–662. [CrossRef]
6. Lin, S.-H.; Sun, W.-H.; Chen, C.-C. Genetic exploration of the role of acid-sensing ion channels. *Neuropharmacology* **2015**, *94*, 99–118. [CrossRef] [PubMed]
7. Yagi, J.; Wenk, H.N.; Naves, L.A.; McCleskey, E.W. Sustained currents through ASIC3 ion channels at the modest pH changes that occur during myocardial ischemia. *Circ. Res.* **2006**, *99*, 501–509. [CrossRef]
8. Deval, E.; Gasull, X.; Noël, J.; Salinas, M.; Baron, A.; Diochot, S.; Lingueglia, E. Acid-sensing ion channels (ASICs): pharmacology and implication in pain. *Pharmacol. Ther.* **2010**, *128*, 549–558. [CrossRef]
9. Jones, N.G.; Slater, R.; Cadiou, H.; McNaughton, P.; McMahon, S.B. Acid-induced pain and its modulation in humans. *J. Neurosci.* **2004**, *24*, 10974–10979. [CrossRef]

10. Dubé, G.R.; Lehto, S.G.; Breese, N.M.; Baker, S.J.; Wang, X.; Matulenko, M.A.; Honoré, P.; Stewart, A.O.; Moreland, R.B.; Brioni, J.D. Electrophysiological and in vivo characterization of A-317567, a novel blocker of acid sensing ion channels. *Pain* **2005**, *117*, 88–96. [CrossRef]
11. Rocha-González, H.I.; Herrejon-Abreu, E.B.; López-Santillán, F.J.; García-López, B.E.; Murbartián, J.; Granados-Soto, V. Acid increases inflammatory pain in rats: effect of local peripheral ASICs inhibitors. *Eur. J. Pharmacol.* **2009**, *603*, 56–61. [CrossRef]
12. Price, M.P.; McIlwrath, S.L.; Xie, J.; Cheng, C.; Qiao, J.; Tarr, D.E.; Sluka, K.A.; Brennan, T.J.; Lewin, G.R.; Welsh, M.J. The DRASIC Cation Channel Contributes to the Detection of Cutaneous Touch and Acid Stimuli in Mice. *Neuron* **2001**, *32*, 1071–1083. [CrossRef]
13. Kang, S.; Jang, J.H.; Price, M.P.; Gautam, M.; Benson, C.J.; Gong, H.; Welsh, M.J.; Brennan, T.J. Simultaneous Disruption of Mouse ASIC1a, ASIC2 and ASIC3 Genes Enhances Cutaneous Mechanosensitivity. *PLoS ONE* **2012**, *7*. [CrossRef]
14. Deval, E.; Noël, J.; Lay, N.; Alloui, A.; Diochot, S.; Friend, V.; Jodar, M.; Lazdunski, M.; Lingueglia, E. ASIC3, a sensor of acidic and primary inflammatory pain. *EMBO J.* **2008**, *27*, 3047–3055. [CrossRef]
15. Sutherland, S.P.; Benson, C.J.; Adelman, J.P.; McCleskey, E.W. Acid-sensing ion channel 3 matches the acid-gated current in cardiac ischemia-sensing neurons. *Proc. Natl. Acad. Sci. USA* **2001**, *98*, 711–716. [CrossRef] [PubMed]
16. Hattori, T.; Chen, J.; Harding, A.M.S.; Price, M.P.; Lu, Y.; Abboud, F.M.; Benson, C.J. ASIC2a and ASIC3 heteromultimerize to form pH-sensitive channels in mouse cardiac dorsal root ganglia neurons. *Circ. Res.* **2009**, *105*, 279–286. [CrossRef] [PubMed]
17. Yan, J.; Edelmayer, R.M.; Wei, X.; De Felice, M.; Porreca, F.; Dussor, G. Dural afferents express acid-sensing ion channels: a role for decreased meningeal pH in migraine headache. *Pain* **2011**, *152*, 106–113. [CrossRef] [PubMed]
18. Izumi, M.; Ikeuchi, M.; Ji, Q.; Tani, T. Local ASIC3 modulates pain and disease progression in a rat model of osteoarthritis. *J. Biomed. Sci.* **2012**, *19*, 77. [CrossRef] [PubMed]
19. Walder, R.Y.; Rasmussen, L.A.; Rainier, J.D.; Light, A.R.; Wemmie, J.A.; Sluka, K.A. ASIC1 and ASIC3 Play Different Roles in the Development of Hyperalgesia Following Inflammatory Muscle Injury. *J. Pain* **2010**, *11*, 210–218. [CrossRef]
20. Baron, A.; Lingueglia, E. Pharmacology of acid-sensing ion channels - Physiological and therapeutical perspectives. *Neuropharmacology* **2015**, *94*, 19–35. [CrossRef]
21. Waldmann, R.; Champigny, G.; Bassilana, F.; Heurteaux, C.; Lazdunski, M. A proton-gated cation channel involved in acid-sensing. *Nature* **1997**, *386*, 173–177. [CrossRef] [PubMed]
22. Li, W.-G.; Yu, Y.; Huang, C.; Cao, H.; Xu, T.-L. Nonproton Ligand Sensing Domain Is Required for Paradoxical Stimulation of Acid-sensing Ion Channel 3 (ASIC3) Channels by Amiloride. *J. Biol. Chem.* **2011**, *286*, 42635–42646. [CrossRef]
23. Yu, Y.; Chen, Z.; Li, W.-G.; Cao, H.; Feng, E.-G.; Yu, F.; Liu, H.; Jiang, H.; Xu, T.-L. A nonproton ligand sensor in the acid-sensing ion channel. *Neuron* **2010**, *68*, 61–72. [CrossRef] [PubMed]
24. Besson, T.; Lingueglia, E.; Salinas, M. Pharmacological modulation of Acid-Sensing Ion Channels 1a and 3 by amiloride and 2-guanidine-4-methylquinazoline (GMQ). *Neuropharmacology* **2017**, *125*, 429–440. [CrossRef]
25. Askwith, C.C.; Cheng, C.; Ikuma, M.; Benson, C.; Price, M.P.; Welsh, M.J. Neuropeptide FF and FMRFamide Potentiate Acid-Evoked Currents from Sensory Neurons and Proton-Gated DEG/ENaC Channels. *Neuron* **2000**, *26*, 133–141. [CrossRef]
26. Li, W.-G.; Yu, Y.; Zhang, Z.-D.; Cao, H.; Xu, T.-L. ASIC3 Channels Integrate Agmatine and Multiple Inflammatory Signals through the Nonproton Ligand Sensing Domain. *Mol. Pain* **2010**, *6*, 88. [CrossRef] [PubMed]
27. Wang, X.; Li, W.-G.; Yu, Y.; Xiao, X.; Cheng, J.; Zeng, W.-Z.; Peng, Z.; Xi Zhu, M.; Xu, T.-L. Serotonin facilitates peripheral pain sensitivity in a manner that depends on the nonproton ligand sensing domain of ASIC3 channel. *J. Neurosci.* **2013**, *33*, 4265–4279. [CrossRef]
28. Diochot, S.; Baron, A.; Rash, L.D.; Deval, E.; Escoubas, P.; Scarzello, S.; Salinas, M.; Lazdunski, M. A new sea anemone peptide, APETx2, inhibits ASIC3, a major acid-sensitive channel in sensory neurons. *EMBO J.* **2004**, *23*, 1516–1525. [CrossRef] [PubMed]
29. Osmakov, D.I.; Kozlov, S.A.; Andreev, Y.A.; Koshelev, S.G.; Sanamyan, N.P.; Sanamyan, K.E.; Dyachenko, I.A.; Bondarenko, D.A.; Murashev, A.N.; Mineev, K.S.; et al. Sea anemone peptide with uncommon β-hairpin structure inhibits acid-sensing ion channel 3 (ASIC3) and reveals analgesic activity. *J. Biol. Chem.* **2013**, *288*, 23116–23127. [CrossRef] [PubMed]

30. Bohlen, C.J.; Chesler, A.T.; Sharif-Naeini, R.; Medzihradszky, K.F.; Zhou, S.; King, D.; Sánchez, E.E.; Burlingame, A.L.; Basbaum, A.I.; Julius, D. A heteromeric Texas coral snake toxin targets acid-sensing ion channels to produce pain. *Nature* **2011**, *479*, 410–414. [CrossRef]
31. Tikhonova, T.B.; Nagaeva, E.I.; Barygin, O.I.; Potapieva, N.N.; Bolshakov, K.V.; Tikhonov, D.B. Monoamine NMDA receptor channel blockers inhibit and potentiate native and recombinant proton-gated ion channels. *Neuropharmacology* **2015**, *89*, 1–10. [CrossRef]
32. Shteinikov, V.Y.; Barygin, O.I.; Gmiro, V.E.; Tikhonov, D.B. Multiple modes of action of hydrophobic amines and their guanidine analogues on ASIC1a. *Eur. J. Pharmacol.* **2019**, *844*, 183–194. [CrossRef] [PubMed]
33. Nagaeva, E.I.; Tikhonova, T.B.; Magazanik, L.G.; Tikhonov, D.B. Histamine selectively potentiates acid-sensing ion channel 1a. *Neurosci. Lett.* **2016**, *632*, 136–140. [CrossRef] [PubMed]
34. Nikolaev, M.; Komarova, M.; Tikhonova, T.; Korosteleva, A.; Potapjeva, N.; Tikhonov, D.B. Modulation of proton-gated channels by antidepressants. *ACS Chem. Neurosci.* **2019**. [CrossRef] [PubMed]
35. Shteinikov, V.Y.; Korosteleva, A.S.; Tikhonova, T.B.; Potapieva, N.N.; Tikhonov, D.B. Ligands of histamine receptors modulate acid-sensing ion channels. *Biochem. Biophys. Res. Commun.* **2017**, *490*, 1314–1318. [CrossRef] [PubMed]
36. Bolshakov, K.V.; Kim, K.H.; Potapjeva, N.N.; Gmiro, V.E.; Tikhonov, D.B.; Usherwood, P.N.R.; Mellor, I.R.; Magazanik, L.G. Design of antagonists for NMDA and AMPA receptors. *Neuropharmacology* **2005**, *49*, 144–155. [CrossRef] [PubMed]
37. Magazanik, L.G.; Buldakova, S.L.; Samoilova, M.V.; Gmiro, V.E.; Mellor, I.R.; Usherwood, P.N. Block of open channels of recombinant AMPA receptors and native AMPA/kainate receptors by adamantane derivatives. *J. Physiol. (Lond.)* **1997**, *505*, 655–663. [CrossRef]
38. Bormann, J. Memantine is a potent blocker of *N*-methyl-D-aspartate (NMDA) receptor channels. *Eur. J. Pharmacol.* **1989**, *166*, 591–592. [CrossRef]
39. Benveniste, M.; Mayer, M.L. Trapping of glutamate and glycine during open channel block of rat hippocampal neuron NMDA receptors by 9-aminoacridine. *J. Physiol. (Lond.)* **1995**, *483 (Pt 2)*, 367–384. [CrossRef]
40. Pereira, V.S.; Hiroaki-Sato, V.A. A brief history of antidepressant drug development: from tricyclics to beyond ketamine. *Acta Neuropsychiatrica* **2018**, *30*, 307–322. [CrossRef]
41. Wagstaff, A.J.; Ormrod, D.; Spencer, C.M. Tianeptine: a review of its use in depressive disorders. *Mol. Diag. Ther.* **2001**, *15*, 231–259. [CrossRef] [PubMed]
42. Arrang, J.M.; Garbarg, M.; Lancelot, J.C.; Lecomte, J.M.; Pollard, H.; Robba, M.; Schunack, W.; Schwartz, J.C. Highly potent and selective ligands for histamine H3-receptors. *Nature* **1987**, *327*, 117–123. [CrossRef]
43. Garbarg, M.; Arrang, J.M.; Rouleau, A.; Ligneau, X.; Tuong, M.D.; Schwartz, J.C.; Ganellin, C.R. S-[2-(4-imidazolyl)ethyl]isothiourea, a highly specific and potent histamine H3 receptor agonist. *J. Pharmacol. Exp. Ther.* **1992**, *263*, 304–310.
44. Alijevic, O.; Hammoud, H.; Vaithia, A.; Trendafilov, V.; Bollenbach, M.; Schmitt, M.; Bihel, F.; Kellenberger, S. Heteroarylguanidines as Allosteric Modulators of ASIC1a and ASIC3 Channels. *ACS Chem. Neurosci.* **2018**, *9*, 1357–1365. [CrossRef]
45. Chen, X.; Kalbacher, H.; Grunder, S. The tarantula toxin psalmotoxin 1 inhibits acid-sensing ion channel (ASIC) 1a by increasing its apparent H+ affinity. *J. Gen. Physiol.* **2005**, *126*, 71–79. [CrossRef]
46. Yoder, N.; Yoshioka, C.; Gouaux, E. Gating mechanisms of acid-sensing ion channels. *Nature* **2018**, *555*, 397–401. [CrossRef] [PubMed]
47. Vullo, S.; Bonifacio, G.; Roy, S.; Johner, N.; Bernèche, S.; Kellenberger, S. Conformational dynamics and role of the acidic pocket in ASIC pH-dependent gating. *Proc. Natl. Acad. Sci. USA* **2017**, *114*, 3768–3773. [CrossRef] [PubMed]
48. Alijevic, O.; Kellenberger, S. Subtype-specific modulation of acid-sensing ion channel (ASIC) function by 2-guanidine-4-methylquinazoline. *J. Biol. Chem.* **2012**, *287*, 36059–36070. [CrossRef] [PubMed]
49. Reeh, P.W.; Steen, K.H. Tissue acidosis in nociception and pain. *Prog. Brain Res.* **1996**, *113*, 143–151.
50. Yan, G.X.; Kléber, A.G. Changes in extracellular and intracellular pH in ischemic rabbit papillary muscle. *Circ. Res.* **1992**, *71*, 460–470. [CrossRef]
51. Street, D.; Bangsbo, J.; Juel, C. Interstitial pH in human skeletal muscle during and after dynamic graded exercise. *J. Physiol.* **2001**, *537*, 993–998. [CrossRef] [PubMed]
52. Woo, Y.C.; Park, S.S.; Subieta, A.R.; Brennan, T.J. Changes in tissue pH and temperature after incision indicate acidosis may contribute to postoperative pain. *Anesthesiology* **2004**, *101*, 468–475. [CrossRef]

53. McVicar, N.; Li, A.X.; Gonçalves, D.F.; Bellyou, M.; Meakin, S.O.; Prado, M.A.; Bartha, R. Quantitative tissue pH measurement during cerebral ischemia using amine and amide concentration-independent detection (AACID) with MRI. *J. Cereb. Blood Flow Metab.* **2014**, *34*, 690–698. [CrossRef] [PubMed]
54. Salinas, M.; Lazdunski, M.; Lingueglia, E. Structural elements for the generation of sustained currents by the acid pain sensor ASIC3. *J. Biol. Chem.* **2009**, *284*, 31851–31859. [CrossRef] [PubMed]
55. Maddox, V.H.; Godefroi, E.F.; Parcell, R.F. The Synthesis of Phencyclidine and Other 1-Arylcyclohexylamines. *J. Med. Chem.* **1965**, *8*, 230–235. [CrossRef]
56. Kalir, A.; Edery, H.; Pelah, Z.; Balderman, D.; Porath, G. 1-Phenylcycloalkylamine derivatives. II. Synthesis and pharmacological activity. *J. Med. Chem.* **1969**, *12*, 473–477. [CrossRef]
57. Geluk, H.W.; Schut, J.; Schlatmann, J.L.M.A. Synthesis and antiviral properties of 1-adamantylguanidine. A modified method for preparing tert-alkylguanidines. *J. Med. Chem.* **1969**, *12*, 712–715. [CrossRef] [PubMed]
58. Thurkauf, A.; De Costa, B.; Yamaguchi, S.; Mattson, M.V.; Jacobson, A.E.; Rice, K.C.; Rogawski, M.A. Synthesis and anticonvulsant activity of 1-phenylcyclohexylamine analogs. *J. Med. Chem.* **1990**, *33*, 1452–1458. [CrossRef]
59. Staruschenko, A.; Dorofeeva, N.A.; Bolshakov, K.V.; Stockand, J.D. Subunit-dependent cadmium and nickel inhibition of acid-sensing ion channels. *Dev. Neurobiol.* **2007**, *67*, 97–107. [CrossRef]
60. Hesselager, M.; Timmermann, D.B.; Ahring, P.K. pH dependency and desensitization kinetics of heterologously expressed combinations of acid-sensing ion channel subunits. *J. Biol. Chem.* **2004**, *279*, 11006–11015. [CrossRef]

MDPI
St. Alban-Anlage 66
4052 Basel
Switzerland
Tel. +41 61 683 77 34
Fax +41 61 302 89 18
www.mdpi.com

*International Journal of Molecular Sciences* Editorial Office
E-mail: ijms@mdpi.com
www.mdpi.com/journal/ijms

Lightning Source UK Ltd.
Milton Keynes UK
UKHW050633110820
367999UK00007B/189